Maritime Welding

Level 2

Trainee Guide

Pearson

Boston Columbus Indianapolis New York San Francisco Amsterdam
Cape Town Dubai London Madrid Milan Munich Paris Montreal Toronto Delhi
Mexico City Sao Paulo Sydney Hong Kong Seoul Singapore Taipei Tokyo

NCCER

President and Chief Executive Officer: Don Whyte
President: Boyd Worsham
Vice President: Steve Greene
Chief Operations Officer: Katrina Kersch
Maritime Welding Project Manager: Chris Wilson
Senior Development Manager: Mark Thomas
Senior Production Manager: Tim Davis
Managing Editor: Natalie Richoux
Desktop Publishing Manager: James McKay
Art Manager: Kelly Saddler
Desktop Publishing Specialists: Gene Page, Eric Caraballoso
Editors: Graham Hack, Jordan Hutchinson

Writing and development services provided by Topaz Publications, Liverpool, NY

Lead Writer/Project Manager: Troy Staton
Desktop Publisher: Joanne Hart
Art Director: Alison Richmond
Permissions Editor: Tonia Burke
Writers: Troy Staton, Carol Herbert, Thomas Burke

Pearson Education, Inc.

Director of Alliance/Partnership Management: Kelly Trakalo
Content Producer: Alexandrina B. Wolf
Assistant Content Producer: Alma Dabral
Digital Content Producer: Jose Carchi
Senior Marketing Manager: Brian Hoehl
Composition: NCCER
Printer/Binder: LSC Communications
Cover Printer: LSC Communications
Text Fonts: Palatino and Univers

Credits and acknowledgments for content borrowed from other sources and reproduced, with permission, in this textbook appear at the end of each module.

ScoutAutomatedPrintCode

Pearson

Perfect Bound ISBN-13: 978-0-13-560378-9
ISBN-10: 0-13-560378-1

Preface

Welding is a high-tech industry that can take you to places all over the world. Welds are everywhere — just look around. From ladders to aircraft carriers; from NASCAR to national defense; from the laboratory to sales and repair; the welding trade impacts virtually every industry.

Few career choices offer as many options for employment and opportunities for growth. According to the Department of Labor, "the basic skills of welding are similar across industries, so welders can easily shift from one industry to another, depending on where they are needed most." Welders will continue to be in high demand since there is a continual need to rebuild aging infrastructure such as highways and buildings. In addition, the creation of new power generation facilities will yield new job prospects for trained and skilled welders.

Of course, one of the places to find a prosperous career as a welder is in the maritime industry. Shipyards and repair facilities are plentiful in coastal areas and on major rivers and lake systems. Maritime welding can be done on land or sea and above or underneath the surface of the water. It includes the work on ships, boats, and off-shore oil rigs. It includes the installation and repair of metal structures and components.

Maritime Welding introduces the fundamentals of the welding trade. The three levels of this curriculum, plus the Maritime Aluminum Welding curriculum, present an apprentice approach and will help you be knowledgeable, safe, and effective on the job.

Maritime Welding

This *Maritime Welding* curriculum is a product of collaboration between NCCER and the National Maritime Education Council (NMEC). Several maritime welding subject matter experts determined which of NCCER's *Welding* modules are suitable to maritime industry training. This edition of *Maritime Welding* can be used for welding training and can be supplemented by instructors with information specific to the maritime industry.

We invite you to visit the NCCER website at **www.nccer.org** for the latest releases, training information, newsletter, and much more. You can also reference the Pearson product catalog online at **www.nccer.org**. Your feedback is welcome. You may e-mail your comments to **curriculum@nccer.org** or send general comments and inquiries to **info@nccer.org**.

NMEC

The National Maritime Education Council (NMEC) is a multi-regional, industry-centered organization working to address critical workforce issues facing its member yards and the maritime industry. Established in 2012 to provide a centralized voice for the industry as it relates to workforce development, NMEC's primary mission is to lead the maritime industry in the development, promotion, and implementation of a national workforce development system that includes standardized craft training, portable credentials, and journey-level assessments.

Through its partnership with NCCER, NMEC is working in collaboration with its member yards to:

- Establish a system of career awareness and recruiting that will allow the industry to speak with one voice and deliver a consistent message to students, student influencers, the underemployed, women, and returning military.
- Establish and implement an industry-wide, comprehensive, standardized training and credentialing system for maritime craft professionals.
- Grow a pipeline of skilled craft professionals trained and credentialed through an industry-driven, uniform process.
- Improve industry outcomes by developing a better skilled, safer, more productive workforce.

NCCER Standardized Curricula

NCCER is a not-for-profit 501(c)(3) education foundation established in 1995 by the world's largest and most progressive construction companies and national construction associations. It was founded to address the severe workforce shortage facing the industry and to develop a standardized training process and curricula. Today, NCCER is supported by hundreds of leading construction and maintenance companies, manufacturers, and national associations. The NCCER Standardized Curricula was developed by NCCER in partnership with Pearson Education, Inc., the world's largest educational publisher.

Some features of the NCCER Standardized Curricula are as follows:

- An industry-proven record of success
- Curricula developed by the industry for the industry
- National standardization providing portability of learned job skills and educational credits
- Compliance with Office of Apprenticeship requirements for related classroom training (*CFR 29:29*)
- Well-illustrated, up-to-date, and practical information

NCCER also maintains a Registry that provides transcripts, certificates, and wallet cards to individuals who have successfully completed a level of training within a craft in the NCCER Standardized Curricula. *Training programs must be delivered by an NCCER Accredited Training Sponsor in order to receive these credentials.*

Special Features

In an effort to provide a comprehensive user-friendly training resource, we have incorporated many different features for your use. Whether you are a visual or hands-on learner, this book will provide you with the proper tools to get started in the welding industry.

Introduction Page

This page is found at the beginning of each module and lists the Objectives, Performance Tasks, and Trade Terms, for that module. The Objectives list the skills and knowledge you will need in order to complete the module successfully. The Performance Tasks give you an opportunity to apply your knowledge to real-world tasks. The list of Trade Terms identifies important terms you will need to know by the end of the module.

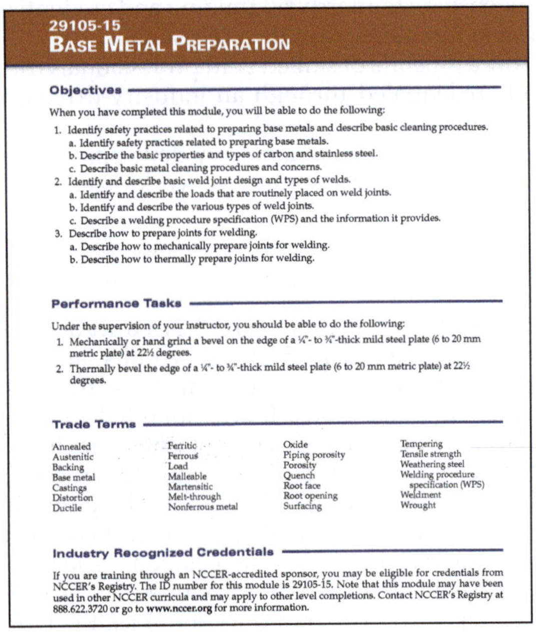

29105-15
BASE METAL PREPARATION

Objectives

When you have completed this module, you will be able to do the following:

1. Identify safety practices related to preparing base metals and describe basic cleaning procedures.
   a. Identify safety practices related to preparing base metals.
   b. Describe the basic properties and types of carbon and stainless steel.
   c. Describe basic metal cleaning procedures and concerns.
2. Identify and describe basic weld joint design and types of welds.
   a. Identify and describe the loads that are routinely placed on weld joints.
   b. Identify and describe the various types of weld joints.
   c. Describe a welding procedure specification (WPS) and the information it provides.
3. Describe how to prepare joints for welding.
   a. Describe how to mechanically prepare joints for welding.
   b. Describe how to thermally prepare joints for welding.

Performance Tasks

Under the supervision of your instructor, you should be able to do the following:

1. Mechanically or hand grind a bevel on the edge of a ¼"- to ¾"-thick mild steel plate (6 to 20 mm metric plate) at 22½ degrees.
2. Thermally bevel the edge of a ¼"- to ¾"-thick mild steel plate (6 to 20 mm metric plate) at 22½ degrees.

Trade Terms

Annealed	Ferritic	Oxide	Tempering
Austenitic	Ferrous	Piping porosity	Tensile strength
Backing	Load	Porosity	Weathering steel
Base metal	Malleable	Quench	Welding procedure specification (WPS)
Castings	Martensitic	Root face	Weldment
Distortion	Melt-through	Root opening	Wrought
Ductile	Nonferrous metal	Surfacing	

Industry Recognized Credentials

If you are training through an NCCER-accredited sponsor, you may be eligible for credentials from NCCER's Registry. The ID number for this module is 29105-15. Note that this module may have been used in other NCCER curricula and may apply to other level completions. Contact NCCER's Registry at 888.622.3720 or go to **www.nccer.org** for more information.

Notes, Cautions, and Warnings

Safety features are set off from the main text in highlighted boxes and organized into three categories based on the potential danger of the issue being addressed. Notes simply provide additional information on the topic area. Cautions alert you of a danger that does not present potential injury but may cause damage to equipment. Warnings stress a potentially dangerous situation that may cause injury to you or a co-worker.

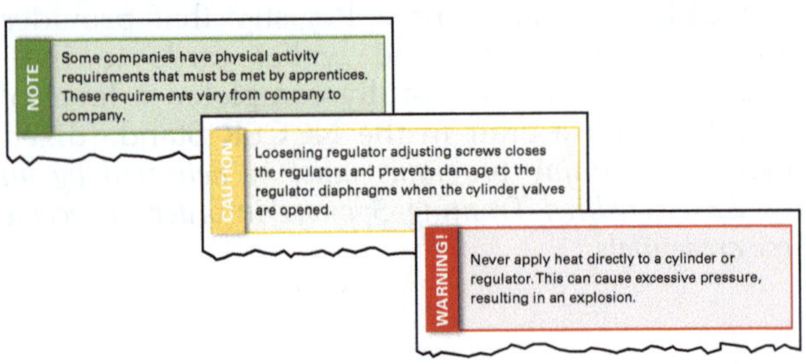

NOTE
Some companies have physical activity requirements that must be met by apprentices. These requirements vary from company to company.

CAUTION
Loosening regulator adjusting screws closes the regulators and prevents damage to the regulator diaphragms when the cylinder valves are opened.

WARNING!
Never apply heat directly to a cylinder or regulator. This can cause excessive pressure, resulting in an explosion.

Special Features

Features present technical tips and professional practices from the construction industry. These features often include real-life scenarios similar to those you might encounter on the job site.

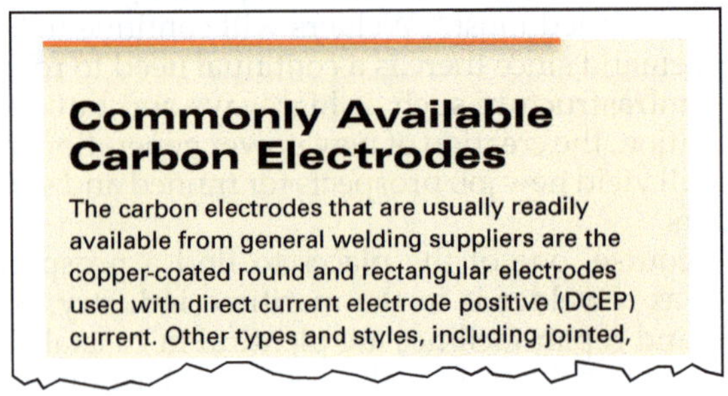

Commonly Available Carbon Electrodes

The carbon electrodes that are usually readily available from general welding suppliers are the copper-coated round and rectangular electrodes used with direct current electrode positive (DCEP) current. Other types and styles, including jointed,

Going Green

Going Green looks at ways to preserve the environment, save energy, and make good choices regarding the health of the planet. Through the introduction of new construction practices and products, you will see how the "greening of the world" has already taken root.

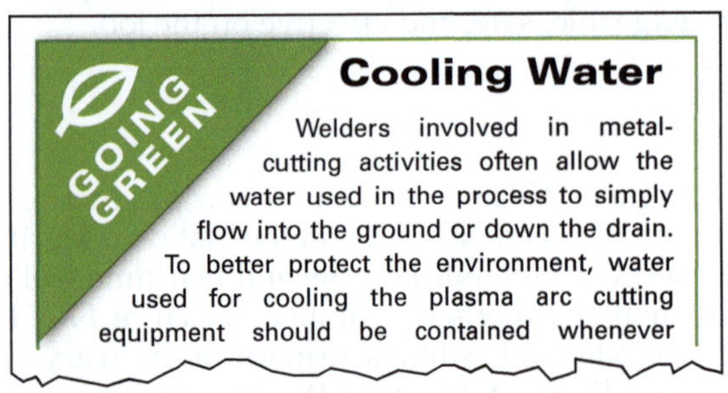

Cooling Water

Welders involved in metal-cutting activities often allow the water used in the process to simply flow into the ground or down the drain. To better protect the environment, water used for cooling the plasma arc cutting equipment should be contained whenever

Did You Know?

The Did You Know? features introduce historical tidbits or modern information about the construction industry. Interesting and sometimes surprising facts about construction are also presented.

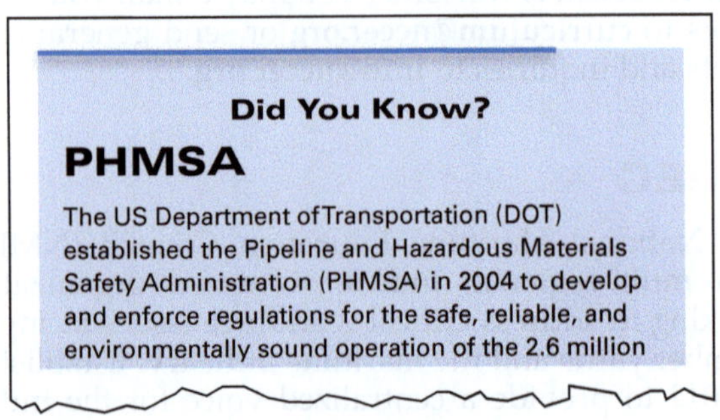

Did You Know?

PHMSA

The US Department of Transportation (DOT) established the Pipeline and Hazardous Materials Safety Administration (PHMSA) in 2004 to develop and enforce regulations for the safe, reliable, and environmentally sound operation of the 2.6 million

Case History

Case History features emphasize the importance of safety by citing examples of the costly (and often devastating) consequences of ignoring safety rules and regulations.

> **Case History**
>
> **Mexico City Refinery**
>
> In Mexico City in 1984, a weld on a liquid propane gas (LPG) tank cracked. The escaping gas cloud drifted toward a nearby residential area until it encountered an ignition source. When the gas ignited, the fireball quickly burned its way back to the leak source. The resulting explosion destroyed the refinery and many residential neighborhoods, leaving 503 people dead and more than 4,000 injured.
>
> **The Bottom Line:** Welds must be made cor-

Color Illustrations and Photographs

Full-color illustrations and photographs are used throughout each module to provide vivid detail. These figures highlight important concepts from the text and provide clarity for complex instructions. Each figure is denoted in the text in *italic type* for easy reference.

Figure 15 Worker in a confined space.

29101-15_F15.EPS

Step-by-Step Instructions

Step-by-step instructions are used throughout to guide you through technical procedures and tasks from start to finish. These steps show you not only how to perform a task but also how to do it safely and efficiently.

> Follow these steps to make an overhead fillet weld:
>
> *Step 1* Tack two plates together to form a T-joint for the fillet weld coupon.
>
> *Step 2* Tack-weld the coupon so it is in the overhead position.

Trade Terms

Each module presents a list of Trade Terms that are discussed within the text, defined in the Glossary at the end of the module, and reinforced with a Trade Terms Quiz. These terms are denoted in the text with **blue bold type** upon their first occurrence. To make searches for key information easier, a comprehensive Glossary of Trade Terms from all modules is found at the back of this book.

> Metals are classified into two basic groups: **ferrous** metals, which are composed mainly of iron, and **nonferrous metals**, which contain very little or no iron. Ferrous metals include all steel, cast iron, **wrought** iron, **malleable** iron, and **ductile**

Section Review

The Section Review features helpful additional resources and review questions related to the objectives in each section of the module.

> **Additional Resources**
>
> *2014 Technical Training Guide*. Current Edition. Cleveland, OH, USA: The Lincoln Electric Company. www.lincolnelectric.com
>
> *Welding Handbook*. Current Edition. Miami, FL: The American Welding Society.
>
> **2.0.0 Section Review**
>
> 1. Which of the following can be quenched without concern?
>   a. Weld test coupons
>   b. Functional welds
>   c. Practice coupons
>   d. Welds done in the 3G position
> 2. On an open-root V-groove weld in the 4G position, the electrode work angle should be _____.
>   a. 30 degrees
>   b. 45 degrees
>   c. 90 degrees
>   d. varied as needed to tie in

Review Questions

Review Questions are provided to reinforce the knowledge you have gained. This makes them a useful tool for measuring what you have learned.

> **Review Questions**
>
> 1. The A-CAC process uses a high-current electric arc between the base metal and the end of a(n) _____.
>   a. carbon-graphite electrode to melt the base metal
>   b. high-frequency pilot arc to melt the base metal
>   c. plasma jet to melt the base metal
>   d. idle arc to melt the base metal
> 2. An A-CAC torch requires compressed air to blow the _____.
>   a. carbon dioxide away from the cut
>   b. molten metal away from the cut
>   c. nitrogen away from the cut
>   d. carbon arc away from the cut
> 3. When cutting and gouging with A-CAC equipment, the jet orifices of the torch head(s) should be located _____.
>   a. beside the electrode
>   b. above the electrode
>   c. ahead of the electrode
>   d. between the electrode and the base metal surface
> 4. To prevent overheating, the electrode tip to the V-groove head should be no farther than _____.
>   a. 1" (2 cm)
>   b. 2" (5 cm)
>   c. 4" (10 cm)
>   d. 6" (15 cm)
> 5. Light-duty gouging or cutting is typically done with a _____.
>   a. 10A welding machine
>   b. 200A welding machine
>   c. 600A welding machine
>   d. 1,200A welding machine
> 6. Heavy-duty cutting can require a torch airflow rate up to _____.
>   a. 20 cfm (566 lpm)
>   b. 35 cfm (991 lpm)
>   c. 40 cfm (1,133 lpm)
>   d. 50 cfm (1,416 lpm)
> 7. Compressed air is supplied to the torch through a hose attached to a fitting on the _____.
>   a. welding machine
>   b. end of the torch cable
>   c. air jet orifices
>   d. workpiece
> 8. Electrodes designed for A-CAC contain a rare-earth, _____.
>   a. magnet metal
>   b. undercurrent ingredient
>   c. amplifier ingredient
>   d. arc-stabilizing metal
> 9. The most common style of A-CAC electrode used for manual cutting or gouging is the _____.
>   a. flat electrode
>   b. round electrode
>   c. half-round electrode
>   d. round-jointed electrode
> 10. When performing A-CAC cutting or gouging, the minimum groove width is controlled by the _____.
>   a. size of the torch
>   b. size of the electrode
>   c. speed of travel
>   d. angle of the electrode holder
> 11. Because A-CAC activities tend to blow molten metal considerable distances, it is important to station a(n) _____.
>   a. observer in the A-CAC work area
>   b. slag trap in the A-CAC work area
>   c. oxygen monitor in the A-CAC work area
>   d. fire watch in the A-CAC work area
> 12. To wash a projection from a metal surface using A-CAC equipment, the electrode is typically used at a push angle of about _____.
>   a. 30 degrees to the base metal surface
>   b. 45 degrees to the base metal surface
>   c. 55 degrees to the base metal surface
>   d. 90 degrees to the base metal surface

NCCER Standardized Curricula

NCCER's training programs comprise more than 80 construction, maintenance, pipeline, and utility areas and include skills assessments, safety training, and management education.

Boilermaking
Cabinetmaking
Carpentry
Concrete Finishing
Construction Craft Laborer
Construction Technology
Core Curriculum: Introductory Craft Skills
Drywall
Electrical
Electronic Systems Technician
Heating, Ventilating, and Air Conditioning
Heavy Equipment Operations
Highway/Heavy Construction
Hydroblasting
Industrial Coating and Lining Application Specialist
Industrial Maintenance Electrical and Instrumentation Technician
Industrial Maintenance Mechanic
Instrumentation
Insulating
Ironworking
Masonry
Millwright
Mobile Crane Operations
Painting
Painting, Industrial
Pipefitting
Pipelayer
Plumbing
Reinforcing Ironwork
Rigging
Scaffolding
Sheet Metal
Signal Person
Site Layout
Sprinkler Fitting
Tower Crane Operator
Welding

Maritime

Maritime Industry Fundamentals
Maritime Pipefitting
Maritime Structural Fitter

Green/Sustainable Construction

Building Auditor
Fundamentals of Weatherization
Introduction to Weatherization
Sustainable Construction Supervisor
Weatherization Crew Chief
Weatherization Technician
Your Role in the Green Environment

Energy

Alternative Energy
Introduction to the Power Industry
Introduction to Solar Photovoltaics
Introduction to Wind Energy
Power Industry Fundamentals
Power Generation Maintenance Electrician
Power Generation I&C Maintenance Technician
Power Generation Maintenance Mechanic
Power Line Worker
Power Line Worker: Distribution
Power Line Worker: Substation
Power Line Worker: Transmission
Solar Photovoltaic Systems Installer
Wind Turbine Maintenance Technician

Pipeline

Control Center Operations, Liquid
Corrosion Control
Electrical and Instrumentation
Field Operations, Liquid
Field Operations, Gas
Maintenance
Mechanical

Safety

Field Safety
Safety Orientation
Safety Technology

Supplemental Titles

Applied Construction Math
Tools for Success

Management

Fundamentals of Crew Leadership
Project Management
Project Supervision

Spanish Titles

Acabado de concreto: nivel uno
Aislamiento: nivel uno
Albañilería: nivel uno
Andamios
Carpintería: Formas para carpintería, nivel tres
Currículo básico: habilidades introductorias del oficio
Electricidad: nivel uno
Herrería: nivel uno
Herrería de refuerzo: nivel uno
Instalación de rociadores: nivel uno
Instalación de tuberías: nivel uno
Instrumentación: nivel uno, nivel dos, nivel tres, nivel cuatro
Orientación de seguridad
Paneles de yeso: nivel uno
Seguridad de campo

Acknowledgments

This curriculum was revised as a result of the farsightedness and leadership of the following sponsors:

Austal USA
Bollinger Shipyards
Claddagh Enterprises
Central Louisiana Technical College
Continental Maritime
Exelon Generation
Gulf States, Inc., Michigan Division
Ingalls Shipbuilding
KBR Industrial Services
Lee College
Lincoln Electric
Northeast Community College
Northland Pioneer College
Robins & Morton
Tecnico Corporation
TIC-The Industrial Company
Toledo Refining Co. LLC
Zachry Industrial, Inc.

This curriculum would not exist were it not for the dedication and unselfish energy of those volunteers who served on the Authoring Team. A sincere thanks is extended to the following:

Tom Ashley
Paul Bennett
Gerald Bickerstaff
Curtis Casey
Bill Cherry
Gerald Guthrie
Rod Hellyer
Frank Johnson
John Knapp
Jason Langford
Jerry McDaniel
Tim Riley
Kyle Savoie
Jason Scales
Dan Sterry
Holley Thomas
Jerry Trainor
Adam Webb

NCCER Partners

American Fire Sprinkler Association
Associated Builders and Contractors, Inc.
Associated General Contractors of America
Association for Career and Technical Education
Association for Skilled and Technical Sciences
Construction Industry Institute
Construction Users Roundtable
Construction Workforce Development Center
Design Build Institute of America
GSSC – Gulf States Shipbuilders Consortium
ISN
Manufacturing Institute
Mason Contractors Association of America
Merit Contractors Association of Canada
NACE International
National Association of Minority Contractors
National Association of Women in Construction
National Insulation Association
National Technical Honor Society
NAWIC Education Foundation
North American Crane Bureau
North American Technician Excellence
Pearson
Pearson Qualifications International
Prov
SkillsUSA®
Steel Erectors Association of America
U.S. Army Corps of Engineers
University of Florida, M. E. Rinker School of Building Construction
Women Construction Owners & Executives, USA

NCCER Business Partners

Contents

Module Nine

GTAW – Plate

Describes how to build pads on carbon steel plate using GTAW and carbon steel filler metal. Also explains how to make multiple-pass GTAW fillet welds on carbon steel plate coupons in the 1F, 2F, 3F, and 4F positions, and how to make GTAW V-groove welds in the 1G, 2G, 3G, and 4G positions. (Module ID 29208; 60 hours)

Module Ten

Physical Characteristics and Mechanical Properties of Metals

Explains physical characteristics, mechanical properties, composition, and classification of common ferrous and nonferrous metals. Identifies the various standard metal forms and structural shapes. Shows how to extract metal information from Welding Procedure Specification (WPS) sheets and Procedure Qualification Records (PQRs). Covers visual inspection, magnetic testing, and X-ray fluorescent spectrometry methods used to identify metals. (Module ID 29203; 7.5 hours)

Module Eleven

Preheating and Postheating of Metals

Explains preheating, interpass temperature control, and postheating procedures that sometimes need to be done to preserve weldment strength, ductility, and weld quality. Covers the equipment used for heat treating metals. (Module ID 29204; 5 hours)

Glossary

Index

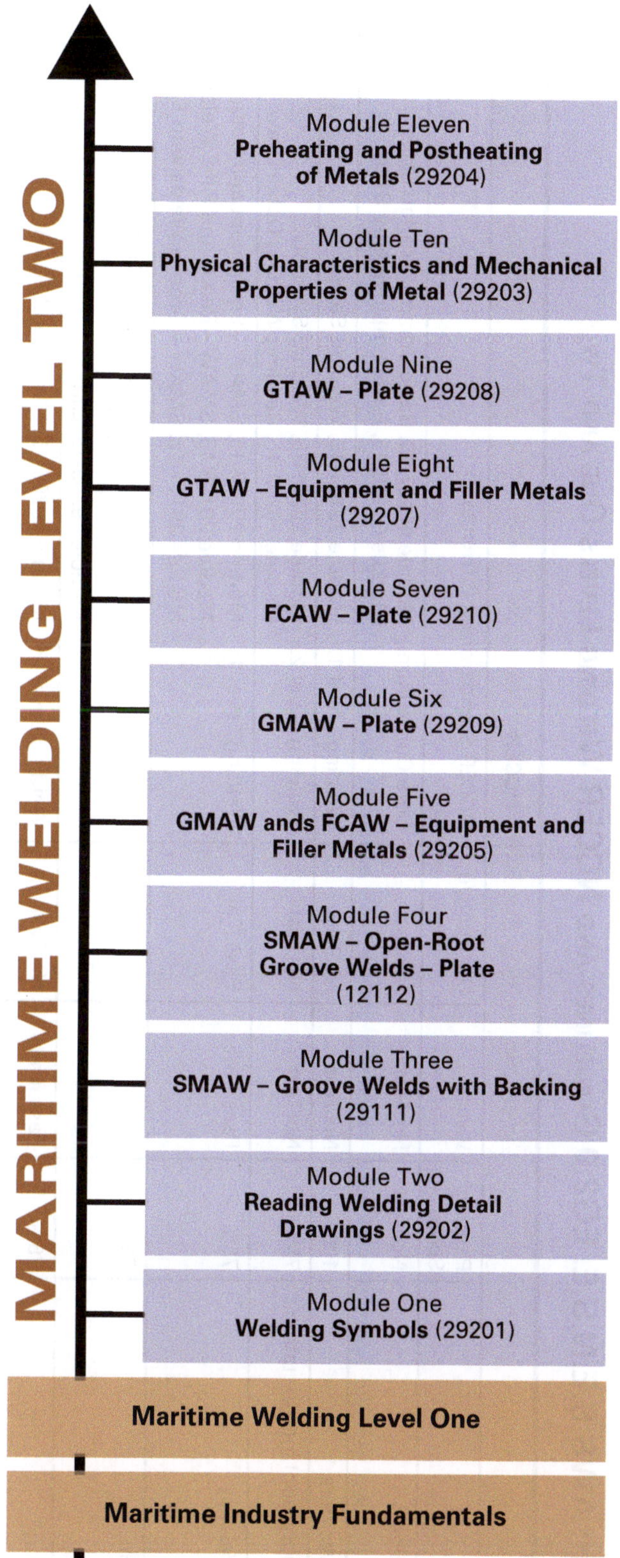

This course map shows all of the modules in *Maritime Welding Level Two*. The suggested training order begins at the bottom and proceeds up. Skill levels increase as you advance on the course map. The local Training Program Sponsor may adjust the training order.

Correlations Chart: AWS S.E.N.S.E. EG2.0 Guidelines and NCCER Welding Levels One and Two

SENSE Guidelines				NCCER
Modules	Passing Score	Visual	Destructive	NCCER Standardized Curricula
1 – Occupational Orientation *	No Test	No Test	No Test	Core Curriculum: Module 00107-15, Basic Communication Skills; Module 00108-15, Basic Employability Skills
2 – Safety and Health of Welders *	100%	No Test	No Test	Welding Level 1: Module 29101-15, Welding Safety
3 – Drawing and Welding Symbol Interpretation *	75%	No Test	No Test	Welding Level 2: Module 29201-15, Welding Symbols; Module 29202-15, Reading Welding Detail Drawings
4 – Shielded Metal Arc Welding	75%	Pass /Fail	Pass /Fail	Welding Level 1: Module 29107-15, SMAW—Equipment and Setup; Module 29109-15, SMAW—Beads and Fillet Welds; Module 29111-15, SMAW—Groove Welds with Backing
5 – Gas Metal Arc Welding	75%	Pass /Fail	No Test	Welding Level 2: Module 29205-15, GMAW and FCAW – Equipment and Filler Materials; Module 29209-15, GMAW – Plate
6 – Flux Cored Arc Welding	75%	Pass /Fail	No Test	Welding Level 2: Module 29205-15, GMAW and FCAW – Equipment and Filler Materials; Module 29210-15, FCAW – Plate
7 – Gas Tungsten Arc Welding	75%	Pass /Fail	No Test	Welding Level 2: Module 29207-15, GTAW – Equipment and Filler Materials; Module 29208-15, GTAW – Plate
8 – Thermal Cutting Process * †	75%	No Test	No Test	
Unit 1 Manual Oxyfuel Gas Cutting (OFC) †		Pass /Fail	No Test	Welding Level 1: Module 29102-15, Oxyfuel Cutting
Unit 2 Mechanized Oxyfuel Gas Cutting (OFC) †		Optional	No Test	Welding Level 1: Module 29105-15, Base Metal Preparation
Unit 3 Manual Plasma Arc Cutting—PAC †		Pass /Fail	No Test	Welding Level 1: Module 29103-15, Plasma Arc Cutting
Unit 4 Manual Air Carbon Arc Cutting †		Optional	No Test	Welding Level 1: Module 29104-15, Air-Carbon Arc Cutting and Gouging
9 – Welding Inspection and Testing *	75%	Pass /Fail	No Test	Welding Level 1: Module 29106-15, Weld Quality

* Required module for Level I Entry Welder Completion (plus one welding process module)

† Completion of Units 1 and 3 minimum

Welding Symbols

Overview

Project drawings and specifications contain numerous symbols that communicate critical information about the welds to be used at various locations throughout the project. Welders must learn this symbolic language so they can properly interpret the symbols and make welds that meet design specifications. This module introduces a wide range of welding symbols, describes how they are structured, and explains the basic rules for applying the information that the symbols convey.

Module 29201

Trainees with successful module completions may be eligible for credentialing through the NCCER Registry. To learn more, go to **www.nccer.org** or contact us at **1.888.622.3720**. Our website has information on the latest product releases and training, as well as online versions of our *Cornerstone* magazine and Pearson's product catalog.

Your feedback is welcome. You may email your comments to **curriculum@nccer.org,** send general comments and inquiries to **info@nccer.org**, or fill in the User Update form at the back of this module.

This information is general in nature and intended for training purposes only. Actual performance of activities described in this manual requires compliance with all applicable operating, service, maintenance, and safety procedures under the direction of qualified personnel. References in this manual to patented or proprietary devices do not constitute a recommendation of their use.

29201 V5

29201

Welding Symbols

Objectives

When you have completed this module, you will be able to do the following:

1. Identify and interpret welding symbols and their structure.
   a. Describe the structure and placement of welding symbols and identify basic symbols.
   b. Identify and interpret size and dimension markings for common types of welds.
   c. Identify and interpret various supplemental symbols.
   d. Identify and interpret less common welding symbols.

Performance Task

Under the supervision of your instructor, you should be able to do the following:

1. Identify and interpret welding symbols on an instructor-provided drawing.

Trade Terms

Arrow line
Countersink
Pitch
Reference line
Weld symbol
Welding symbol

Industry Recognized Credentials

If you are training through an NCCER-accredited sponsor, you may be eligible for credentials from NCCER's Registry. The ID number for this module is 29201. Note that this module may have been used in other NCCER curricula and may apply to other level completions. Contact NCCER's Registry at 888.622.3720 or go to **www.nccer.org** for more information.

Contents

Topics to be presented in this module include:

Figures and Tables

Figures and Tables

Section One

1.0.0 Welding Symbols

Objective

Identify and interpret welding symbols and their structure.

a. Describe the structure and placement of welding symbols and identify basic symbols.
b. Identify and interpret size and dimension markings for common types of welds.
c. Identify and interpret various supplemental symbols.
d. Identify and interpret less common welding symbols.

Performance Task

1. Identify and interpret welding symbols on an instructor-provided drawing.

Trade Terms

Arrow line: The line drawn at an angle from the reference line (either end or both ends) to an arrowhead at the location of the weld.

Countersink: A hole with tapered sides and a wider opening that allows a flat-head fastener to seat flush to the surface of the material in which the hole is made.

Pitch: The center-to-center distance between welds.

Reference line: The horizontal line in the center of the welding symbol from which all elements of the welding symbol are referenced. The reference line is one of the most important elements of the welding symbol.

Weld symbol: A graphic character connected to the reference line of a welding symbol specifying the weld type.

Welding symbol: A graphical representation of the specifications for producing a welded joint; includes a reference line and arrow line and many also include a weld symbol.

Welding symbols are used on drawings, project specifications, and welding procedure specifications (WPS) to convey the design specifications for welds. A series of symbols is used to indicate the joint configuration and weld type, location, size, and length of weld required. Welders must be able to properly interpret welding symbols to ensure that the welds they make will meet the design specifications.

Most companies use the symbols that have been standardized by the American Welding Society (AWS) in *AWS A2.4, Standard Symbols for Welding, Brazing, and Nondestructive Examination.* Most companies also have a site quality standard that provides guidelines and examples of how welding symbols are to be used. Always refer to the site-specific quality standard when interpreting welding symbols.

1.1.0 Welding Symbol Construction

The base for all welding symbols is the horizontal reference line which has an arrow at one end. The arrow line, which can be on either side of the reference line, points to the location to which the welding symbol applies (*Figure 1*). The welding symbol describes the type of weld, its size, and its surface finish.

The opposite end of the reference line, called the tail, is used for information that aids in making the weld but does not have its own special place on the symbol (*Figure 2*). For instance, the tail may be used to indicate the welding and cutting processes, the reference to a note, or the welding procedure or electrode type to be used. When a reference is not required, the tail is omitted, as shown in *Figure 1.*

In the case of a T-joint, the arrow points to the two sides of the joint, and welding may be accomplished on either side. The welding symbol distinguishes between the two sides of a joint by using the arrow line and the spaces above and below the reference line. The side of the joint to which the arrow points is called the arrow side, and its weld is made in accordance with the instructions below the reference line. The opposite side of the joint is referred to as the other side,

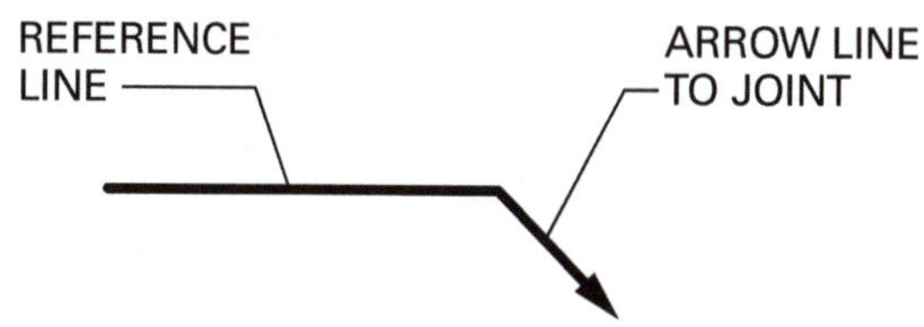

Figure 1 Horizontal reference line.

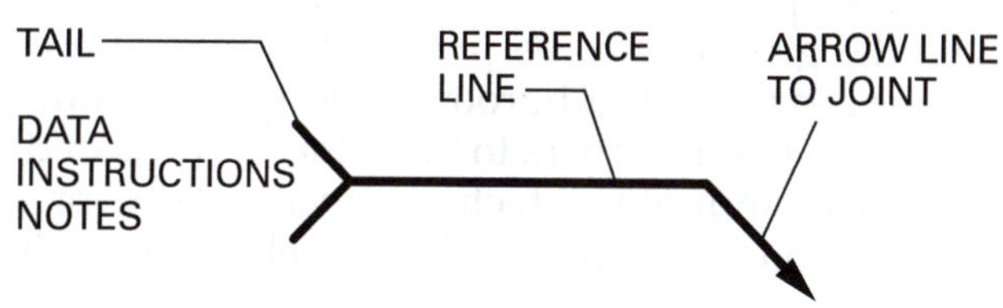

Figure 2 Tail-side information

Welding and Cutting Processes

The *Appendixes* in this module contain tables listing AWS abbreviations for various welding and cutting processes. Abbreviations from all three tables can be used in tail references, or in single-line or multiple-line welding symbols.

and its weld is made in accordance with the instructions found above the reference line. *Figure 3* shows examples of these symbols.

Regardless of which end of the reference line the arrow line is on, information on the reference line is always read from left to right. *Figure 4* shows a welding symbol base and the numerous elements of a welding symbol.

The **weld symbols** used to identify the type of weld to be made signify a basic type of weld or joint preparation (*Figure 5*). It is important to note the difference between welding symbols and weld symbols. A weld symbol is a graphic character connected to the reference line of a welding symbol, specifying the weld type. A welding symbol graphically represents the specifications for producing a welded joint; it includes any weld symbols as well as the additional components. All welding symbols include a reference line and arrow line, at a minimum. Many also include a weld symbol.

1.1.1 Symbols for Welds

The type of weld used is determined by the way the members to be welded are positioned to form the joint. The five basic types of weld joints are the butt, corner, tee, lap, and edge joints (*Figure 6*).

1.1.2 Location of Weld Symbols

When the weld symbol is on both the top and bottom of the reference line, the information applies to both sides of the joint or surface to which the arrow is pointing. Symbols appear on both sides of the reference line when a double-fillet or double-groove weld is specified. *Figure 7* shows symbols for arrow-side, other-side, and both sides for fillet and V-groove welds.

The shape of the groove weld symbol indicates how the groove is to be prepared. In the case of the bevel-groove and J-groove weld symbols, only one of the two members to be welded is prepared. The arrow indicates which member is to be prepared by breaking toward that member (*Figure 8*).

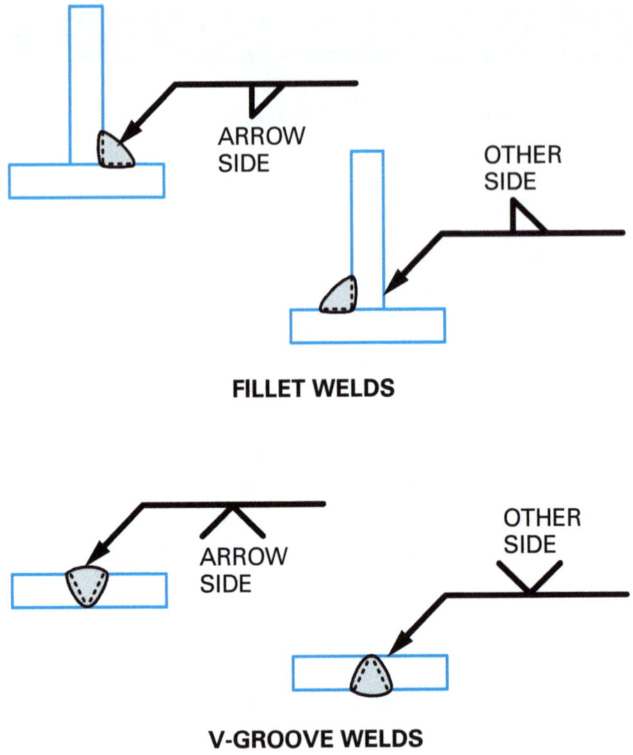

Figure 3 Arrow-side and other-side significance.

Notice also that the vertical leg of the weld symbol is shown drawn to the left of the slanted/curved leg(s). Regardless of whether the symbol is for a fillet, bevel-groove, J-groove, or flare bevel-groove weld, the vertical leg is always drawn to the left.

1.1.3 Combining Weld Symbols

When more than one type of weld is to be made on the same joint, it is necessary to combine weld symbols (*Figure 9*). A typical combination weld is made by adding a fillet weld to a single or double bevel groove weld in a T-joint. The fillet weld symbol is always placed on top of the groove weld symbol, just as it would be on the actual weld.

1.1.4 Multiple Reference Lines

When more than one welding operation must be performed in a certain sequence, two or more reference lines may be used to indicate the sequence.

Think About It

Breaking the Arrow

Are the symbol representations shown in *Figure 8* the only ways a breaking arrow can be shown?

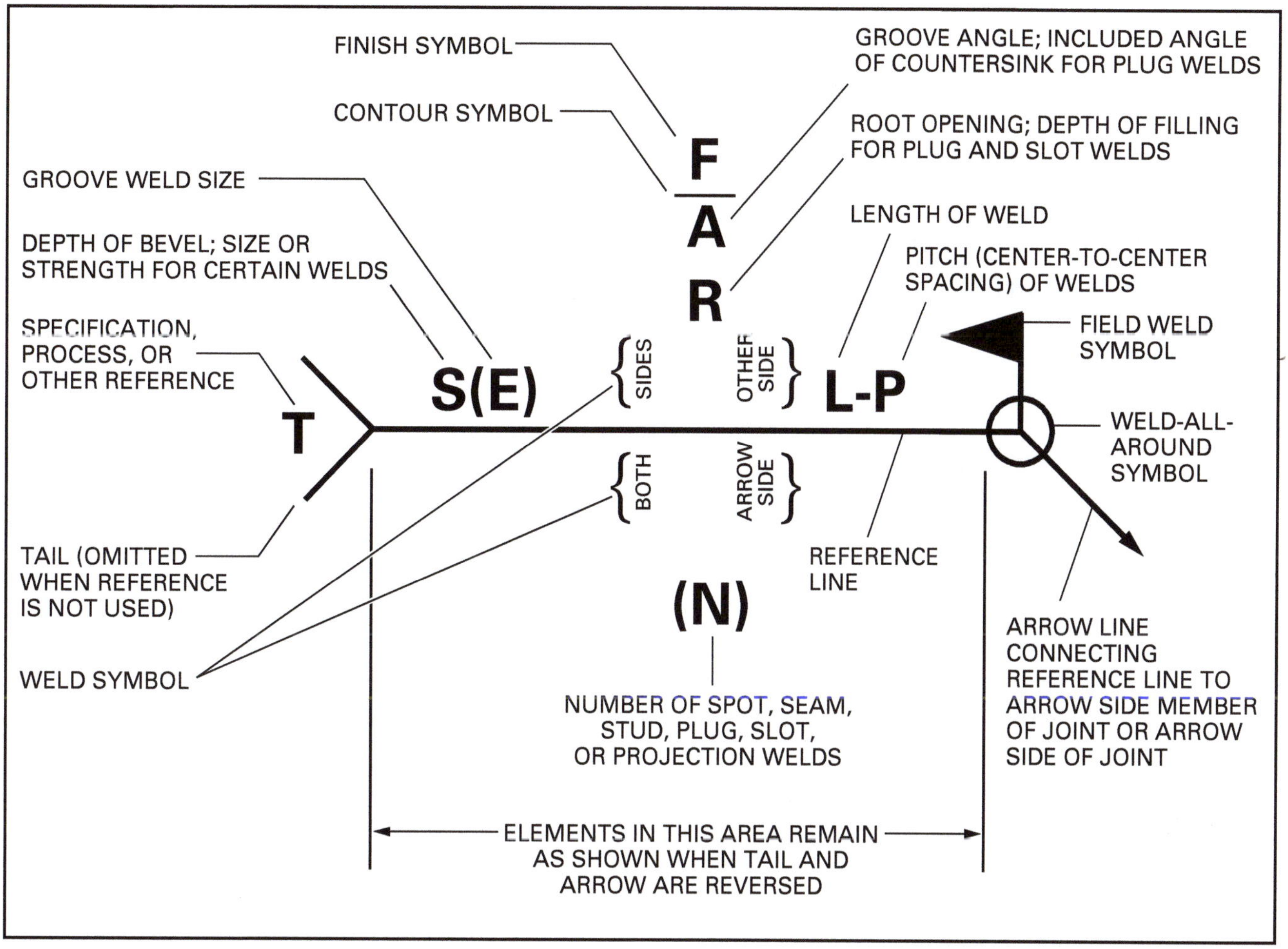

Figure 4 Standard location of elements of a welding symbol.

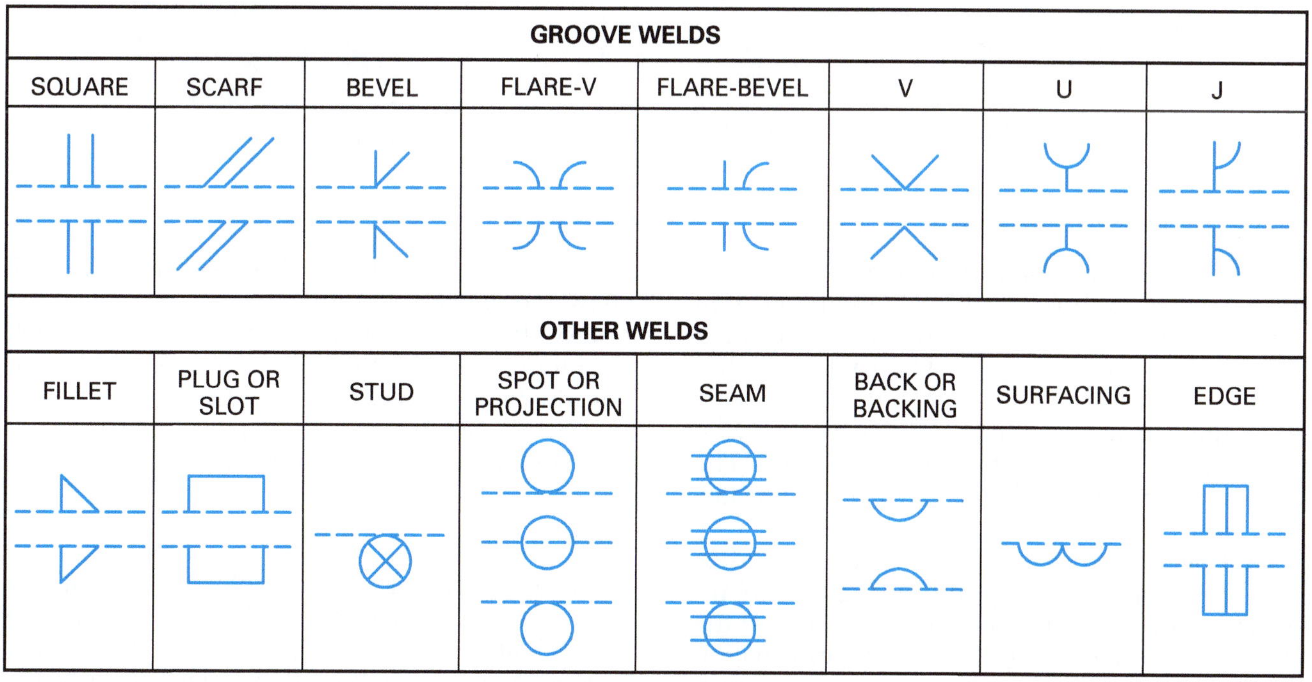

Figure 5 Basic weld symbols.

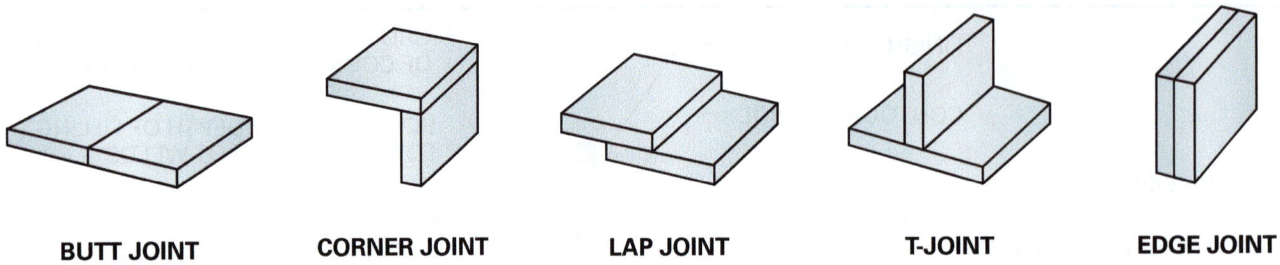

Figure 6 Basic types of weld joints.

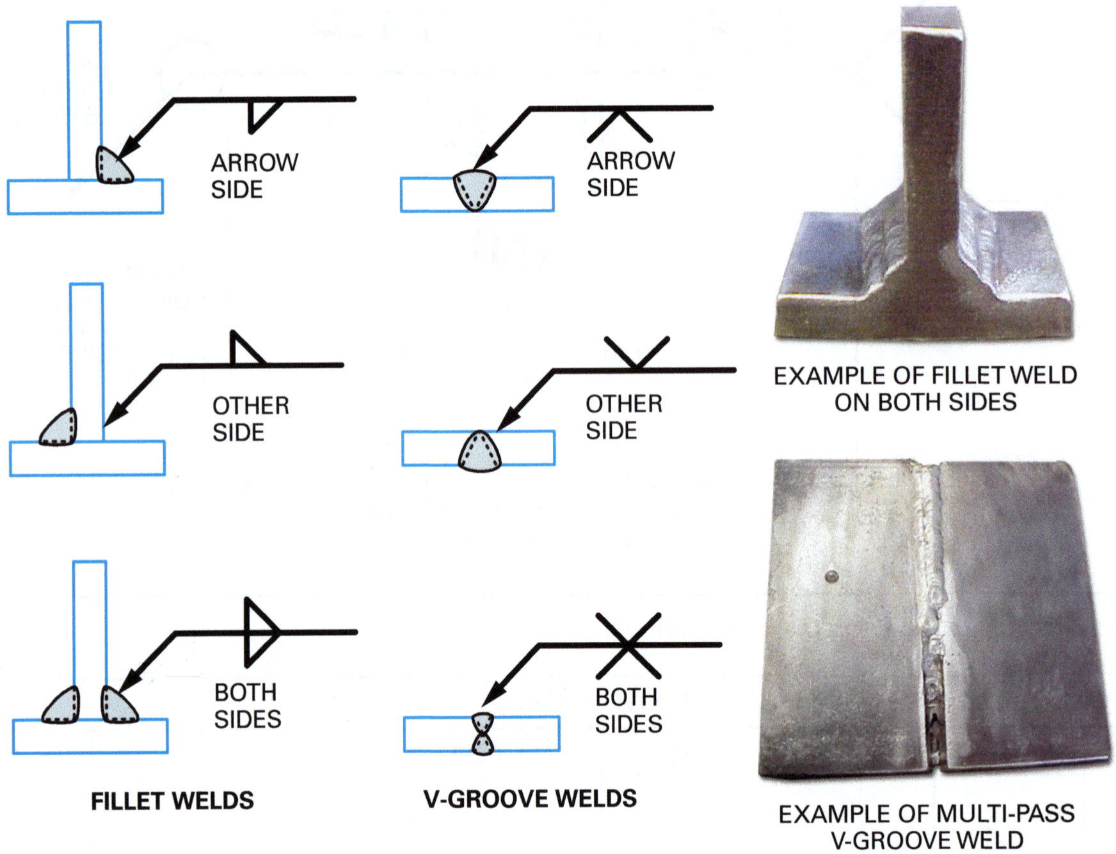

Figure 7 Fillet and V-groove welds.

Figure 10 illustrates the typical sequences of operation when multiple reference lines are used. In each sequence, the first operation to be performed will be closest to the arrow. The first operation must be shown on the reference line that is nearest to the arrow. Subsequent operations may be shown sequentially on other reference lines. The last operation to be performed will be the one farthest from the arrow.

Multiple reference lines may also be used to show data supplementary to welding symbol information that is on the reference line nearest the arrow. As shown in *Figure 11*, test information may be shown on a second or third line, away from the arrow.

When required, the weld-all-around symbol, shown in *Figure 12(A)*, must be placed at the junction of the arrow line and reference line for each operation to which it applies. The field weld symbol shown in *Figure 12(B)* may also be used in this manner. Supplemental symbols are discussed in more detail later in this module.

1.2.0 Sizing and Dimensioning Welds

Unless defined in a drawing note, the size data for a weld is always shown to the left of the symbol to which it applies. For example, the size of the fillet weld shown in *Figure 13* is ¼ inch. The length of the weld is always shown to the right of the symbol to which the length applies. An example of this is seen in *Figure 14*, where the length is 6 inches.

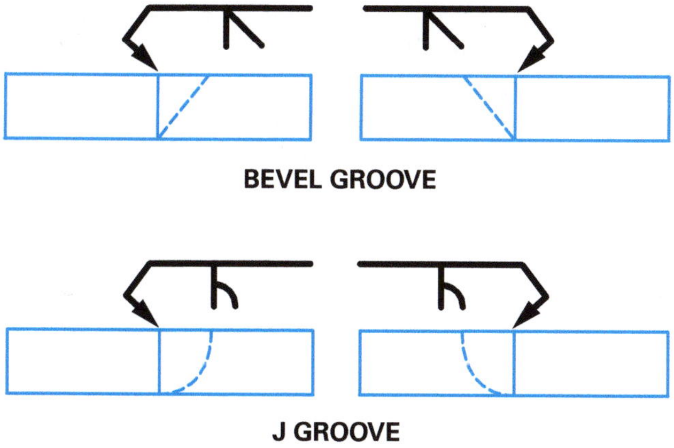

NOTE: DASHED LINES INDICATE PREPARATION SPECIFIED BY WELD SYMBOL

Figure 8 Bevel and J-groove weld symbols.

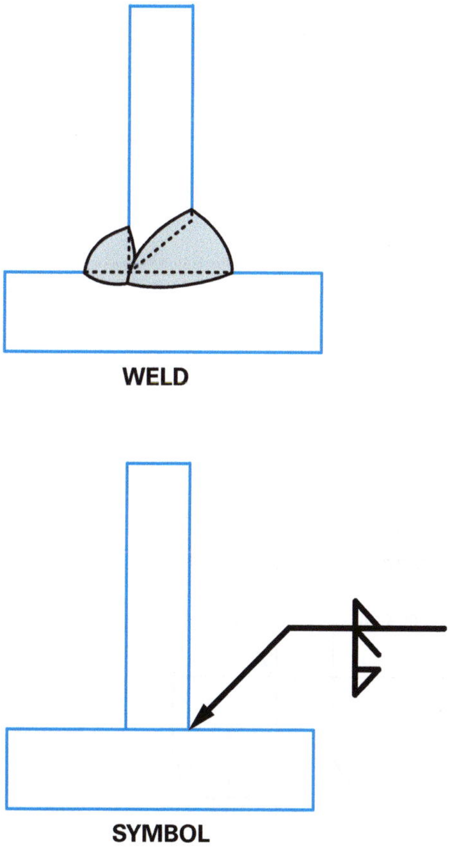

Figure 9 Combination weld symbol.

Dimensions can be given as fractions, decimals, or metric measurements. The following sections explain how to size and dimension the following weld types:

- Fillet
- Groove
- Plug
- Slot

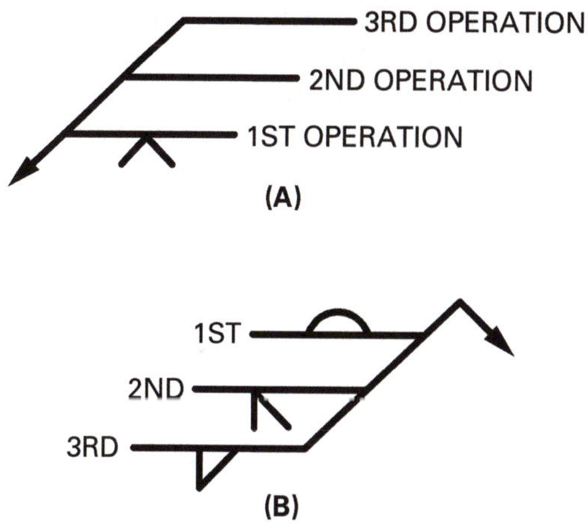

Figure 10 Multiple reference lines.

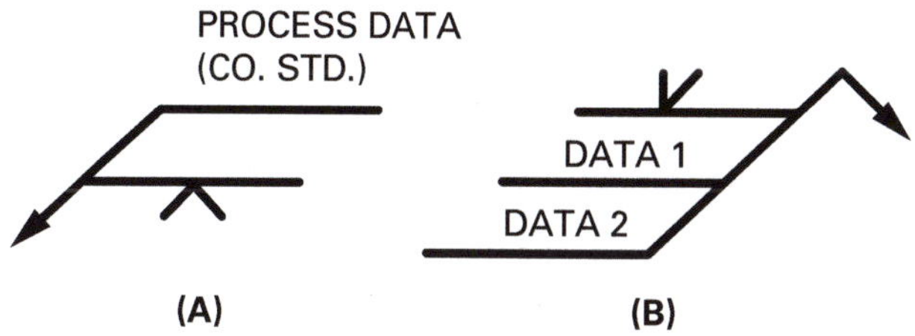

Figure 11 Supplementary data.

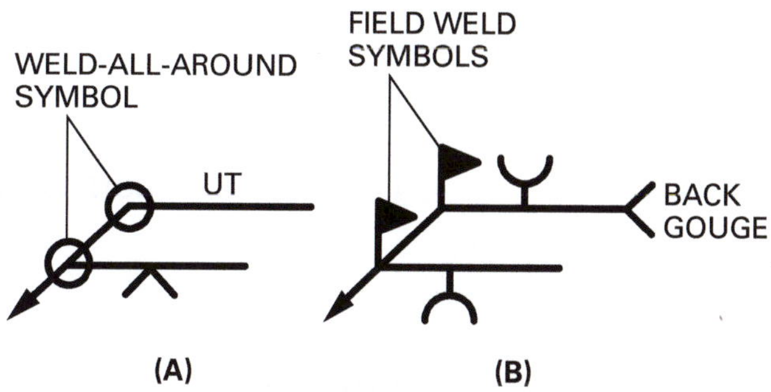

Figure 12 Typical placement of supplemental symbols.

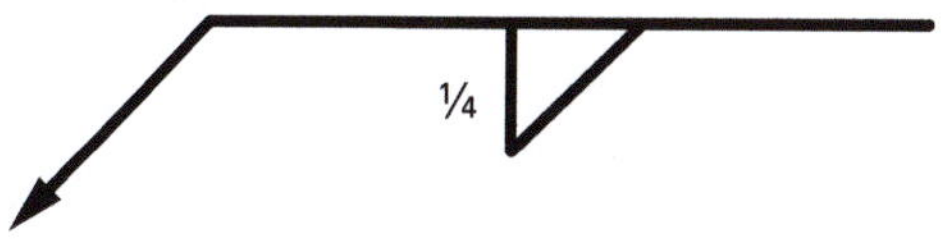

Figure 13 Sizing welds

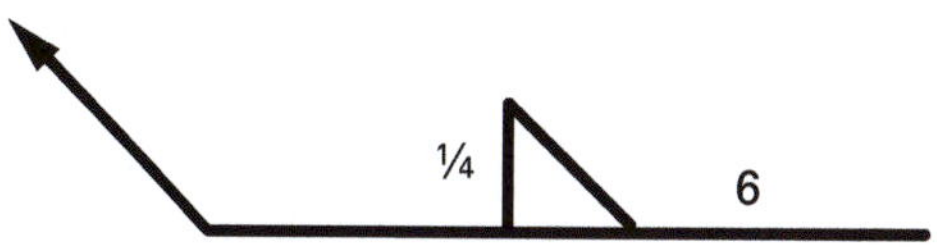

Figure 14 Dimensioning welds.

1.2.1 Sizing Fillet Welds

A fillet weld is a weld made in a corner. It is roughly triangular in shape and joins two welding surfaces that are perpendicular to each other, such as in a lap, T-, or corner joint. The size of a fillet weld is the length of the legs (sides) of the largest triangle that can be inscribed within the fillet weld cross-section. *Figure 15* shows the profiles of convex and concave welds.

The lengths of the legs on a typical fillet weld are equal, so only one dimension is given. Equal-leg fillet welds are sized by placing the size to the left of the fillet weld symbol or as defined in a drawing note.

If the fillet weld has unequal legs, the size of each leg is given in parentheses or brackets to the

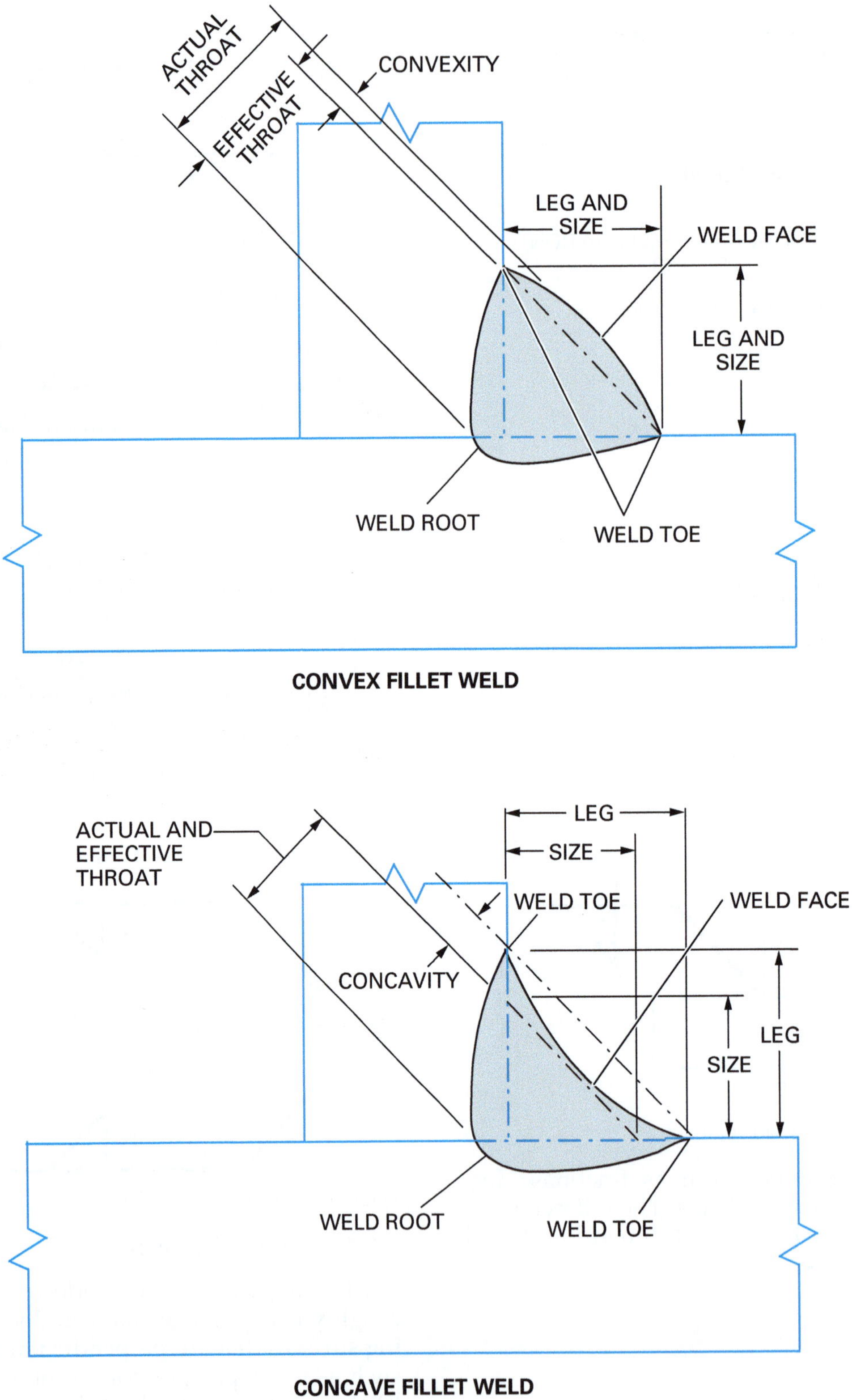

Figure 15 Profiles of fillet welds.

Nondestructive Examination Symbols

Nondestructive examination symbols may appear on the same drawings as (and look very similar to) welding symbols. Inspectors use nondestructive examination symbols to examine joints once welding is complete. The nondestructive methods of inspection are defined by letter codes on the reference line. For instance, VT indicates visual testing, UT indicates ultrasonic testing, RT indicates radiographic testing, and PT indicates penetrant testing.

NDT symbols and weld symbols may be combined by using two reference lines. In contrast to the arrow for welding symbols, the arrow for NDT always points to the weld that is to be tested, and never to a weld on the other side of the joint. If a double fillet weld were to be tested, for example, this would be shown with a separate arrow on each side of the joint.

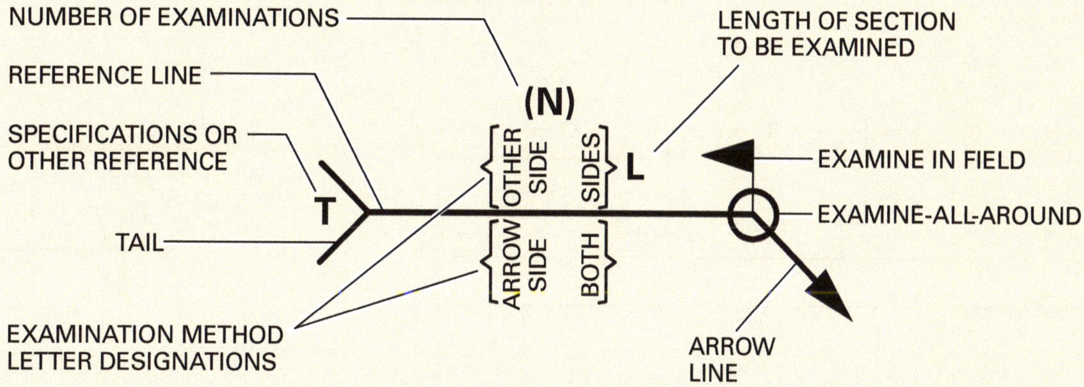

left of the fillet weld symbol with a drawing or explanation of leg orientation. An example of this is shown in *Figure 16* where the unequal legs are ¼ inch (6.4 mm) and ½ inch (12.7 mm).

1.2.2 Dimensioning Fillet Welds

The length of a fillet weld is shown to the right of the fillet weld symbol. If the fillet weld is to be an intermittent fillet weld, the length (length of each segment) and the **pitch** (center-to-center distance between welds) are shown with a dash. The dash appears between the length, which is always shown first, and the pitch, which is always shown second. *Figure 17* shows a length of 2 inches (5.1 cm) and a pitch of 6 inches (15.2 cm). If fillet welds are required on both sides of the joint, they can be back-to-back or staggered. If the welds are back-to-back, as shown in *Figure 17(A)*, the fillet weld symbols are aligned evenly on both sides of the reference line, as shown in *Figure 17(B)*. If the welds are to be staggered, as shown in *Figure 17(C)*, the fillet weld symbols above and below the reference line are offset, as shown in *Figure 17(D)*.

Unequal Fillet Weld Legs

The only practical use of unequal fillet weld legs is on metal of unequal thickness. Unequal legs do not make the weld stronger; the thinner material prevails. An unequal fillet leg weld helps distribute loads and eliminates incomplete fusion discontinuities that would be caused by making a small weld on heavy material without preheat.

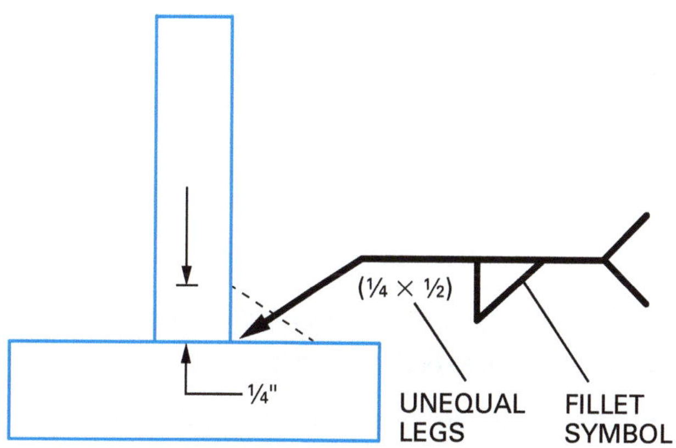

Figure 16 Sizing fillet welds.

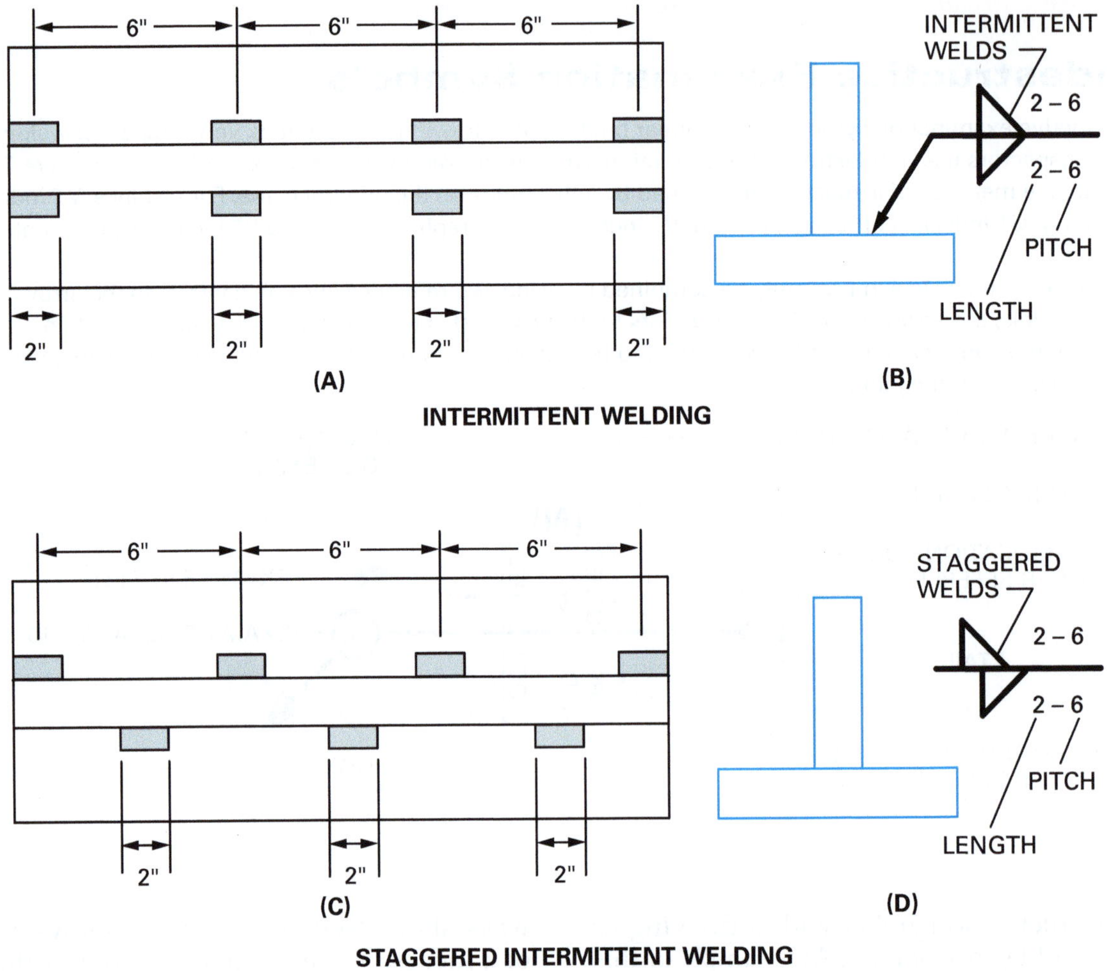

INCHES	CENTIMETERS
2	5.1
6	15.2

Figure 17 Dimensioning fillet welds.

When there is no dimension to the right of the fillet weld symbol, the weld extends the full length of the joint.

1.2.3 Sizing Groove Welds

A groove weld is a weld that is made in the groove formed by two beveled pieces of metal (*Figure 18*).

Think About It

Location of Unequal Fillet Weld Legs

How does a welder determine which side of a fillet weld has the longer leg?

Several factors must be considered when sizing groove welds. These factors apply to all groove weld types.

Groove preparation is the depth to which the groove extends into the base metal. When the groove weld extends all the way through the joint, as shown in *Figure 19(A)*, no preparation size needs to be shown. When the groove weld extends only part of the way through the joint, as shown in *Figure 19(B)*, the depth of the groove is shown to the left of the symbol. *Figure 19(B)* shows sizing a groove with a depth of ¼ inch (6.4 mm).

For partial joint penetration (PJP) groove welds, both the depth of the groove and weld size (joint penetration) are shown, as in *Figure 20*. The depth of the groove is shown first and the weld size is shown next, in parentheses. In

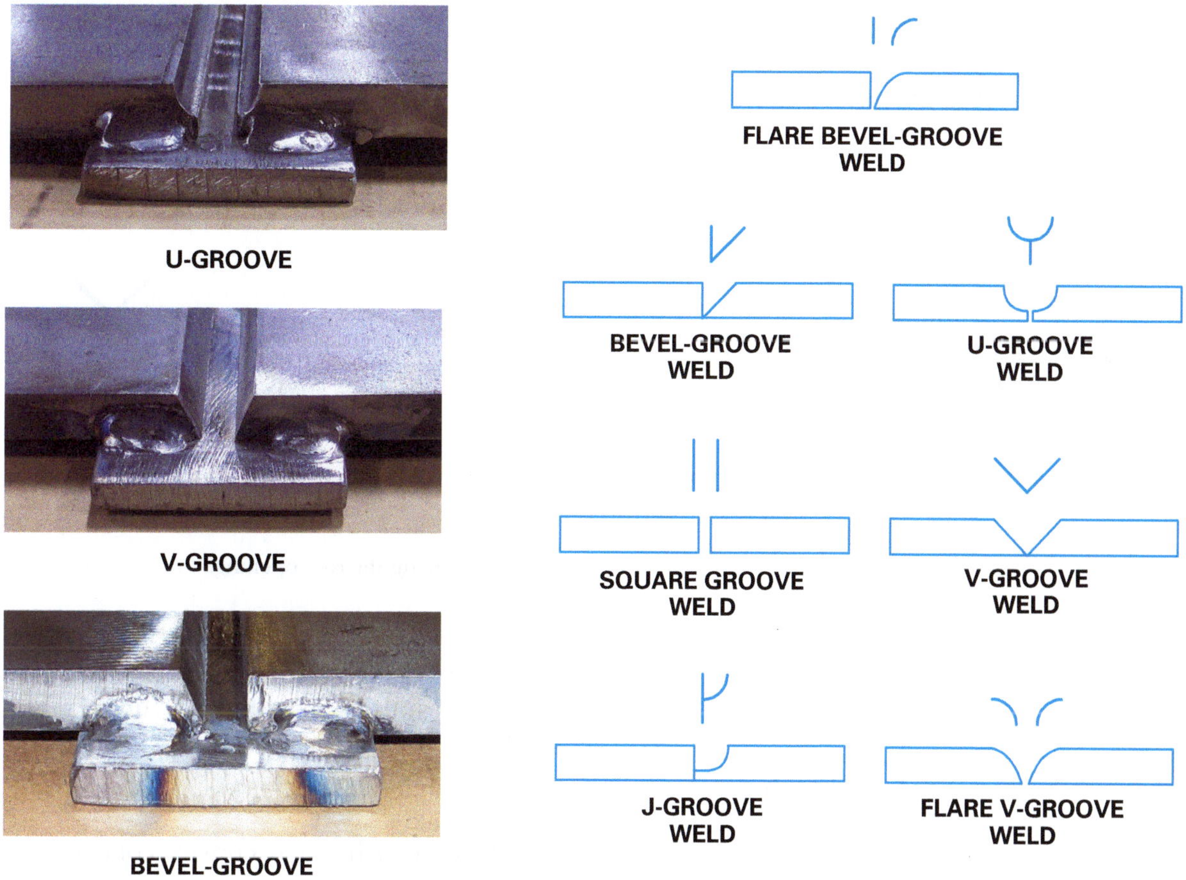

Figure 18 Types of groove welds.

this example, the groove depth is 0.25 inch (6.4 mm) and the weld size is 0.385 inch (9.8 mm). For complete joint penetration (CJP) groove welds, the weld metal extends through the joint thickness.

If a root opening is required, the root opening dimension of a groove weld will be shown inside the symbol. In *Figure 21*, the root opening is $\frac{1}{8}$ inch (3.6 mm) and is shown on only the arrow-side symbol because the opening is common to both grooves.

The angle for the type of groove is shown above or below the welding symbol, depending on whether the symbol is above or below the reference line. In *Figure 22*, the groove angle is shown as 45 degrees and is shown only by the arrow-side symbol because the groove is the same on both sides. The bevel angle for groove preparation on each workpiece is $22\frac{1}{2}$ degrees.

1.2.4 Dimensioning Groove Welds

The length of a groove weld is shown to the right of the groove-weld symbol. If there is no dimension on the right of the groove-weld symbol, the weld extends the full length of the joint.

1.2.5 Sizing and Dimensioning Plug Welds

A plug weld is a weld made in a circular hole in one member of a joint, fusing that member to another member. The size of a plug weld, which is the diameter of the plug, is shown to the left of the welding symbol. The number of plug (and slot) welds can be shown below the symbol. If the plug is not completely filled, the depth of the fill is given inside the plug weld symbol, as shown in *Figure 23(A)*. If the plug weld is to be countersunk, the angle of the **countersink** is shown below or above the weld symbol, depending on which side of the reference line the symbol is on, as shown in *Figure 23(B)*. The pitch, shown in *Figure 23(C)*, represents the center-to-center spacing of the plugs shown in *Figure 23(B)* to the right of the plug weld symbol. *Figure 23* shows a composite view of sizing and dimensioning plug welds.

1.2.6 Sizing and Dimensioning Slot Welds

The only time depth of fill is used with a slot weld symbol is when the slots are partially filled. The partial depth of fill is shown in the symbol (*Figure 24*). All other information, such as length, width, spacing, angle of countersink, and location of slots, is shown in a special detail on the print.

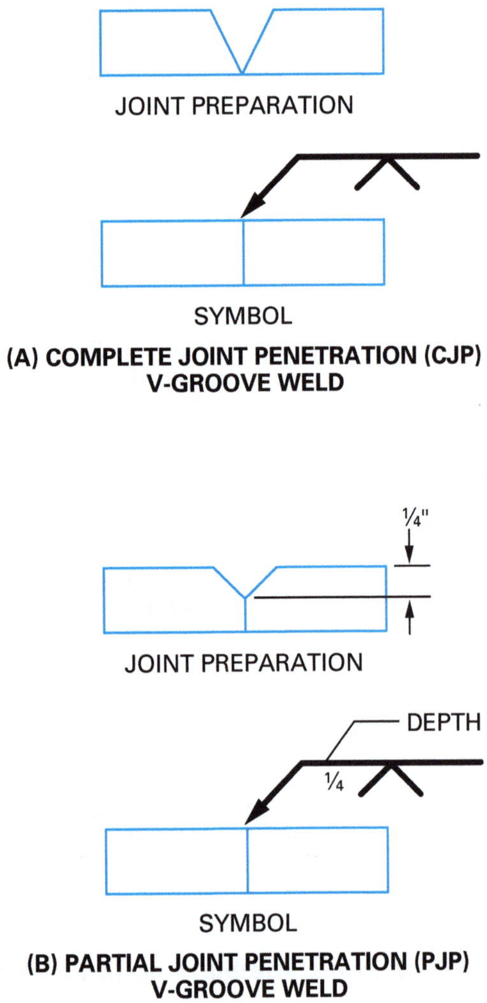

Figure 19 Sizing groove welds.

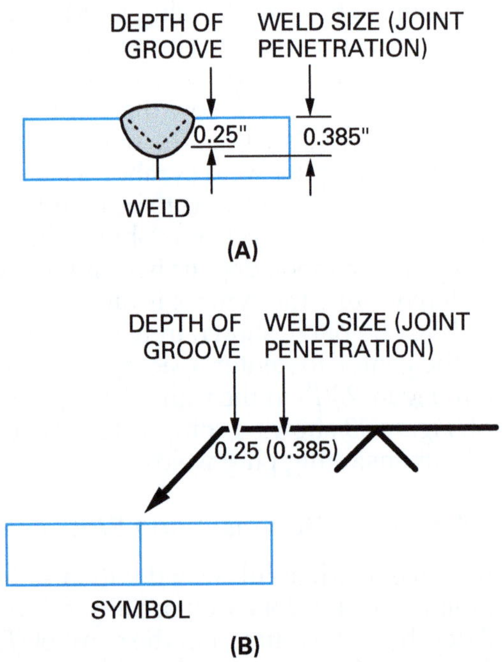

Figure 20 Sizing partial joint preparation (PJP) welds.

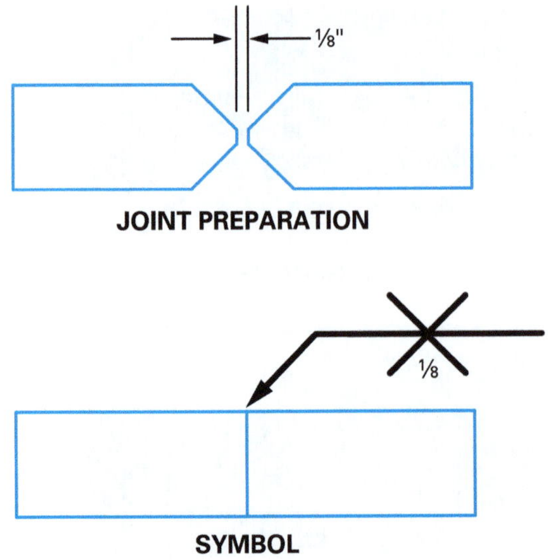

Figure 21 Sizing the root opening.

1.3.0 Supplemental Symbols

The following supplemental symbols are used to convey special instructions that apply to the welding symbol:

- Weld-all-around
- Field weld
- Contour finish (flush, convex, concave)

Figure 25 shows these and other supplemental symbols.

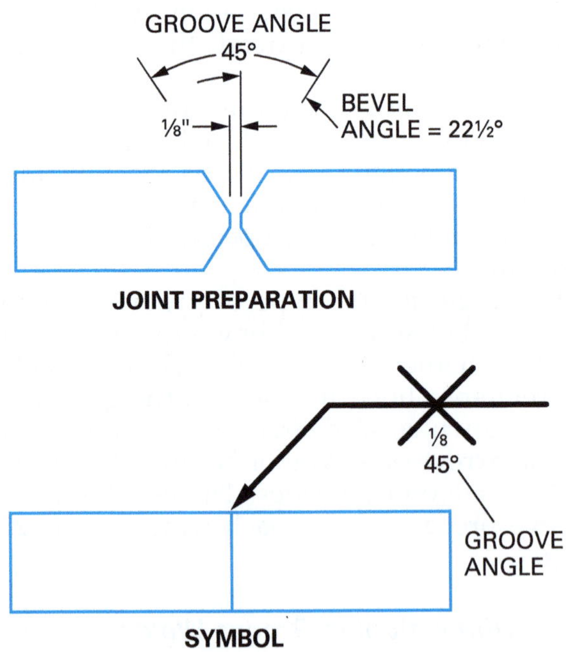

Figure 22 Specifying the groove angle.

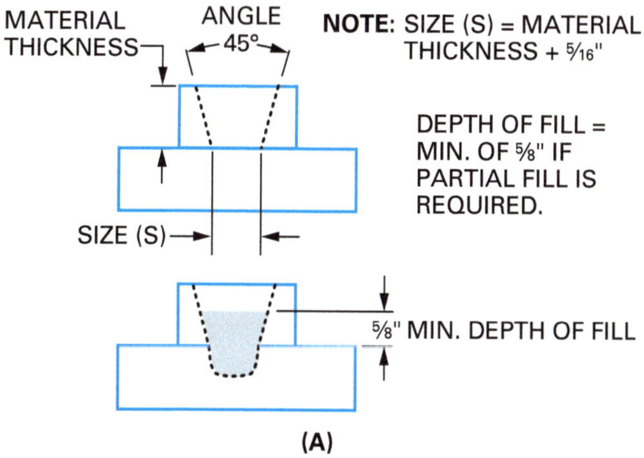

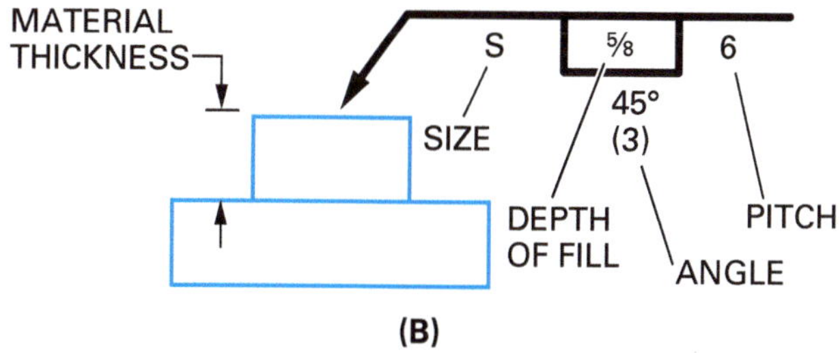

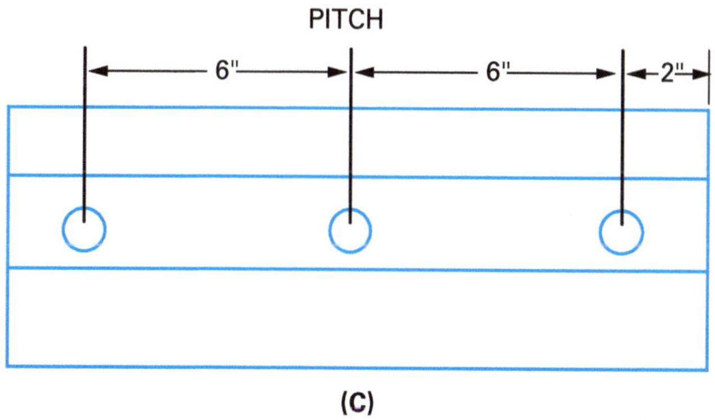

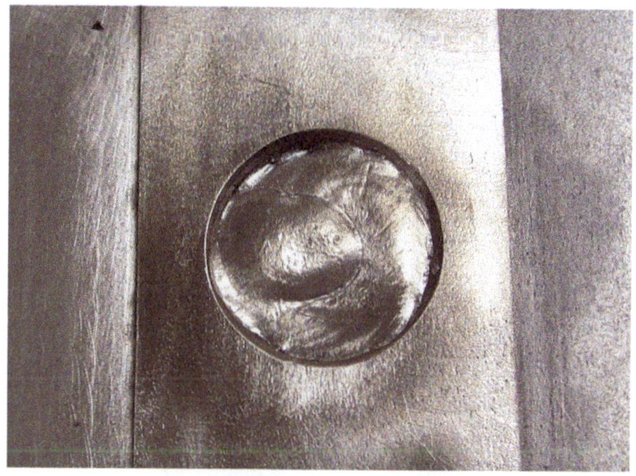

TYPICAL PLUG WELD

INCHES	MILLIMETERS
5⁄16	7.9
5⁄8	15.9
2	50.8
6	152.4

Figure 23 Sizing and dimensioning plug welds.

1.3.1 Weld-All-Around

When a weld is to extend completely around the joint, such as when welding a channel to a base plate, a small circle is shown where the arrow line meets the reference line. *Figures 25* and *26* both show examples of the symbol placement. *Figure 26* also shows the intended result of a typical weld-all-around symbol.

1.3.2 Field Weld

When the weld indicated by the welding symbol is to be made at a location other than that of the initial construction, a small flag is placed where the arrow line meets the reference line. The flag can be pointing either left or right. It can also point either up and down. *Figure 27* shows two examples of field welding symbols.

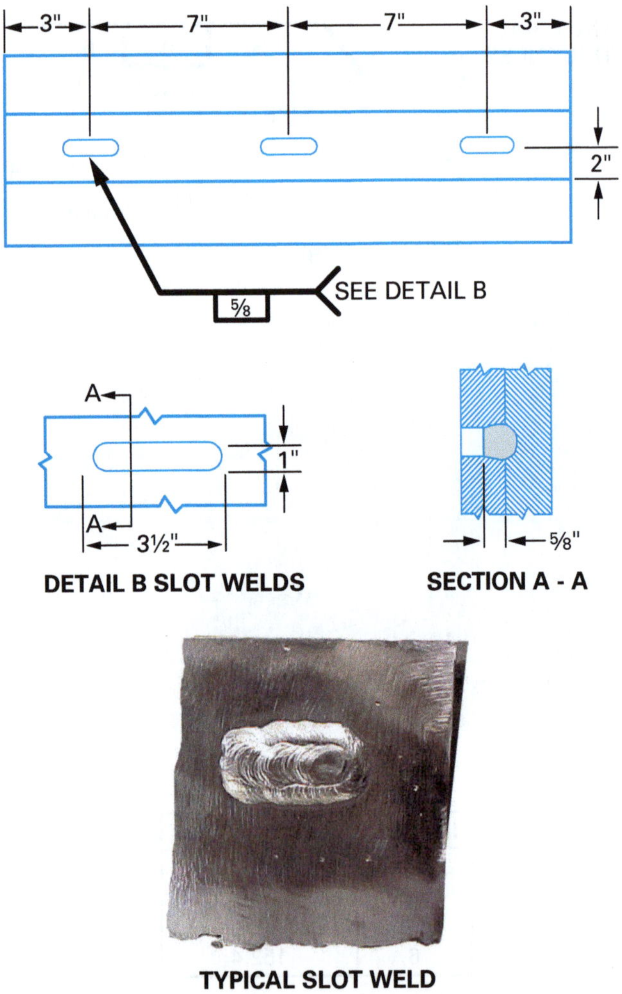

INCHES	MILLIMETERS
5⁄8	15.9
1	25.4
2	50.8
3	76.2

Figure 24 Sizing slot welds.

1.3.3 Contour Finish

When the face of the weld must have a finished shape that is not its normal as-welded condition, a finish symbol that indicates a flush, convex, or concave surface is placed adjacent to the welding symbol. The method used to finish the weld face is specified by a letter, such as M for machining (*Figure 28*). *Table 1* shows the related finishing methods with their letter symbols.

The Letters TYP

The letters TYP (meaning typical) are placed in the tail of the welding symbol to indicate multiple joints of the same configuration. This eliminates duplication of the same welding symbol on the drawing.

1.4.0 Other Symbols

Other, less common weld symbols are shown in *Figure 29*. They include the following:

- Backing and spacer
- Back or backing
- Melt-through
- Surfacing
- Edge
- Spot
- Seam

If a consumable insert is required, details are provided with the symbol.

1.4.1 Backing and Spacer Symbols

Backing is used at the root of the weld to control melt-through. The backing can be metal strips or shapes, flux-coated tape, fiberglass-coated tape, or ceramic tape. The backing symbol is placed opposite a groove weld symbol on the reference line. The material, size, and shape of the backing are specified in the tail of the welding symbol. Spacers are metal strips or shapes that are placed at the root of the joint to control burn-through. They are used most often on double-groove joint preparations. The spacer symbol is centered on the reference line. The joint weld symbol is modified to project from the corners of the spacer symbol. The material and size of the spacer are noted in the tail of the welding symbol. *Figure 30* includes illustrations of typical backing and spacer symbols.

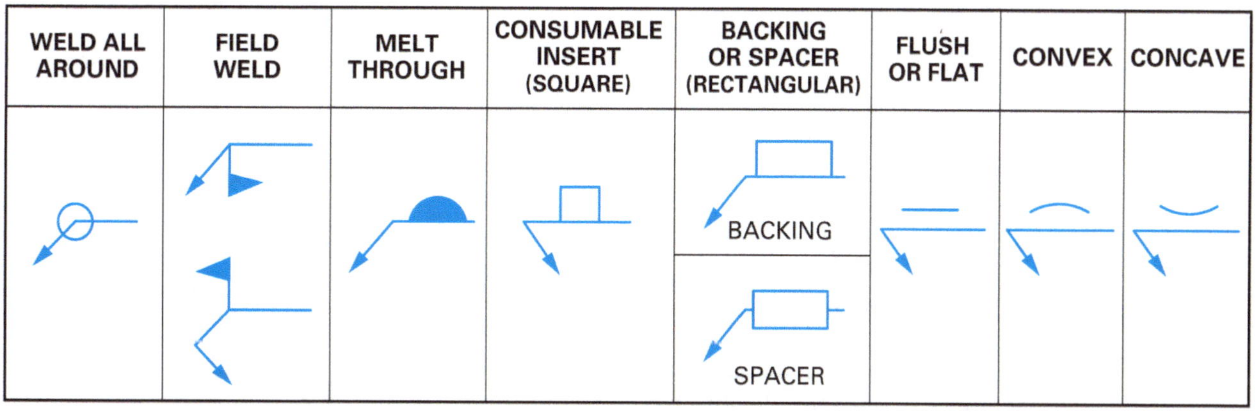

INCHES	MILLIMETERS
3 ½	88.9
7	177.8

Figure 25 Supplemental symbols.

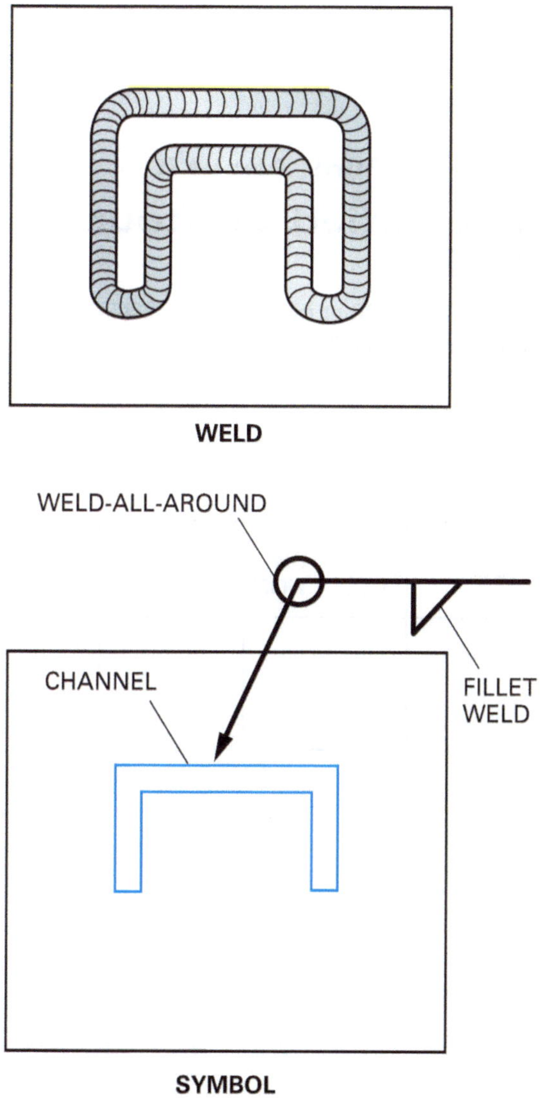

Figure 26 Weld-all-around symbol.

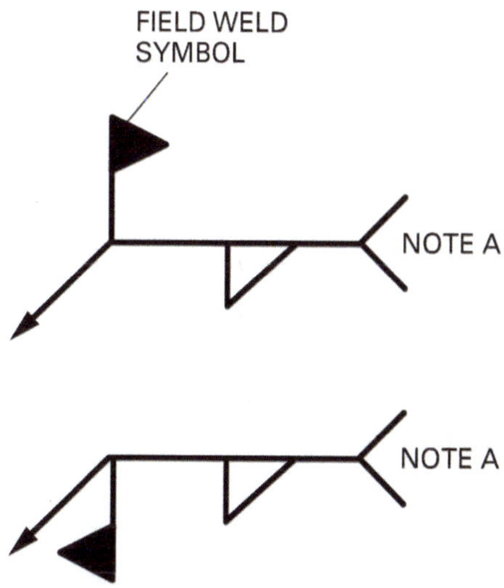

Figure 27 Field welding symbols.

1.4.2 Back or Backing Weld Symbols

A back weld is made after the groove weld. A backing weld is made before the groove weld. The same symbol is used to indicate either a back or a backing weld (*Figure 31*). The back or the backing weld symbol is placed opposite a groove weld symbol on the reference line. A note in the tail of the welding symbol indicates which weld is to be made.

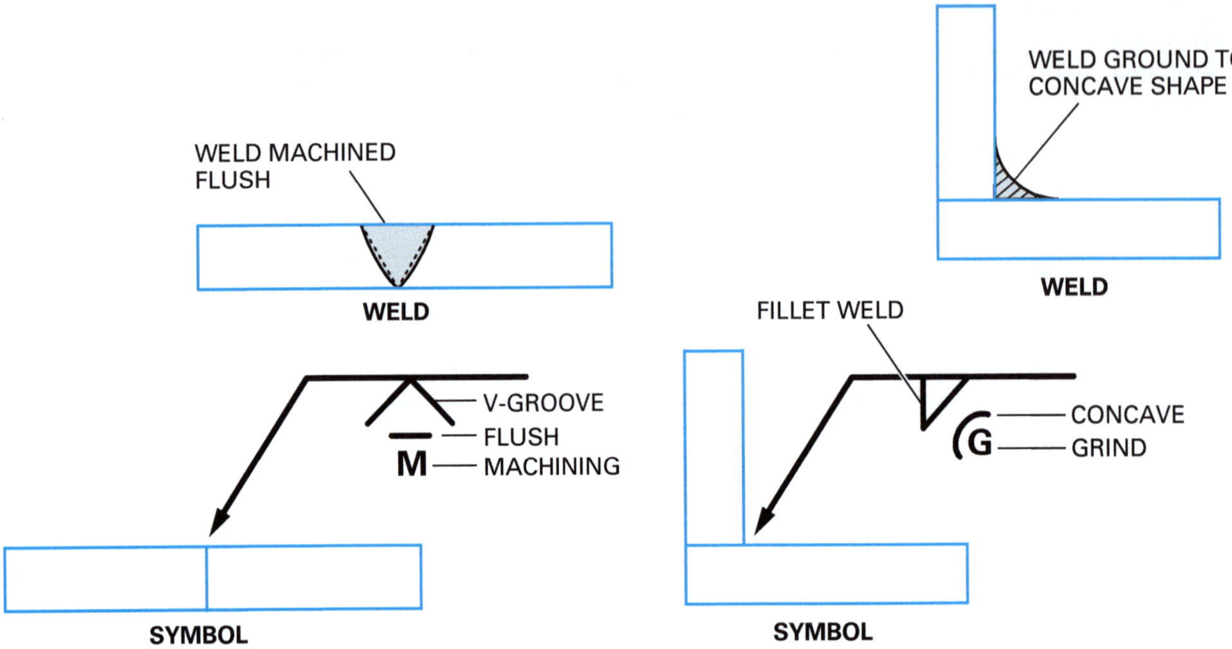

Figure 28 Examples of contour finishes and symbols.

Table 1 Finishing Methods

Method	Letter
Chipping	C
Grinding	G
Hammering	H
Machining	M
Rolling	R
Unspecified	U

Think About It

Unspecified Contour Finishes

Refer to *Table 1*. What might an unspecified (U) contour finish mean?

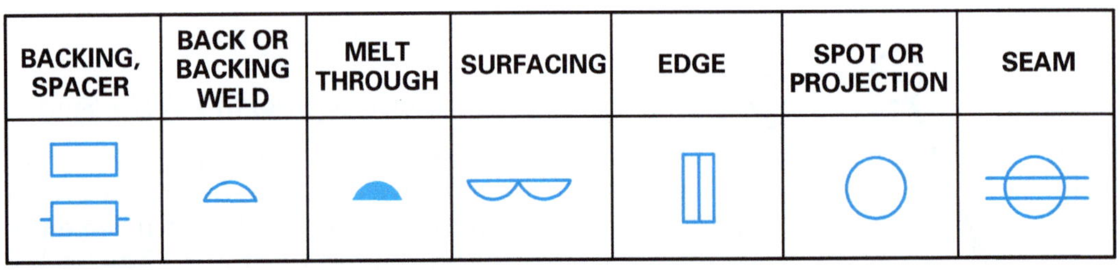

Figure 29 Other symbols.

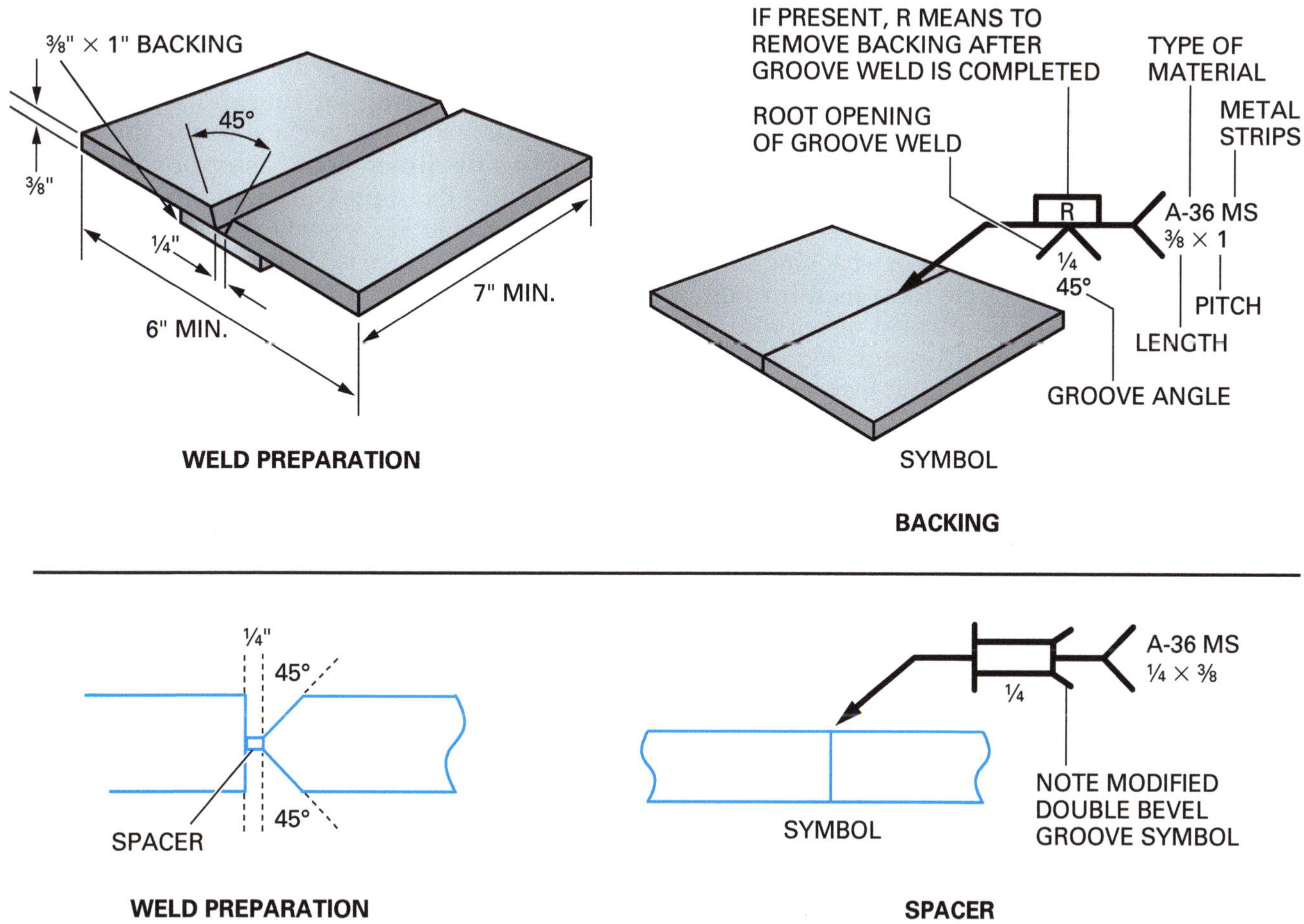

INCHES	MILLIMETERS
¼	6.4
⅜	9.5
1	25.4
6	152.4
7	177.8

Figure 30 Backing and spacer symbols.

An Alternate Welding Symbol

The welding symbol shown in *Figure 31* could also be presented as a multiple line symbol. Using a multiple line symbol eliminates the need for the tail information showing the back weld. The multiple line symbol indicates that the first operation is the groove weld and the second operation, by definition, can only be a back weld rather than a backing weld.

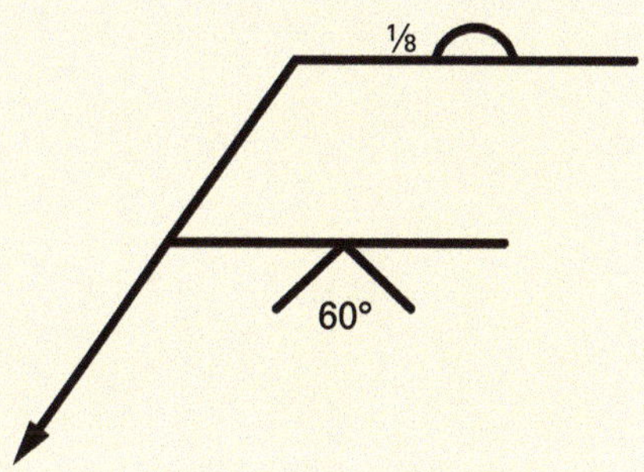

1.4.3 Melt-Through Symbols

When complete penetration is required from one side of a joint and root reinforcement is desired, a melt-through symbol (*Figure 32*) can be used to specify the amount of melt-through. The symbol is placed opposite a groove weld symbol on the reference line. A melt-through symbol is similar in shape to the semi-circle of a back or a backing symbol. However, the semi-circle for a melt-through symbol is solidly filled in, instead of remaining blank. The amount of melt-through required is placed to the left of the melt-through symbol.

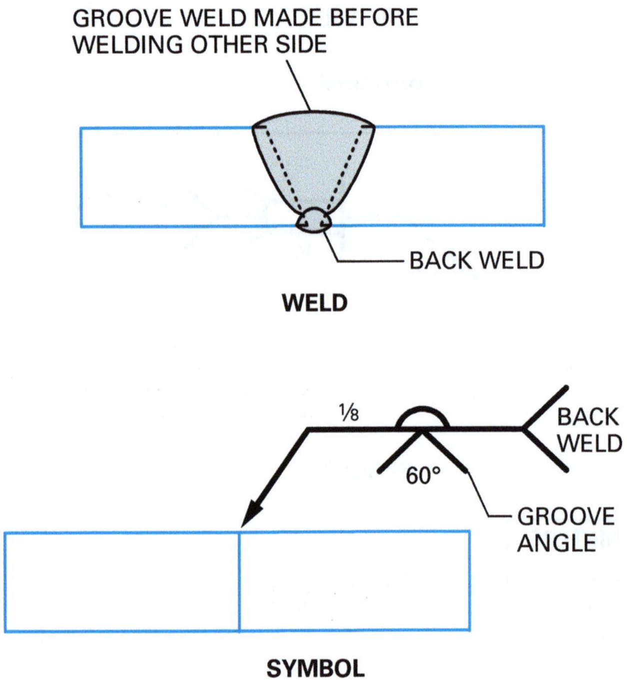

Figure 31 Back or backing weld symbol.

1.4.4 Surfacing Weld Symbols

The AWS defines surfacing as "the application by welding, brazing, or thermal spraying of a layer of material to a surface to obtain desired properties or dimensions, as opposed to making a joint." Examples of surfacing welds are shown in *Figure 33*. A surfacing weld symbol is used when a surface is to be built up with weld. The depth of buildup is indicated to the left of the symbol.

When less than the entire surface is to be built up, the surface to be built up is dimensioned on the blueprint in a detail drawing.

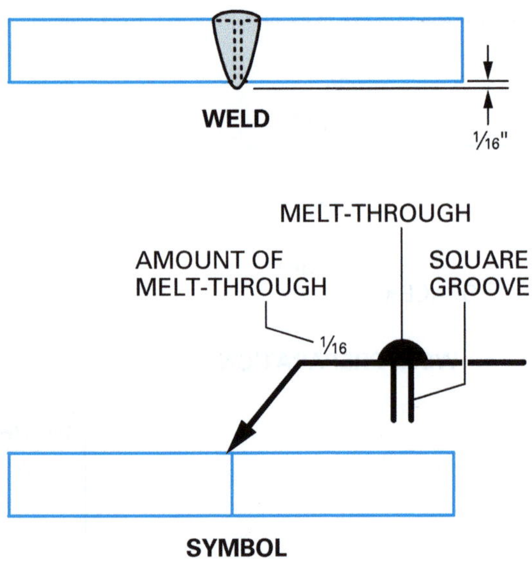

Figure 32 Melt-through symbol.

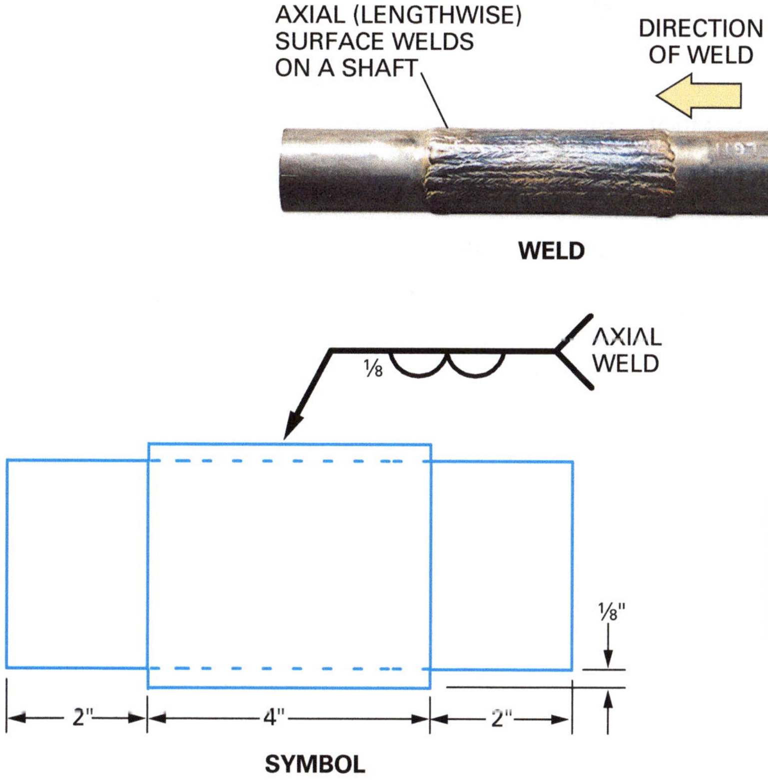

INCHES	MILLIMETERS
1/8	3.2
2	50.8
4	101.6

Figure 33 Surfacing welds and symbols.

1.4.5 Edge Weld Symbols

The AWS defines an edge weld as "a joint between the edges of two or more parallel or nearly parallel members." Edge welds are commonly used on sheet metal and can be flanged butt joints or flanged T-joints. *Figure 34* shows a variety of edge welds, symbols, and dimensioning.

1.4.6 Spot Weld Symbols

Spot welding (*Figure 35*) is a process used to join sheet metal using a series of small spots. There are two types: resistance spot welding (RSW) and arc spot welding. RSW uses electrical resistance heating and clamping pressure to fuse sheet metal panels. Arc spot welding involves either gas metal arc welding (GMAW) or gas tungsten arc welding (GTAW). The process used to make the spot weld is specified in the tail. The size of a spot weld is indicated to the left of the symbol. The pitch is indicated to the right of the symbol. The number of spot welds is defined in parentheses beneath the symbol. For arrow-side spot welds, the circle is placed on the bottom of the reference line. When the spot weld is placed on the other side, the circle is placed on top of the reference line. When there is no side significance (such as with a resistance spot), the circle is centered on the reference line. The drawings will define the starting or ending locations for the spot welds.

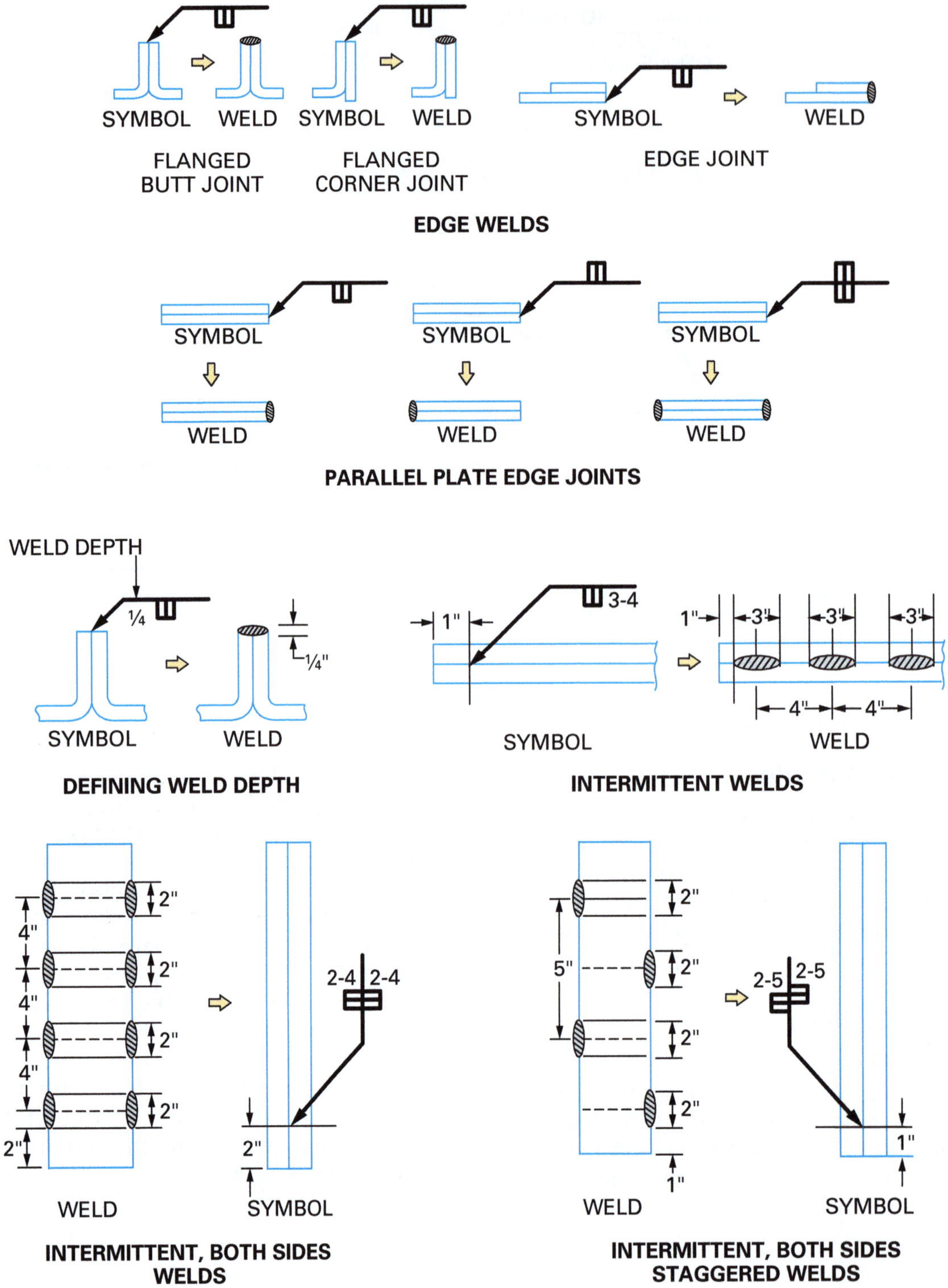

INCHES	MILLIMETERS
¼	6.4
1	25.4
2	50.8
3	76.2
4	101.6
10	254

Figure 34A Edge welds and symbols (1 of 2).

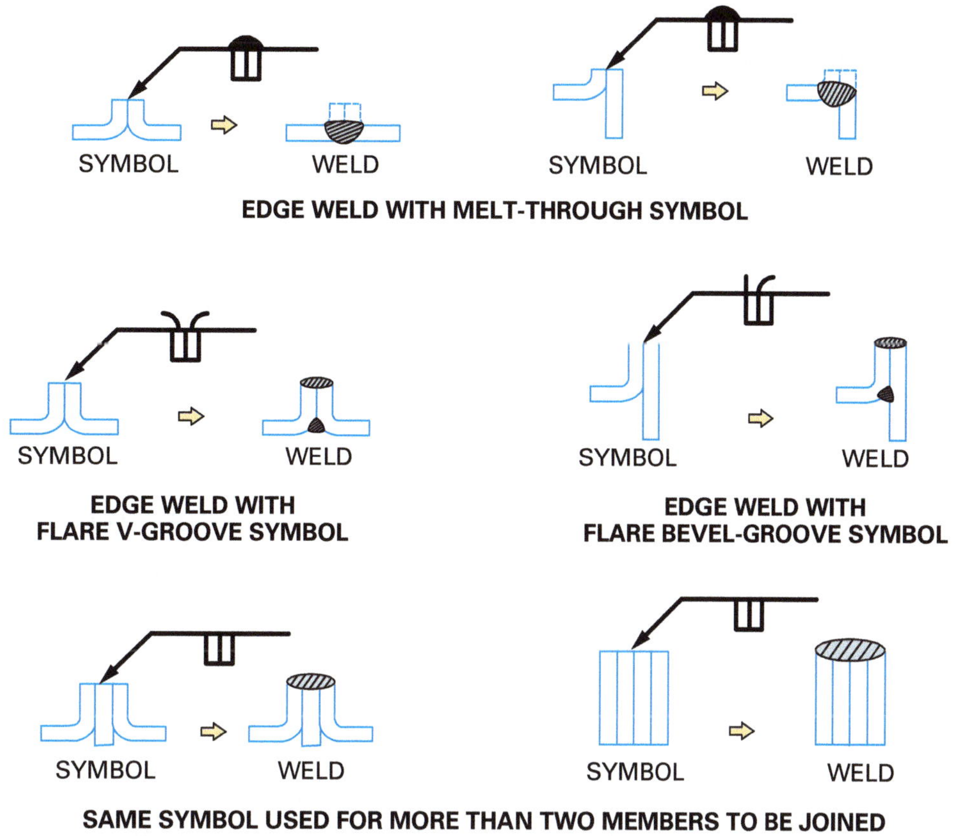

Figure 34B Edge welds and symbols (2 of 2).

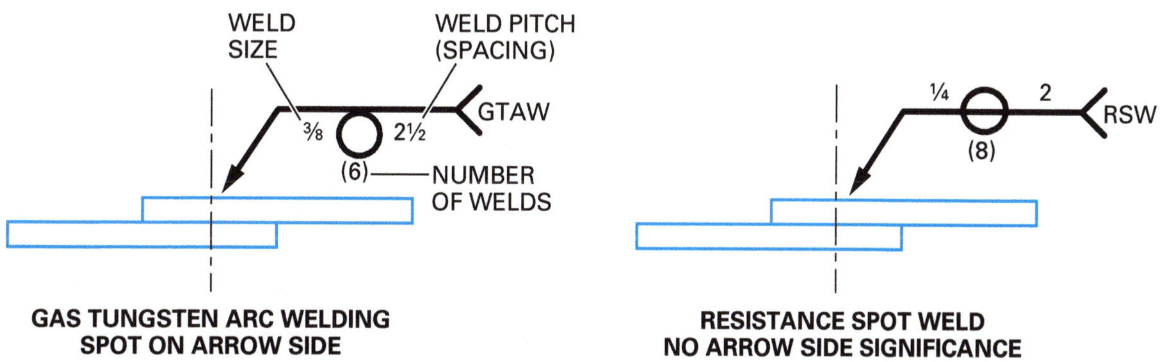

Figure 35 Spot weld symbols.

Projection Welding

Projection welding (PW) is a type of spot welding in which a number of welds are performed simultaneously with one welding machine. Because of tooling costs, projection welding is primarily used in high-production applications.

Prior to welding, one of the members to be joined is passed through a machine that presses projections at specified locations on the member. This member is joined to another member in a welding machine by pressing the two members together and applying current through the projections. The resistance at the points of contact of the projections causes the members to heat, melt, and fuse together while under pressure. This results in welded members that are fused closely together. The side without the projections has minimal surface marking as a result of the weld.

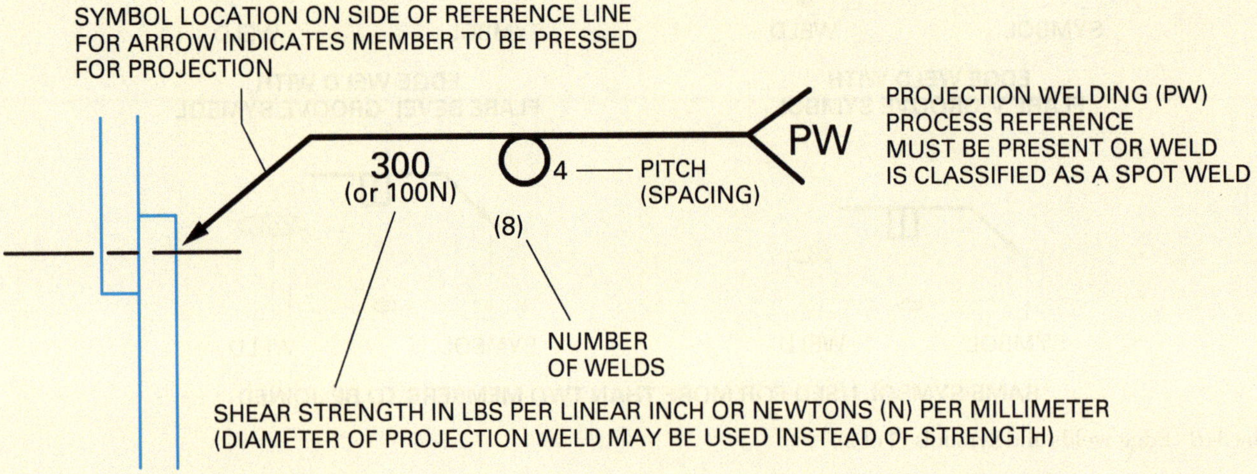

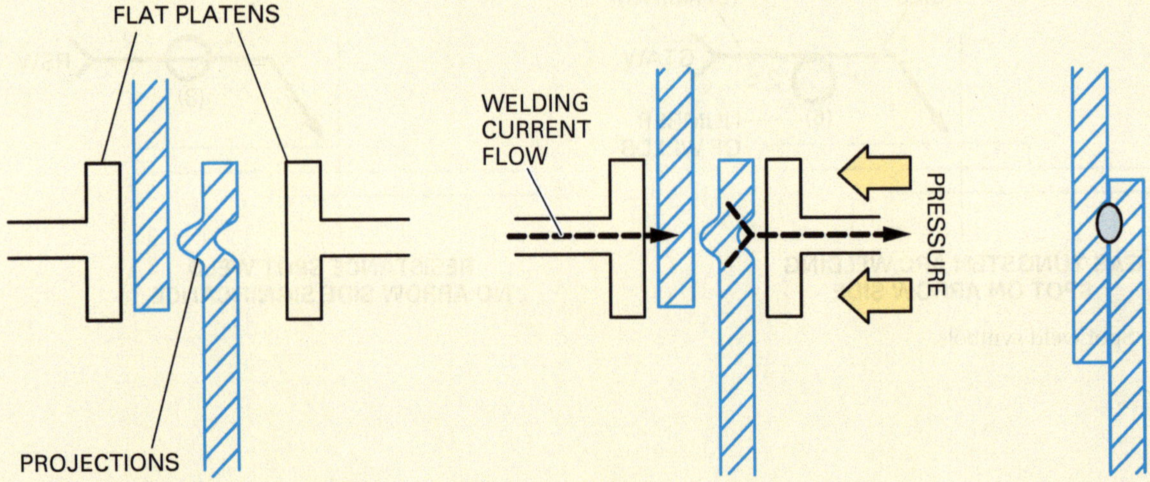

STEP 1 ONE MEMBER IS PRESSED TO CREATE PROJECTIONS AND BOTH MEMBERS ARE PLACED BETWEEN FLAT PLATENS IN WELDING MACHINE

STEP 2 MEMBERS HEAT AND MELT AT PROJECTIONS UNDER PRESSURE

STEP 3 COMPLETED WELD

1.4.7 Seam Weld Symbols

Seam welding is used to join two pieces of sheet metal in a continuous seam or in a series of overlapping spot welds. Several processes can be used to make a seam weld. The most common are GMAW, GTAW, and resistance seam welding (RSEW). The process used to make the seam weld is specified in the tail. The size of a seam weld (width) or strength of the weld is given to the left of the symbol. The length and pitch are given on the right of the symbol for intermittent welds. For welds that are less than the full length of the joined members, the drawing details the starting point and the length of the weld.

For arrow-side seam welds, the circle is placed on the bottom of the reference line, as shown in *Figure 36(A)*. When the seam weld is placed on the other side, the circle is placed on top of the reference line. When there is no side significance, such as with a resistance weld seam, the circle is centered on the reference line, as shown in *Figure 36(B)*. Flush contour symbols can be applied to seam weld symbols for other than resistance weld seams. For parallel seams, a drawing note detailing the spacing and number of seams is included in the tail.

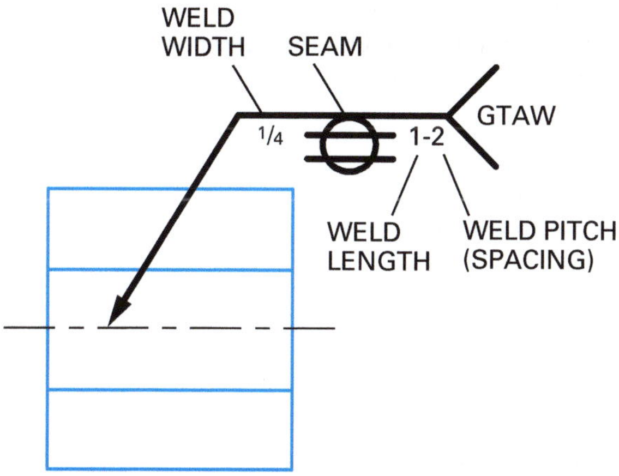

(A) GAS TUNGSTEN ARC WELD SEAM ON ARROW SIDE

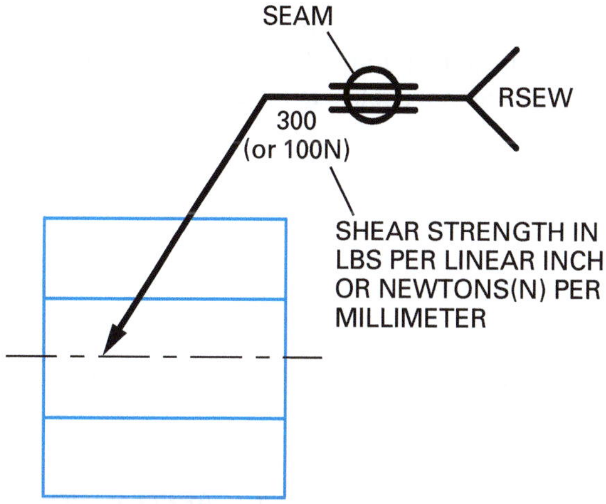

(B) RESISTANCE SEAM WELD NO ARROW SIDE SIGNIFICANCE

Figure 36 Seam weld symbols.

Around the World

AWS and ISO Welding Symbol Standards

AWS A2.4, Standard Symbols for Welding, Brazing, and Nondestructive Examination is the welding symbols standard used in the United States. AWS welding symbols comply with the requirements of the American National Standards Institute (ANSI) and are designated ANSI/AWS. The AWS system of welding symbols has long been used in the oil industry, which is a global enterprise. Today, the AWS system is used by approximately half of the world's welding industry.

The other widely used welding symbols standard is *ISO 2553, Welded, Brazed, and Soldered Joints – Symbolic Representation on Drawings.* The ISO system was designed by the International Organization for Standardization (ISO).

The AWS and the ISO systems are very similar. For example, the reference line and arrow system in the ISO system are the same as in the AWS system. There are, however, differences that welders need to be aware of in order to interpret the symbols correctly. The most significant difference, as shown in *Figure A*, is that the ISO system uses a dashed identification line that is not used in the AWS system. The dashed identification line, which may be drawn above or below the solid reference line, is used to indicate the other side of the joint. Information that applies to the arrow side of the joint is placed on the solid reference line. Information that applies to the other side of the joint is placed on the dashed line.

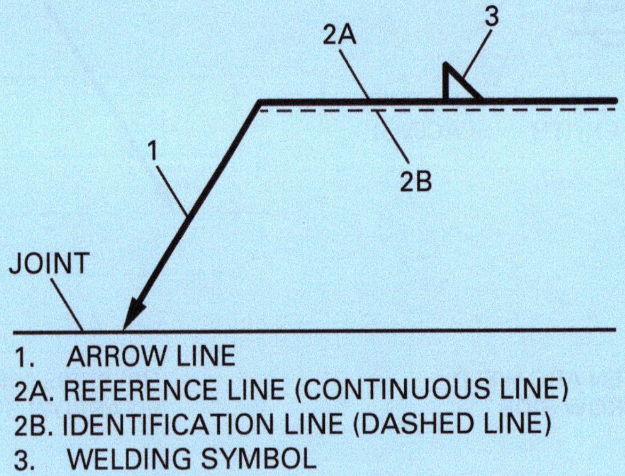

(A) ISO WELDING SYMBOL SYSTEM

Figure B uses fillet welding symbols as an example to further illustrate typical differences between AWS and ISO welding symbols.

Some countries that have extensive international trade relationships have actually used one system of welding symbols in their own country and symbols from the other system as needed to satisfy the requirements of their overseas customers. With the publication of *ISO 2553-2013*, the ISO has sought to provide a set of standardized welding symbols that recognizes both the AWS and the ISO welding symbol systems and can be applied on a worldwide basis.

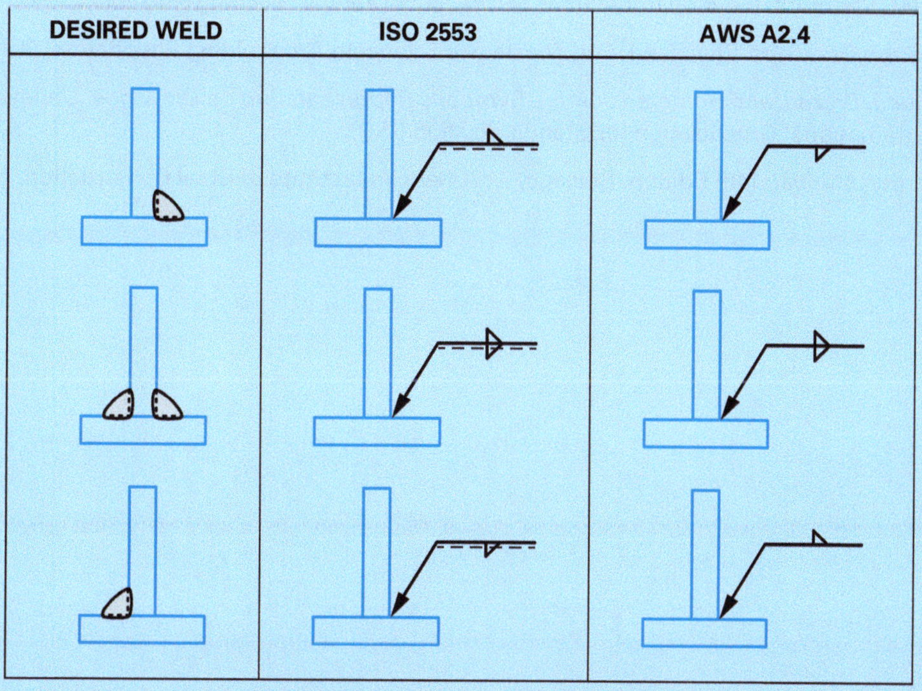

(B) COMPARISON OF ISO AND AWS SYMBOLS

Additional Resources

ASTM A325, Standard Specification for Structural Bolts, Steel, Heat Treated, 120/105 ksi Minimum Tensile Strength. West Conshohocken, PA: ASTM International.

AWS A2.4, Standard Symbols for Welding, Brazing, and Nondestructive Examination. Miami, FL: American Welding Society.

AWS A3.0, Standard Definitions; Including Terms for Adhesive Bonding, Brazing, Soldering, Thermal Cutting, and Thermal Spraying. Miami, FL: American Welding Society.

AWS D1.1/D1.1M, Structural Welding Code Steel. Miami, FL: American Welding Society.

How to Read Shop Drawings. Cleveland, OH: The James F. Lincoln Arc Welding Foundation, 2012.

ISO 2553, Welded, Brazed, and Soldered Joints—Symbolic Representation on Drawings. Geneva, Switzerland: International Organization for Standardization (ISO).

Steel Construction Manual. 14th Edition. Chicago, IL: American Institute of Steel Construction, 2011.

1.0.0 Section Review

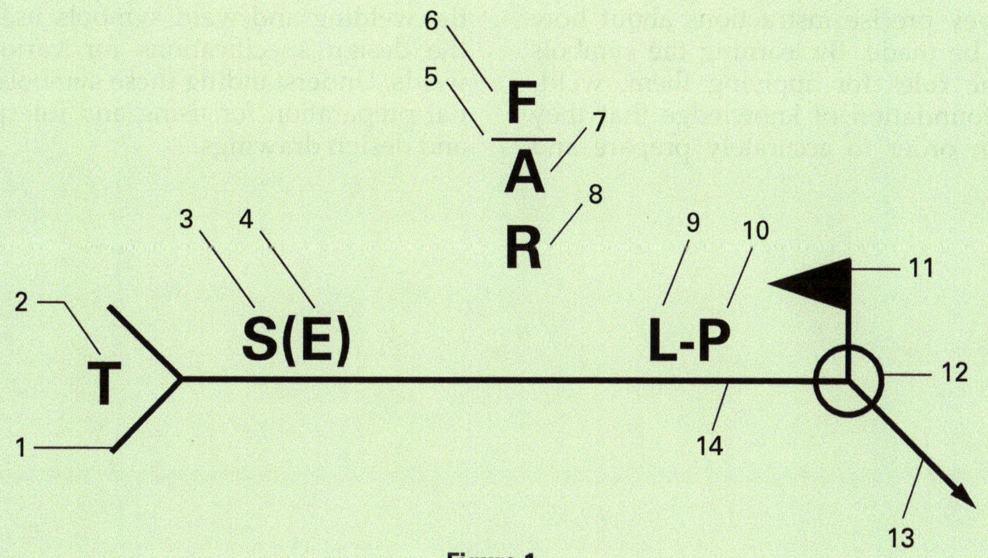

Figure 1

1. Match each of the terms listed below to the number that identifies the corresponding welding symbol shown on *Section Review Question Figure 1.*
   a. Arrow line = ____
   b. Groove weld size = ____
   c. Field weld symbol = ____
   d. Specification or process = ____
   e. Reference line = ____

2. A weld made in a corner is a ____.
   a. slot weld
   b. fillet weld
   c. plug weld
   d. groove weld

3. Which one of the symbols in *Section Review Question Figure 2* is a supplemental symbol?
   a. Symbol A
   b. Symbol B
   c. Symbol C
   d. Symbol D

4. The symbol in *Section Review Question Figure 3* that indicates that complete penetration from one side of a joint and root reinforcement are required is ____.
   a. Symbol A
   b. Symbol B
   c. Symbol C
   d. Symbol D

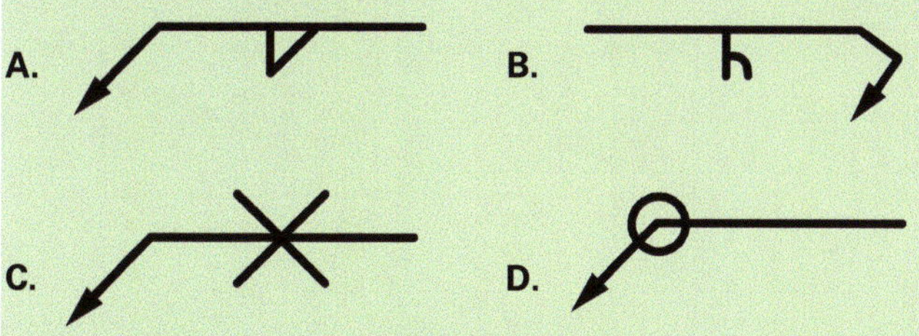

Figure 2

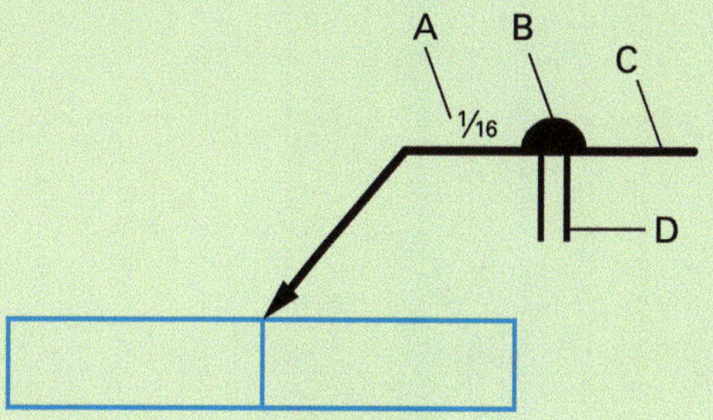

Figure 3

SUMMARY

Welding symbols are a graphic language that is used to convey precise instructions about how a weld is to be made. By learning the symbols and the basic rules for applying them, welders build a foundation of knowledge that they must have in order to accurately prepare and make the required welds. This module covered the welding and weld symbols used to convey the design specifications for various types of welds. Understanding these symbols is an essential preparation for using and interpreting WPS and design drawings.

Review Questions

1. The welding symbol describes the type of weld, its size, and its ____.
   a. surface finish
   b. tail references
   c. cutting process
   d. depth of field
2. The side of the joint or surface that is opposite the side to which the arrow of a reference line is pointing is called the ____.
   a. arrow side
   b. other side
   c. tail joint
   d. countersink
3. Regardless of which end of the reference line the arrow line is on, information on the reference line is always read from ____.
   a. top to bottom
   b. bottom to top
   c. left to right
   d. right to left

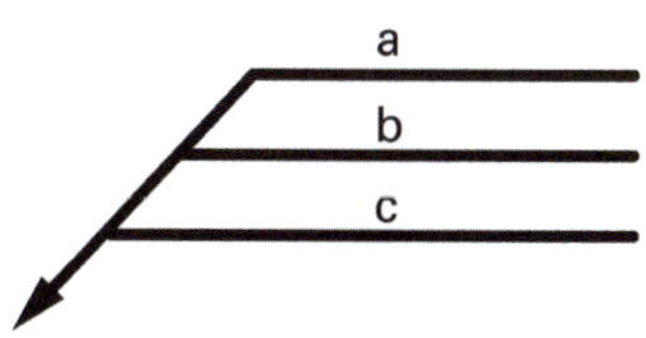

Figure 1

4. In the welding symbol shown in *Review Question Figure 1*, the operation to be conducted first is ____.
   a. Operation A
   b. Operation B
   c. Operation C
   d. not critical

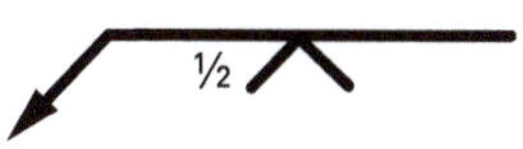

Figure 2

5. In the welding symbol shown in *Review Question Figure 2*, ½ represents the ____.
   a. size of the weld
   b. width of the weld
   c. length of the weld
   d. depth of preparation
6. In relation to the fillet weld symbol, the fillet weld size is placed ____.
   a. above the symbol
   b. below the symbol
   c. to the left of the symbol
   d. to the right the symbol

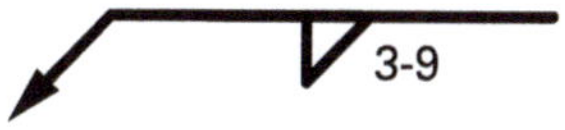

Figure 3

7. In the welding symbol shown in *Review Question Figure 3*, the number 9 represents the weld ____.
   a. pitch
   b. depth
   c. width
   d. length
8. In relation to the plug weld symbol, the pitch of a plug weld is shown ____.
   a. above the symbol
   b. below the symbol
   c. to the left of the symbol
   d. to the right of the symbol
9. The field weld symbol is a(n) ____.
   a. number
   b. letter
   c. arrow
   d. flag
10. When the face of the weld must have a finished shape, a finish symbol is placed adjacent to the welding symbol, along with a letter indicating ____.
    a. size of the weld
    b. method of finish
    c. smoothness of the finish
    d. width of the finished area

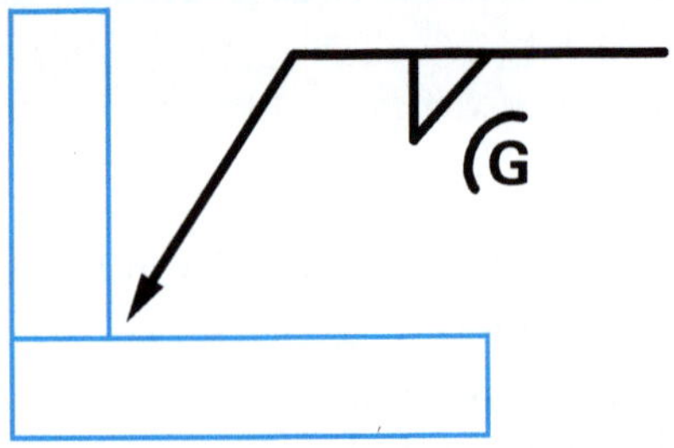

Figure 4

11. The finish of the weld face in *Review Question Figure 4* is to be achieved by ____.
    a. filleting
    b. grinding
    c. chipping
    d. reaming

12. When backing is to be used at the root of the weld to control melt-through, the backing symbol is placed opposite which weld symbol on the reference line?
    a. tail
    b. field
    c. spacer
    d. groove

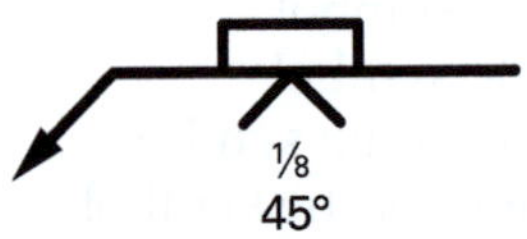

Figure 5

13. In the weld symbol shown in *Review Question Figure 5*, the numbers inside the symbol represent the ____.
    a. root opening and groove angle
    b. groove angle and depth
    c. root opening and pitch
    d. pitch and angle

Figure 6

14. The type of weld symbol shown above the reference line in *Review Question Figure 6* is a ____.
    a. spot weld
    b. back weld
    c. surfacing weld
    d. melt-through

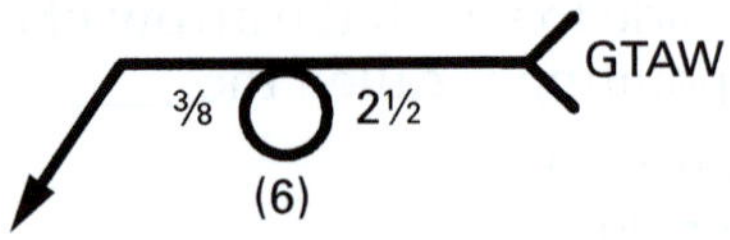

Figure 7

15. The type of weld represented by the weld symbol in *Review Question Figure 7* is a ____.
    a. spot weld
    b. backing weld
    c. surfacing weld
    d. melt-through

Appendix A

Letter Designations for Welding Processes			
AAW	Air Acetylene Welding	GTAW	Gas Tungsten Arc Welding
AB	Adhesive Bonding	GTAW-P	Gas Tungsten Arc Welding–Pulsed Arc
AHW	Atomic Hydrogen Welding	HPW	Hot Pressure Welding
AW	Arc Welding	IB	Induction Brazing
B	Brazing	IRB	Infrared Brazing
BB	Block Brazing	IW	Induction Welding
BMAW	Bare Metal Arc Welding	LBW	Laser Beam Welding
CAW	Carbon Arc Welding	OAW	Oxyacetylene Welding
CAW-G	Gas Carbon Arc Welding	OHW	Oxyhydrogen Welding
CAW-T	Twin Carbon Arc Welding	PAW	Plasma Arc Welding
CEW	Coextrusion Welding	PEW	Percussion Welding
CW	Cold Welding	PGW	Pressure Gas Welding
DP	Dip Brazing	PW	Projection Welding
DFB	Diffusion Brazing	RB	Resistance Brazing
DFW	Diffusion Welding	ROW	Roll Welding
EBW	Electronic Beam Welding	RSEW	Resistance Seam Welding
EBW-HV	Electronic Beam Welding–High Vacuum	RSEW-HF	Resistance Seam Welding–High Frequency
EBW-MV	Electronic Beam Welding–Medium Vacuum	RSEW-I	Resistance Seam Welding–Induction
EBW-NV	Electronic Beam Welding–Nonvacuum	RSW	Resistance Spot Welding
EGW	Electrogas Welding	RW	Resistance Welding
ESW	Electroslag Welding	SAW	Submerged Arc Welding
EXW	Explosion Welding	SAW-S	Series Submerged Arc Welding
FB	Furnace Brazing	SMAW	Shielded Metal Arc Welding
FCAW	Flux Cored Arc Welding	SSW	Solid-State Welding
FLB	Flow Brazing	SW	Stud Arc Welding
FLOW	Flow Welding	TB	Torch Brazing
FOW	Forge Welding	TCAB	Twin Carbon Arc Brazing
FRW	Friction Welding	TW	Thermite Welding

Appendix B

Letter Designations for Cutting Processes			
AC	Arc Cutting	LOC	Oxygen Lance Cutting
AOC	Oxygen Arc Cutting	MAC	Metal Arc Cutting
CAC	Carbon Arc Cutting	OC	Oxygen Cutting
CAC-A	Air Carbon Arc Cutting	OC-F	Flux Cutting
EBC	Electron Beam Cutting	OFC	Oxyfuel Gas Cutting
GMAC	Gas Metal Arc Cutting	OFC-A	Oxyacetylene Cutting
GTAC	Gas Tungsten Arc Cutting	OFC-H	Oxyhydrogen Cutting
LBC	Laser Beam Cutting	OFC-N	Oxynatural Gas Cutting
LBC-A	Laser Beam Cutting–Air	OFC-P	Oxypropane Cutting
LBC-EV	Laser Beam Cutting–Evaporative	OC-P	Metal Powder Cutting
LBC-IG	Laser Beam Cutting–Inert Gas	PAC	Plasma Arc Cutting
LBC-O	Laser Beam Cutting–Oxygen	SMAC	Shielded Metal Arc Cutting

Appendix C

Letter Designations for Applying Welding Processes	
AU	Automatic
ME	Mechanized
MA	Manual
SA	Semi-Automatic

Trade Terms Introduced in This Module

Arrow line: The line drawn at an angle from the reference line (either end or both ends) to an arrowhead at the location of the weld.

Countersink: A hole with tapered sides and a wider opening that allows a flat-head fastener to seat flush to the surface of the material in which the hole is made.

Pitch: The center-to-center distance between welds.

Reference line: The horizontal line in the center of the welding symbol from which all elements of the welding symbol are referenced. The reference line is one of the most important elements of the welding symbol.

Weld symbol: A graphic character connected to the reference line of a welding symbol specifying the weld type.

Welding symbol: A graphical representation of the specifications for producing a welded joint; includes a reference line and arrow line and many also include a weld symbol.

Additional Resources

This module is intended as a thorough resource for task training. The following reference works are suggested for further study.

ASTM A325, Standard Specification for Structural Bolts, Steel, Heat Treated, 120/105 ksi Minimum Tensile Strength. West Conshohocken, PA: ASTM International.

AWS A2.4, Standard Symbols for Welding, Brazing, and Nondestructive Examination. Miami, FL: American Welding Society.

AWS A3.0, Standard Definitions; Including Terms for Adhesive Bonding, Brazing, Soldering, Thermal Cutting, and Thermal Spraying. Miami, FL: American Welding Society.

AWS D1.1/D1.1M, Structural Welding Code Steel. Miami, FL: American Welding Society.

How to Read Shop Drawings. Cleveland, OH: The James F. Lincoln Arc Welding Foundation, 2012.

ISO 2553, Welded, Brazed, and Soldered Joints—Symbolic Representation on Drawings. Geneva, Switzerland: International Organization for Standardization (ISO).

Steel Construction Manual. 14th Edition. Chicago, IL: American Institute of Steel Construction, 2011.

Figure Credits

The Lincoln Electric Company, Cleveland, OH, USA, Module opener

AWS A2:2007, Figure 2, Reproduced with permission from the American Welding Society (AWS), Miami, FL, USA, Figure 4

Topaz Publications, Inc., Figures 7 (photos), 24 (photo), 33 (photo)

Specialty Welding & Fabricating of New York, Inc., Figures 18 (photos), 23 (photos)

Section Review Answer Key

Answer	Section Reference	Objective
Section One		
1. a. 13 b. 4 c. 11 d. 2 e. 14	1.1.0	1a
2. b	1.2.1	1b
3. d	1.3.0	1c
4. b	1.4.3	1d

NCCER CURRICULA — USER UPDATE

NCCER makes every effort to keep its textbooks up-to-date and free of technical errors. We appreciate your help in this process. If you find an error, a typographical mistake, or an inaccuracy in NCCER's curricula, please fill out this form (or a photocopy), or complete the online form at **www.nccer.org/olf**. Be sure to include the exact module ID number, page number, a detailed description, and your recommended correction. Your input will be brought to the attention of the Authoring Team. Thank you for your assistance.

Instructors – If you have an idea for improving this textbook, or have found that additional materials were necessary to teach this module effectively, please let us know so that we may present your suggestions to the Authoring Team.

NCCER Product Development and Revision

13614 Progress Blvd., Alachua, FL 32615

Email: curriculum@nccer.org
Online: www.nccer.org/olf

❑ Trainee Guide ❑ Lesson Plans ❑ Exam ❑ PowerPoints Other ____________

Craft / Level: Copyright Date:

Module ID Number / Title:

Section Number(s):

Description:

Recommended Correction:

Your Name:

Address:

Email: Phone:

Reading Welding Detail Drawings

Overview

Reading and interpreting the information contained on construction drawings is an essential skill for welders. Incorrectly interpreting this information can result in producing work that fails to meet design specifications, building codes, and safety standards. This module focuses on how to read and interpret assembly and detail drawings. It introduces the various object views and basic drawing elements that are used in welding detail drawings. It also explains how to interpret dimensional information, notes, and bills of materials.

Module 29202

Trainees with successful module completions may be eligible for credentialing through the NCCER Registry. To learn more, go to **www.nccer.org** or contact us at **1.888.622.3720**. Our website has information on the latest product releases and training, as well as online versions of our *Cornerstone* magazine and Pearson's product catalog.

Your feedback is welcome. You may email your comments to **curriculum@nccer.org,** send general comments and inquiries to **info@nccer.org**, or fill in the User Update form at the back of this module.

This information is general in nature and intended for training purposes only. Actual performance of activities described in this manual requires compliance with all applicable operating, service, maintenance, and safety procedures under the direction of qualified personnel. References in this manual to patented or proprietary devices do not constitute a recommendation of their use.

29202 V5

29202
Reading Welding Detail Drawings

Objectives

When you have completed this module, you will be able to do the following:

1. Describe welding detail drawings and identify basic drawing elements and features.
   a. Describe the object views used to depict welding details.
   b. Identify basic drawing elements related to welding detail drawings.
2. Identify and explain how to interpret dimensional information, notes, and a bill of materials.
   a. Identify and explain how to interpret dimensional information.
   b. Identify and explain how to interpret notes and a bill of materials.

Performance Task

Under the supervision of your instructor, you should be able to do the following:

1. Draw or sketch a welding drawing based on an instructor-provided image or object.

Trade Terms

Blind hole
Counterbore
Elevation
Isometric view
Mark
Orthographic projection
Scale
Through hole
Tolerance

Industry Recognized Credentials

If you are training through an NCCER-accredited sponsor, you may be eligible for credentials from NCCER's Registry. The ID number for this module is 29202. Note that this module may have been used in other NCCER curricula and may apply to other level completions. Contact NCCER's Registry at 888.622.3720 or go to **www.nccer.org** for more information.

Contents

Topics to be presented in this module include:

Figures

Section One

1.0.0 Basic Welding Detail Drawings

Objective

Describe welding detail drawings and identify basic drawing elements and features.

a. Describe the object views used to depict welding details.
b. Identify basic drawing elements related to welding detail drawings.

Performance Task

1. Draw or sketch a welding drawing based on an instructor-provided image or object.

Trade Terms

Isometric view: A drawing depicting a three-dimensional view of an object.

Orthographic projection: A method of developing multiple flat views of an object by projecting lines out from the object at right angles from the face.

Engineers, designers, and drafters use drawings, also called prints or blueprints, to convey information. There are many categories of drawings, including civil, structural, mechanical, piping, electrical, and instrumentation. Welders must be able to correctly interpret drawings so that the design specifications and safety factors shown in the drawings will be included in the finished product. Altering or failing to comply with a drawing could result in a violation of federal or state standards or building codes, which could lead to fines and other penalties. More importantly, a deviation from the drawings could create an unsafe structure or product, which might result in death or injury. To avoid these problems, welders must be able to read and interpret drawings accurately. This module explains how to read and interpret assembly and detail drawings.

Each category contains several types of drawings. The most common drawing used by welders is the welding detail drawing, which generally depicts the weldment that must be fabricated and the pieces to be joined by welding. The parts are drawn assembled, as they will appear after welding, except that the welds are not shown. Welding symbols show how the pieces are to be joined. *Figure 1* is an example of a small bracket that may need to be fabricated on the job.

Figure 2 shows how the same bracket would appear on a welding detail drawing.

NOTE

The performance task for this module is to draw or sketch a welding detail drawing based on a given image or object. An example of such an image and its corresponding drawing is provided in the *Appendix* of this module.

When the object to be fabricated is too complicated for one drawing, the object may be shown on an assembly drawing as well as one or more detail drawings. The assembly drawing (*Figure 3*) shows the completed and assembled object. The main purpose of the assembly drawing is to show how the parts are assembled in relation to one another. Detail drawings of each part are then provided (*Figure 4*). The detail drawings contain all the information necessary to make the part.

NOTE

Letters and leader lines identify the parts of the object in an assembly drawing. These same letters are then used to label the parts in the corresponding detail drawing.

1.1.0 Object Views

Drawings must clearly show the shape of an object. To do this, most drawings have more than one view of the object. The most commonly used views are the following:

- Isometric
- Multiview
- Section

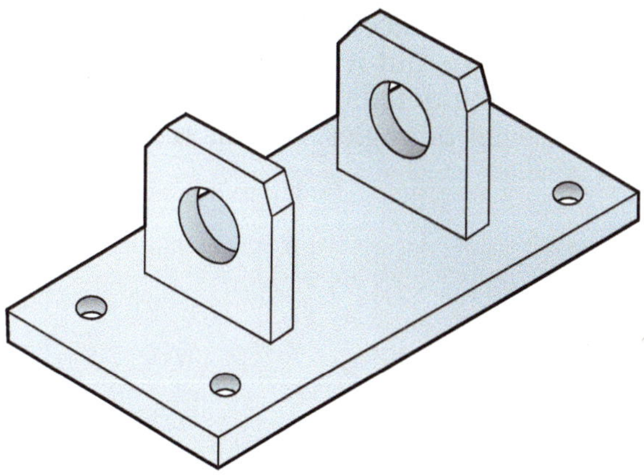

Figure 1 Bracket.

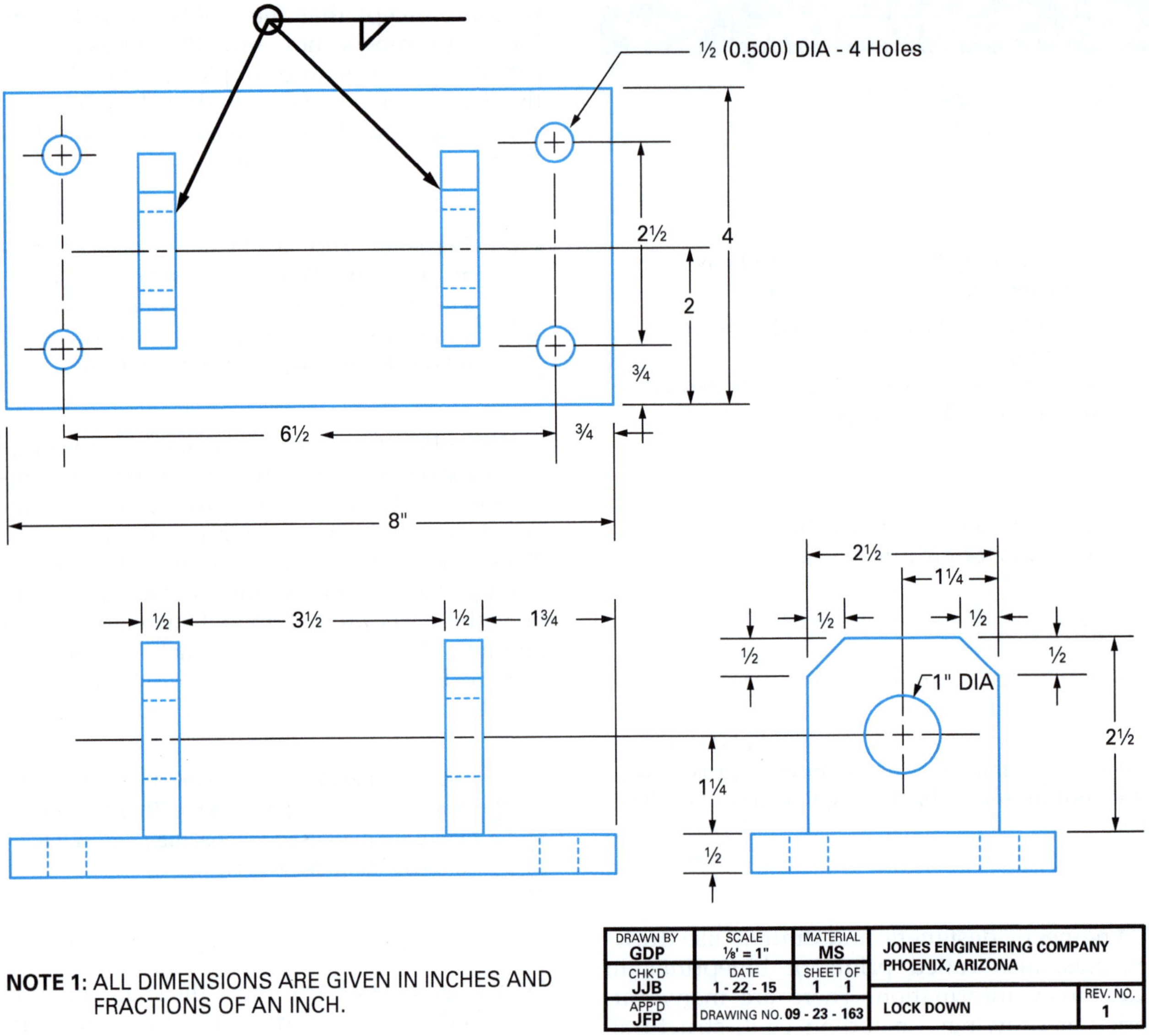

(A) IMPERIAL MEASUREMENTS

Figure 2A Welding detail drawing. (1 of 2)

1.1.1 Isometric View

The **isometric view** shows a three-dimensional picture of an object (*Figure 5*). The isometric (pictorial) view is an excellent way to show what an object looks like, but it is a difficult view to use when providing details, such as dimensions, hole locations, sizes, or angles. For this reason, the isometric view is generally not used on detail drawings.

1.1.2 Multiview

Multiviews are developed by a process called **orthographic projection**, and they are generally used on detail drawings. Several views, each from a different direction, are developed and shown on the same drawing. The views are designed as if the object had been placed inside a transparent box. Each view is then expanded by projecting points at right angles from the object onto a surface of the transparent box. If the sides of the box were folded open, a flat, multiview drawing would result, as shown in *Figure 6*.

Isometric Views

An isometric drawing of an object shows the object at 30-degree angles. This causes the object to appear on paper similar to the way the real, three-dimensional object appears.

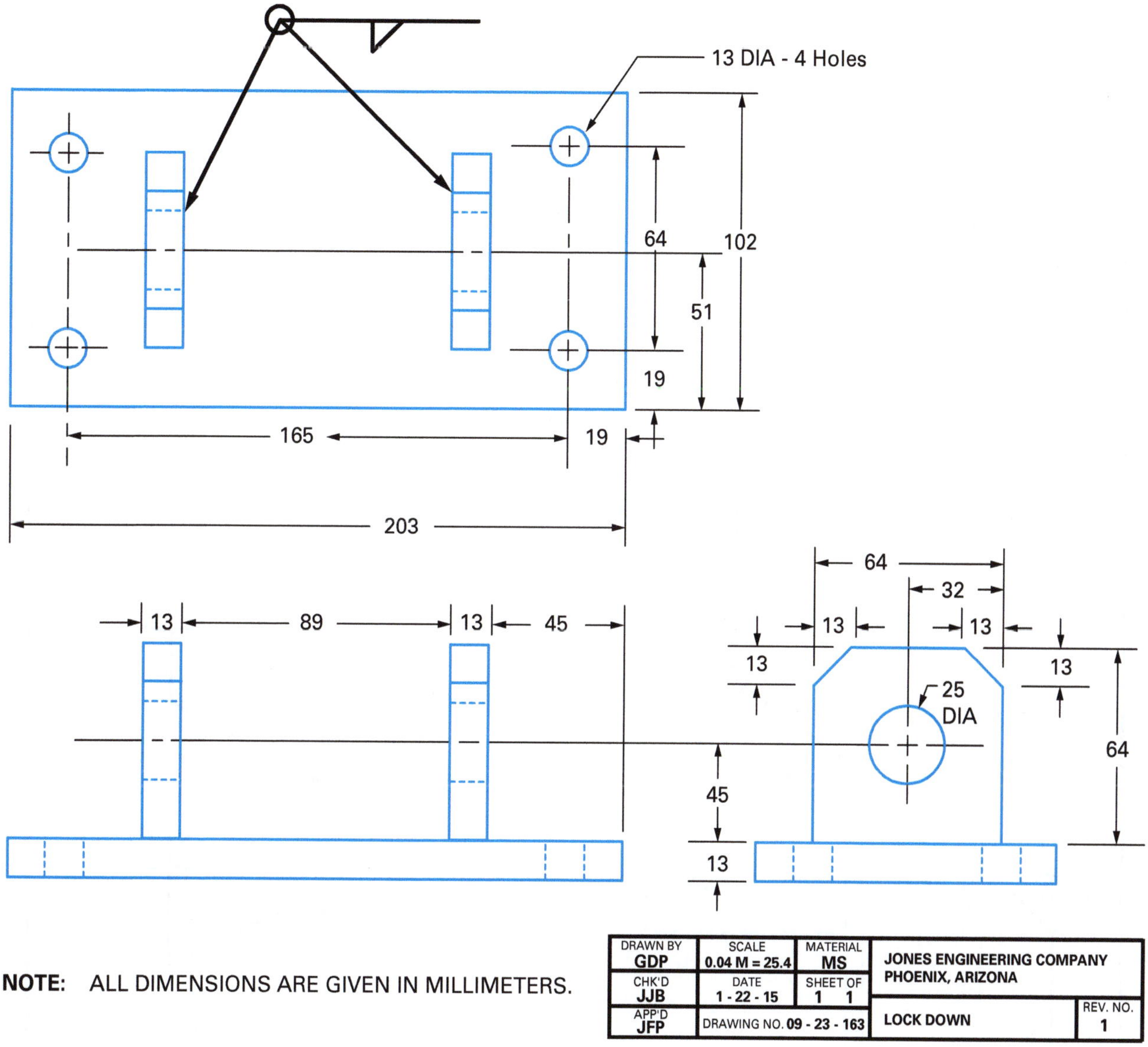

(B) METRIC MEASUREMENTS

Figure 2B Welding detail drawing. (2 of 2)

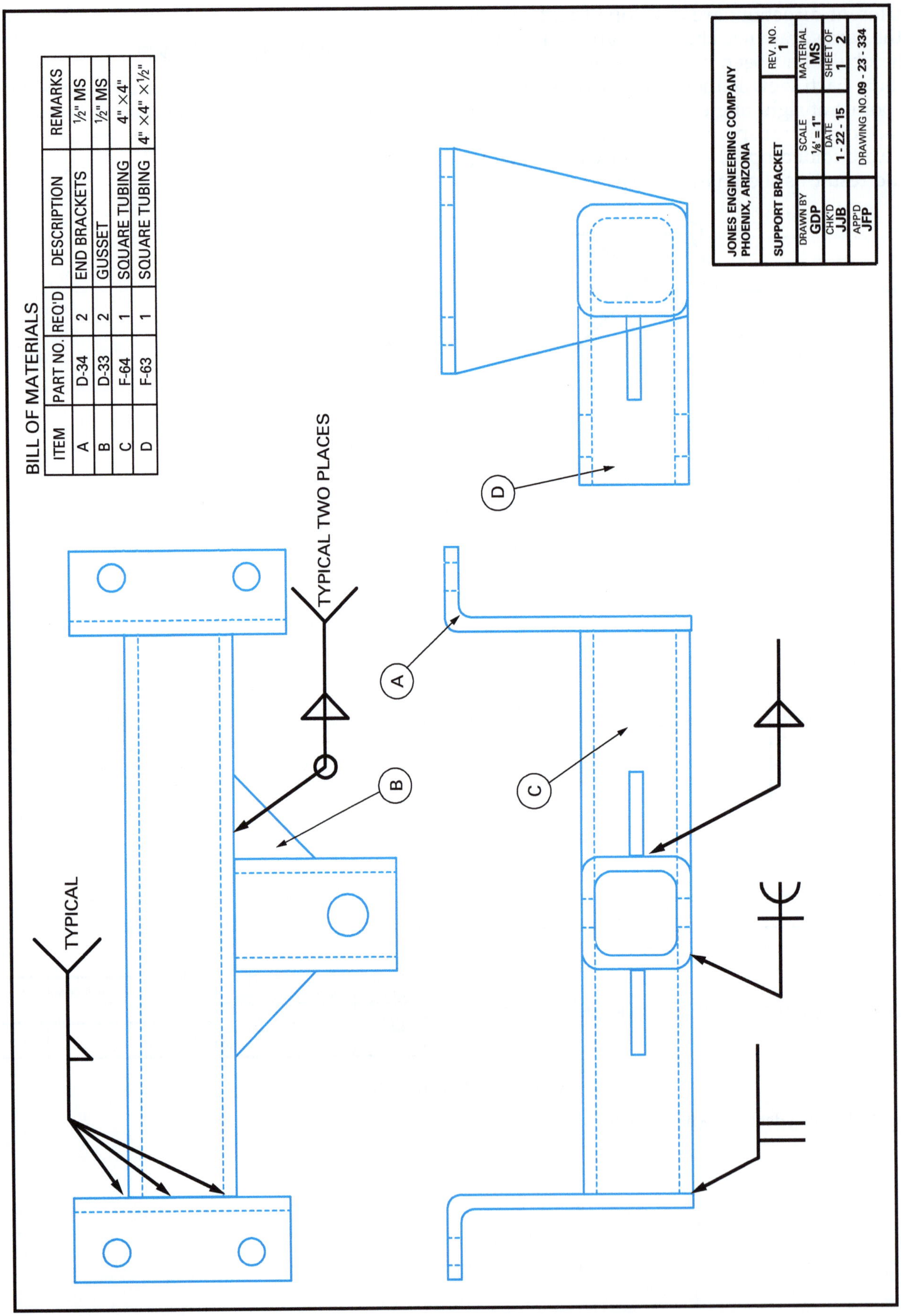

ITEM	PART NO.	REQ'D	DESCRIPTION	REMARKS
A	D-34	2	END BRACKETS	½" MS
B	D-33	2	GUSSET	½" MS
C	F-64	1	SQUARE TUBING	4" × 4"
D	F-63	1	SQUARE TUBING	4" × 4" × ½"

Figure 3 Assembly drawing.

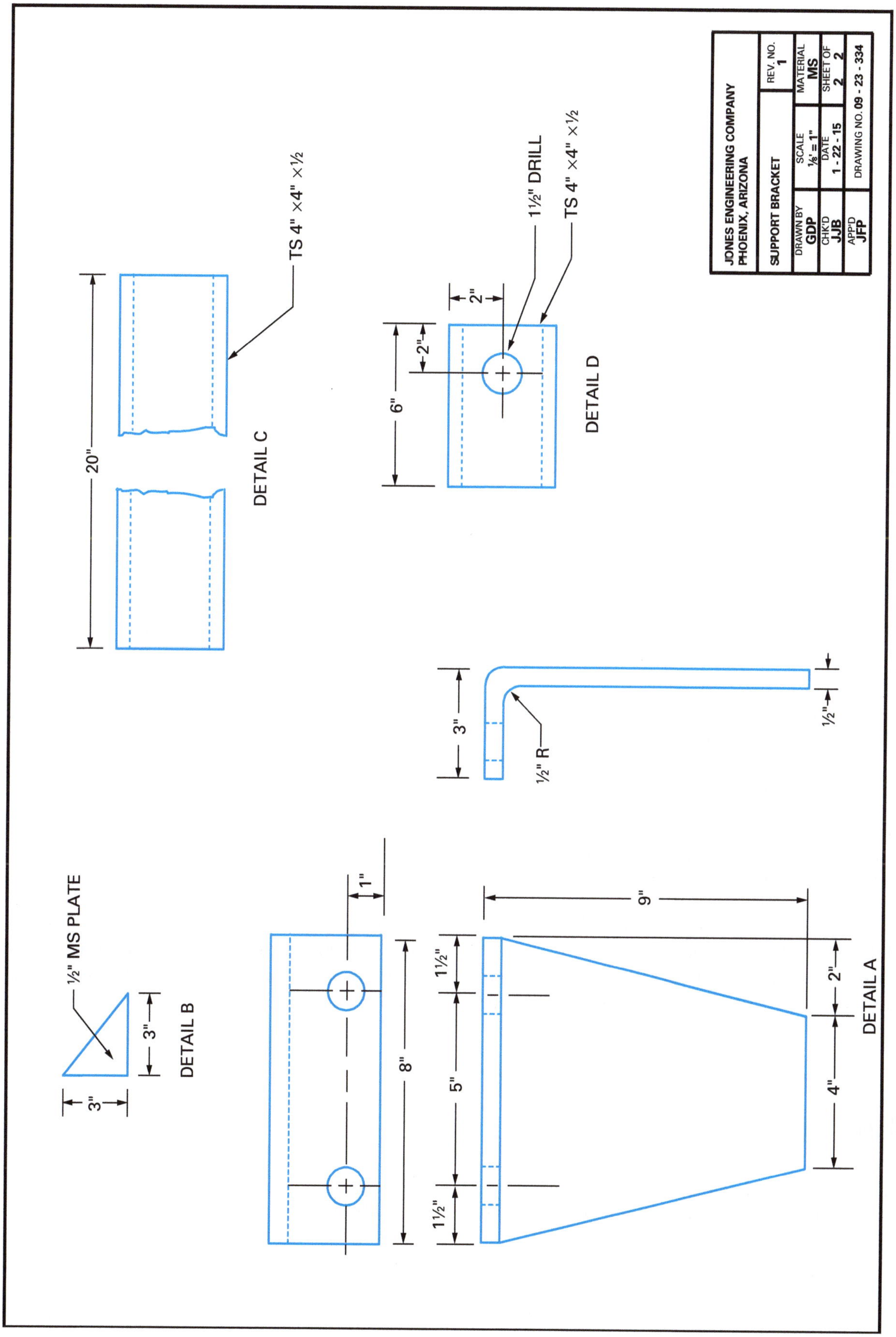

Figure 4 Detail drawing.

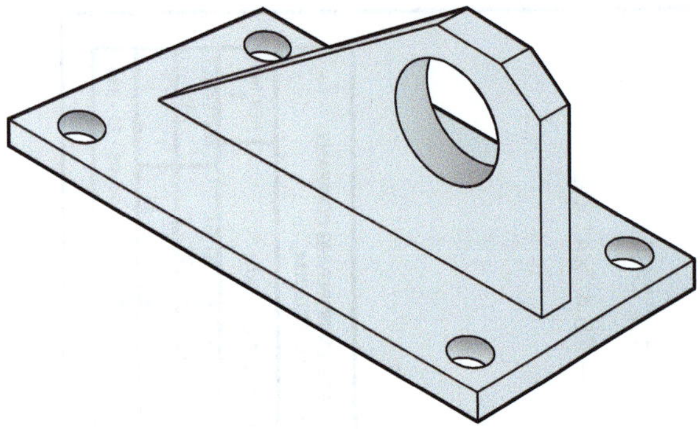

Figure 5 Isometric view

Each view in an orthographic projection has a set position in relation to the other views. In addition, the views are aligned vertically and horizontally. All sides of an object can be shown, although the views used most frequently are the front, top, and right side. The front view is selected as the view that best represents the shape of the object. All other views are positioned in reference to the front view. *Figure 7* shows the relationship of possible views used with orthographic projection.

Figure 8 shows the relationship of views on a welding detail drawing.

1.1.3 Section

A full section, or cross-section, is a view of an object that is used to show internal details that would not normally be visible (*Figure 9*). A section line is used on one of the complete views to indicate where the part has been cut or sectioned. The arrowheads on the section line point in the direction from which the sectioned object is to be viewed. Because there can be more than one section view, each one is identified by placing a letter next to the arrowheads. The sectioned view is then labeled with the letters.

A drawing shows a more detailed view of a part by showing different sections of the same part.

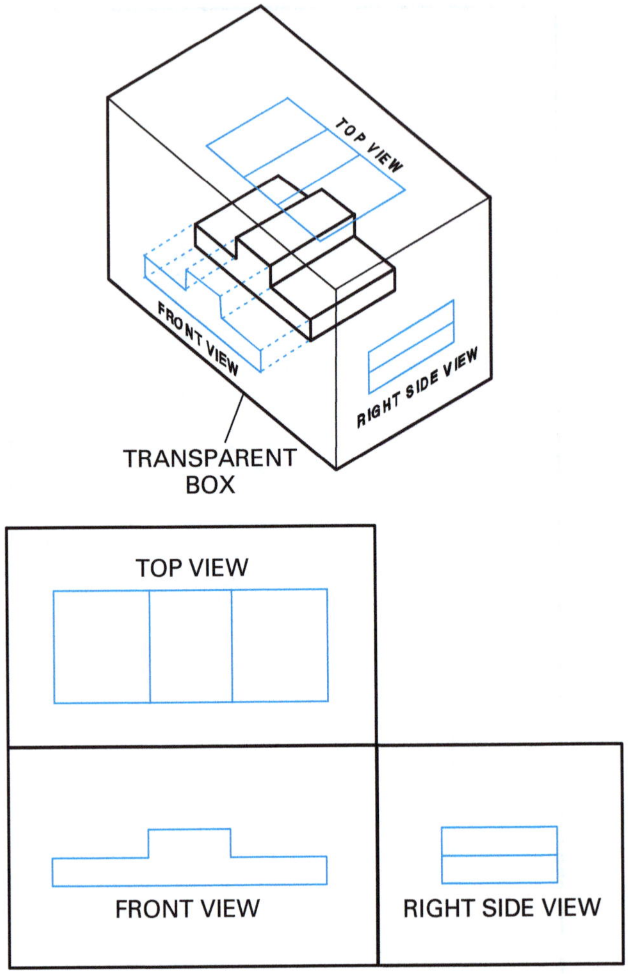

Figure 6 Orthographic projection.

Each of these sections is identified by pairs of letters located near the points of the cutting plane line.

1.2.0 Drawing Components

Welding drawings have several components, including the following: lines; material symbols; solid round breaks; pipe, or tubing, breaks; and revolved sections.

Blueprints

Blueprints got their name from an obsolete chemical process that was used to develop chemically treated print paper, which turned paper blue. The lines produced on the paper were white against the blue background. Drawings are now made using other methods and are typically black on white.

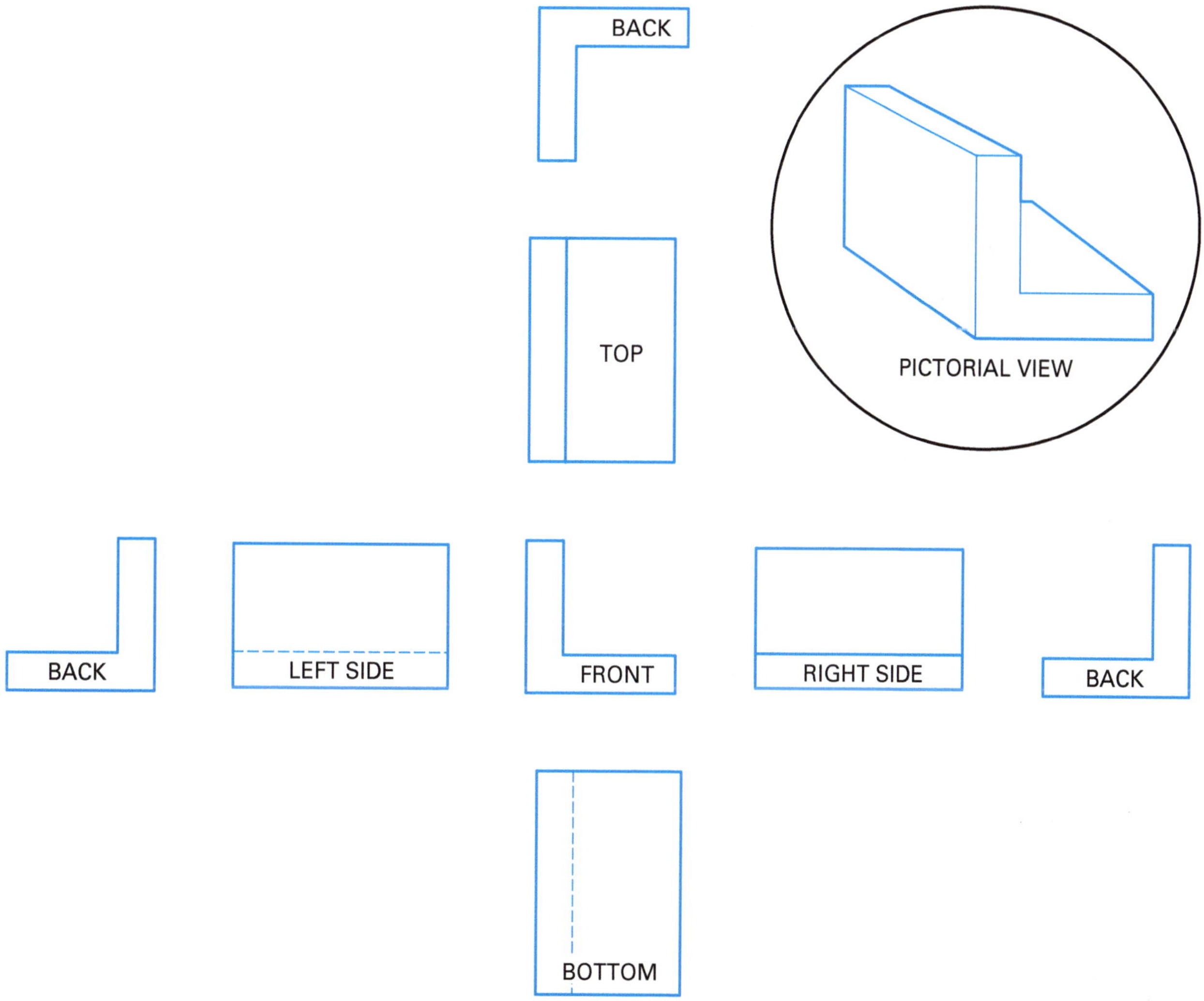

Figure 7 Relationship of possible views.

1.2.1 Lines

Many different types of lines are used on drawings, and each type has its own special meaning. The lines are drawn to a standard that specifies thickness (thick, medium, or thin) and the way the lines are interpreted or broken. The more commonly used lines are shown in *Figure 10* and are described as follows:

- *Object line* – A thick, solid line used to show all visible edges of a part; also called a visible line.
- *Hidden line* – A medium, broken line made up of evenly spaced dashes; used to show edges that are not visible.
- *Center line* – A thin, broken line made up of short and long dashes; used to show the center of holes or other objects.
- *Dimension line* – A thin line with arrowheads that shows the extent of a dimension; the line is broken by the dimension number.
- *Extension line* – A thin line used with dimension lines; it extends from the point on the object that the dimension applies to.
- *Leader line* – A thin, horizontal line with an angled arrowhead at one end; used with a note or dimension. The arrowhead points directly to the surface of the note or dimension to which the line applies.
- *Cutting plane line* – A line used to show where an imaginary cut has been made through an object. It has a heavy line that can be made up of either a series of one long and two short dashes or a series of short dashes. Short lines with arrowheads are on both ends of the line. The arrowheads point in the direction in which the section should be viewed.
- *Short break line* – A thick, solid, irregular line drawn freehand to indicate that part of the object has been removed or broken away; used to conserve space or show internal detail.
- *Long break line* – A thin, solid line with zigzags inserted at several places; used to show where a section of an object has been removed to conserve space on the drawing.

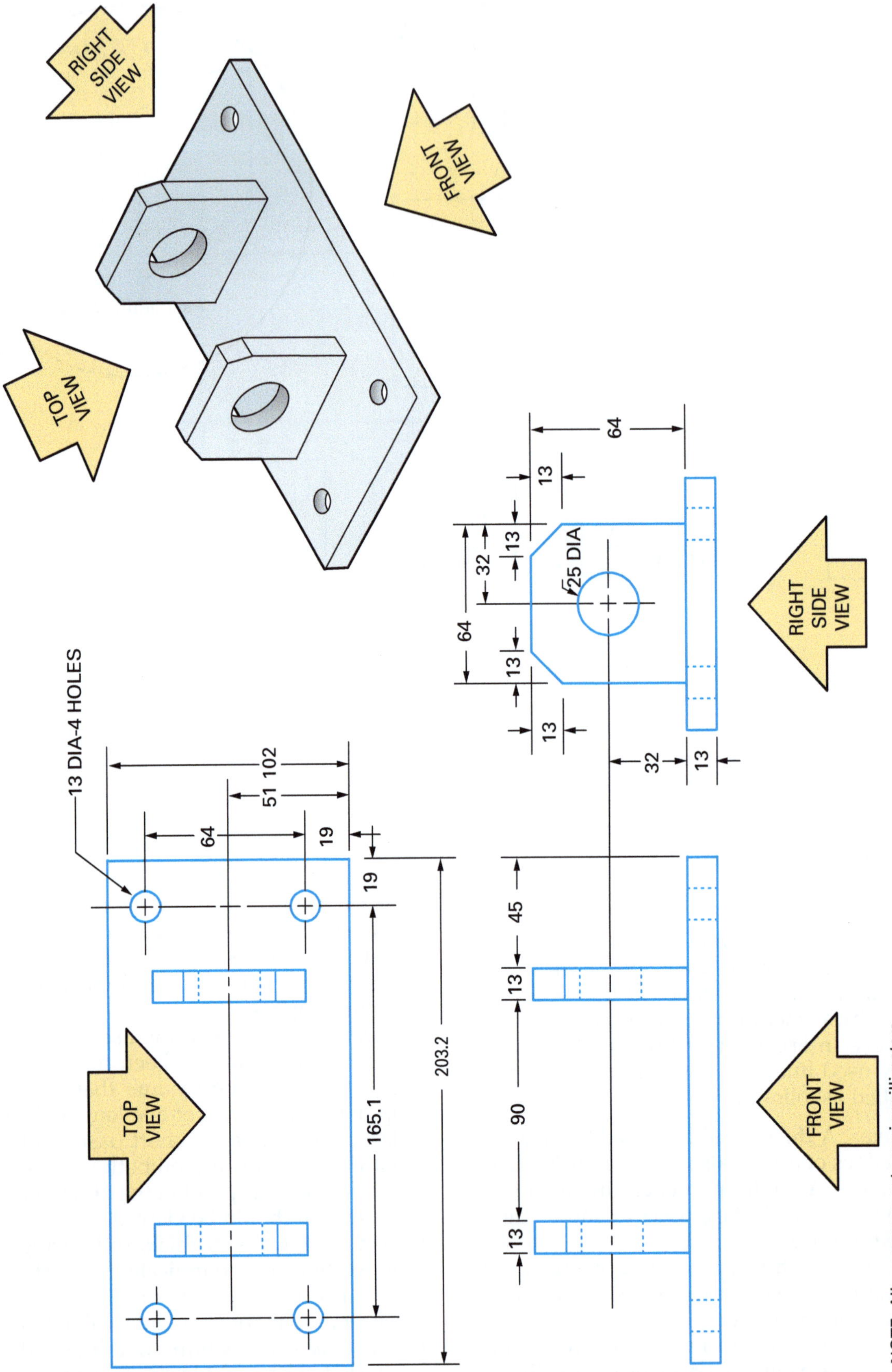

Figure 8 Relationship of views.

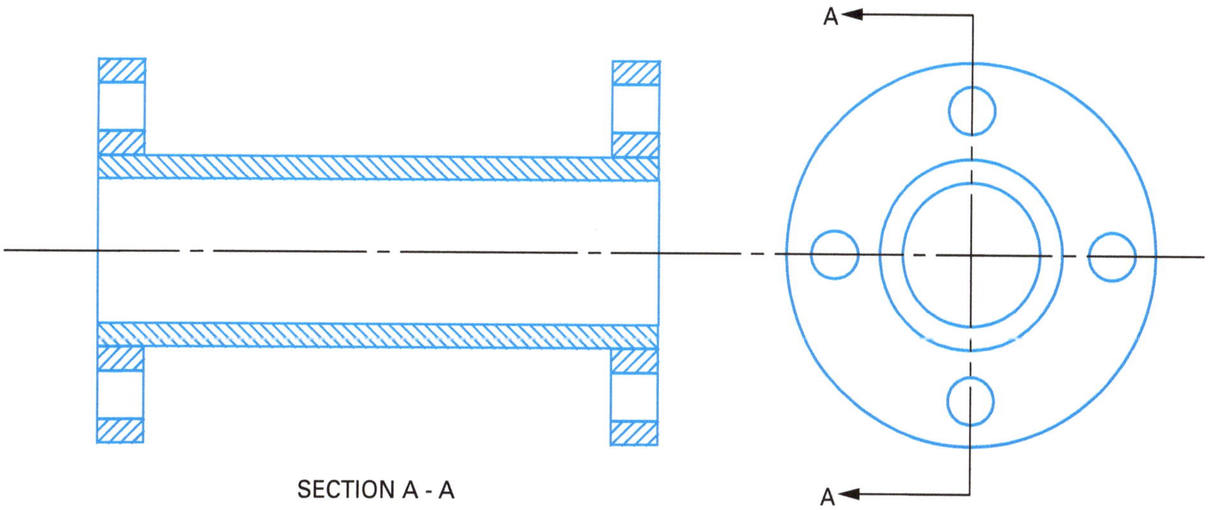

INCHES	MILLIMETERS
6 ½	165
8	204

Figure 9 Section view.

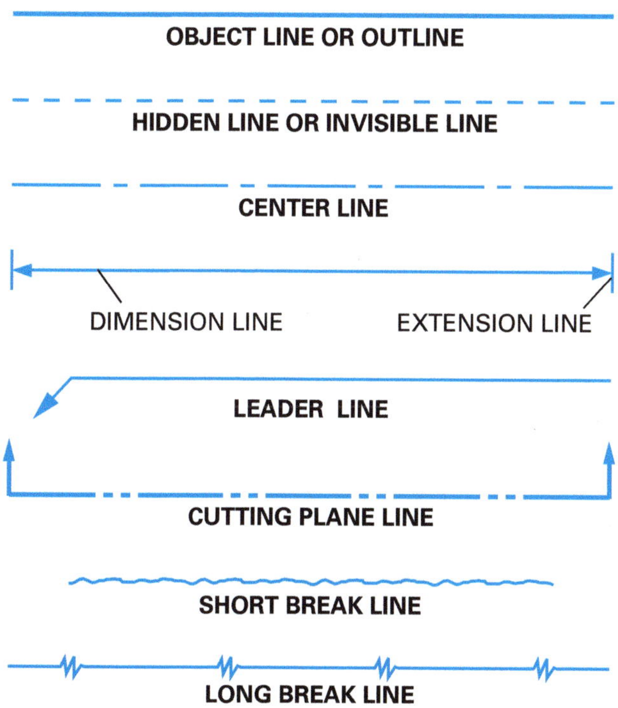

Figure 10 Common drawing lines.

1.2.2 Material Fill Symbols

Material symbols are a series of thin lines, either solid or a combination of solid and broken, that are arranged in a pattern to represent various types of materials. The symbols are used to identify internal surfaces that have been exposed by a cutting plane line or a short cutaway view. *Figure 11* shows common material symbols used to represent various metals.

1.2.3 Bar Stock and Pipe Breaks

A bar stock or pipe break is represented by thick, curved lines that loop back on themselves (*Figure 12*). The approach is the same regardless of the shape of the stock. The loop is filled with a material symbol, such as cross-hatching, to indicate an internal surface. Breaks are used to show where a section of an object has been removed to conserve space on the drawing.

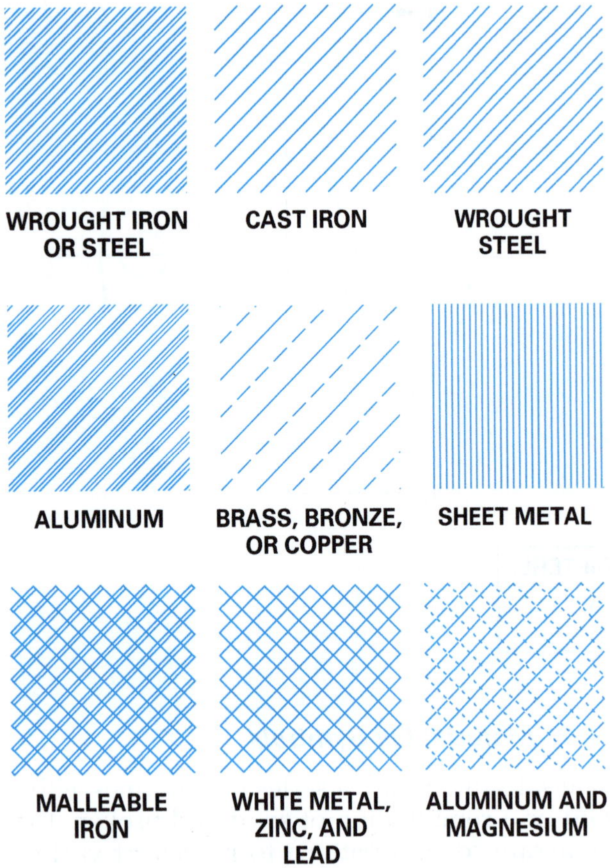

Figure 11 Material symbols.

1.2.4 Revolved Sections

Revolved sections are similar to solid round breaks and pipe, or tubing, breaks except that the shape of the object that is broken is shown in the break as a section view (*Figure 13*).

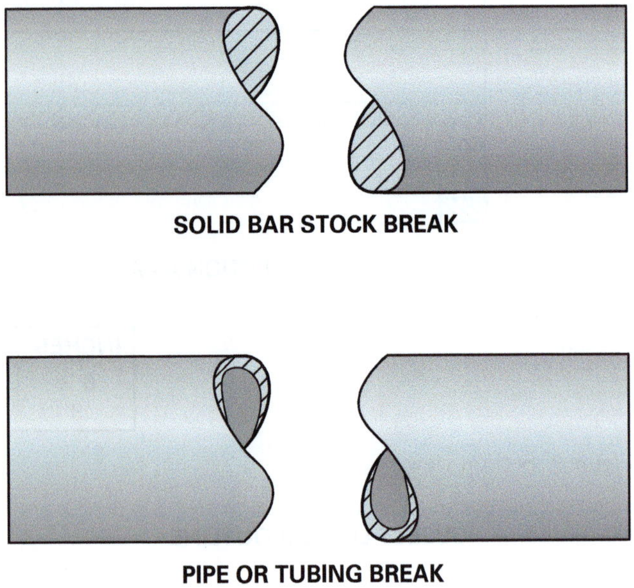

Figure 12 Bar stock and pipe breaks.

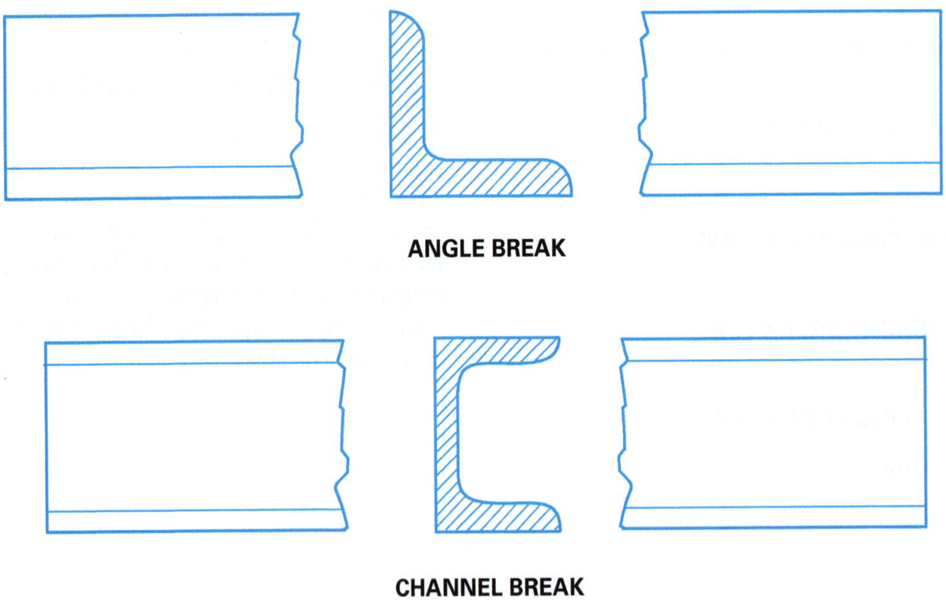

Figure 13 Revolved sections.

Material Symbols

Do not select materials based upon material symbols shown on a drawing. These symbols only represent the described material and do not provide specific information about the material. For example, there are two or three symbols that represent steel, but there are many different types of steel, each with specific characteristics. Always use the drawing notes, material list, or a bill of materials to determine the exact materials that are required.

Additional Resources

ASTM A325, Standard Specification for Structural Bolts, Steel, Heat Treated, 120/105 ksi Minimum Tensile Strength. West Conshohocken, PA: ASTM International.

AWS A2.4, Standard Symbols for Welding, Brazing, and Nondestructive Examination. Miami, FL: American Welding Society.

AWS A3.0, Standard Welding Terms and Definitions. Miami, FL: American Welding Society.

How to Read Shop Drawings. Cleveland, OH: The James F. Lincoln Arc Welding Foundation, 2012.

Steel Construction Manual. Fourteenth Edition. Chicago, IL: American Institute of Steel Construction, 2011.

1.0.0 Section Review

1. The most common drawing used by welders is the ____.
   a. section view
   b. isometric view
   c. assembly drawing
   d. welding detail drawing

2. In a drawing, a solid round break and a pipe, or tubing, break are used to show where a section of an object has been ____.
   a. broken
   b. repaired
   c. inserted
   d. removed

Section Two

2.0.0 Dimensions, Notes, and Materials

Objective

Identify and explain how to interpret dimensional information, notes, and a bill of materials.

a. Identify and explain how to interpret dimensional information.
b. Identify and explain how to interpret notes and a bill of materials.

Performance Task

1. Draw or sketch a welding drawing based on an instructor-provided image or object.

Trade Terms

Blind hole: A drilled hole that does not go all the way through the object.

Counterbore: To enlarge a portion of a drilled hole so that a socket head fastener will seat flush or below the surface.

Elevation: A height measurement above sea level or other identified point; also, the vertical sides of an object in an orthographic view.

Mark: A system of letters and numbers used with structural steel to identify the type of member and where it is placed.

Scale: A method of drawing an object larger or smaller than its actual size while keeping the proportions the same.

Through hole: A drilled hole that goes completely through an object.

Tolerance: The allowance over or under the specified dimension, value, or other parameter.

Welding detail drawings include precise dimensional information for the various pieces that make up an object. Additional details about the construction of the object are often provided in notes and a bill of materials. To fabricate an object that meets design specifications, welders must be able to read and correctly interpret the dimensional information, general or local notes, and the bill of materials.

2.1.0 Dimensions

Dimensions identify sizes. Drawings must have dimensions if the object shown is to be fabricated. The drawing must include all dimensions in correct relationship to one another in order to give a complete size description of the object. The two basic methods for dimensioning a part are conventional dimensioning and baseline dimensioning.

In conventional dimensioning (*Figure 14*), the overall dimensions of length, width, and height are given. Dimensions A and B are not given because they are easily found by adding and/or subtracting the given dimensions. If they were given, dimensions A and B would be identified in parentheses as reference dimensions. A reference dimension, such as the 1¾ inch dimension in parentheses on the drawing, is obtained by adding or subtracting other dimensions found on the drawing. If the overall size of a dimensioned part is not given, then those dimensions that make up the total must be added to find the overall size. Reference dimensions help to speed up the reading of drawings.

In baseline dimensioning (*Figure 15*), all dimensions originate from baselines. When imperial measurements are used, baseline dimensions are usually expressed as three-place decimals for accuracy. Metric baseline dimensions are also normally expressed as three-place decimals, depending upon the chosen unit of measure.

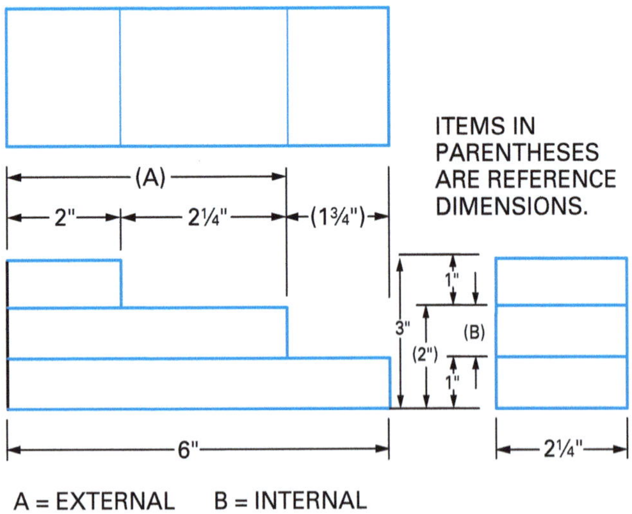

Figure 14 Conventional dimensioning.

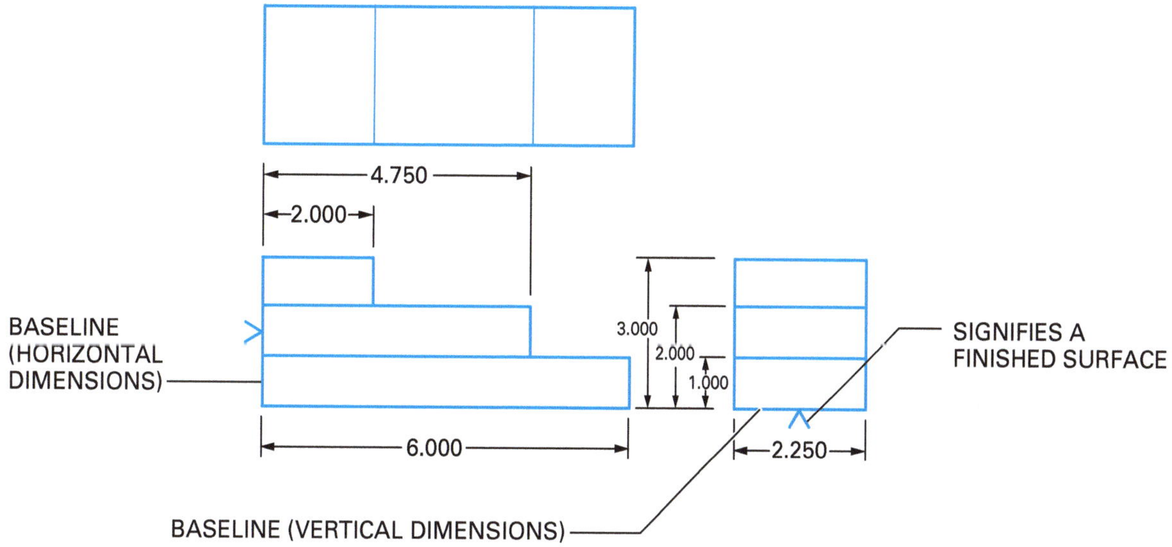

Figure 15 Baseline dimensioning.

NOTE

Baseline dimensions on metric drawings are never fractional because, unlike imperial measurements, metric measurements are always shown as decimals. When working with drawings that have baseline dimensions, always check to make sure which system of measurement (metric or imperial) the drawing is based on.

The baselines are typically machined surfaces that are indicated by the symbol V. Baselines may also originate from center lines or from a critical edge. The advantage of baseline dimensioning is that it eliminates the cumulative error that often occurs with conventional dimensioning. Baseline dimensioning does not require calculations to find a dimension. Most welding drawings that require hand assembly and welding do not use three-place decimal dimensioning because of the difficulty in maintaining tight **tolerances**.

Think About It

Dimension Calculation

What is the reference dimension of a section whose overall part dimension is 4⅞ inches? The part consists of three sections. Two of the sections are 1⅞ inches and 1⅛ inches, respectively.

2.1.1 Scale

Most drawings are made in **scale**. This means that the drawing can be larger, the same size, or smaller than the actual object, as long as the proportions in the drawing are the same as the proportions of the actual object. Most objects shown in welding detail drawings are too large to fit on a standard drawing, so they are drawn smaller than the actual object's size. The scale used on the drawing is given in the title block in the form of an equation. Many different scales can be used, for example:

- ¼ inch = 1 inch
- ⅛ inch = 1 inch
- ¼ inch = 1 foot

The figure to the left of the equal sign in a scale equation represents the measure used for the drawing, while the figure to the right of the equal sign corresponds to the measure used for the object. For example, if the scale ¼ inch = 1 inch is used, an actual surface 3 inches long would be represented on the drawing by a line ¾ inches long (¼ inch for each actual inch). If the scale ¼ inch = 1 foot is used, an actual surface 4 feet long would be shown on the drawing by a line 1 inch long. *Figure 16* shows how scale is used to reduce the size of a drawing.

On drawings that are dimensioned in metric units, the scale is stated as a ratio. Actual size would be stated as 1:1. A reduction ratio would be stated as 1:2, 1:5, 1:10, etc. An increasing ratio would be stated as 2:1, 5:1, etc. The drawing title block will usually specify the applicable scale, unless more than one scale is used on the drawing.

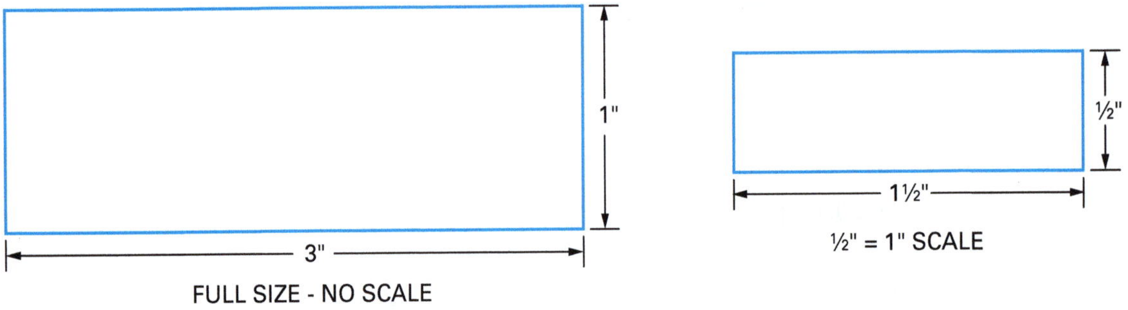

Figure 16 Drawing in scale.

2.1.2 Size and Locations Dimensions

The most common dimensions are size and location. Size dimensions provide the measurements for the overall size of the object. Location dimensions provide measurements from one feature of an object to another part, hole, break, or component. Examples of both dimension types are shown in *Figure 17.*

2.1.3 Hole Dimensions

Drilled hole dimensions are shown with a leader line and a note (*Figure 18*). The note gives the hole size and the number of holes to be drilled, if more than one. If the hole is to be drilled only partway through the object, which is called a **blind hole**, then the depth will also be given. If the depth is not given, then the hole goes completely through the object and is called a **through hole**.

Holes may also be **counterbored** or countersunk. A hole that is counterbored has a square shoulder that is used to recess socket head fasteners. A hole that is countersunk is tapered from the surface to the drilled hole size and is used for flat-head fasteners. The required dimensions for countersinking or counterboring holes are added to the note (*Figure 19*).

2.1.4 Angle and Bevel Dimensions

Angles and bevels are common on welding detail drawings because they are used with weld joint preparation. They can be specified with a leader line and a note, or they can be specified using extension and dimension lines (*Figure 20*). The size can be given as angular, linear, or a combination of angular and linear dimensions.

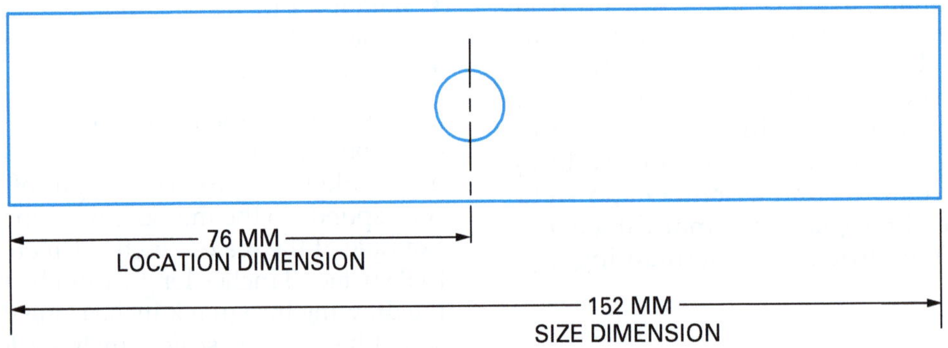

Figure 17 Size and location dimensions.

Drawing Measurements

Be sure to use only dimensions provided on the drawing. If a dimension is needed but is not provided, add or subtract the dimensions that were furnished to obtain the needed dimension. Avoid measuring directly from a drawing (called scaling a drawing). Measuring directly from a drawing almost always results in errors. If you cannot determine a measurement by adding or subtracting given dimensions, check with your supervisor.

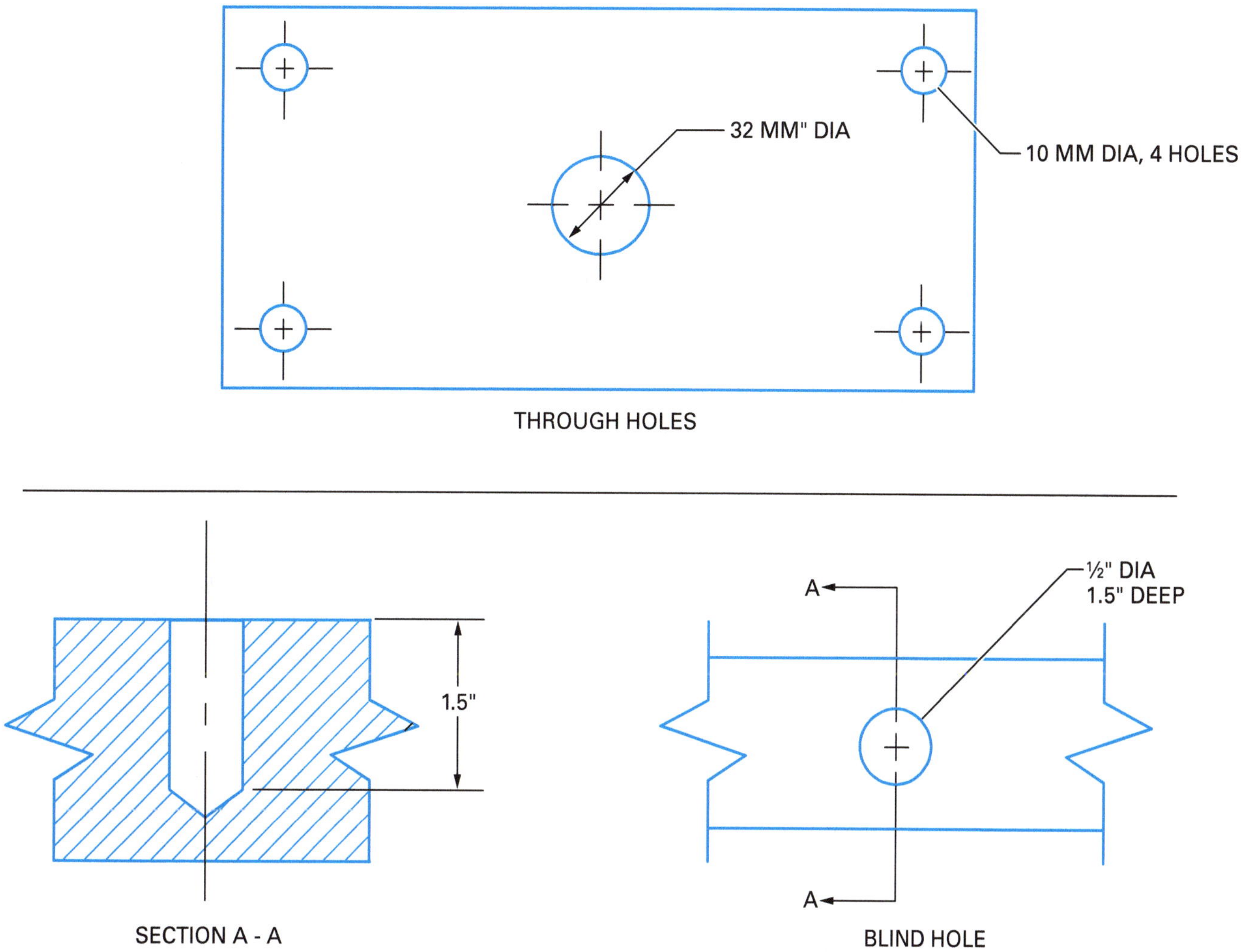

Figure 18 Dimensioning holes.

Structural Steel and Pipe Sizes

Be aware that structural steel and pipe sizes, if given on drawings, are usually nominal (approximate) sizes only. Structural steel is classified as hot rolled or cold rolled. Hot-rolled varieties are somewhat oversized and have rounded corners and edges. The sizes of cold-rolled types are more precise, and the corners and edges are square. Size specifications for some structural shapes including S-beams (formerly called I-beams), H-beams, and wide flange (WF) beams are nominal.

Pipe is sometimes sized through the application of a schedule number as part of a specification. The eleven schedules commonly used are: 5, 10, 20, 30, 40, 50, 60, 80, 100, 120, and 160. The schedule numbers, which represent the strength of the pipe, refer to wall thickness. Pipe schedule tables provide the nominal wall thickness, weight, and outside diameter for each size and type of pipe.

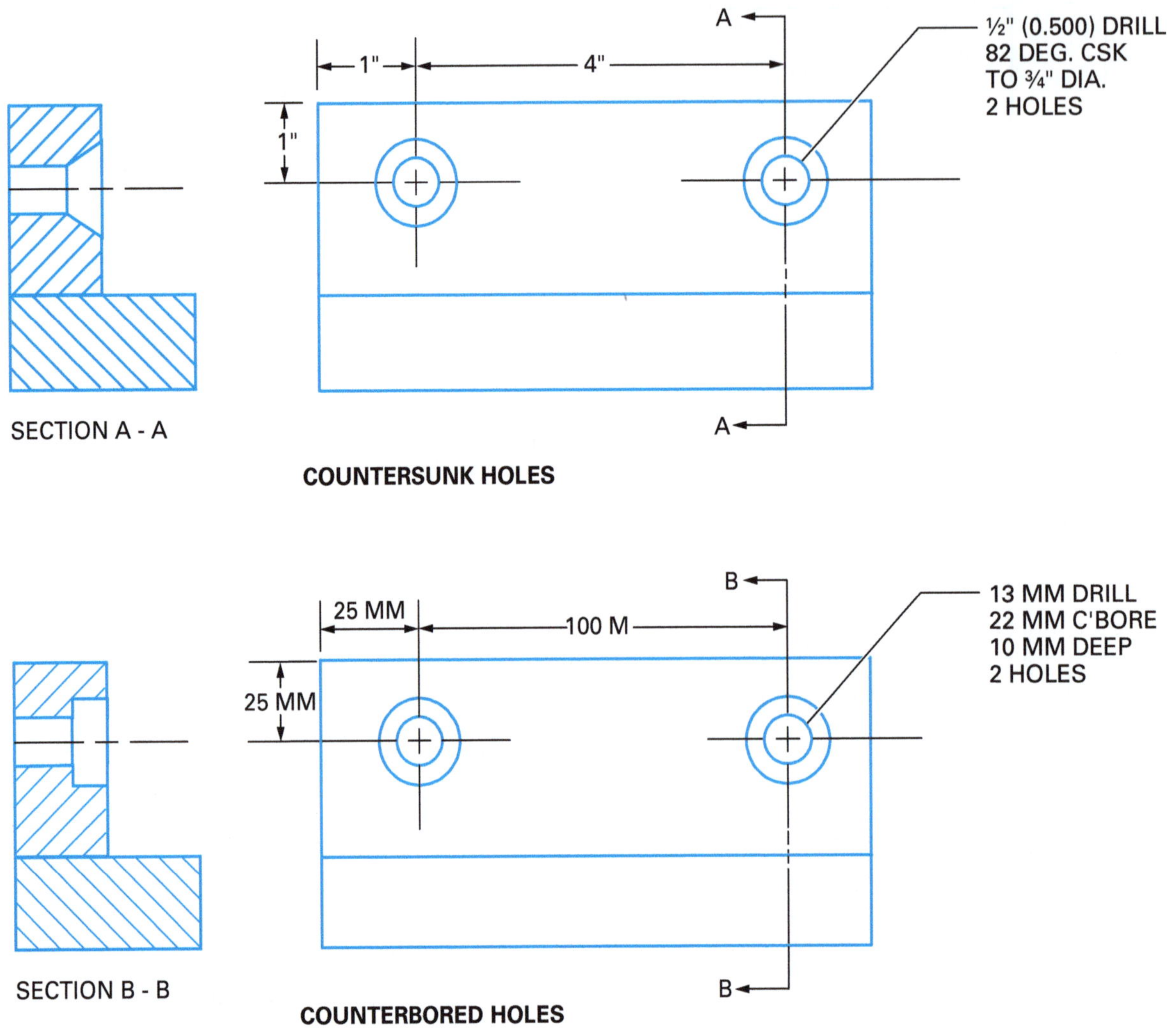

Figure 19 Dimensioning counterbores and countersinks.

2.1.5 Radius and Arc Dimensions

When the ends, sides, or corners of an object are to be rounded, a radius or arc is shown and dimensioned (*Figure 21*). Typically a combination of center lines, radius, extension, and dimension lines are used. To obtain the arc required, the radius line may start outside the physical shape of the object.

2.1.6 Tolerances

Tolerance is an important part of dimensioning because it identifies the variance in size that is allowable on a part. Depending on its intended use, a part can have a large or a small tolerance. If a part is fabricated outside the allowable tolerance, the part will have to be reworked or scrapped at an additional cost. On the other hand, if a part with a large tolerance is fabricated to very fine tolerances, this will generally add to the cost since parts with large tolerances typically can be fabricated faster than parts with smaller tolerances. For these reasons, it is important to identify the tolerances on a drawing and then work accordingly.

Tolerances can be given following the dimension, as a note, or in a table as part of the title block. They are generally given as a plus (+) or minus (−) quantity. For example, if the dimension 10 inches was followed by the tolerance, + or − ⅛ inch, the object could be 9⅞ inches through 10⅛ inches.

The plus and minus tolerances for a size can be different. For example, a dimension of 12 inches, + ⅛ inch and −0, indicates that the object can be from 12 inches through 12⅛ inches. *Figure 22* shows three ways of indicating tolerances.

Think About It

Cut Tolerance

What are the minimum and maximum allowable sizes of an 18-inch cut that has to be made with a tolerance of 6¼ inches?

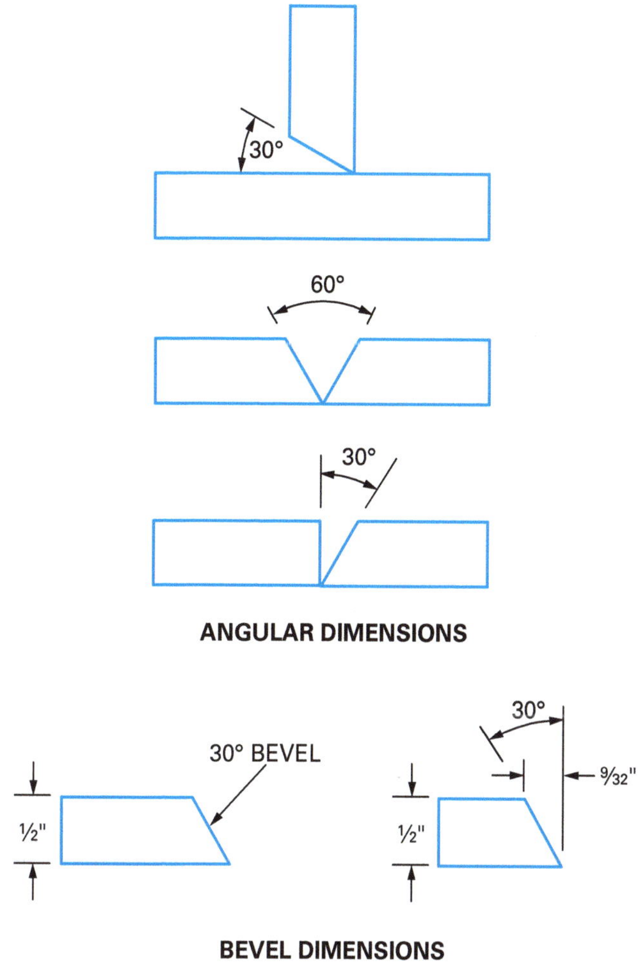

Figure 20 Specifying angles and bevels.

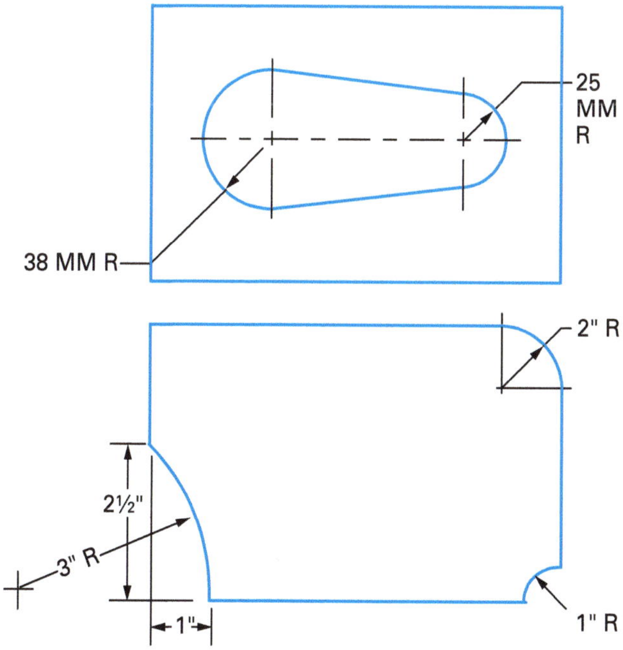

Figure 21 Dimensioning radius and arcs.

2.2.0 Notes and Bill of Materials

Additional information that can be found on many drawings includes notes and a bill of materials. These forms of documentation provide data that are not given elsewhere on the drawing.

2.2.1 Notes

A note is an explanatory or critical comment concerning the details of construction. Two types of notes are found on drawings: general notes and local (specific) notes (*Figure 23*). A general note applies to the drawing in its entirety, and it is placed in an area away from the views so that it will be noticed. On large jobs, general notes are placed on a separate drawing. On smaller jobs, general notes are placed on detail drawings, where they refer to the entire drawing. General notes are grouped together and numbered.

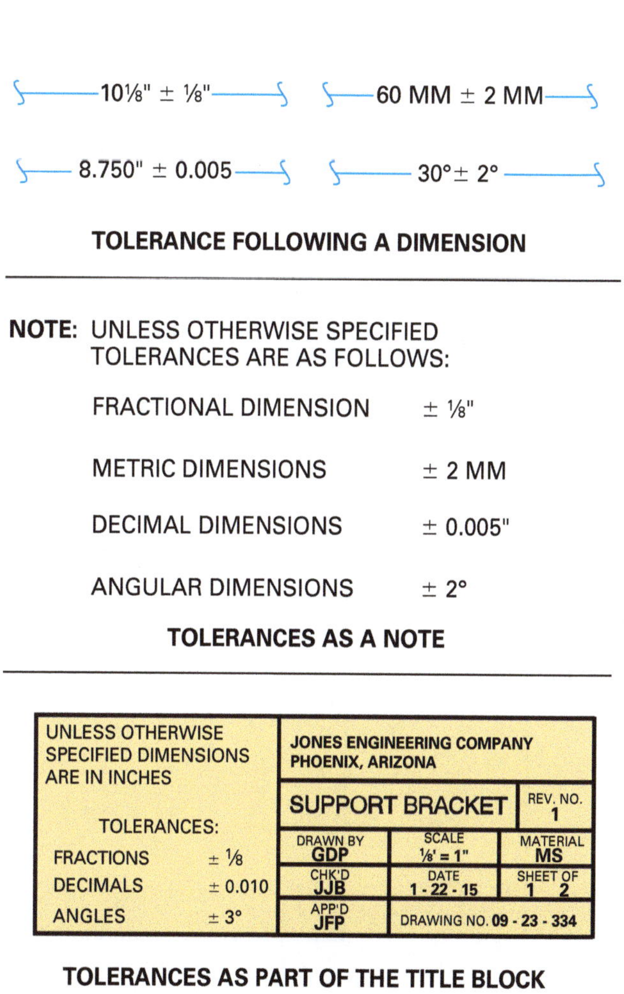

Figure 22 Dimensioning tolerances.

Items typically found in general notes include the following:

- References to specifications, procedures, or other notes
- Identification of work responsibility
- Surface finish requirements (smoothness, type of finish)
- Special handling requirements
- Material requirements such as specification, grade, and class
- Special assembly requirements

A local note applies to a particular part of an object. It is placed near one of the views that represent the part. Local notes are used for items such as the following:

GENERAL NOTES - STEEL

1. Design, fabrication, and erection of structural steel shall be in accordance with the American Institute of Steel Construction (AISC) *Steel Construction Manual*, Thirteenth Edition (2007), unless otherwise modified on the drawings or in the specifications.
2. Material shall meet the requirements of the following specifications unless otherwise noted:

STRUCTURAL STEEL	- ASTM	A 36 - 08
HIGH STRENGTH BOLTS	- ASTM	A325 - 07a, ⅞ INCH DIA
WELD STEEL	- AWS	D 1.1 - 08, CLASS E70

3. Shop and field paint shall be as called for in the specifications.
4. Steel to be encased in concrete shall not be painted.
5. Hot dipped galvanized surfaces that have been damaged by welding, cutting, burning, shearing, or other damage incurred during transit or erection shall be repaired with Z.R.C. Cold Galvanizing Compound as manufactured or approved equivalent. Repair shall be in strict accordance with the manufacturer's recommendations.
6. All shop connections shall be welded or made with high strength bolts unless noted specifically.
7. Erectors shall provide all temporary shoring & bracing needed for stability until structure is completed.

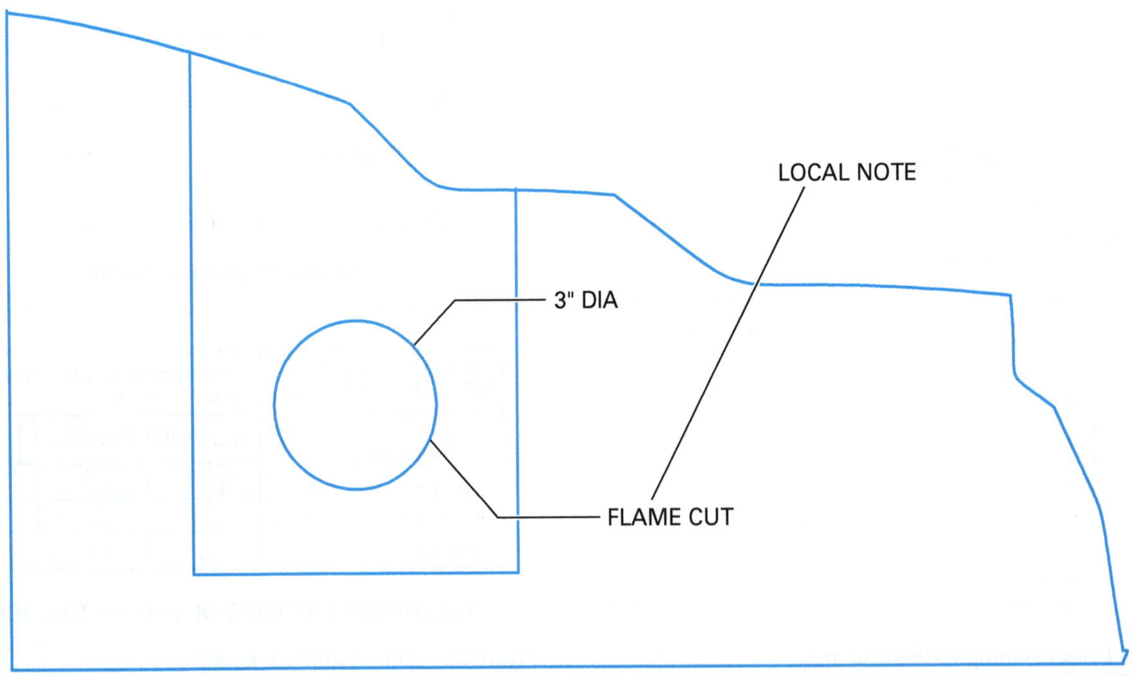

Figure 23 General notes and local notes.

- Thread fastener type and size
- Location of special items (parts to be supplied by others)
- Reference to details on another drawing
- **Elevations**
- Special treatments or finishes
- Drilled hole size

A specification is a note that provides a detailed description of requirements for the type and size of the parts or material to be used. For instance, it may specify the type and size of electrode and/or the welding process to be used. In *Figure 24*, the letter A refers to Specification A. In this case, Specification A in the tail of the arrow indicates that No. 30, ⅛-inch copper rod is to be used for the weld. Specifications are usually located near the views to which they refer. If there are many specifications, they are attached on a separate sheet of paper and referenced to the drawing.

2.2.2 Bill of Materials

A bill of materials, also called a list of material, parts list, or schedule of parts, is a list of all the parts required to fabricate an object. It generally identifies each part with a number or letter, sometimes called a **mark**. The bill of materials includes a description of the part, the quantity needed for each assembly, and the part number. Some bills of materials may also include size and dimensions for the part. This is often the case when the parts being described are structural shapes. *Figure 25* shows a portion of a drawing that includes a bill of materials. Not every drawing, however, will have a bill of materials. For larger projects, the bill of materials alone may require several drawing sheets.

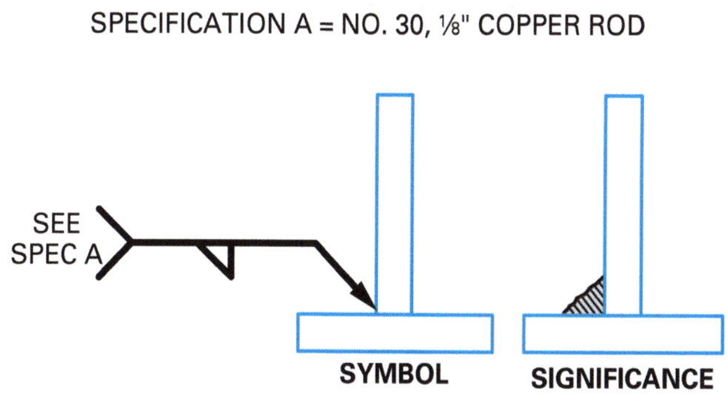

Figure 24 Application of a specification

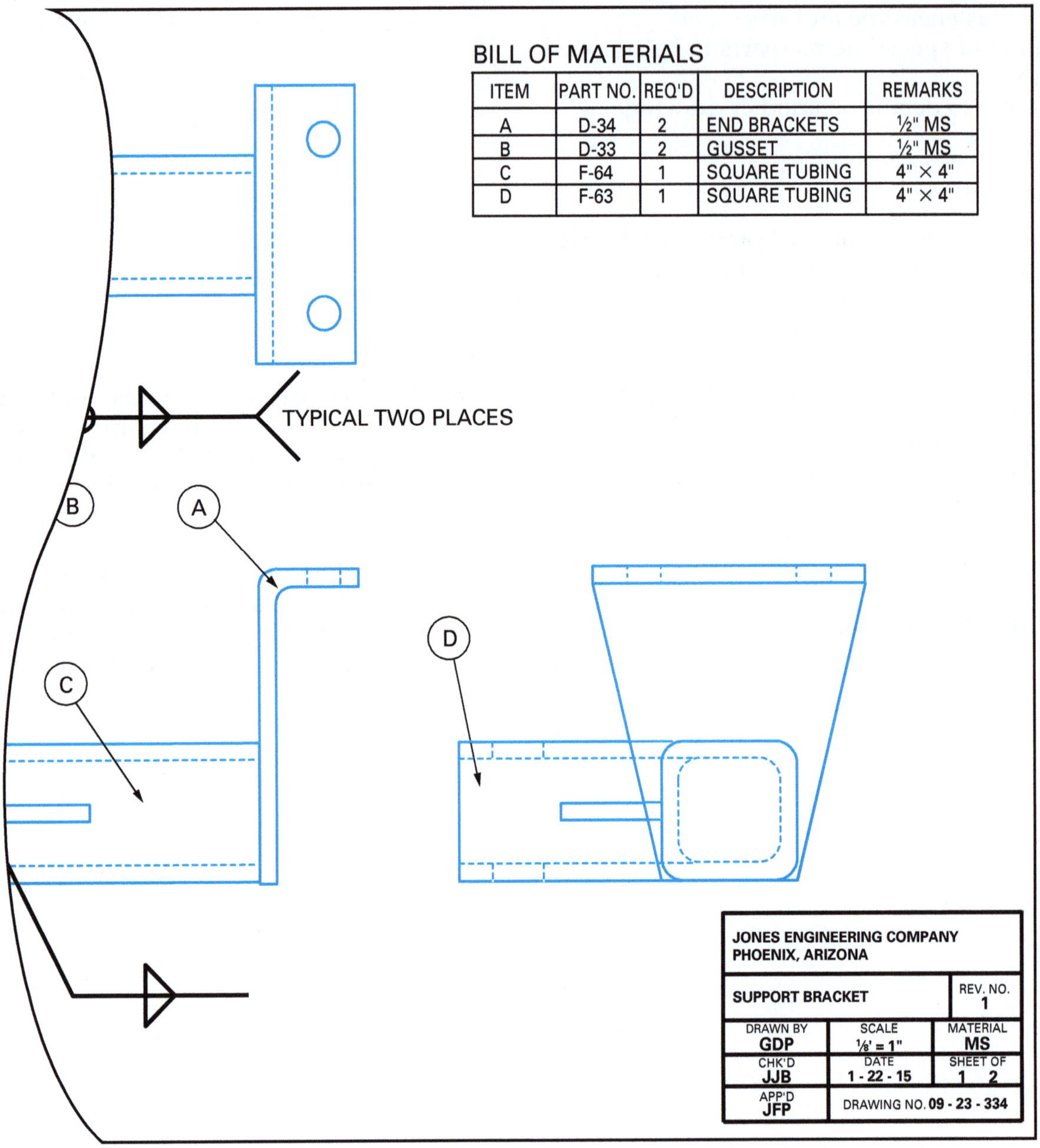

Figure 25 Bill of materials.

Additional Resources

ASTM A325, Standard Specification for Structural Bolts, Steel, Heat Treated, 120/105 ksi Minimum Tensile Strength. West Conshohocken, PA: ASTM International.

AWS A2.4, Standard Symbols for Welding, Brazing, and Nondestructive Examination. Miami, FL: American Welding Society.

AWS D1.1/D1.1M, Structural Welding Code Steel. Miami, FL: American Welding Society.

AWS A3.0, Standard Welding Terms and Definitions. Miami, FL: American Welding Society.

How to Read Shop Drawings. Cleveland, OH: The James F. Lincoln Arc Welding Foundation, 2012.

Steel Construction Manual. Fourteenth Edition. Chicago, IL: American Institute of Steel Construction, 2011.

Welding Symbols on Drawings. E. N. Gregory and A. A. Armstrong. Boca Raton, FL: CRC Press, 2005.

2.0.0 Section Review

1. A hole that will only be drilled part of the way through an object is called a ____.
   a. counterbored hole
   b. countersunk hole
   c. blind hole
   d. through hole

2. Items such as identification of work responsibility and special assembly requirements are given in ____.
   a. classification schedules
   b. specification notes
   c. welding parts lists
   d. general notes

Summary

Engineers, designers, and draftsmen use welding detail drawings to precisely convey information to welders in the field. To correctly interpret this information, welders must be able to accurately read welding detail drawings.

Reading and interpreting welding detail drawings requires welders to have an understanding of lines, fills, views, and dimensioning. It is also important for welders to be able to read and interpret drawing notes and the bill of materials.

Review Questions

1. What type of weld is used to join parts 1F and 1B of the object shown in *Review Question Figure 1*?
   a. Concave
   b. Convex
   c. Groove
   d. Fillet

2. When an object to be fabricated is too complicated for one drawing, the complete and assembled object may be shown on one or more detail drawings plus a(n) ____.
   a. general note
   b. bill of materials
   c. structural print
   d. assembly drawing

3. On a typical drawing, a thick, solid line used to show all visible edges of a part is called a(n) ____.
   a. object line
   b. center mark
   c. leader line
   d. hidden section

4. A series of thin lines that are either solid, or solid and broken, and are arranged in a pattern to indicate various types of substances, such as metals, is called a(n) ____.
   a. material symbol
   b. cut-away projection
   c. cutting plane
   d. extension grid

5. How tall is part 1C of the object shown in *Figure 1*?
   a. 2"
   b. 3$\frac{13}{16}$"
   c. 4"
   d. 6"

6. Which part or parts of the object shown in *Figure 1* must be drilled?
   a. 1A
   b. 1F
   c. 1A and 1D
   d. 1B and 1C

7. Angles and bevels are common on welding detail drawings because they are used with ____.
   a. fillet welds
   b. groove welds
   c. oxyfuel cuts
   d. weld joint preparation

8. What degree of angle is used for the bevel groove weld joining 1D and 1C in *Figure 1*?
   a. 60 degrees
   b. 50 degrees
   c. 30 degrees
   d. 20 degrees

9. The tolerances used for the dimensions of a drawing can be given (shown) following the dimension, as a note, or in the ____.
   a. title block
   b. scale factor
   c. drawing title
   d. object drawing itself

10. How many pieces make up the weldment shown in *Figure 1*?
    a. 4
    b. 5
    c. 6
    d. 8

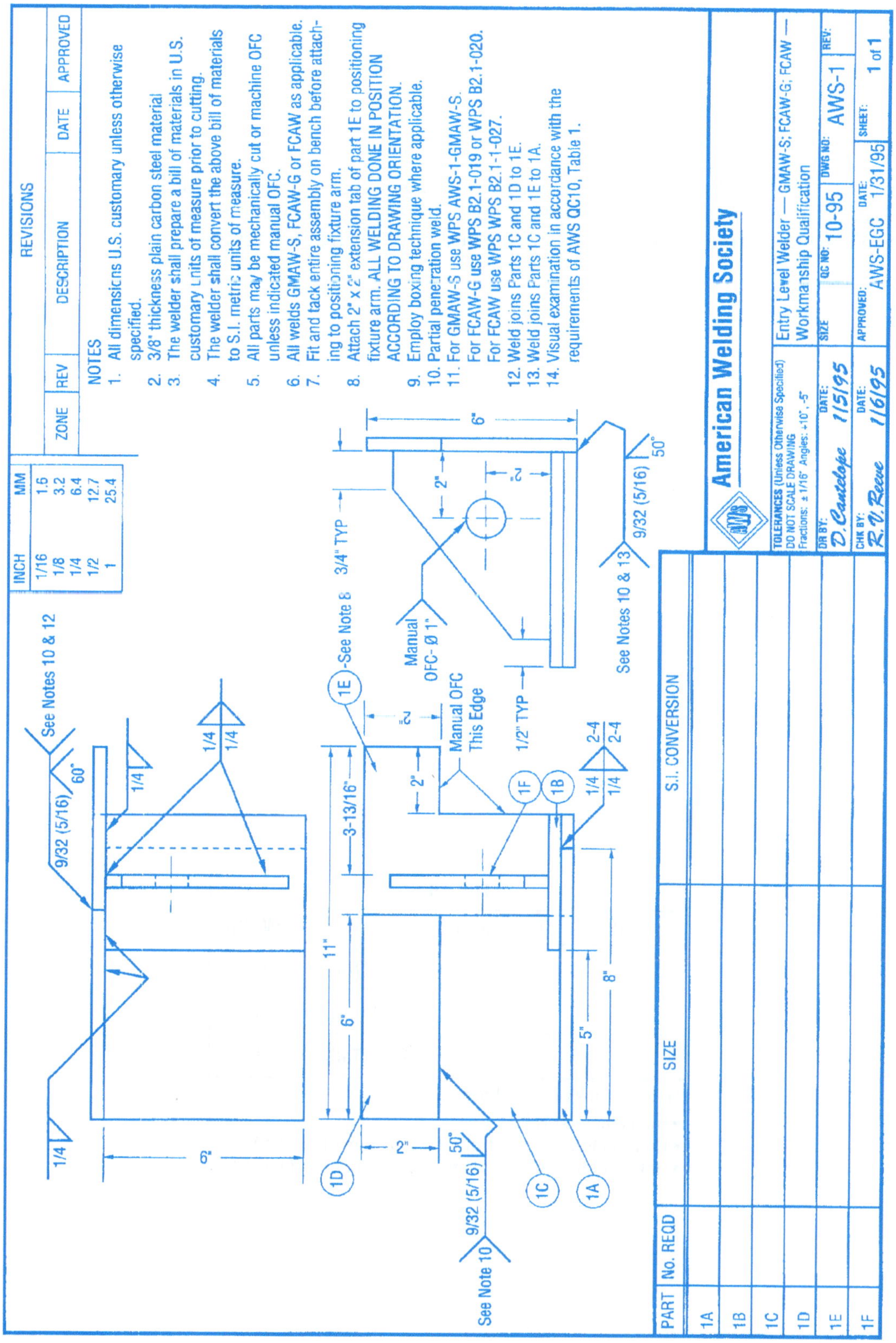
REVISIONS
ZONE
REV
DESCRIPTION
DATE
APPROVED
INCH MM
1/16 1.5
1/8 3.2
1/4 6.4
1/2 12.7
1 25.4
NOTES
1. All dimensions U.S. customary unless otherwise specified.
2. 3/8" thickness plain carbon steel material
3. The welder shall prepare a bill of materials in U.S. customary units of measure prior to cutting.
4. The welder shall convert the above bill of materials to S.I. metric units of measure.
5. All parts may be mechanically cut or machine OFC unless indicated manual OFC.
6. All welds GMAW-S, FCAW-G or FCAW as applicable.
7. Fit and tack entire assembly on bench before attaching to positioning fixture arm.
8. Attach 2" x 2" extension tab of part 1E to positioning fixture arm. ALL WELDING DONE IN POSITION ACCORDING TO DRAWING ORIENTATION.
9. Employ boxing technique where applicable.
10. Partial penetration weld.
11. For GMAW-S use WPS AWS-1-GMAW-S.
For FCAW-G use WPS B2.1-019 or WPS B2.1-020.
For FCAW use WPS B2.1-027.
12. Weld joins Parts 1C and 1D to 1E.
13. Weld joins Parts 1C and 1E to 1A.
14. Visual examination in accordance with the requirements of AWS QC10, Table 1.
See Notes 10 & 12
9/32 (5/16)
60°
1/4
1/4
1/4
6"
2"
50°
See Note 10
1D
1C
1A
1E
See Note 8
3/4" TYP
Manual OFC-Ø 1"
Manual OFC This Edge
1/2" TYP
1F
1B
2-4
2-4
3-13/16"
11"
6"
5"
8"
2"
6"
See Notes 10 & 13
50°
PART
No. REQD
SIZE
S.I. CONVERSION
1A
1B
1C
1D
1E
1F
American Welding Society
TOLERANCES (Unless Otherwise Specified)
DO NOT SCALE DRAWING
Fractions: ± 1/16" Angles: +10°, -5°
Entry Level Welder — GMAW-S; FCAW-G; FCAW — Workmanship Qualification
DR BY: D. Cantelope
DATE: 1/5/95
SIZE
QC NO: 10-95
DWG NO: AWS-1
REV:
CHK BY: R.V. Reeve
DATE: 1/6/95
APPROVED: AWS-EGC
DATE: 1/31/95
SHEET: 1 of 1

Figure 1

Performance Task

The Performance Task for this module is to draw or sketch a three-view detail drawing based on a given image or object. The following is an example of such an image. The three-view detail drawing representing the completed Performance Task is also provided here for trainee review.

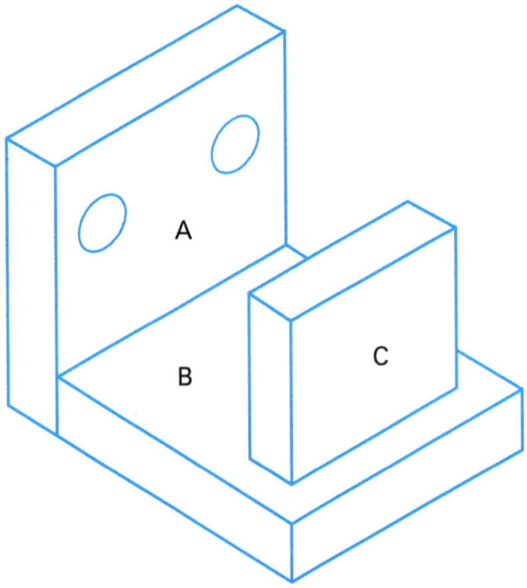

DETAILS			
Part	**Thickness**	**Length**	**Width**
A	¾"	3½" (Tall)	3½"
B	¾"	4" (Long)	3½"
C	¾"	2½" (Tall)	2½"

Hole Dimensions and Location

- Hole diameter is ¾"
- Hole centers are 1¼" from the top edge of block "A"
- Hole centers are ¾" from the outside edges of block "A"

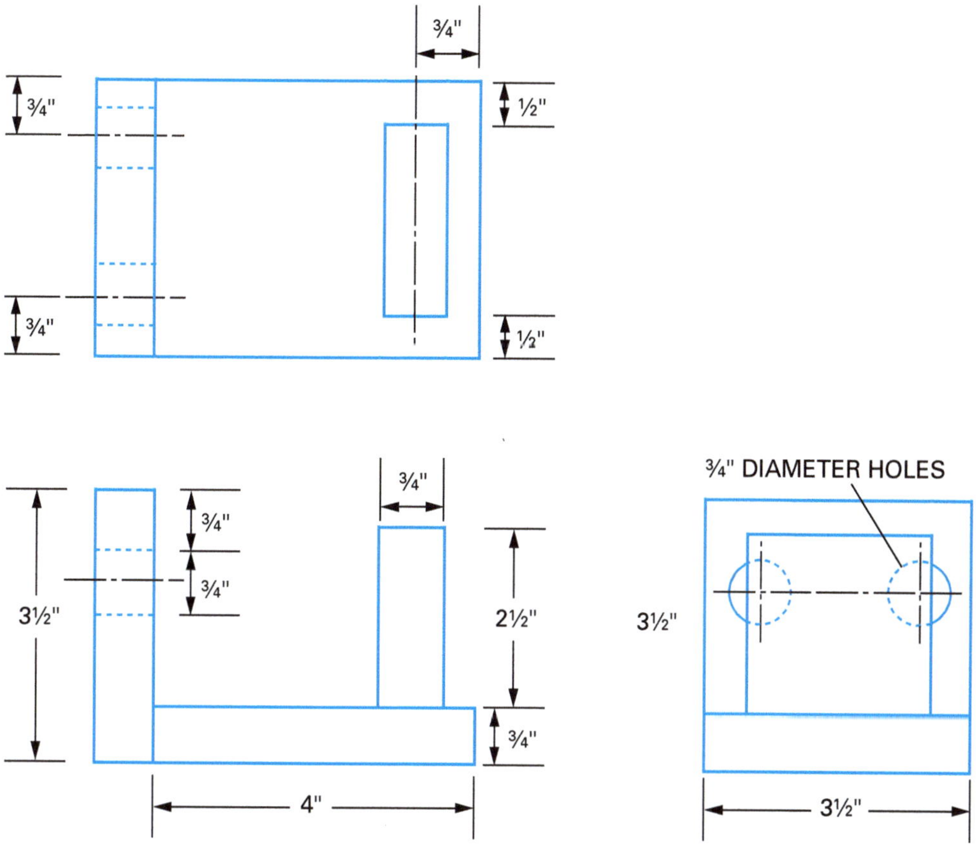

RESULT OF REDRAWING THE OBJECT USING THREE ORTHOGRAPHIC VIEWS

Trade Terms Introduced in This Module

Blind hole: A drilled hole that does not go all the way through the object.

Counterbore: To enlarge a portion of a drilled hole so that a socket head fastener will seat flush or below the surface.

Elevation: A height measurement above sea level or other identified point; also, the vertical sides of an object in an orthographic view.

Isometric view: A drawing depicting a three-dimensional view of an object.

Mark: A system of letters and numbers used with structural steel to identify the type of member and where it is placed.

Orthographic projection: A method of developing multiple flat views of an object by projecting lines out from the object at right angles from the face.

Scale: A method of drawing an object larger or smaller than its actual size while keeping the proportions the same.

Through hole: A drilled hole that goes completely through an object.

Tolerance: The allowance over or under the specified dimension, value, or other parameter.

Additional Resources

This module is intended as a thorough resource for task training. The following reference works are suggested for further study.

ASTM A325, Standard Specification for Structural Bolts, Steel, Heat Treated, 120/105 ksi Minimum Tensile Strength. West Conshohocken, PA: ASTM International.

AWS A2.4, Standard Symbols for Welding, Brazing, and Nondestructive Examination. Miami, FL: American Welding Society.

AWS A3.0, Standard Welding Terms and Definitions. Miami, FL: American Welding Society.

AWS D1.1/D1.1M, Structural Welding Code Steel. Miami, FL: American Welding Society.

How to Read Shop Drawings. Cleveland, OH: The James F. Lincoln Arc Welding Foundation, 2012.

Steel Construction Manual. Fourteenth Edition. Chicago, IL: American Institute of Steel Construction, 2011.

Welding Symbols on Drawings. E. N. Gregory and A. A. Armstrong. Boca Raton, FL: CRC Press, 2005.

Figure Credits

The Lincoln Electric Company, Cleveland, OH, USA, Module opener

AWS EG2.0-95, Figure 2, Reproduced with permission from the American Welding Society (AWS), Miami, FL, USA, RQ01

Section Review Answer Key

Answer	Section Reference	Objective
Section One		
1. d	1.0.0	1a
2. d	1.2.3	1b
Section Two		
1. c	2.1.3	2a
2. d	2.2.1	2b

NCCER CURRICULA — USER UPDATE

NCCER makes every effort to keep its textbooks up-to-date and free of technical errors. We appreciate your help in this process. If you find an error, a typographical mistake, or an inaccuracy in NCCER's curricula, please fill out this form (or a photocopy), or complete the online form at **www.nccer.org/olf**. Be sure to include the exact module ID number, page number, a detailed description, and your recommended correction. Your input will be brought to the attention of the Authoring Team. Thank you for your assistance.

Instructors – If you have an idea for improving this textbook, or have found that additional materials were necessary to teach this module effectively, please let us know so that we may present your suggestions to the Authoring Team.

NCCER Product Development and Revision

13614 Progress Blvd., Alachua, FL 32615

Email: curriculum@nccer.org

Online: www.nccer.org/olf

❑ Trainee Guide ❑ Lesson Plans ❑ Exam ❑ PowerPoints Other ____________

Craft / Level: Copyright Date:

Module ID Number / Title:

Section Number(s):

Description:

Recommended Correction:

Your Name:

Address:

Email: Phone:

SMAW – Groove Welds with Backing

Overview

There are a number of different groove styles used in welding. The standard V-groove is one of the most popular. Completing groove welds in the various positions requires attention to detail and an understanding of work, drag, and push angles to ensure a quality weld. This module explores the differences in technique used to complete V-groove welds with backing, and provides a significant amount of time to develop these essential skills.

Module 29111

Trainees with successful module completions may be eligible for credentialing through the NCCER Registry. To learn more, go to **www.nccer.org** or contact us at **1.888.622.3720**. Our website has information on the latest product releases and training, as well as online versions of our *Cornerstone* magazine and Pearson's product catalog.

Your feedback is welcome. You may email your comments to **curriculum@nccer.org**, send general comments and inquiries to **info@nccer.org**, or fill in the User Update form at the back of this module.

This information is general in nature and intended for training purposes only. Actual performance of activities described in this manual requires compliance with all applicable operating, service, maintenance, and safety procedures under the direction of qualified personnel. References in this manual to patented or proprietary devices do not constitute a recommendation of their use.

29111 V5

29111
SMAW – Groove Welds with Backing

Objectives

When you have completed this module, you will be able to do the following:

1. Identify various types of groove welds and describe how to prepare for groove welding.
   a. Identify various types of groove welds and define related terms.
   b. Describe how to prepare for groove welding.
2. Describe the technique required to produce various groove welds.
   a. Describe the technique required to produce groove welds in the 1G and 2G positions.
   b. Describe the technique required to produce groove welds in the 3G and 4G positions.

Performance Tasks

Under the supervision of your instructor, you should be able to do the following:

1. Safely set up arc welding equipment for making groove welds.
2. Make flat welds with backing on V-groove joints using E7018 electrodes.
3. Make horizontal welds with backing on V-groove joints using E7018 electrodes.
4. Make vertical welds with backing on V-groove joints using E7018 electrodes.
5. Make overhead welds with backing on V-groove joints using E7018 electrodes.

Industry Recognized Credentials

If you are training through an NCCER-accredited sponsor, you may be eligible for credentials from NCCER's Registry. The ID number for this module is 29111. Note that this module may have been used in other NCCER curricula and may apply to other level completions. Contact NCCER's Registry at 888.622.3720 or go to **www.nccer.org** for more information.

Contents

Topics to be presented in this module include:

Figures

Section One

1.0.0 Groove Welds

Objective

Identify various types of groove welds and describe how to prepare for groove welding.

a. Identify various types of groove welds and define related terms.
b. Describe how to prepare for groove welding.

Performance Task

1. Safely set up arc welding equipment for making groove welds.

Different welding applications require differently shaped welds at the point where two metal parts are to be joined. This module explains groove welds and the various techniques and arc welding equipment used to make them.

A groove weld is made in an opening in a part or between two parts. Groove welds can be made in all five of the basic types of joints as well as on the surface of a part. There are several different groove styles. The name of the groove is based on its shape. Some must be prepared, while others appear as a result of fitting together two or more parts to be welded.

If a groove only goes part of the way through the joint, welding in the groove results in what is called a partial joint penetration (PJP) weld. If the groove allows for welding the complete thickness of the base metals, it is called a complete joint penetration (CJP) weld.

1.1.0 Typical Groove Weld Styles

Groove weld styles include the square, V, U, bevel, J, flare V, and flare bevel. *Figure 1* shows these typical single-groove weld styles.

For a part whose shape must be changed to a shape required for a specific groove, you need to start with an object having a squared (90-degree) end or edge, such as the square groove objects in *Figure 1*. Grinders are normally used to bevel metal parts. Beveling the parts means removing enough material to form an incline, such as the bevel groove image in *Figure 1*. Beveling the metal

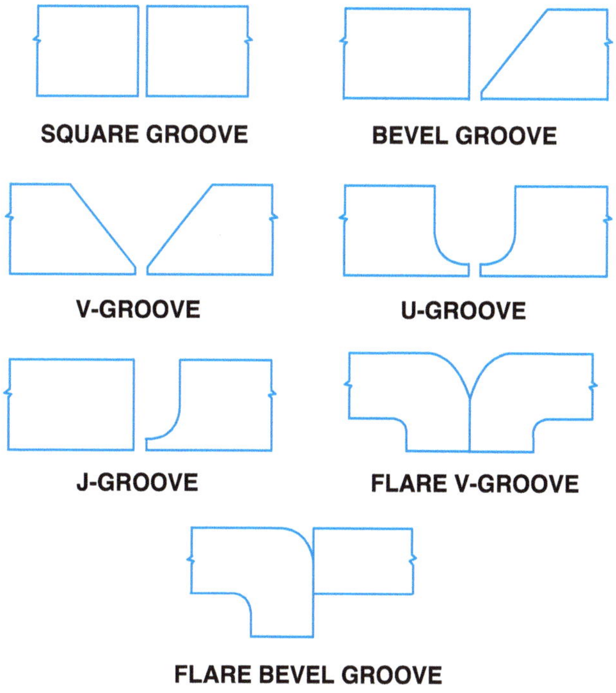

Figure 1 Single groove weld styles.

serves two purposes. First, it cleans all of the areas the weld must contact. Second, it creates a greater or wider space, which requires more weld to fill the joint. The bevel angle must be sufficient to allow access to the root, but not so large that it requires an excessive amount of filler metal to fill it.

1.1.1 Double-Groove Welds

V-, bevel, U-, and J-grooves can be used and welded on one side of a joint or on both sides. If they are used on both sides of a joint, they are called double-groove welds (*Figure 2*).

> **NOTE**
> Throughout this module, both metric and Imperial units have been provided for most measurements. Weld symbols are generally provided in one measurement system or another, but not both. For this reason, metric values have not been placed alongside Imperial values in weld symbols.

1.1.2 Combination Groove and Fillet Welds

Groove welds are often used in combination with fillet welds to maintain the strength of the welded joint and use less weld metal. *Figure 3* shows a number of groove and fillet weld combinations.

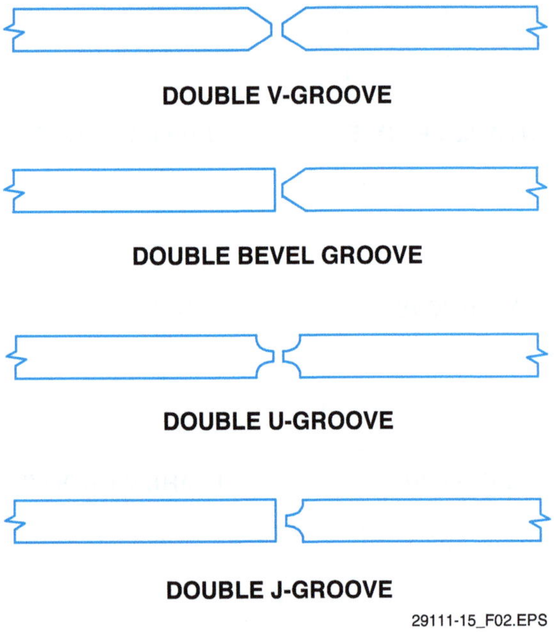

Figure 2 Double groove weld styles.

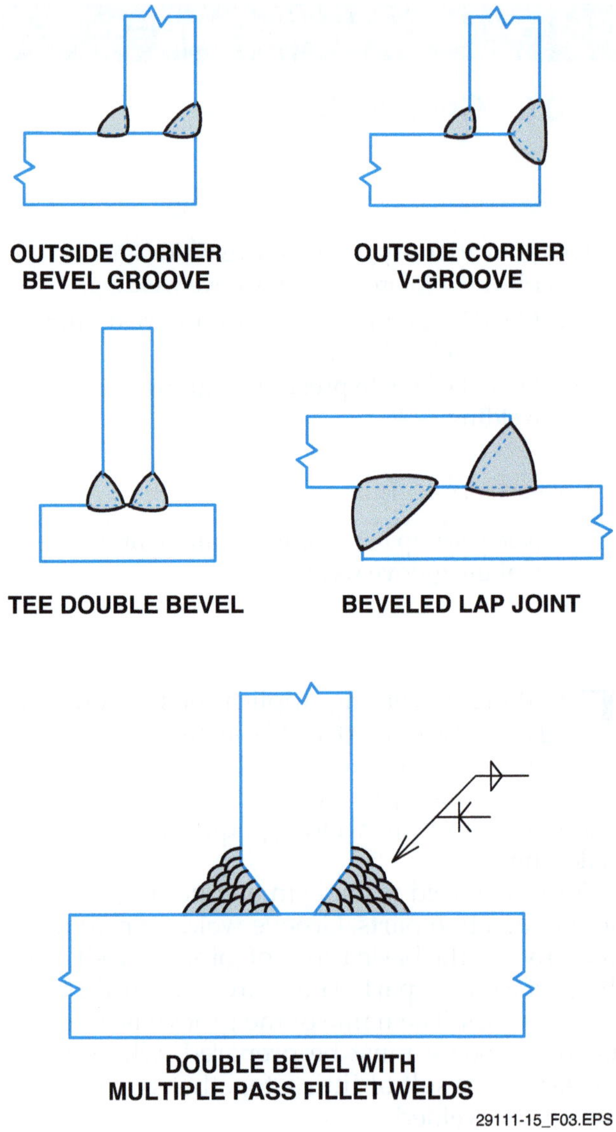

Figure 3 Groove and fillet weld combinations.

1.1.3 Groove Weld Terms

Part of a welder's certification includes the proper understanding of related welding terms. The terms associated with a groove are the following:

- *Bevel angle* – The angle formed between the prepared edge of a member and a line perpendicular to the surface of the member.
- *Bevel face* – The prepared surface of a bevel edge shape.
- *Bevel radius* – The radius used to form the shape of a J- or U-groove.
- *Depth of bevel* – The distance the joint preparation extends into the base metal (used for partial joint penetration welds).
- *Groove angle* – The included (total) angle between the groove faces of a weld groove.
- *Groove face* – Any surface in a weld groove prior to welding. See *bevel face, root face,* and *bevel radius.*
- *Joint root* – The portion of a joint where the members approach closest to each other.
- *Root face* – The portion of the edge of a part to be joined by a groove weld that has not been beveled or grooved.
- *Root opening* – The separation between the members to be joined at the root of the joint.

Figure 4 shows examples of grooves and their named parts.

The terms used to describe features of a groove weld are as follows:

- *Face reinforcement* – Reinforcement of the weld by excess weld metal on the side from which the weld was made.
- *Groove weld size* – The joint penetration (depth of joint preparation plus root penetration). The size can never be greater than the base metal thickness.
- *Weld face* – The exposed surface of the weld.
- *Weld root* – The point at which the back of the weld extends the farthest into the weld joint.
- *Weld toe* – The points where the face of the weld metal and the base metal meet.

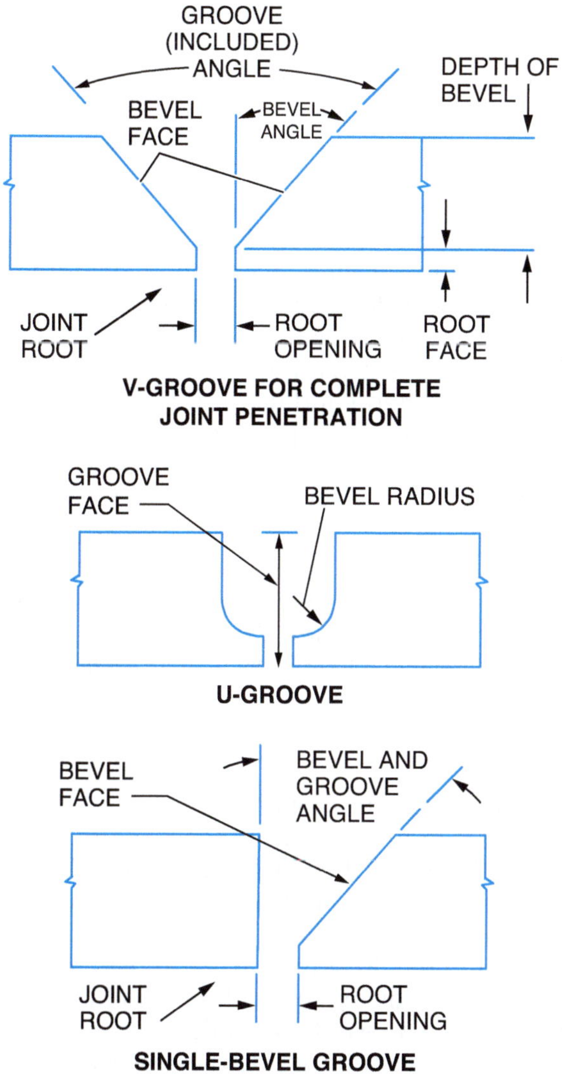

Figure 4 Grooves.

1.1.4 Backings

Although single groove welds can be done without backing, single groove welds with backing are used for high-strength, high-quality welds. Backing also makes welding the root easier. Thick metal pieces that are to be joined must be beveled so that the weld will penetrate the full thickness. The simplest form of backing is a strip of metal matching the base metal being welded. Other types of backing include dissimilar metals, ceramic, and coated tapes that can be removed when the welding has been completed.

Figure 5 shows some V-groove welds with backing. Although a weld joint done with an inert gas on the opposite side of the weld has no physical, solid backing material, it is still considered a joint with backing. The other two backings shown have a true, solid backing.

This module will explain the preparation and welding of V-groove welds with metal backing using low-hydrogen electrodes in all positions. Other types of groove welds can be performed in the same manner. Performance demonstrations will require that the welds meet the qualification standards in the applicable code for visual and destructive testing.

> **NOTE**
> The dimensions and specifications in this module are designed to be representative of codes in general and are not specific to any particular code. Always follow the proper codes for your site.

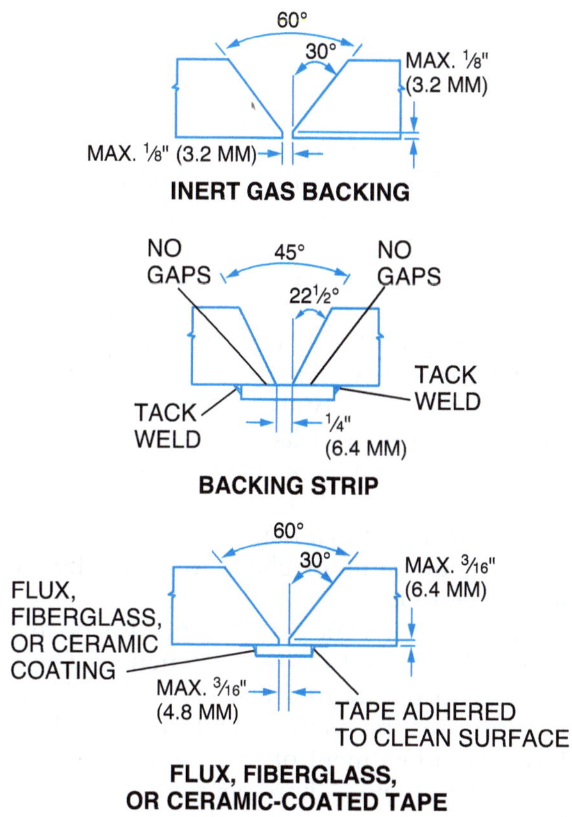

Figure 5 V-groove welds with backings.

1.2.0 Welding Preparations

Before welding can take place, the area has to be made ready, the welding equipment must be set up, and the metal to be welded must be prepared. The following sections explain how to set up arc welding equipment for welding.

1.2.1 Safety Practices

The following is a summary of safety procedures and practices that must be observed when cutting or welding. Keep in mind that this is a summary. Complete safety coverage is provided in the *Welding Safety* module. If you have not completed that module, do so before continuing. Above all, be sure to wear appropriate protective clothing and equipment when welding or cutting.

To maintain safety and prevent injury, it is essential that you wear the appropriate protective clothing and equipment when cutting or welding metals. Be sure to follow these guidelines during all phases of cutting or welding:

- Always use safety glasses along with a full face shield or a helmet. The glasses, face shield, or helmet lens must have the proper light-reducing tint for the type of welding or cutting being performed. Never directly or indirectly view an electric arc without using a properly tinted lens.
- Wear proper protective leather and/or flame retardant clothing along with welding gloves that will protect you from flying sparks and molten metal as well as heat.
- Wear high-top safety shoes or boots. Make sure that the tongue and lace area of the footwear will be covered by a pant leg. If the tongue and lace area is exposed, or if the footwear must be protected from burn marks, wear leather spats under your pants or chaps to cover the front top of the footwear.
- Wear a 100-percent cotton cap with no mesh material included in its construction. The bill of the cap points to the rear. If a hard hat is required for the environment, use one that allows the attachment of rear deflector material and a face shield. A hard hat with a rear deflector is generally preferred when working overhead, and may be required by some employers and job sites.

WARNING!

Do not wear a cap with a button in the middle. The conductive metal button beneath the fabric represents a safety hazard.

- Wear a face shield over snug-fitting cutting goggles or safety glasses for gas welding or cutting. Either the face shield or the lenses of the welding goggles must be an approved shade for the application. For electric arc welding or cutting, wear safety gasses and a welding hood with the correct tinted lens.
- Do not open your welding hood lens and expose unprotected eyes when grinding or cleaning a weld. Either use a hood that maintains a protective lens when the shaded lens is opened, a hood with an auto-shading lens, or remove the hood and use a full-face shield.
- Wear earplugs to protect ear canals from sparks. Wear hearing protection to protect against the consistent sound of the torch.

WARNING!

Ear protection is essential to protect ears from the noise of the torch. Other personal protective equipment (PPE) must be worn to protect the operator from hot metal and slag.

Welding activities involve the use of fire or extreme heat to melt metal. Whenever fire is used, it must be controlled and contained. Welding activities are often performed on vessels that may once have contained flammable or explosive materials. Residues from those materials can catch fire or explode when a welder begins work on such a vessel. The following are fire and explosion prevention guidelines associated with welding:

- Never carry matches or gas-filled lighters in your pockets. Sparks can cause the matches to ignite or the lighter to explode, causing serious injury.
- Always comply with all site and/or employer requirements for a hot-work permit and a fire watch.
- Never use oxygen to blow dust or dirt off clothing. The oxygen can remain trapped in the fabric for a time. If a spark hits the oxygen in the fabric, the clothing can burn rapidly and violently out of control.
- Make sure that any flammable material in the work area is moved or shielded by a fire-resistant covering.
- Approved fire extinguishers must be available before any heating, welding, or cutting operations are attempted. Make sure the extinguisher is charged, the inspection tag is valid, and any individual that may be required to operate it knows how to do so.

- Never release a large amount of oxygen or use oxygen in place of compressed air. The presence of oxygen around flammable materials or sparks can cause rapid and uncontrolled combustion. Keep oxygen away from oil, grease and other petroleum products.
- Never release a large amount of fuel gas, especially acetylene. Methane and propane are heavier than air and tend to migrate to and concentrate in low areas. As a result, they can ignite at a considerable distance from the release point. Acetylene is lighter than air but is even more dangerous than methane; when mixed with air or oxygen, it will explode at much lower concentrations than any other common fuel gas.
- To prevent fires, maintain a neat and clean work area, and make sure that any metal scrap or slag is cold before disposal.

Before cutting containers such as tanks or barrels, check to see if they have contained any explosive, hazardous, or flammable materials, including petroleum products, citrus products, or chemicals that decompose into toxic fumes when heated. As a standard practice, always clean and then fill any tanks or barrels with water, or purge them with a flow of inert gas such as nitrogen to displace any oxygen.

WARNING!

Welding or cutting must never be performed on drums, barrels, tanks, vessels, or other containers until they have been emptied and cleaned thoroughly, eliminating all flammable materials and all substances (such as detergents, solvents, greases, tars, or acids) that might produce flammable, toxic, or explosive vapors when heated. Do not assume that a container that has held combustibles is clean and safe until proven so by proper tests. Do not weld in places where dust or other combustible particles are suspended in air or where explosive vapors are present.

Containers must be cleaned by steam cleaning, flushing with water, or washing with detergent until all traces of the material have been removed.

WARNING!

Clean containers only in well-ventilated areas. Vapors can accumulate during cleaning, causing explosions or injury.

Vapors and fumes tend to rise in the air from their sources. Welders often have to work above welding areas where fumes are being created. Welding fumes can be harmful. Good work area ventilation helps to remove the vapors and protect the welder. Always perform cutting or welding operations in a well-ventilated area. Cutting or welding operations involving zinc or cadmium materials or coatings result in toxic fumes. For long-term cutting or welding of such materials, always wear an approved full-face, supplied-air respirator (SAR) that uses breathing air supplied- from outside the work area. For occasional, very short-term exposure, a HEPA-rated or metal-fume filter may be used on a standard respirator.

1.2.2 Preparing the Welding Area

To practice welding, a welding table, bench, or stand is needed (*Figure 6*). The welding surface must be steel, and provisions must be made for placing welding coupons out of position.

To set up the area for welding, follow these steps:

Step 1 Make sure that the area is properly ventilated. Make use of doors, windows, and fans.

Step 2 Check the area for fire hazards. Remove any flammable materials before proceeding.

29111-15_F06.EPS

Figure 6 Welding station with vacuum system.

Step 3 Know the location of the nearest fire extinguisher. Do not proceed unless the extinguisher is charged and you know how to use it.

Step 4 Position a welding table near the welding machine, or vice versa.

Step 5 Set up flash shields around the welding area.

1.2.3 Preparing the Weld Coupons

If possible, the weld coupons should be ⅜" thick carbon steel to conform with AWS limited-thickness test coupon requirements. If this size is not readily available, ¼" to 1" thick steel can be used for practice welds. If you are working with metric metals, a 10 mm thickness is preferred. If not readily available, 6 to 25 mm plate can be used for practice. Clean the steel plate prior to welding by using a wire brush or grinder to remove heavy mill scale or corrosion.

WARNING!

Be extremely careful when using wire brush attachments on powered tools. Always ensure that the rotational speed rating of the wire brush matches or exceeds the rated speed of the powered tool. Loose wires from the brush and debris tend to fly off and embed into anything or anyone near the work. Make sure that all appropriate PPE is worn when wire brushing or grinding with all tools, but be especially cautious with powered tools.

For each weld coupon, you will need two pieces of plate 3" × 7" (7.6 cm × 17.8 cm) and one piece of backing. The backing should be ¼" × 1" × 8" (6 mm × 2.5 cm × 20.3 cm). The backing is 1" (2.5 cm) longer than the coupons, so that it can extend past them ½" (12.7 mm) on each end of the joint. Cut the coupons into rectangles, with the long sides beveled at approximately 22½ degrees. A 37½-degree bevel can also be used, subject to your instructor's choice. The backing is not beveled.

Note that this 3" × 7" size is the minimum size allowed for an AWS test coupon. However, for the purpose of practice and material conservation, your instructor may select a smaller size.

GOING GREEN

Mill Scale and Corrosion Debris

To reduce air pollution in general and to cut down on respiratory problems in the workplace, make sure that all mill scale and corrosion debris ground from the steel is properly extracted from the work area and contained as waste material.

Follow these steps to prepare the V-groove with metal backing weld coupon:

Step 1 Check the bevel. There should be no root face and no dross, and the bevel angle should be approximately 22½ or 37½ degrees.

Step 2 Obtain clamps large enough to hold the coupons to the backing strips.

Step 3 Lay the backing strip on a flat surface and then position the two beveled coupons on it so that their root opening is ¼" (6.4 mm) wide. *Figure 7* shows coupons set up for a V-groove with metal backing.

Step 4 Clamp the beveled strips to the backing strip, making sure to keep the root opening at ¼" (6.4 mm). Also, make sure that the backing strip is held tight against the back of the beveled strips with the clamps.

Step 5 Turn the strips over and place the tack welds on the backside of the joint in the lap formed by the backing strip and the beveled strips. Use a tack in the center and one on each end, applied to each half of the coupon. The result is six tack welds (*Figure 8*).

Step 6 Remove the electrode from the electrode holder and discard as soon as you stop welding to prevent arcing to nearby surfaces, and place the stub in a proper container.

NOTE

At this point, the strips can be positioned into any of the four welding positions as required.

Backing Strip

When the backing strip is ½" to 1" (1.3 to 2.5 cm) longer at each end of the weld groove, it allows the welder to start and stop the bead outside the weld groove.

GOING GREEN

Welding Coupons

To reduce landfill waste, collect used welding coupons and other scrap metals on behalf of your organization.

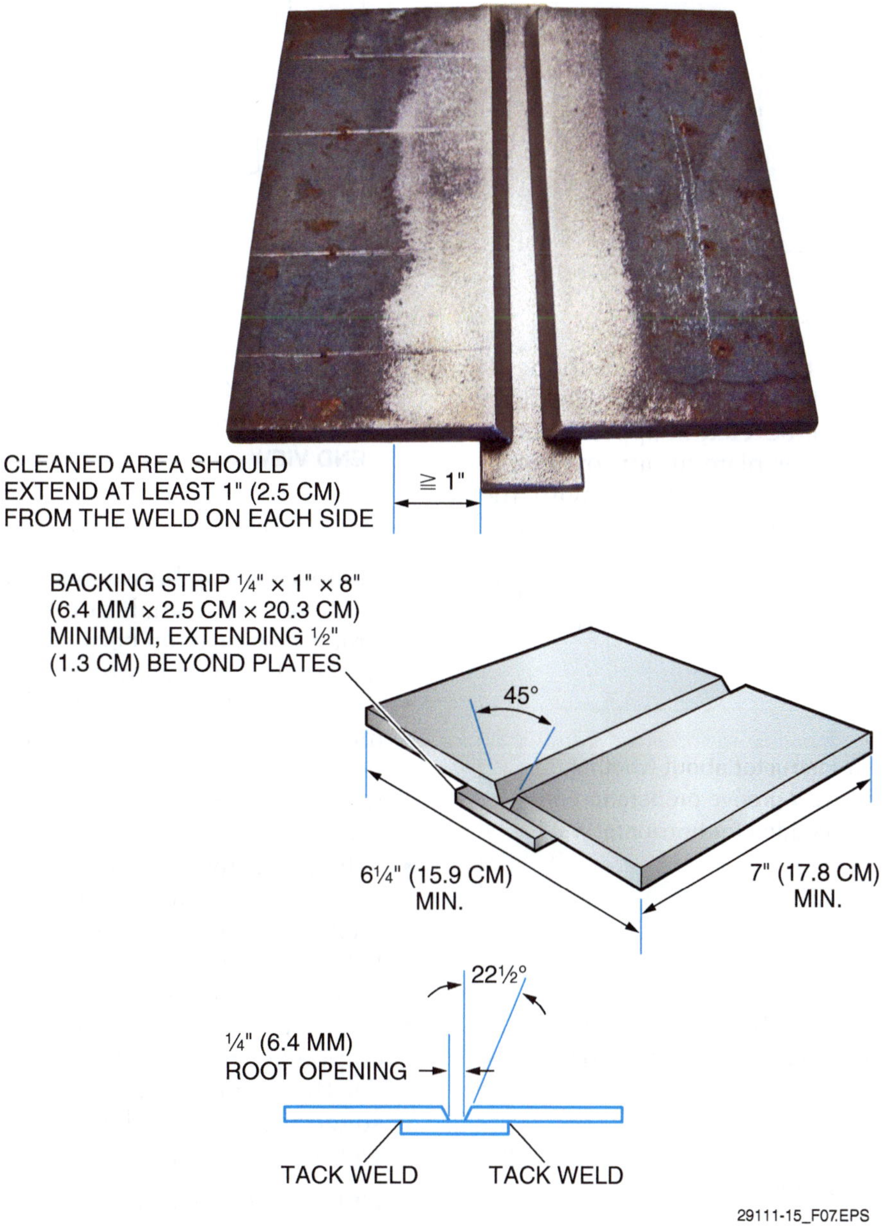

Figure 7 V-groove with metal backing weld coupon.

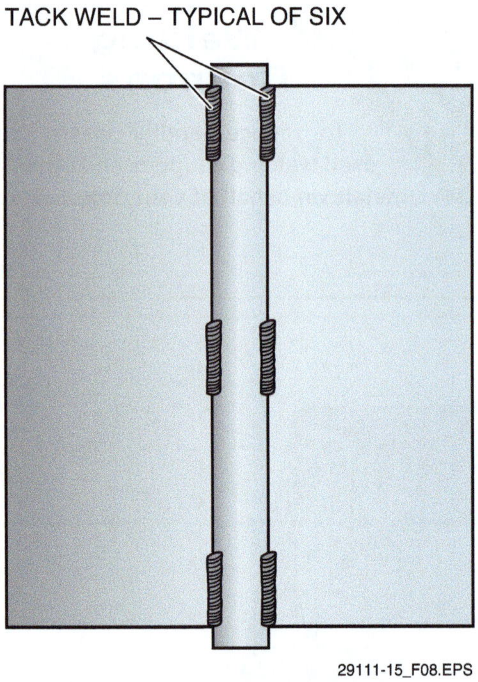

Figure 8 Coupon tack welds.

When welding in the horizontal position, an alternate joint preparation can be used. The alternate joint has one plate beveled at approximately 45 degrees and the other plate at approximately 90 degrees. The plates are positioned with the beveled plate above the 90-degree plate and a ¼" (6.4 mm) root opening.

Figure 9 is a diagram showing the alternate horizontal weld coupons. Degrees shown are approximate.

> **NOTE**
>
> Check with your instructor about whether to use the standard V-groove preparation or the alternate preparation for horizontal weld coupons.

1.2.4 Electrodes

Obtain only a small quantity of the electrodes to be used. The welding exercises in this module use ³⁄₃₂" (2.4 mm), ⅛" (3.2 mm), or ⁵⁄₃₂" (4.0 mm) E7018 electrodes.

Keep the following general rules in mind when running low-hydrogen electrodes:

- Remember not to whip low-hydrogen electrodes.
- Maintain a short arc because a low-hydrogen weld relies on its slag to protect the molten metal. It does not have a heavy gaseous shield.

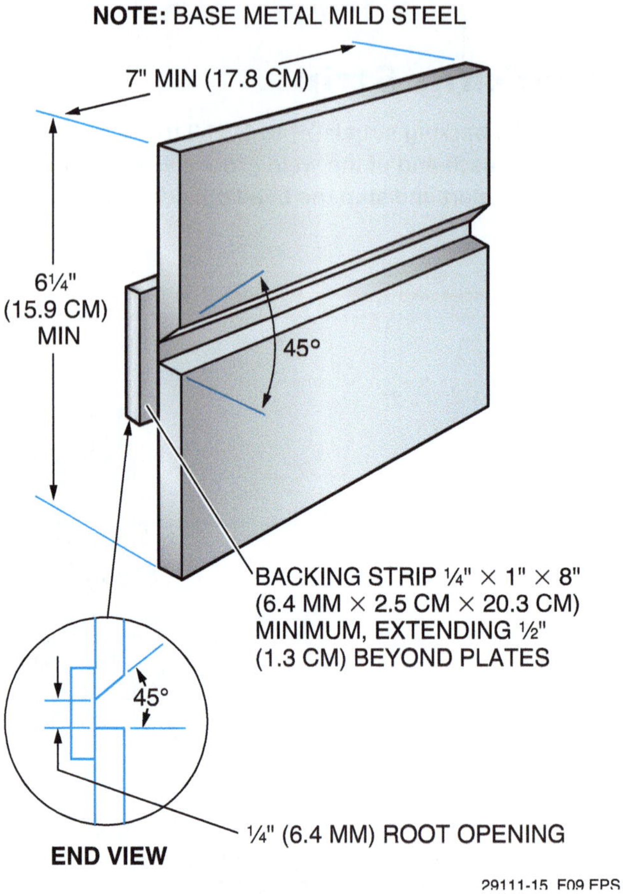

Figure 9 Alternate horizontal weld coupons.

- Remove all slag between passes.
- Remove only a small number of electrodes at a time from the oven because low-hydrogen electrodes absorb moisture quickly.
- Restart by striking the arc ahead of the crater, move quickly back into the crater, and then proceed as usual. This technique welds over the arc strike, eliminating porosity and giving a smoother restart.
- When running stringer beads, use a slight side-to-side motion with a small pause at the weld toe to tie-in and flatten the bead.
- When running weave beads, use a slight pause at the weld toe to tie-in and flatten the bead.
- Do not use low-hydrogen electrodes with chipped or missing flux.
- Set the amperage in the lower portion of the suggested range for better puddle control when welding vertically.

Obtain only the electrodes to be used for a particular welding exercise at one time. Have some type of pouch or rod holder in which to store the electrodes to prevent them from becoming damaged. Never store electrodes loose on a table.

They may fall on the floor where they can become a tripping hazard or become damaged. Some type of metal container or bucket must also be available to discard hot electrode stubs.

WARNING!

Do not throw electrode stubs on the floor. They roll easily and could cause someone to slip and fall.

1.2.5 Preparing the Welding Machine

Welders can expect to find different types and makes of welding machines in welding shops and on jobs in the field. Many machines will not have operator manuals with them. As a trainee, and even as a trained welder, you must be able to recognize the different types of welding machines and to figure out how to safely set up and operate them. Select a welding machine to use and then follow these steps to set it up for welding:

Step 1 Verify that the welding machine can be used for DC welding.

Step 2 Check the area for proper ventilation.

Step 3 Verify the location of the power disconnect for the unit.

Step 4 Set the polarity to direct current electrode positive (DCEP).

Conserving Material

Steel for practice welding is expensive and difficult to obtain. Every effort should be made to conserve and not waste the material that is available. Reuse weld coupons until all surfaces have been welded upon by cutting the weld coupon apart and reusing the pieces. Use material that cannot be cut into weld coupons to practice running beads.

Step 5 Connect the clamp of the workpiece lead to the workpiece.

Step 6 Set the amperage for the electrode type and size to be used. Typical settings are as follows:

Electrode	Size	Amperage
E7018	3⁄32" (2.4 mm)	70A to 100A
E7018	1⁄8" (3.2 mm)	100A to 150A
E7018	5⁄32" (4.0 mm)	130A to 200A

NOTE

Amperage recommendations vary by manufacturer, welding position, current type, and electrode brand. Refer to the Welding Procedure Specification (WPS) as the primary source of information, or to the manufacturer's recommendations for the electrode being used if a WPS is not in use.

Step 7 Check to be sure the electrode holder is not grounded.

Step 8 Turn on the welding machine.

Figure 10 shows DC and AC/DC welding machines with their various parts labeled.

Electrode Housekeeping

A floor littered with electrode stubs is an accident waiting to happen. Not only does throwing electrode stubs on the floor or ground create a hazard, it can also result in disciplinary action from the employer. In some cases, a worker may even lose his or her job for this infraction.

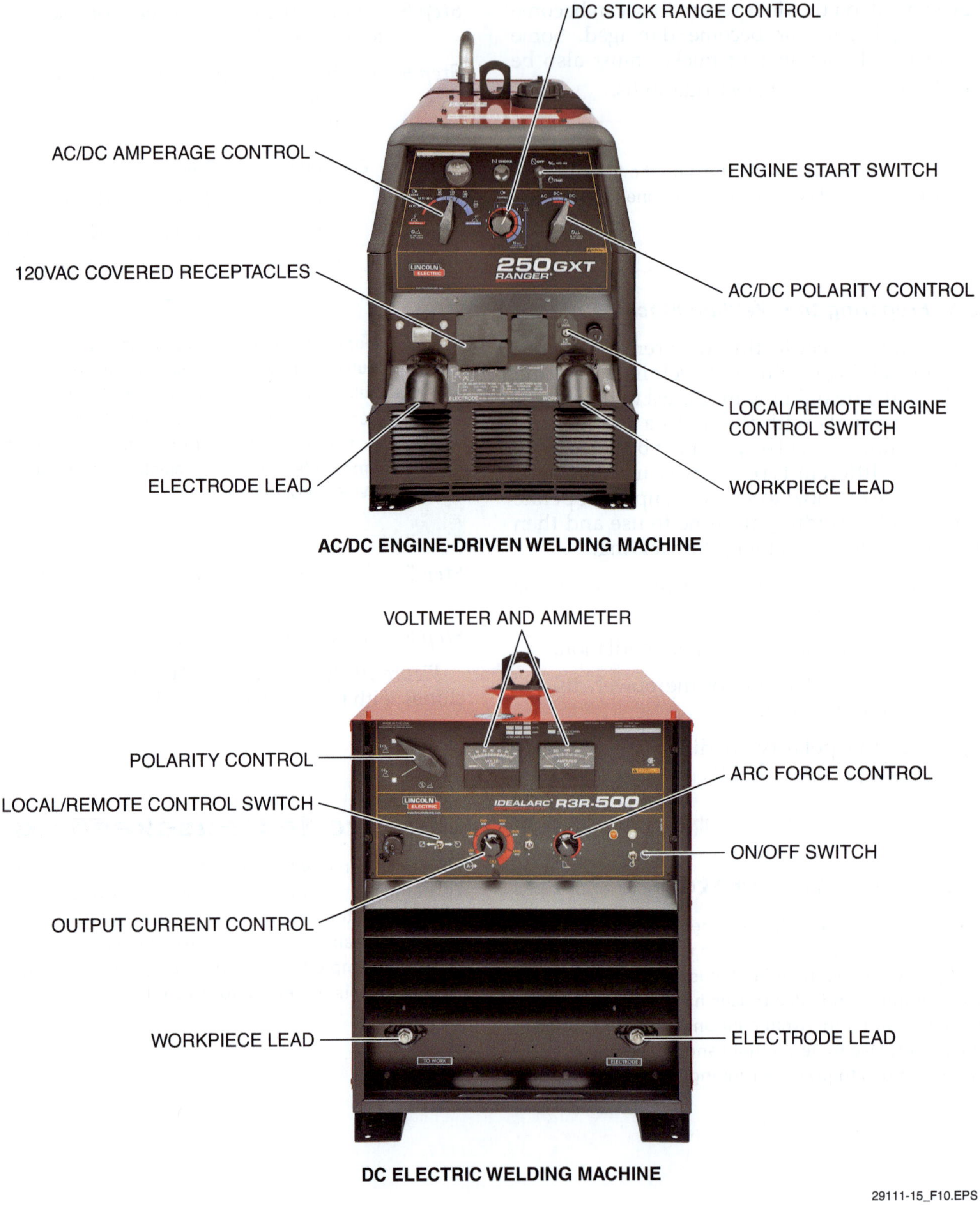

Figure 10 DC and AC/DC welding machines.

Lash

To ensure good welding results, remember LASH.

L *(length of arc)* –The distance between the electrode and the base metal (usually one times the electrode diameter).

A *(angle)* –Two angles are critical:

- Travel angle –The longitudinal angle of the electrode in relation to the axis of the weld joint
- Work angle –The traverse angle of the electrode in relation to the axis of the weld joint

S *(speed)* –Travel speed is measured in inches per minute (IPM). The width of the weld will determine if the travel speed is correct.

H *(heat)* – Controlled by the amperage setting and dependent upon the electrode diameter, base metal type, base metal thickness, and the welding position.

Additional Resources

2014 Technical Training Guide. Current Edition. Cleveland, OH. USA: The Lincoln Electric Company. **www.lincolnelectric.com**

Welding Handbook. Current Edition. Miami, FL: The American Welding Society.

1.0.0 Section Review

1. The angle formed between the prepared edge of a member and a line perpendicular to the surface of the member is called the _____.
   a. bevel radius
   b. bevel angle
   c. groove angle
   d. groove face

2. Welding operations involving zinc or cadmium create _____.
   a. heavy slag
   b. stronger welds
   c. heavy weld spatter
   d. toxic fumes

SECTION TWO

2.0.0 V-GROOVE WELDS WITH BACKING

Objective

Describe the technique required to produce various groove welds.

a. Describe the technique required to produce groove welds in the 1G and 2G positions.
b. Describe the technique required to produce groove welds in the 3G and 4G positions.

Performance Tasks

2. Make flat welds with backing on V-groove joints using E7018 electrodes.
3. Make horizontal welds with backing on V-groove joints using E7018 electrodes.
4. Make vertical welds with backing on V-groove joints using E7018 electrodes.
5. Make overhead welds with backing on V-groove joints using E7018 electrodes.

The V-groove with backing weld is a common groove weld normally made with low-hydrogen electrodes. The backing can be of dissimilar material, steel, or a backing weld run with E7018. Depending on the code or procedure, low-hydrogen electrodes may be used for the root pass on an open root weld. However, if low-hydrogen electrodes are used on an open root, back gouging and back welding may be required. Using a steel backing eliminates the need for back gouging, and is a fast and effective way to prepare a joint for welding. It also makes the root pass easier to run.

The V-groove with steel backing procedure is also the standard AWS qualification test for plate welding. The most common test is the limited-thickness qualification (up to ¾", or 19 mm), which requires the use of ⅜" (10 mm) carbon steel plate and welding with low-hydrogen electrodes.

> **Backing**
>
> Backing can either be left in place or removed after welding. If the backing is to be removed, an R is inserted in the welding symbol. Backing that is to be removed is usually made of a different material than the workpiece, to keep it from being welded into the joint.

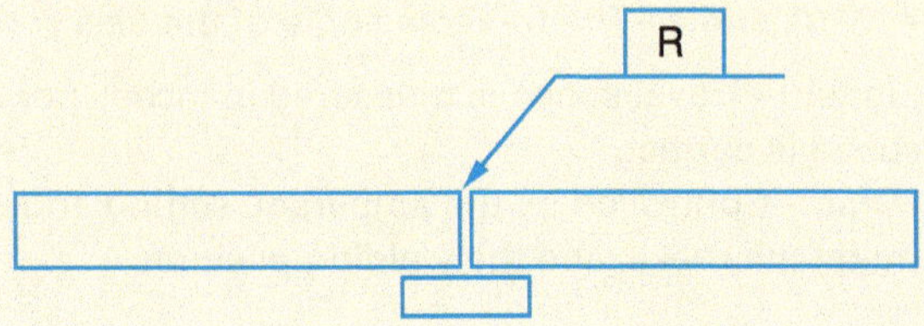

29111-15_SA01.EPS

2.0.1 Groove Weld Positions

Groove welds can be made in all positions. The weld position for plate is determined by the weld and the orientation of the workpiece (*Figure 11*). Groove weld positions for plate are flat, or 1G; horizontal, or 2G; vertical, or 3G; and overhead, or 4G. The G represents a groove weld. In the 1G and 2G positions, the weld axis can be inclined up to 15 degrees. Any weld axis inclination for the other positions varies with the rotational position of the weld face as specified in AWS standards.

2.0.2 Acceptable and Unacceptable Groove Weld Profiles

Groove welds should be made with slight reinforcement and a gradual transition to the base metal at each toe. Groove welds must not have excessive reinforcement, underfill, undercut, or overlap. If a groove weld has any of these defects, it is unacceptable. *Figure 12* shows examples of acceptable and unacceptable groove weld profiles.

> **NOTE**
>
> Refer to your site's WPSs for specific requirements on groove welds. The information in this module is provided as a general guideline only. The site WPS or quality specifications must be followed for all welds. Check with your supervisor if you are unsure of the specifications for your application.

2.1.0 Groove Welding—1G and 2G Positions

The 1G and 2G positions will be practiced first. The following sections describe the types of beads applied, as well as the techniques used to ensure an acceptable weld.

2.1.1 Practicing Flat V-Groove Welds with Backing (1G Position)

Practice flat V-groove welds using $\frac{3}{32}$", $\frac{1}{8}$", or $\frac{5}{32}$" (2.4, 3.2, or 4 mm) E7018 electrodes. Stringer beads or weave beads can be used. When using weave beads, keep the electrode at a 90-degree work angle with a 10- to 15-degree drag angle. When using stringer beads, use a 10- to 15-degree drag angle, but adjust the electrode work angle to tie in the weld on one side or the other as needed. Pay particular attention at the termination of the weld to fill the crater. If the coupon gets too hot between passes, you may have to quench it in water.

CAUTION

Quenching with water is only done on practice coupons. Never cool testing coupons or functional welds with water. Rapid cooling with water can cause weld cracks and affect the mechanical properties of the base metal.

WARNING!

Use pliers and wear gloves to handle hot practice coupons and quench. Steam will rise off the coupon quickly and can burn or scald unprotected hands. Do not cool practice coupons in a sink used for washing hands, or in a water fountain. Use a bucket or bin to avoid clogging the sink with slag and dirt.

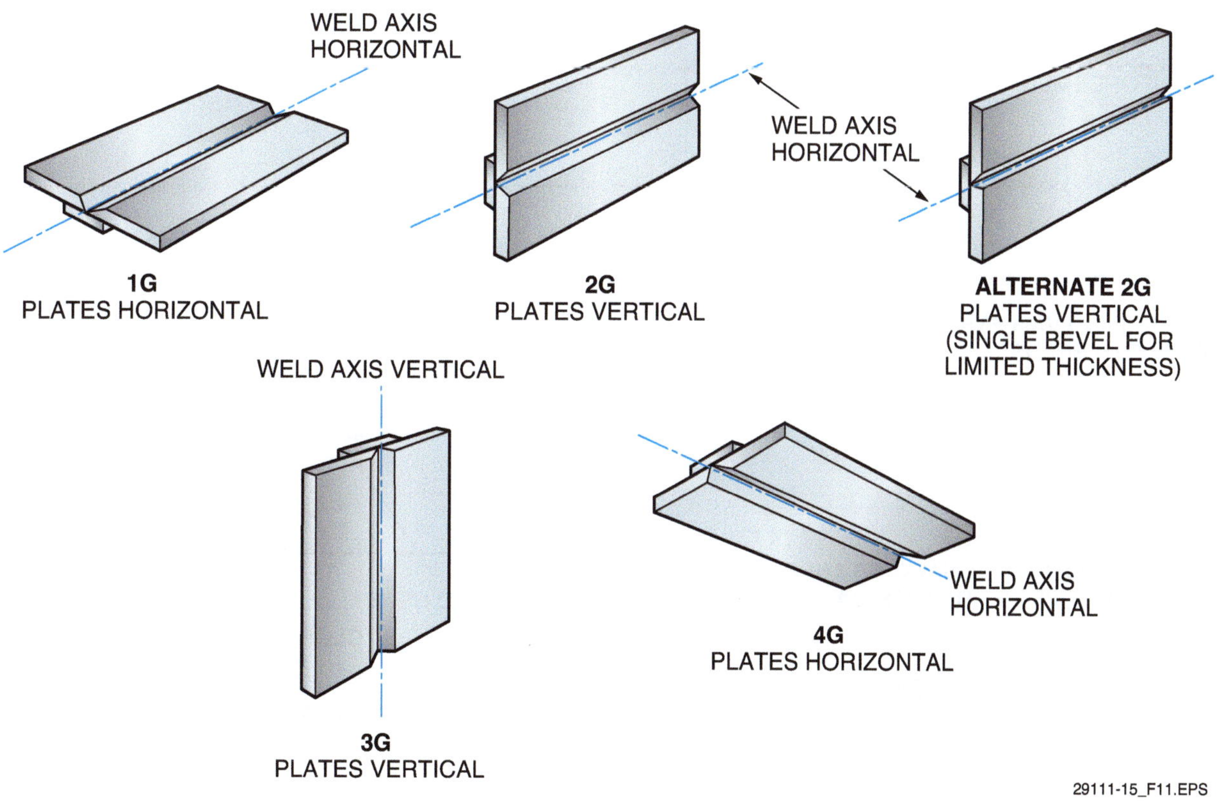

Figure 11 Groove-weld positions for plate.

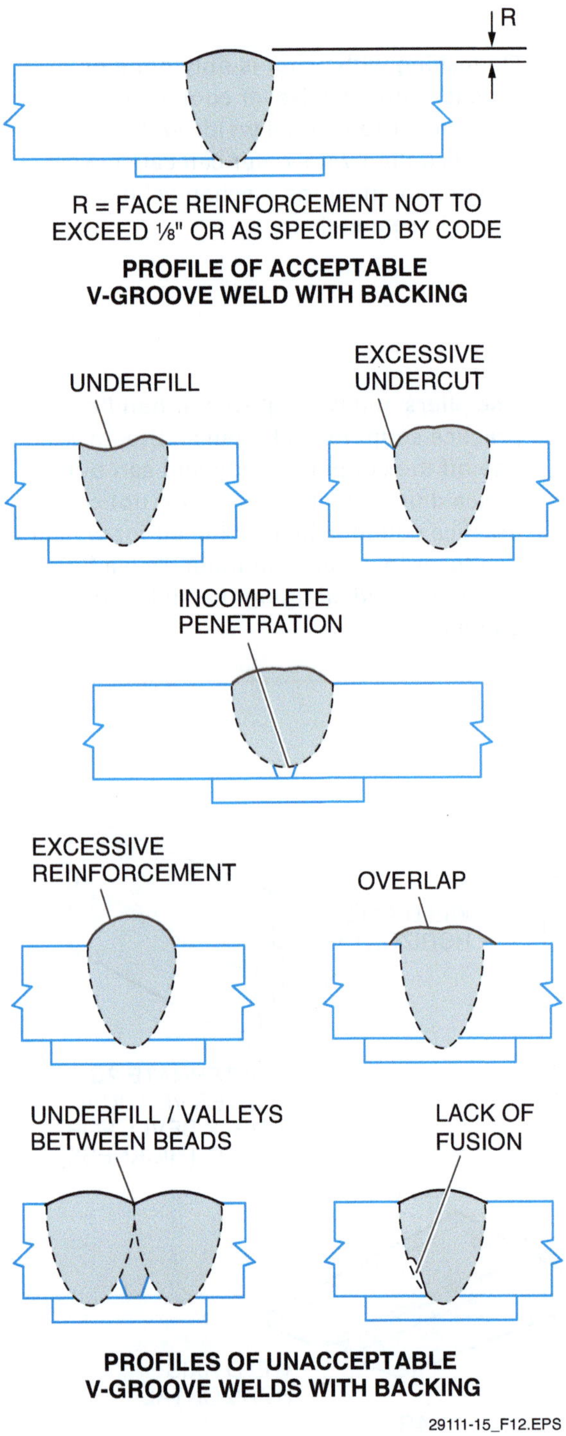

Figure 12 Acceptable and unacceptable profiles of V-groove welds with backing.

Follow these steps to practice V-groove welds with metal backing in the flat position:

Step 1 Tack weld the practice coupon together (*Figure 13*).

Step 2 Position the weld coupon flat on the welding table.

Step 3 Run the root pass using E7018 electrodes. *Figure 14* shows the root, filler, and cover pass sequences and work angles. Degrees shown are approximate.

Step 4 Chip, clean, and brush the weld between all passes.

Step 5 Run the fill and cover passes, cleaning between each bead, to complete the weld using E7018 electrodes. Use stringer or weave beads as directed by your instructor.

Figure 13 Tack welding the practice coupon.

Step 6 Remove the electrode stub from the electrode holder and discard as soon as you stop welding to prevent arcing to nearby surfaces, and place the stub in a proper container.

Figure 15 shows a flat V-groove weld with typical filler and cover passes.

2.1.2 Horizontal Welds (2G Position)

Horizontal welds can be made with or without backing. This section explains welding horizontal beads and making horizontal V-groove welds with backing.

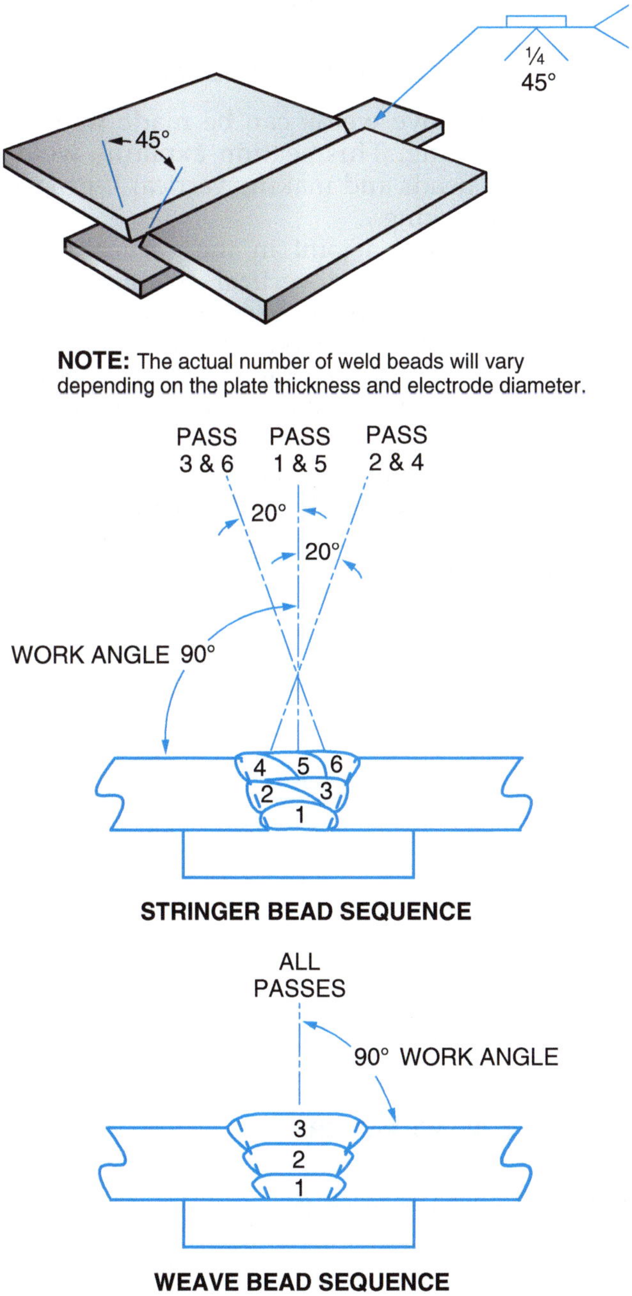

29111-15_F14.EPS

Figure 14 Multiple-pass 1G weld sequences and work angles.

29111-15_F15.EPS

Figure 15 V-groove weld filler and cover pass.

Before welding a V-groove in the horizontal position, practice running horizontal stringer beads by building a pad in the horizontal position. Use 3/32", 1/8", or 5/32" E7018 electrodes. The electrode drag angle should be 10 to 15 degrees, and the work angle should be 0 degrees. To control the weld puddle, the electrode work angle can be dropped slightly, but no more than 10 degrees. Follow these steps to practice welding beads in the horizontal position:

Step 1 Tack weld a flat plate welding coupon in the horizontal position.

Step 2 Run the first pass along the bottom of the coupon using E7018 electrodes.

Step 3 Clean the weld bead with a chipping hammer and wire brush between all passes.

Step 4 Weld the second bead just above the first bead.

Step 5 Continue running beads to complete the pad. *Figures 16* and *17* show the building of a pad in the horizontal position. Degrees shown are approximate.

Step 6 Remove the electrode from the electrode holder and discard as soon as you stop welding to prevent arcing to nearby surfaces, and place the stub in a proper container.

Practice horizontal V-groove welds using 3/32", 1/8", or 5/32" E7018 electrodes and stringer beads. Use a 10- to 15-degree drag angle, but adjust the electrode work angle to tie-in the weld as needed. Pay particular attention at the termination of the weld to fill the crater. The root pass is done using a weave bead.

Follow these steps to practice welding V-groove welds with metal backing in the horizontal position:

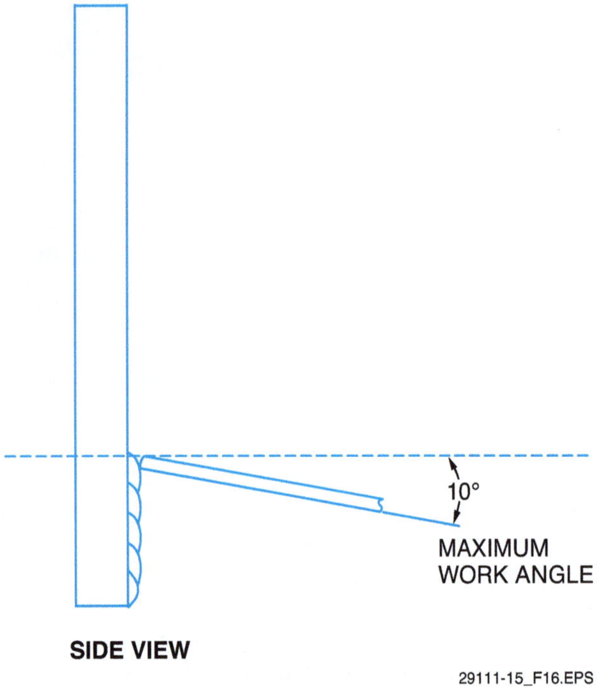

Figure 16 Work angle in the horizontal position.

Step 1 Tack weld the practice coupon together following the example given earlier. Use the standard or alternate horizontal weld coupon as directed by your instructor.

Step 2 Tack weld the weld coupon in the horizontal position.

Step 3 Run the root pass. Note that the first pass, or root pass, is critical because it needs to tie in the backing plate to each of the beveled test coupons (*Figure 18*). A welder's ability to weave in this bead will determine the quality of the root pass.

Step 4 Clean the weld bead with a chipping hammer and wire brush between all passes.

Step 5 Run the fill and cover passes, cleaning the weld between passes, to complete the weld.

Step 6 Remove the electrode from the electrode holder and discard as soon as you stop welding to prevent arcing to nearby surfaces, and place the stub in a proper container.

Figure 19 shows a horizontal V-groove weld with metal backing. Degrees shown are approximate.

Figure 17 A pad built in the horizontal position.

2.2.0 Groove Welding – 3G and 4G Positions

The vertical and overhead welds, positions 3G and 4G respectively, can be more challenging than the 1G and 2G positions.

2.2.1 Vertical Welds (3G Position)

Vertical V-groove welds can be made with or without backing. This section explains welding vertical beads and making vertical V-groove welds with backing.

Before welding a joint in the vertical position, practice running vertical stringer beads by building a pad in the vertical position. Use 3⁄32", 1⁄8", or 5⁄32" E7018 electrodes. Set the amperage in the lower part of the range for better puddle control. The electrode should be at a 0- to 10-degree push angle with a 90-degree work angle. Use a slight side-to-side motion to control the weld puddle. The width of the stringer beads should be no more than three times the electrode diameter.

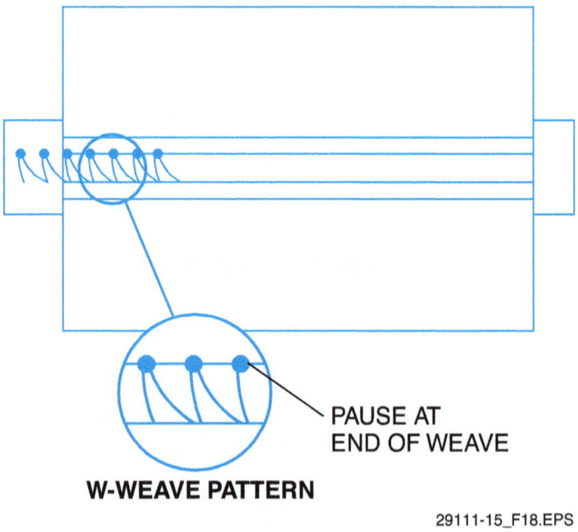

Figure 18 Weaving the root pass.

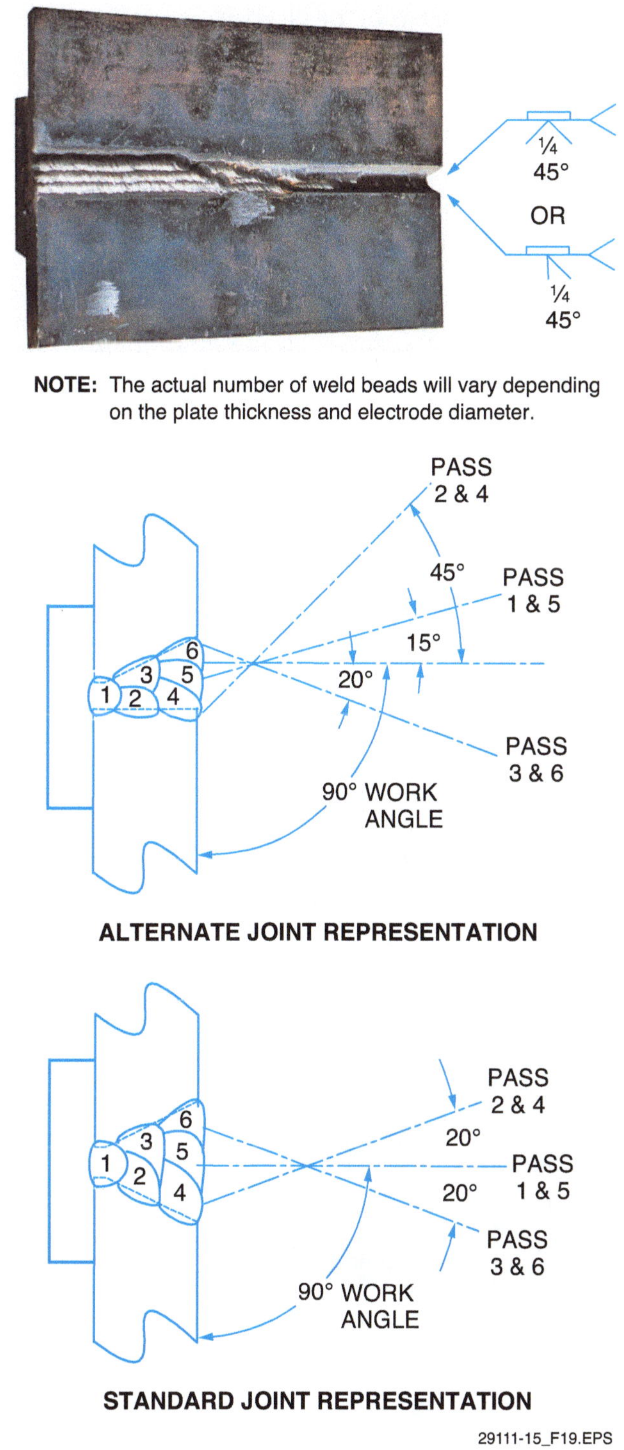

Figure 19 Multiple-pass 2G weld sequences and work angles.

Follow these steps to practice welding beads in the vertical position:

Step 1 Tack weld a flat plate welding coupon in the vertical position.

Step 2 Run the first pass along the vertical edge of the coupon using E7018 electrodes.

Step 3 Clean the weld bead with a chipping hammer and wire brush between all passes.

Step 4 Continue running beads to complete the pad. Clean the weld between each pass. *Figure 20* is a diagram showing the travel angle in the vertical position. Degrees shown are approximate.

Step 5 Remove the electrode from the electrode holder and discard as soon as you stop welding to prevent arcing to nearby surfaces, and place the stub in a proper container.

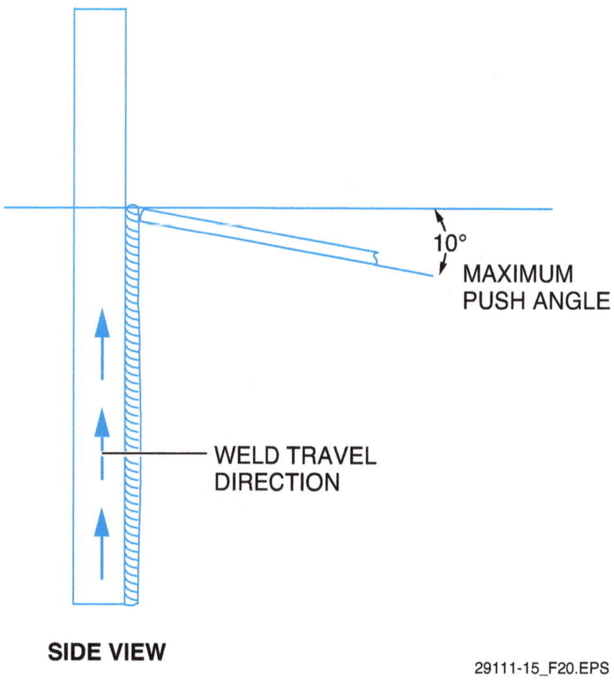

Figure 20 Travel direction and push angles in the vertical (3G) position.

Practice vertical V-groove welds using 3⁄32", 1⁄8", or 5⁄32" E7018 electrodes with stringer beads and/or weave beads. Set the amperage in the lower part of the range for better puddle control. Use a 0- to 10-degree push angle, but adjust the electrode work angle to tie in the weld on one side or the other as needed. If using a weave bead, pause slightly at each toe to penetrate and fill the crater to prevent undercut. Pay particular attention at the termination of the weld to fill the crater.

Follow these steps to practice welding V-groove welds with metal backing in the vertical position:

Step 1 Tack weld the practice coupon together following the example given earlier.

Step 2 Tack weld the weld coupon in the vertical position.

Step 3 Run the root pass with a slight oscillation, pausing at each side.

Step 4 Clean the weld bead with a chipping hammer and wire brush between all passes.

Step 5 Run the fill and cover passes, cleaning each pass, to complete the weld.

Step 6 Remove the electrode from the electrode holder and discard as soon as you stop welding to prevent arcing to nearby surfaces, and place the stub in a proper container.

Figure 21 shows a vertical V-groove weld with metal backing. Degrees shown are approximate.

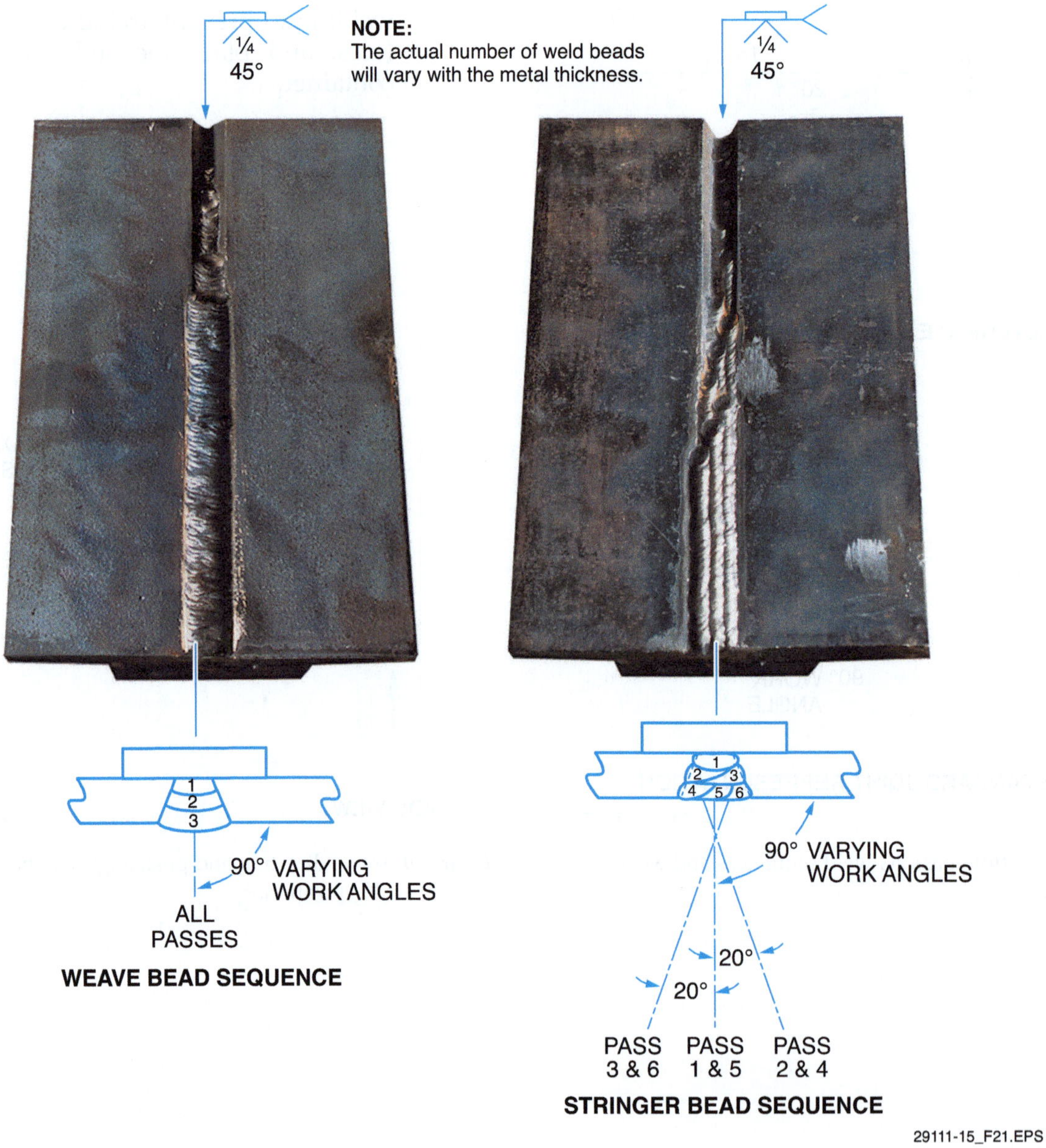

Figure 21 Multiple-pass 3G weld sequences and work angles.

2.2.2 Overhead Welds (4G Position)

Overhead V-groove welds (4G) can be made with or without backing. This section explains welding overhead beads and making overhead V-groove welds with backing.

Before welding a V-groove in the overhead position, practice running overhead stringer beads by building a pad in the overhead position. Use 3⁄32", 1⁄8", or 5⁄32" E7018 electrodes. The electrode should have a 90-degree work angle and a 10- to 15-degree drag angle.

> **WARNING!**
> Try not to stand directly under the coupons while welding because hot slag and molten metal can cause very severe burns. Always wear the proper PPE and any additional clothing required for overhead welding.

Follow these steps to practice welding beads in the overhead position:

Step 1 Tack weld the welding coupon in the overhead position.

Step 2 Run the first pass along the edge of the coupon using E7018 electrodes.

Step 3 Clean the weld bead with a chipping hammer and wire brush between all passes.

Step 4 Continue running and cleaning the beads to complete the pad.

Step 5 Remove the electrode from the electrode holder and discard as soon as you stop welding to prevent arcing to nearby surfaces, and place the stub in a proper container.

Figure 22 is a diagram showing how a pad is built in the overhead position.

Practice making overhead V-groove welds using 3⁄32", 1⁄8", or 5⁄32" E7018 electrodes and stringer beads. Use a 10- to 15-degree drag angle, but adjust the electrode work angle to tie in the weld as needed. Pay particular attention at the termination of the weld to fill the crater.

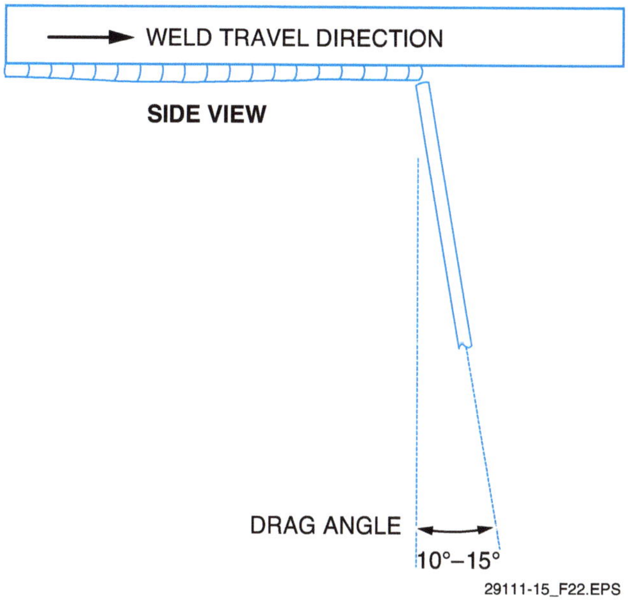

Figure 22 Building a pad in the overhead position.

Follow these steps to practice welding V-groove welds with metal backing in the overhead position:

Step 1 Tack weld the practice coupon together following the example given earlier.

Step 2 Tack weld the weld coupon in the overhead position.

Step 3 Run the root pass.

Step 4 Clean the weld bead with a chipping hammer and wire brush between all passes.

Step 5 Run the fill and cover passes, cleaning after each pass, to complete the weld.

Step 6 Remove the electrode from the electrode holder and discard as soon as you stop welding to prevent arcing to nearby surfaces, and place the stub in a proper container.

Figure 23 shows an overhead V-groove weld with metal backing. Degrees shown are approximate.

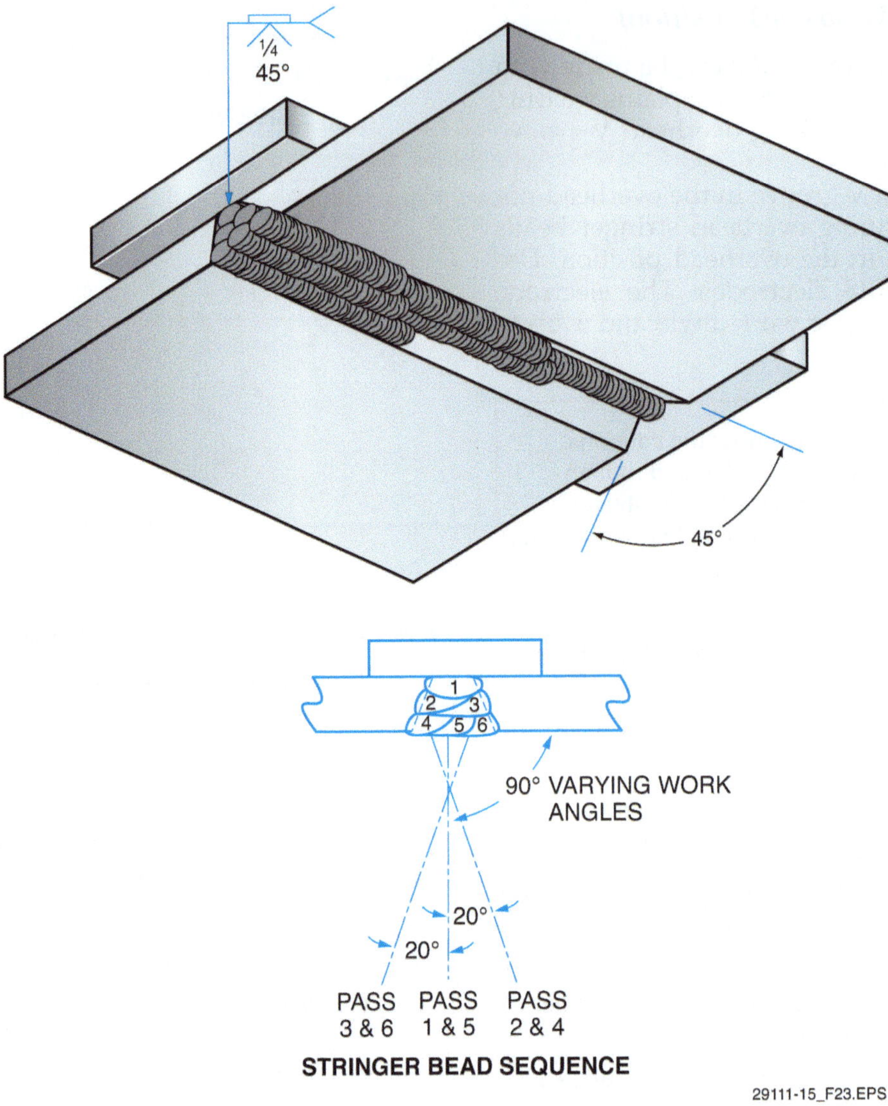

Figure 23 Multiple-pass 4G weld sequences and work angles.

Using Your Bifocals and Trifocals

29111-15_SA02.EPS

Many welders have used what are known as cheater lenses in welding hoods for years to help sharpen their vision when the hood is down. A cheater lens is simply a magnification lens. Although many welders wear prescription glasses with bifocal or trifocal lenses, it was impractical to use the glasses due to the design of the hood. Adjustment of the hood's position in front of the face was nonexistent, and the lens size restricted vision.

Newer welding hoods, like the Viking model shown here, allow welders to make use of their bifocal and trifocal prescription glasses. The hood distance from the front of the face can be adjusted in the headgear. This allows the hood to be tilted down farther, even to the chest. The welder can then tilt the head back and forth slightly, using the different focal lengths of the lenses, without losing sight of the task. The significantly larger lens also ensures that the task at hand remains in the welder's field of view.

Additional Resources

2014 Technical Training Guide. Current Edition. Cleveland, OH. USA: The Lincoln Electric Company. **www.lincolnelectric.com**

Welding Handbook. Current Edition. Miami, FL: The American Welding Society.

2.0.0 Section Review

1. Which of the following can be quenched without concern?
   a. Weld test coupons
   b. Functional welds
   c. Practice coupons
   d. Welds done in the 3G position

2. When V-groove welding overhead, you are welding in the _____.
   a. 1G position
   b. 2F position
   c. 3F position
   d. 4G position

Summary

The ability to make groove welds on plate with metal backing using low-hydrogen electrodes is an essential welding skill. These skills are necessary to progress to more difficult pipe welding procedures in the future. Several groove-weld styles and combination styles, including V-grooves with backing, were covered in this module.

Proper preparation of the equipment, base metal, and backing is important for making quality groove welds. Welds in the vertical and overhead positions are typically required for welder certification on plate.

Groove welds are made in the flat (1G), horizontal (2G), vertical (3G), and overhead (4G) positions. It is important to commit these position designations to memory. Practice these welds until you can consistently produce acceptable welds.

Review Questions

1. If a groove for a weld allows for the entire thickness of the base metals to be welded, it is called a(n) _____.
   a. partial joint penetration (PJP) weld
   b. complete joint penetration (CJP) weld
   c. axial joint penetration (AJP) weld
   d. subsurface joint penetration (SJP) weld

2. The portion of a joint where the two separate components to be welded are the closest to each other is called the _____.
   a. joint root
   b. depth of bevel
   c. groove radius
   d. weld toe

3. When preparing welding coupons for V-groove welds with backing, the backing should extend beyond the end of the coupons _____.
   a. ⅛" (3.2 mm)
   b. ¼" (6.4 mm)
   c. ½" (12.7 mm)
   d. ¾" (19.1 mm)

4. Which of the following is an *incorrect* statement about running low-hydrogen electrodes?
   a. You should maintain a short arc.
   b. You should take only a few electrodes from the oven at a time.
   c. You should always whip the electrode.
   d. You should not use electrodes with missing flux.

5. The V-groove weld with backing is a common weld normally made with _____.
   a. fast-fill electrodes
   b. fill-freeze electrodes
   c. fast-freeze electrodes
   d. low-hydrogen electrodes

6. A way to eliminate the need for back gouging, and to quickly and effectively prepare a joint for welding, is to use _____.
   a. tape backing
   b. steel backing
   c. a backing weld run with E7018 electrodes
   d. a backing weld run with fast-freeze electrodes

7. When using weave beads on flat V-groove welds, keep the electrode at a 10- to 15-degree drag angle and a work angle of _____.
   a. 0 degrees
   b. 45 degrees
   c. 60 degrees
   d. 90 degrees

8. When practicing horizontal (2G) V-groove welding using stringer beads, the work angle should be _____.
   a. 0 degrees
   b. 45 degrees
   c. 90 degrees
   d. adjusted as needed to ensure tie-in

9. When using stringer beads on V-groove welds in the vertical (3G) position, keep the electrode at a _____.
   a. 0-degree push angle at all times
   b. 0- to 10-degree push angle
   c. 45-degree push angle
   d. 90-degree push angle

10. Before welding a V-groove in the overhead position, you should _____.
   e. change to E6010 electrodes
   a. increase the welding current significantly
   b. build a pad with stringer beads in the horizontal position
   c. change to AC welding current

Appendix

Performance Accreditation Tasks

The American Welding Society (AWS) School Excelling through National Skills Standards Education (SENSE) program is a comprehensive set of minimum Standards and Guidelines for Welding Education programs. The following performance accreditation tasks are aligned with and designed around the SENSE program.

The Performance Accreditation Tasks (PATs) correspond to and support the learning objectives in *AWS EG2.0, Guide for the Training and Qualification of Welding Personnel: Entry-Level Welder*.

Note that in order to satisfy all learning objectives in *AWS EG2.0*, the instructor must also use the PATs contained in the second level of the NCCER Welding curriculum.

PATs 1 through 4 correspond to *AWS EG2.0, Module 4—Shielded Metal Arc Welding*, Key Indicators 3, 4, and 6.

PATs provide specific acceptable criteria for performance and help to ensure a true competency-based welding program for students.

The following tasks are designed to develop your competency in preparing base metal. Practice each task until you are thoroughly familiar with the procedure.

As you complete each task, take it to your instructor for evaluation. Do not proceed to the next task until instructed to do so by your instructor.

V-GROOVE WELDS WITH BACKING IN THE FLAT (1G) POSITION

Using 3⁄32", 1⁄8", or 5⁄32" (2.4, 3.2, or 4.0 mm) E7018 electrodes, make a V-groove weld with steel backing on carbon steel plate in the flat position as indicated.

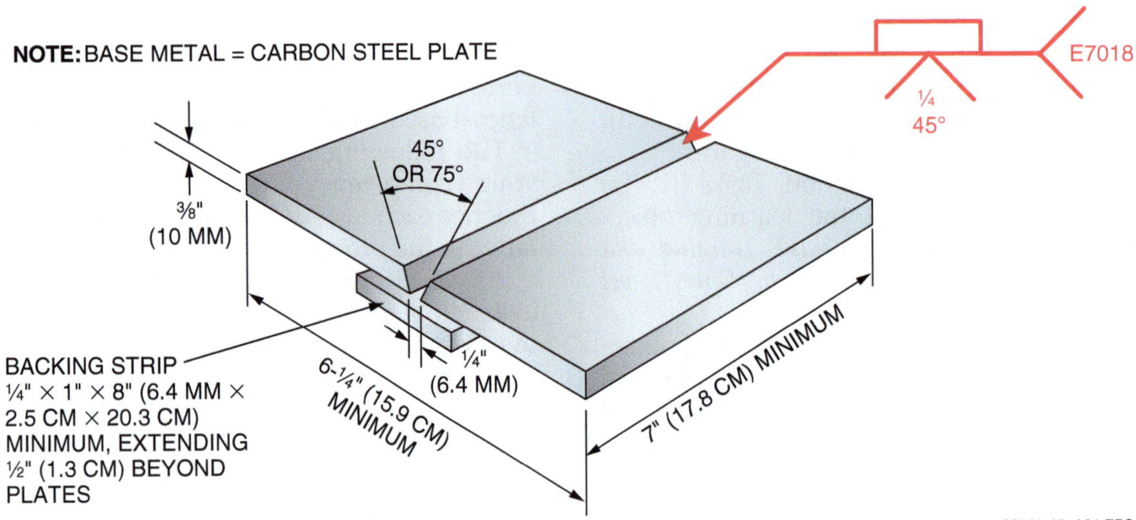

29111-15_A01.EPS

Inspection Hold Points:

- Fit-up __________
- Root pass __________
- Final __________

Criteria for Acceptance:

- No arc strikes outside the weld area __________
- Uniform rippled appearance on the bead face with no valley between the beads and acceptable tie-in __________
- Craters and restarts filled to the full cross section of the weld __________
- Uniform weld size ±1⁄16" (1.6 mm) __________
- Acceptable weld profile in accordance with *AWS D1.1* __________
- Smooth transition with complete fusion at the toes of the weld __________
- No porosity __________
- No overlap __________
- No excessive undercut __________
- No inclusions __________
- No cracks __________

Note to Instructor: Make sure all trainees complete each pass before starting another.

V-GROOVE WELDS WITH BACKING IN THE HORIZONTAL (2G) POSITION

Using 3⁄32", 1⁄8", or 5⁄32" (2.4, 3.2, or 4.0 mm) E7018 electrodes, make a V-groove weld with steel backing on carbon steel plate in the horizontal position as indicated.

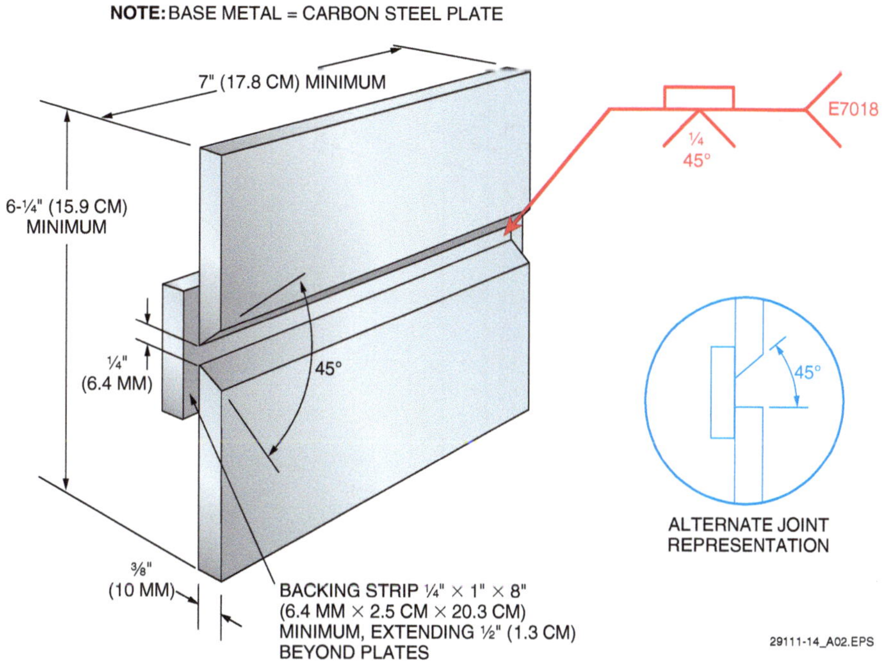

Inspection Hold Points:

- Fit-up ____________
- Root pass ____________
- Final ____________

Criteria for Acceptance:

- No arc strikes outside the weld area ____________
- Uniform rippled appearance on the bead face with no valley between the beads and acceptable tie-in ____________
- Craters and restarts filled to the full cross section of the weld ____________
- Uniform weld size ±1⁄16" (1.6 mm) ____________
- Acceptable weld profile in accordance with *AWS D1.1* ____________
- Smooth transition with complete fusion at the toes of the weld ____________
- No porosity ____________
- No excessive undercut ____________
- No overlap ____________
- No inclusions ____________
- No cracks ____________
- Acceptable guided bend test results per *AWS QC-10* ____________

V-GROOVE WELDS WITH BACKING IN THE VERTICAL (3G) POSITION

Using 3/32", 1/8", or 5/32" (2.4, 3.2, or 4.0 mm) E7018 electrodes, make a V-groove weld with steel backing on carbon steel plate in the vertical position as indicated.

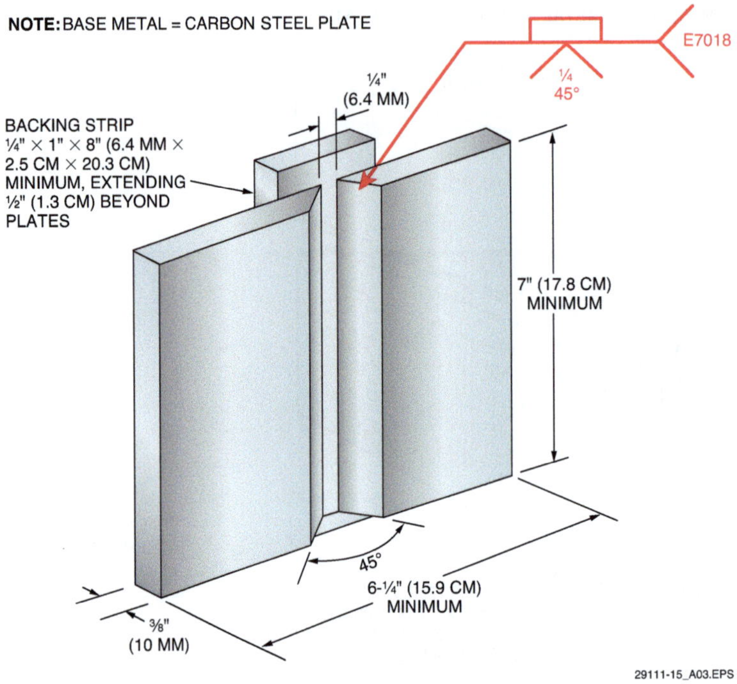

Inspection Hold Points:

- Fit-up ________
- Root pass ________
- Final ________

Criteria for Acceptance:

- No arc strikes outside the weld area ________
- Uniform rippled appearance on the bead face with no valley between the beads and acceptable tie-in ________
- Craters and restarts filled to the full cross section of the weld ________
- Uniform weld size ±1/16" (1.6 mm) ________
- Acceptable weld profile in accordance with *AWS D1.1* ________
- Smooth transition with complete fusion at the toes of the weld ________
- No porosity ________
- No excessive undercut ________
- No overlap ________
- No inclusions ________
- No cracks ________
- Acceptable guided bend test results per *AWS QC-10* ________

V-GROOVE WELDS WITH BACKING IN THE OVERHEAD (4G) POSITION

Using ³⁄₃₂", ⅛", or ⁵⁄₃₂" (2.4, 3.2, or 4.0 mm) E7018 electrodes, make a V-groove weld with steel backing on carbon steel plate in the overhead position as indicated.

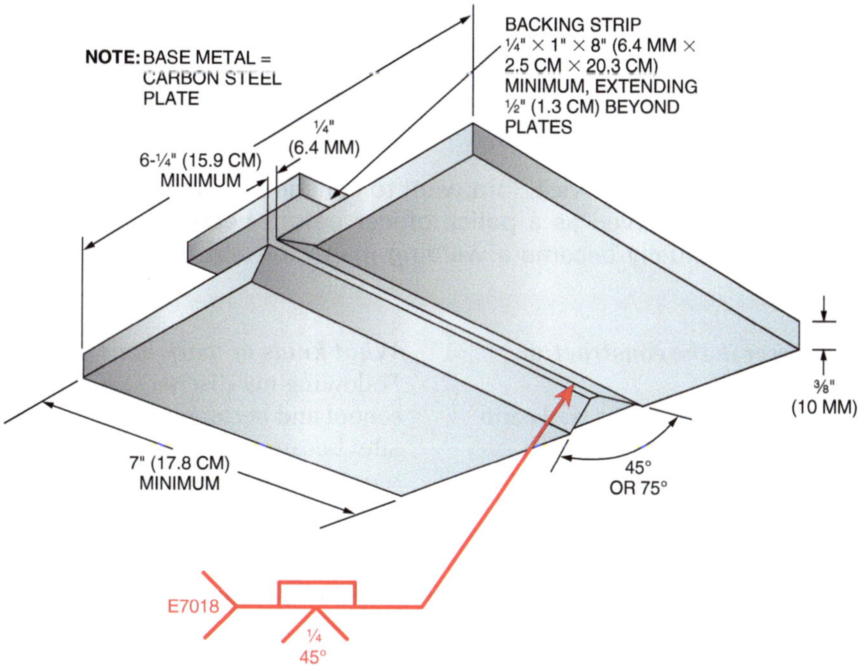

29111-15_A04.EPS

Inspection Hold Points:

- Fit-up ______
- Root pass ______
- Final ______

Criteria for Acceptance:

- No arc strikes outside the weld area ______
- Uniform rippled appearance on the bead face with no valley between the beads and acceptable tie-in ______
- Craters and restarts filled to the full cross section of the weld ______
- Uniform weld size ±¹⁄₁₆" (1.6 mm) ______
- Acceptable weld profile in accordance with *AWS D1.1* ______
- Smooth transition with complete fusion at the toes of the weld ______
- No porosity ______
- No excessive undercut ______
- No overlap ______
- No inclusions ______
- No cracks ______
- Acceptable guided bend test results per *AWS QC-10* ______

Cornerstone of Craftsmanship

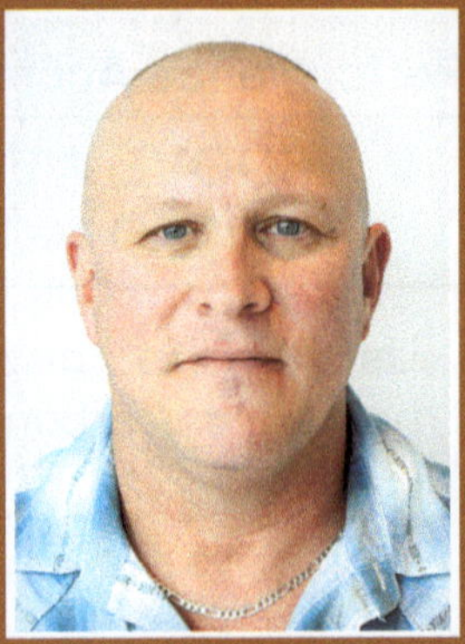

Tom Ashley

Robins and Morton, Birmingham, AL
Quality Assurance/Quality Control Manager

Following his discharge from military service, Tom went to college and trained and worked as a paramedic and firefighter. He subsequently served as a police officer before becoming a welder. That was the end of the career changes. He eventually became a welding inspector and then moved into management as manager of QA/QC.

How did you choose a career in the construction industry?
My father gave me a strong push to learn a trade and suggested that I consider a career in construction.

Who inspired you to enter the industry?
In addition to my father, my grandfather was a strong inspiration.

What types of training have you been through?
I did three years of welding school while serving a pipefitting apprenticeship.

How important are education and training in construction?
Education and training are essential. A person can learn to run beads without formal training, but will not achieve the level of knowledge and skill that lead to a well-paying, secure career. A welder who is well-trained will create a finished product that complies with industry standards.

How important are NCCER credentials to your career?
NCCER credentials provide the means to evaluate a person's knowledge and ability and provide a prescription for additional training. If a craft worker earns a Certified Plus credential, for example, you can be assured that he or she meets the journey-level standards of the craft.

How has training/construction impacted your life?
Training has given me the satisfaction of seeing a project progress from concept to completion. Being involved in multiple projects also provides some very rewarding challenges.

What kinds of work have you done in your career?
Following my discharge form the military, I went to school and became a firefighter and paramedic, then later became a police officer. None of those jobs gave me the satisfaction I needed, so I decided to become a welder.

Tell us about your present job.
I'm QA/QC manager for the Power and Industrial Division of my company. In that position, I supervise inspectors in all disciplines and train and coordinate other QA/QC managers.

What do you enjoy most about your job?
I enjoy working in a variety of disciplines because every day is different and every day presents a new challenge.

What factors have contributed most to your success?
I have to say it's my willingness to learn all disciplines, keeping an open mind, good communication skills, and a strong work ethic.

Would you suggest construction as a career to others? Why?
Absolutely! A construction career can be very rewarding because you are only limited by your desire and level of commitment. Many leaders and business owners in the construction industry started out on the ground floor and worked their way up.

What advice would you give to those new to the welding field?
Stay focused on your goals, get all the training you can, and listen to the advice of those with more experience.

Tell us an interesting career-related fact or accomplishment.
I started out as a public servant with no intention of entering the construction industry. Once I did, I came to love the field and I realized that it's what I'm best at.

How do you define craftsmanship?
Craftsmanship is taking pride in your work and completing it safely. It's creating a product that meets applicable specifications and standards, as well as expectations for its intended use and reliability. lines, boilers—you name it. I've performed work in accordance with a wide variety of welding codes.

Additional Resources

This module presents thorough resources for task training. The following resource material is suggested for further study.

2014 Technical Training Guide. Current Edition. Cleveland, OH. USA: The Lincoln Electric Company. **www.lincolnelectric.com**

Welding Handbook. Current Edition. Miami, FL: The American Welding Society.

Figure Credits

The Lincoln Electric Company, Cleveland, OH, USA, Module Opener, Figures 6, 10, 17, SA02

Topaz Publications, Inc., Figures 7 (photo), 19 (photo), 21 (photo)

Zachry Industrial, Inc., Figures 13, 15

Section Review Answer Key

Answer	Section Reference	Objective
Section One		
1. b	1.1.3	1a
2. d	1.2.1	1b
Section Two		
1. c	2.1.1	2a
2. d	2.2.2	2b

NCCER CURRICULA — USER UPDATE

NCCER makes every effort to keep its textbooks up-to-date and free of technical errors. We appreciate your help in this process. If you find an error, a typographical mistake, or an inaccuracy in NCCER's curricula, please fill out this form (or a photocopy), or complete the online form at **www.nccer.org/olf**. Be sure to include the exact module ID number, page number, a detailed description, and your recommended correction. Your input will be brought to the attention of the Authoring Team. Thank you for your assistance.

Instructors – If you have an idea for improving this textbook, or have found that additional materials were necessary to teach this module effectively, please let us know so that we may present your suggestions to the Authoring Team.

NCCER Product Development and Revision

13614 Progress Blvd., Alachua, FL 32615

Email: curriculum@nccer.org
Online: www.nccer.org/olf

❑ Trainee Guide ❑ Lesson Plans ❑ Exam ❑ PowerPoints Other ______________

Craft / Level: Copyright Date:

Module ID Number / Title:

Section Number(s):

Description:

Recommended Correction:

Your Name:

Address:

Email: Phone:

SMAW – Open-Root Groove Welds – Plate

OVERVIEW

SMAW groove welds in plate can be done with or without a backing. When a backing is not used, there is no material on the opposite side of the weld to prevent molten metal from passing through the groove. There is also less mass involved in the welded pieces. These differences, together with the small differences in technique and fit-up dimensions, make the process different from welding plate with a backing. This module focuses on those differences and on the practicing of SMAW on open-root grooves in plate.

Module 29112

Trainees with successful module completions may be eligible for credentialing through the NCCER Registry. To learn more, go to **www.nccer.org** or contact us at **1.888.622.3720**. Our website has information on the latest product releases and training, as well as online versions of our *Cornerstone* magazine and Pearson's product catalog.

Your feedback is welcome. You may email your comments to **curriculum@nccer.org**, send general comments and inquiries to **info@nccer.org**, or fill in the User Update form at the back of this module.

This information is general in nature and intended for training purposes only. Actual performance of activities described in this manual requires compliance with all applicable operating, service, maintenance, and safety procedures under the direction of qualified personnel. References in this manual to patented or proprietary devices do not constitute a recommendation of their use.

29112 V5

29112
SMAW – Open Root Groove Welds – Plate

Objectives

When you have completed this module, you will be able to do the following:

1. Identify various types of groove welds and describe how to prepare for groove welding.
   a. Identify various types of groove welds and define related terms.
   b. Describe how to prepare the work area and plate for groove welding.
2. Describe the technique required to produce various open V-groove welds.
   a. Describe the technique required to produce open V-groove welds in the 1G and 2G positions.
   b. Describe the technique required to produce open V-groove welds in the 3G and 4G positions.

Performance Tasks

Under the supervision of your instructor, you should be able to do the following:

1. Make open V-groove welds with E6010 and E7018 electrodes in the following positions:
   - Flat (1G) position
   - Horizontal (2G) position
   - Vertical (3G) position
   - Overhead (4G) position

Trade Terms

Feather
Root face

Industry Recognized Credentials

If you are training through an NCCER-accredited sponsor, you may be eligible for credentials from NCCER's Registry. The ID number for this module is 29112. Note that this module may have been used in other NCCER curricula and may apply to other level completions. Contact NCCER's Registry at 888.622.3720 or go to **www.nccer.org** for more information.

Contents

Topics to be presented in this module include:

Figures

Section One

1.0.0 Open-Groove Welds

Objective

Identify various types of groove welds and describe how to prepare for groove welding.

a. Identify various types of groove welds and define related terms.
b. Describe how to prepare the work area and plate for groove welding.

Trade Term

Root face: The portion of the edge of a part to be joined by a groove weld that has not been beveled or grooved; previously referred to as the land.

This module covers open V-groove welds, which are commonly used for welding plate. The open V-groove weld is easy to prepare because it only requires single-face beveling of the matching members. Without backing, however, welding the root pass requires practice to perfect. This module focuses on V-groove welds without a backing.

This module will explain the joint preparation and performance of open-root V-groove welds on plate using E6010 and E7018 electrodes in all positions. It will also prepare the welder to advance to the more difficult pipe welding modules. Performance demonstrations will require that the welds meet the qualification standards in the *ASME Boiler and Pressure Vessel Code, Section IX – Welding and Brazing Qualifications* for visual and destructive testing. Always remember that quality control begins with the welder.

1.1.0 Typical Groove Styles

A groove weld is made in an opening in a part or between two parts. Groove welds can be made in all five of the basic types of joints as well as on the surface of a part. There are several different groove styles, the name of the groove is derived from its shape. Some must be prepared; others appear as a result of fitting together the parts to be welded together.

Note that, if a groove only goes part of the way through the joint, welding in the groove results in what is called a partial joint penetration (PJP) weld. If the groove allows for welding the complete thickness of the base metals, it is called a complete joint penetration (CJP) weld.

1.1.1 *Single- and Double-Groove Welds*

Groove weld styles include the square, V, U, bevel, J, flare V, and flare bevel. *Figure 1* shows typical single-groove weld styles.

V-, bevel, U-, and J-grooves can be used and welded on one side of a joint or on both sides. If they are used on both sides of a joint, they are called double-groove welds (*Figure 2*).

NOTE

Throughout this module, both metric and Imperial units have been provided for most measurements. Weld symbols are generally provided in one measurement system or another, but not both. For this reason, metric values have not been placed alongside Imperial values in weld symbols.

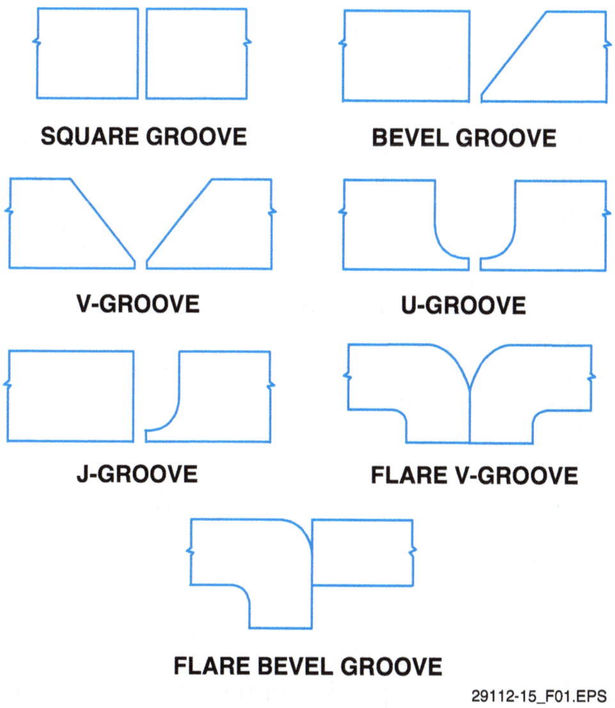

Figure 1 Single-groove weld styles.

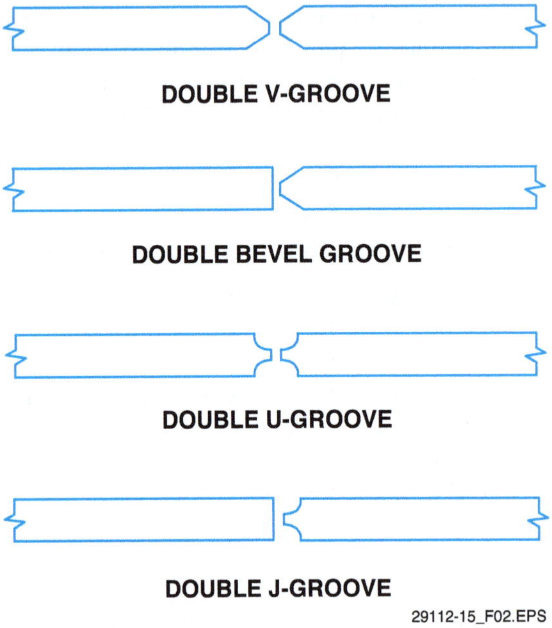

Figure 2 Double-groove weld styles.

1.1.2 Combination Groove and Fillet Welds

Groove welds are often used in combination with fillet welds to maintain the strength of the welded joint and use less weld metal. *Figure 3* shows two groove and fillet weld combinations.

1.1.3 Groove Weld Terms

Part of a welder's certification includes the proper understanding of related welding terms. The terms associated with a groove are the following:

- *Bevel angle* – The angle formed between the prepared edge of a member and a line perpendicular to the surface of the member.
- *Bevel face* – The prepared surface of a bevel edge shape.
- *Bevel radius* – The radius used to form the shape of a J- or U-groove.
- *Depth of bevel* – The distance the joint preparation extends into the base metal (used for partial joint penetration welds).
- *Groove angle* – The included (total) angle between the groove faces of a weld groove.
- *Groove face* – Any surface in a weld groove prior to welding. See *bevel face,* **root face**, and *bevel radius.*
- *Joint root* – The portion of a joint where the members approach closest to each other.
- *Root face* – The portion of the edge of a part to be joined by a groove weld that has not been beveled or grooved; previously referred to as the land.
- *Root opening* – The separation between the members to be joined at the root of the joint.

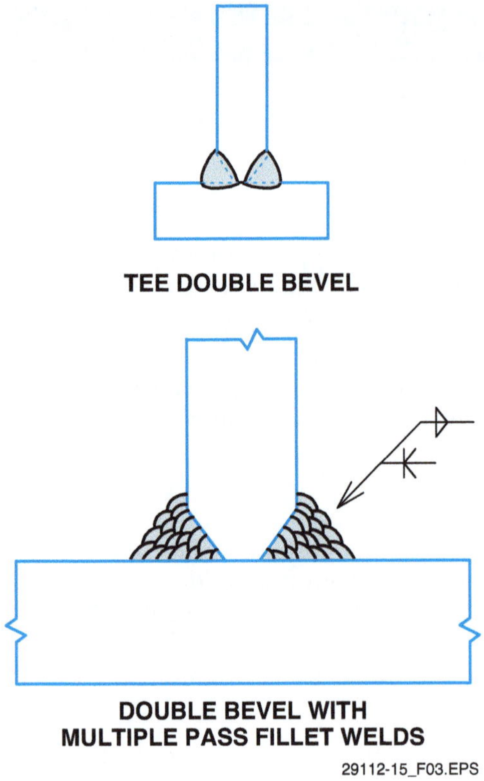

Figure 3 Groove and fillet weld combinations.

Figure 4 shows examples of grooves and their named parts.

The terms used to describe features of a groove weld are as follows:

- *Face reinforcement* – Reinforcement of a joint done from the same side as that from which the weld was made.
- *Groove weld size* – The joint penetration (depth of joint preparation plus root penetration). The size can never be greater than the base metal thickness.
- *Weld face* – The exposed surface of the weld.
- *Weld root* – The point at which the back of the weld extends the farthest into the weld joint.
- *Weld toe* – The points where the face of the weld metal and the base metal meet.

Figure 5 shows the common dimensions of an open-root V-groove weld.

> **NOTE**
>
> The dimensions and specifications in this module are designed to be representative of codes in general and are not specific to any particular code. Always follow the Welding Procedure Specification (WPS) and proper codes for your site.

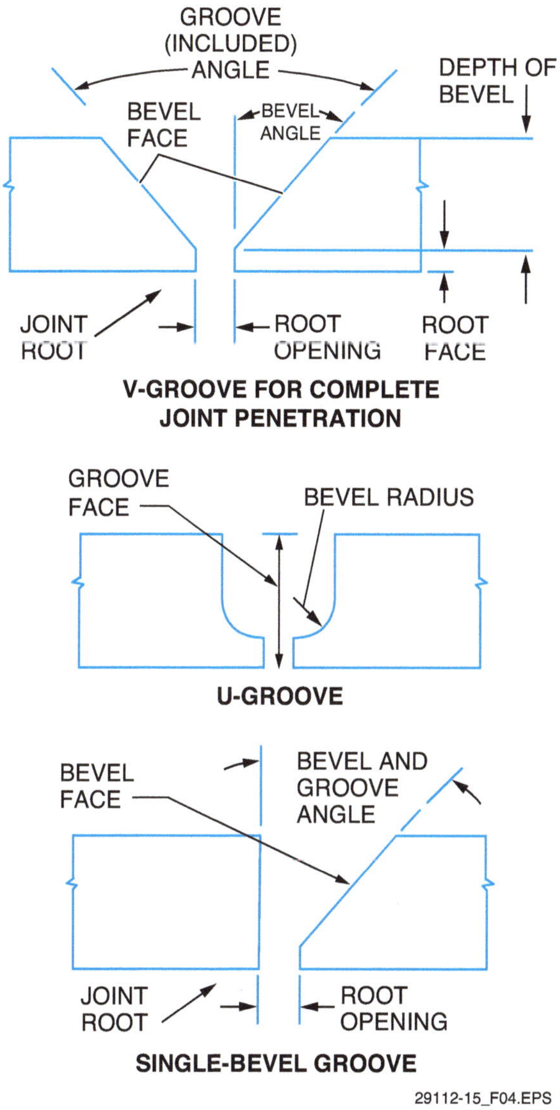

Figure 4 Grooves.

1.2.0 Welding Preparations

Before welding can take place, the area has to be made ready, the welding equipment must be set up, and the metal to be welded must be prepared. The following sections explain how to set up equipment for SMAW.

1.2.1 Safety Practices

The following is a summary of safety procedures and practices that must be observed when cutting or welding; complete safety coverage is provided in the *Welding Safety* module. If you have not completed that module, do so before continuing. Above all, be sure to wear appropriate protective clothing and equipment when welding or cutting.

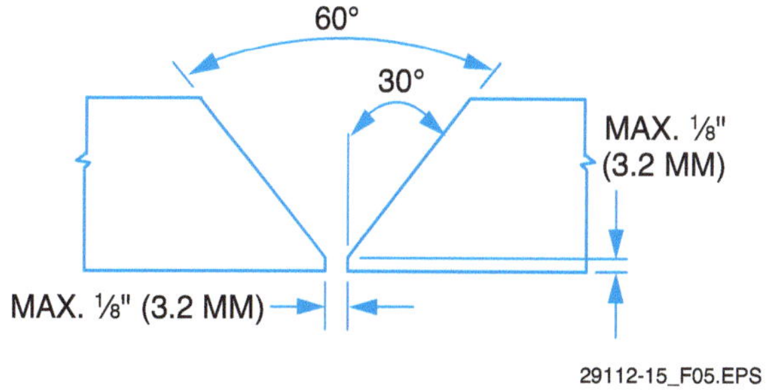

Figure 5 Common open-root V-groove dimensions.

To maintain safety and prevent injury, it is essential that you wear the appropriate protective clothing and equipment when cutting or welding metals (*Figure 6*). Be sure to follow these guidelines during all phases of cutting or welding:

- Always use safety glasses along with a full face shield or a helmet. The glasses, face shield, or helmet lens must have the proper light-reducing tint for the type of welding or cutting being performed. Never directly or indirectly view an electric arc without using a properly tinted lens.
- Wear proper protective leather and/or flame retardant clothing along with welding gloves that will protect you from flying sparks and molten metal as well as heat.
- Wear high-top safety shoes or boots. Make sure that the tongue and lace area of the footwear will be covered by a pant leg. If the tongue and lace area is exposed, or if the footwear must be protected from burn marks, wear leather spats under your pants or chaps to cover the front top of the footwear.

Figure 6 Welder working in full protective equipment.

- Wear a 100-percent cotton cap with no mesh material included in its construction. The bill of the cap points to the rear. If a hard hat is required for the environment, use one that allows the attachment of rear deflector material and a face shield. A hard hat with a rear deflector is generally preferred when working overhead, and may be required by some employers and job sites.

WARNING!

Do not wear a cap with a button in the middle. The conductive metal button beneath the fabric represents a safety hazard.

- Wear a face shield and snug-fitting safety glasses for gas welding or cutting. Either the face shield or the lenses of the welding glasses must be an approved shade for the application. For electric arc welding or cutting, wear safety glasses and a welding hood with the correct tinted lens.
- Wear earplugs to protect ear canals from sparks. Wear hearing protection to protect against the consistent sound of the torch.

WARNING!

Ear protection is essential to protect ears from the noise of the torch. Other personal protective equipment (PPE) must be worn to protect the operator from hot metal and slag.

- Do not open your welding hood lens and expose unprotected eyes when grinding or cleaning a weld. Use either a hood that maintains a protective lens when the shaded lens is opened, a hood with an auto-shading lens, or remove the hood and use a full-face shield.

Welding activities involve the use of fire or extreme heat to melt metal. Whenever fire is used, it must be controlled and contained. Welding activities are often performed on vessels that may once have contained flammable or explosive materials. Residues from those materials can catch fire or explode when a welder begins work on such a vessel. The following are fire and explosion prevention guidelines associated with welding:

- Never carry matches or gas-filled lighters in your pockets. Sparks can cause the matches to ignite or the lighter to explode, causing serious injury.
- Always comply with all site and/or employer requirements for a hot-work permit and a fire watch.
- Never use oxygen to blow dust or dirt off clothing. The oxygen can remain trapped in the fabric for a time. If a spark hits the oxygen in the fabric, the clothing can burn rapidly and violently.
- Make sure that any flammable material in the work area is moved or shielded by a fire-resistant covering.
- Approved fire extinguishers must be available before attempting any heating, welding, or cutting operations. Make sure the extinguisher is charged, the inspection tag is valid, and any individual that may be required to operate it knows how to do so.
- Never release a large amount of oxygen or use oxygen in place of compressed air. The presence of oxygen around flammable materials or sparks can cause rapid and uncontrolled combustion. Keep oxygen away from oil, grease and other petroleum products.
- Never release a large amount of fuel gas, especially acetylene. Methane and propane are heavier than air and tend to migrate to and concentrate in low areas. As a result, they can ignite at a considerable distance from the release point. Acetylene is lighter than air but is even more dangerous than methane; when mixed with air or oxygen, it will explode at much lower concentrations than any other common fuel gas.
- To prevent fires, maintain a neat and clean work area, and make sure that any metal scrap or slag is cold before disposal.

Before cutting or welding containers such as tanks or barrels, find out if they contained any explosive, hazardous, or flammable materials, including petroleum products, citrus products, or chemicals that decompose into toxic fumes when heated. Proper procedures for cutting or welding hazardous containers are described in the *American Welding Society (AWS) F4.1, Safe Practices for the Preparation of Containers and Piping for Welding and Cutting,* and *ANSI Z49.1.* As a standard practice, always clean and then fill any tanks or barrels with water, or purge them with a flow of inert gas such as nitrogen to displace any oxygen.

WARNING!

Welding or cutting must never be performed on drums, barrels, tanks, vessels, or other containers until they have been emptied and cleaned thoroughly, eliminating all flammable materials and all substances (such as detergents, solvents, greases, tars, or acids) that might produce flammable, toxic, or explosive vapors when heated. Do not assume that a container that has held combustibles is clean and safe until proven so by proper tests. Do not weld in places where dust or other combustible particles are suspended in air or where explosive vapors are present.

Containers must be cleaned by steam cleaning, flushing with water, or washing with detergent until all traces of the material have been removed.

WARNING!

Clean containers only in well-ventilated areas. Vapors can accumulate during cleaning, causing explosions or injury.

After cleaning the container, fill it with water or a purging gas, such as carbon dioxide, argon, or nitrogen to displace the explosive fumes. Air, which contains oxygen, is displaced from inside the container by the water or inert gas. Without oxygen, combustion cannot take place.

A water-filled vessel is the best alternative. When using water, position the container to minimize the air space. When using an inert gas, provide a vent hole so the inert gas can push the air and other vapors out to the atmosphere. Keep in mind, though, that even these precautions do not guarantee the absence of flammable materials inside. For that reason, these types of activities should not be done without proper supervision and the use of proper testing methods.

GOING GREEN

Welding Water

To better protect the environment, send any water used in a welding process, or water used to clean tanks, to a waste treatment facility instead of simply allowing it to run into a storm drain or onto the ground.

Vapors and fumes tend to rise in the air from their sources. Welders often have to work above welding areas where fumes are being created. Welding fumes can be harmful. Good work area ventilation helps to remove the vapors and protect the welder. Always perform cutting or welding operations in a well-ventilated area. Cutting or welding operations involving zinc or cadmium materials or coatings result in toxic fumes. For long-term cutting or welding of such materials, always wear an approved full-face, supplied-air respirator (SAR) that uses breathing air supplied from outside the work area. For occasional, very short-term exposure, a HEPA-rated or metal fume filter may be used on a standard respirator.

1.2.2 Preparing the Welding Area

To practice welding, a welding table, bench, or stand is needed (*Figure 7*). The welding surface must be steel, and provisions must be made for placing welding coupons out of position.

To set up the area for welding, follow these steps:

Step 1 Make sure that the area is properly ventilated. Make use of doors, windows, and fans.

Step 2 Check the area for fire hazards. Remove any flammable materials before proceeding.

29112-15_F07.EPS

Figure 7 Welding station with vacuum system.

Step 3 Know the location of the nearest fire extinguisher. Do not proceed unless the extinguisher is charged and you know how to use it.

Step 4 Position a welding table near the welding machine, or vice versa.

Step 5 Set up welding curtains around the welding area.

1.2.3 Preparing the Weld Coupons

If possible, the weld coupons should be ⅜" (9.6 mm) thick carbon steel. If this size is not readily available, ¼" to ¾" (6.4 to 19.1 mm) thick steel can be used for practice welds. Open-root welds generally have a bevel angle of 30 degrees. Clean the steel plate prior to welding by using a wire brush or grinder to remove heavy mill scale or corrosion.

WARNING!

Be extremely careful when using wire brush attachments on powered tools. Always ensure that the rotational speed rating of the wire brush matches or exceeds the rated speed of the powered tool. Loose wires from the brush and debris tend to fly off and embed into anything or anyone near the work. Make sure that all appropriate PPE is worn when wire brushing or grinding with all tools, but be especially cautious with powered tools.

Prepare weld coupons to practice the following welds:

- *Stringer bead welds* – The coupons can be any size or shape that is easily handled.
- *Open-root V-groove welds* – As shown in *Figure 8*, cut the metal into 3" × 7" (7.6 cm × 17.8 cm) rectangles with a bevel of 30 degrees on one of the long sides. Grind up to a ⅛" (3.2 mm) root face on the bevel, or leave without a root face. Your instructor may have a preference.

NOTE

The welding codes allow the root face and opening on open-root welds to be 0" to ⅛" (0 to 3.2 mm). Adjust the root face and root opening as needed when you start your welding practices.

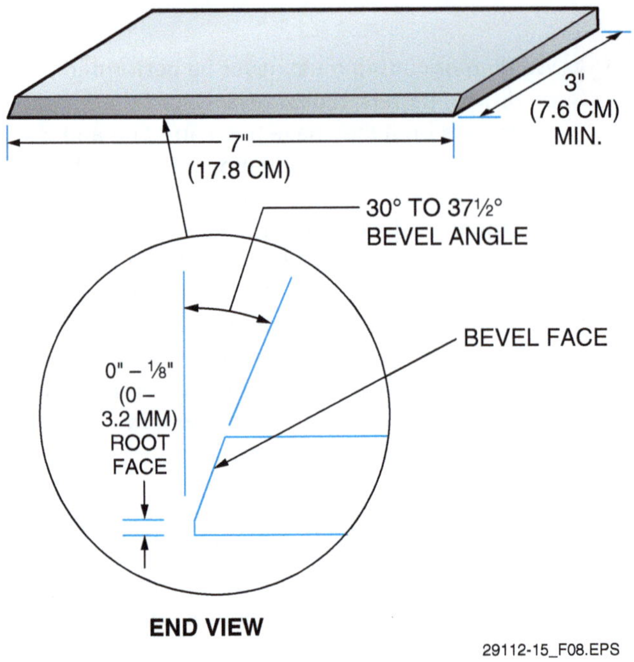

Figure 8 Metal cut for open-root V-groove weld coupons.

Follow these steps to prepare each V-groove weld coupon:

Step 1 Check the bevel face. There should be no dross and a 0" to ⅛" (0 to 3.2 mm) root face, and the bevel angle should be 30 degrees. An example of this is shown in *Figure 9*.

Step 2 Remove any burrs from the edges of the coupon.

Step 3 Align the beveled strips on a flat surface with a 0" to ⅛" (0 to 3.2 mm) root opening and tack weld them on each end. Use three to four ¼" (6.4 mm) long tack welds.

1.2.4 Electrodes

Obtain a small quantity of the electrodes to be used. For the welding exercises in this module, E6010 and E7018 electrodes will be used.

GOING GREEN

Vapor Extraction Systems

To better protect the environment, perform any welding activities under a vapor extraction system that can filter the welding fumes before they reach the atmosphere.

> **NOTE**
> The electrode sizes are recommendations only. Other sizes may be substituted depending on site conditions. Check with your instructor for the electrode sizes to use.

For E6010 electrodes, use a whipping motion when depositing the stringer bead to control the weld puddle, then remove all slag between passes.

For E7018 electrodes, keep the following in mind:

- Remove only a small number of electrodes at a time from the oven because low-hydrogen electrodes absorb moisture quickly.
- Do not whip low-hydrogen electrodes.
- Maintain a short arc because a low-hydrogen weld relies on its slag to protect the molten metal. It does not have a heavy gaseous shield.
- Remove all slag between passes.
- Do not use low-hydrogen electrodes with chipped or missing flux.

For either E6010 or E7108 electrodes, observe the following:

- When running stringer beads, use a slight side-to-side motion and a small pause at the weld toe to tie-in and flatten the bead.
- When running weave beads, use a slight pause at the weld toes to tie-in to the base metal.
- Set the amperage in the lower portion of the suggested range for better puddle control when welding vertically.

Obtain only the electrodes to be used for a particular welding exercise at one time. Use some type of pouch or rod holder to store the electrodes in to prevent them from becoming damaged.

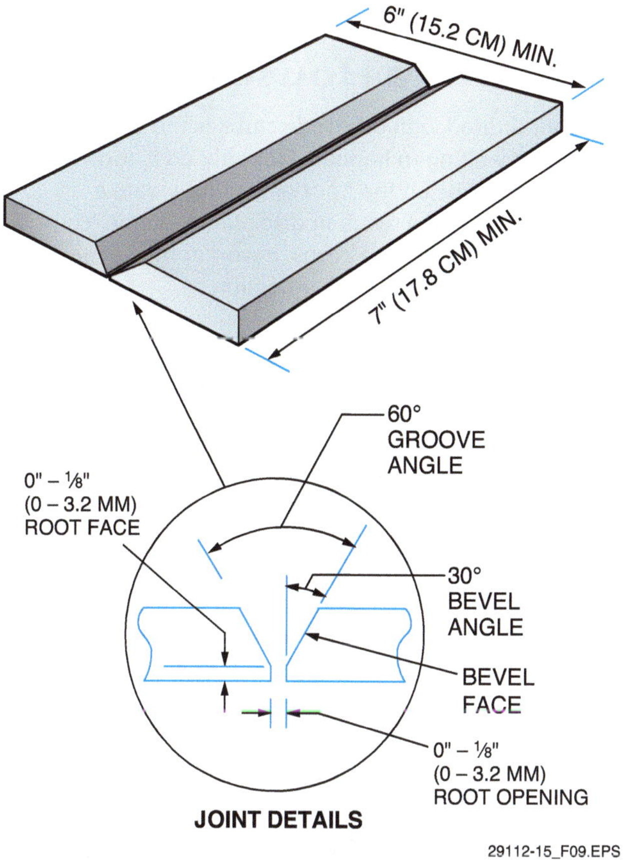

Figure 9 Open-root V-groove weld coupon.

Never store electrodes loose on a table. They may fall on the floor where they can become a tripping hazard or become damaged. A pail or other metal container must also be available to discard hot electrode stubs.

> **WARNING!**
> Do not throw electrode stubs on the floor. They roll easily and could cause someone to slip and fall.

GOING GREEN

Recycle Metal

Steel for practice welding is expensive and difficult to obtain. Make every effort to conserve and not waste the material available. Reuse weld coupons until all surfaces have been welded upon by cutting the weld coupon apart and reusing the pieces. Use material that cannot be cut into weld coupons to practice running beads. When beveling plates for coupons, bevel both of the long edges to help conserve material.

Salvage any coupons or scrap metal that is no longer usable for welding practice and place it in a recycling bin. Almost all metals are recyclable and valuable. One technical school, for example, collected donated steel, used it for cutting and welding practice, and then sold all their recyclable metals to fund their school's annual picnic. The bottom line is that metal no longer usable for cutting or welding practice can be recycled and the proceeds are put to good use.

Electrode Housekeeping

A floor littered with electrode stubs is the site of an accident waiting to happen. Not only does throwing electrode stubs on the floor or ground create a hazard, it can also result in disciplinary action from the employer. In some cases, a worker may even lose his or her job for this infraction.

1.2.5 Preparing the Welding Machine

Welders can expect to find different types and makes of welding machines in welding shops and on jobs in the field. Many machines will have not have the operator manual with them. As a trainee, and even as a trained welder, you must be able to recognize the different types of welding machines and figure out how to safely set up and operate them. *Figure 10* shows typical welding machines.

Select a welding machine to use. Proceed with the following welding set-up steps:

Step 1 Verify that the welding machine can be used for DC (direct current) welding.

Step 2 Check the area for proper ventilation.

Step 3 Verify the location of the power disconnect for the welding machine.

Step 4 Set the polarity to direct current electrode positive (DCEP).

Step 5 Connect the clamp of the workpiece lead to the workpiece.

Running E7018 Electrodes

Moving an electrode back and forth makes a weave bead. Many different patterns can be used to make a weave bead, including circles, crescents, and zigzags. The pause at the edges will also flatten out the weld, giving it the proper profiles.

Step 6 Set the amperage for the electrode type and size to be used. Typical settings are:

Electrode	Size	Amperage
E6010	3/32" (2.4 mm)	40A to 80A
E6010	1/8" (3.2 mm)	70A to 130A
E6010	5/32" (4.0 mm)	90A to 165A
E7018	3/32" (2.4 mm)	65A to 100A
E7018	1/8" (3.2 mm)	110A to 160A
E7018	5/32" (4.0 mm)	130A to 200A

NOTE

Amperage recommendations can vary by manufacturer, position, current type, and electrode brand. Follow the WPS precisely when applicable; the WPS should always be the first source of weld information. Otherwise, consult the welding machine manufacturer's recommendations for the electrode being used.

Step 7 Check to be sure the electrode holder is not grounded to a conductive surface.

Step 8 Turn on the welding machine.

Lash

To ensure good welding results, remember LASH.

L *(length of arc)* – The distance between the electrode and the base metal (usually one times the electrode diameter).

A *(angle)* – Two angles are critical:

- Travel angle – The longitudinal angle of the electrode in relation to the axis of the weld joint
- Work angle – The traverse angle of the electrode in relation to the axis of the weld joint

S *(speed)* – Travel speed is measured in inches per minute (IPM). The width of the weld will determine if the travel speed is correct.

H *(heat)* – Controlled by the amperage setting and dependent upon the electrode diameter, base metal type, base metal thickness, and the welding position.

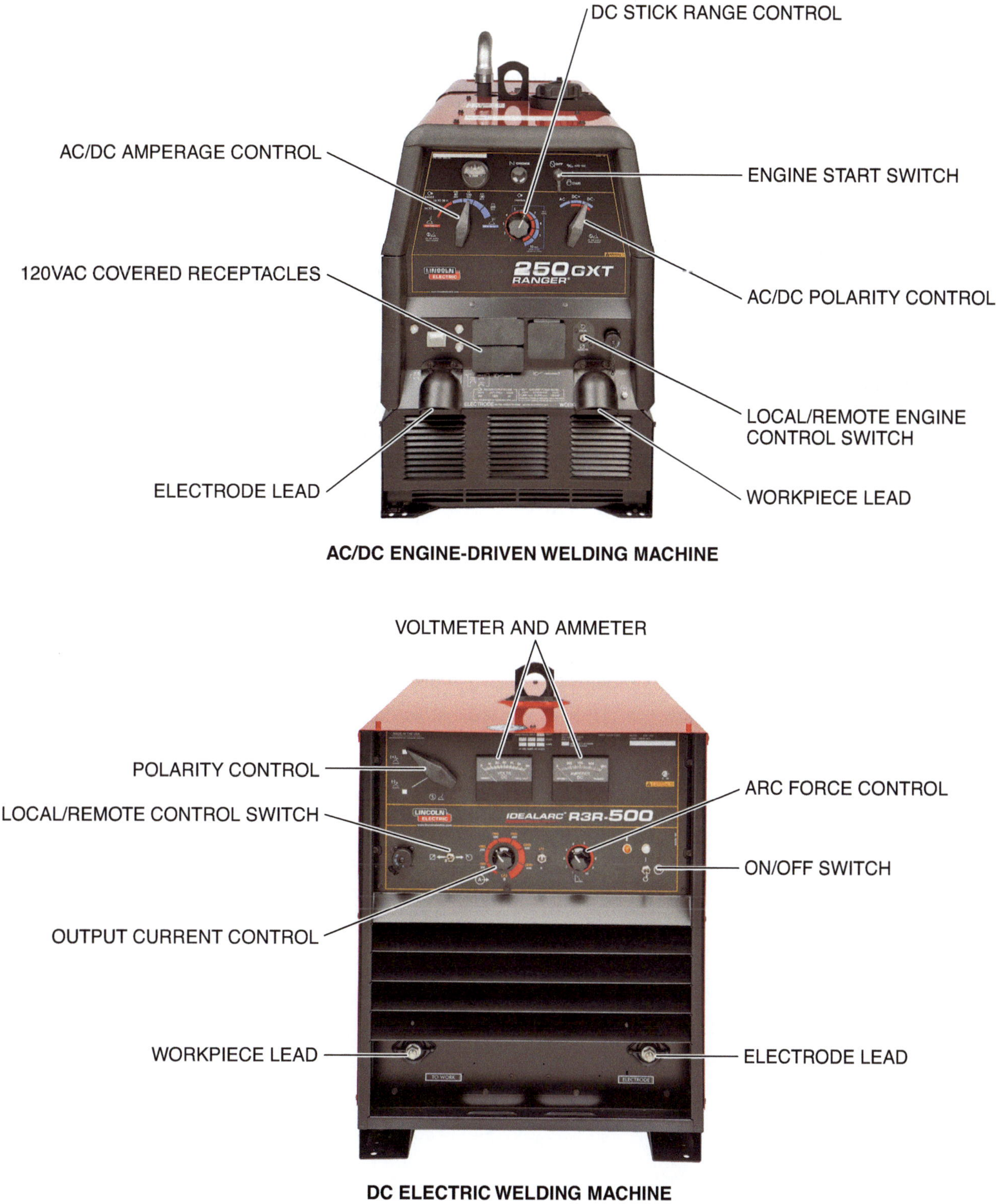

29112-15_F10.EPS

Figure 10 AC/DC and DC-only welding machines.

Additional Resources

2014 Technical Training Guide. Current Edition. Cleveland, OH. USA: The Lincoln Electric Company. **www.lincolnelectric.com**

Welding Handbook. Current Edition. Miami, FL: The American Welding Society.

1.0.0 Section Review

1. The radius used to form the shape of a J- or U-groove is called the _____.
   a. bevel radius
   b. root radius
   c. groove radius
   d. face radius

2. Before welding on previously used containers such as tanks or barrels, it is best to clean and fill them with water or to purge them with _____.
   a. compressed air
   b. acetylene
   c. oxygen
   d. an inert gas

Section Two

2.0.0 Open-Root V-Groove Welds

Objective

Describe the technique required to produce various open V-groove welds.

a. Describe the technique required to produce open V-groove welds in the 1G and 2G positions.
b. Describe the technique required to produce open V-groove welds in the 3G and 4G positions.

Performance Tasks

1. Make open V-groove welds with E6010 and E7018 electrodes in the following positions:
   - Flat (1G) position
   - Horizontal (2G) position
   - Vertical (3G) position
   - Overhead (4G) position

Trade Term

Feather: The process of grinding a tack weld to a tapered edge.

The open-root V-groove weld is a common groove weld normally made on both plate and pipe. Practicing the open-root V-groove welds on plate will prepare the welder to make the more difficult pipe welds in the future.

The performance qualification requirements for this module for visual and destructive testing are based on the *ASME Boiler and Pressure Vessel Code, Section IX, Welder Certification Test.*

2.0.1 Root Pass

The most difficult part of making an open-root V-groove weld is the root pass. The root pass is made from the V-groove side of the joint and must have complete penetration, but not an excessive amount of root reinforcement. The penetration is controlled with a technique called "running a keyhole." A keyhole is a hole made when the root faces of two plates are melted away by the arc. The molten metal flows to the back side of the keyhole, forming the weld. The keyhole should be about ⅛" to 3⁄16" (3.2 to 4.8 mm) in diameter, and centered between the two coupons. If the keyhole is larger, there will be excess root reinforcement. If the keyhole closes up, there will be a lack of penetration. Root reinforcement should be flush to ⅛" (3.2 mm).

The keyhole is controlled by using fast-freeze E6010 electrodes and by whipping the electrode. Move the electrode forward about 3⁄16" to ¼" (4.8 to 6.4 mm) and then back about ⅛" (3.2 mm) and pause. Use care not to increase the arc length when moving the electrode. If the keyhole starts to grow, decrease the pause and increase the forward length of the whip. If the keyhole starts to close up, pause slightly longer and decrease the length of the whip. Other factors that affect the keyhole size are the amperage setting, root face size, and root opening. All these factors can be adjusted within the parameters allowed by the WPS, welding codes, and/or your site quality procedures.

After the root pass has been run, it should be cleaned and inspected. Inspect the root face for excess buildup or undercut, as shown in *Figure 11*, which could trap slag when the filler pass is run. Remove excess buildup or undercut with a hand grinder by grinding the face of the root pass with the edge of a grinding disk. Use care not to grind through the root pass or widen the groove.

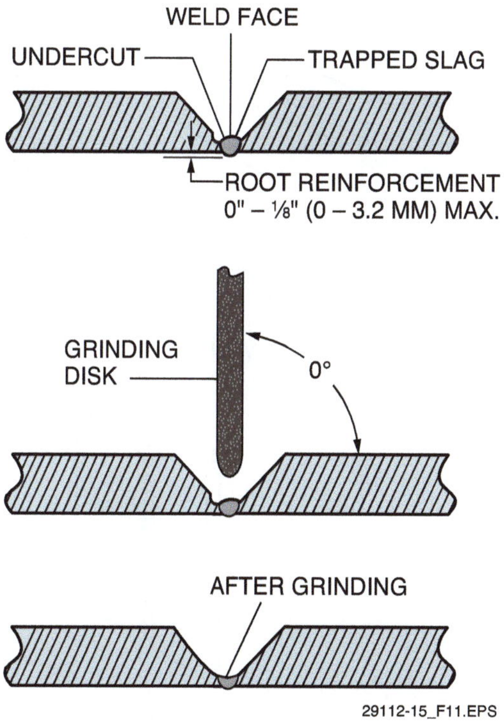

Figure 11 Grinding a root pass.

2.0.2 Groove Weld Positions

As illustrated in *Figure 12*, groove welds can be made in all positions. The weld position for plate is determined by the axis of the weld and the orientation of the workpiece. Groove weld positions for plate are flat, or 1G (groove); horizontal, or 2G; vertical, or 3G; and overhead, or 4G.

Grinding is required more often when low-hydrogen electrodes are used for the filler passes. This is because low-hydrogen electrodes have shallow penetrating characteristics and will not burn out slag inclusions as readily as E6010 electrodes. However, some structural codes do not allow the grinding of welds for any reason. In the 1G and 2G positions, the weld axis can be inclined up to 15 degrees. Any weld axis inclination for the other positions varies with the rotational position of the weld face as specified in AWS standards.

2.0.3 Acceptable and Unacceptable Groove Weld Profiles

Groove welds should be made with slight reinforcement and a gradual transition to the base metal at each toe. As shown in *Figure 13*, groove welds must not have excess reinforcement, underfill, excessive undercut, or overlap. If a groove weld has any of these defects, it is unacceptable.

2.1.0 Open V-Groove Welding – 1G and 2G Positions

The flat (1G) and horizontal (2G) positions are the easiest and best positions to begin practicing open-root V-groove welds.

2.1.1 Practicing Flat Open-Root V-Groove Welds (1G Position)

Practice flat open-root V-groove welds, as shown in *Figure 14*, using E6010 and E7018 electrodes. Use weave beads and keep the electrode angle 90 degrees to the plate surface (0-degree work angle) with a 10- to 15-degree drag angle. When using stringer beads, use a 10- to 15-degree drag angle, but adjust the electrode work angle to tie in and fill the bead. Pay particular attention at the termination of the weld to fill the crater. If the practice coupon gets too hot between passes, you may have to quench it in water.

WARNING!

Use pliers to handle the hot practice coupons. Wear gloves when quenching the practice coupon in water. Steam will rise off the coupon and can burn or scald unprotected hands.

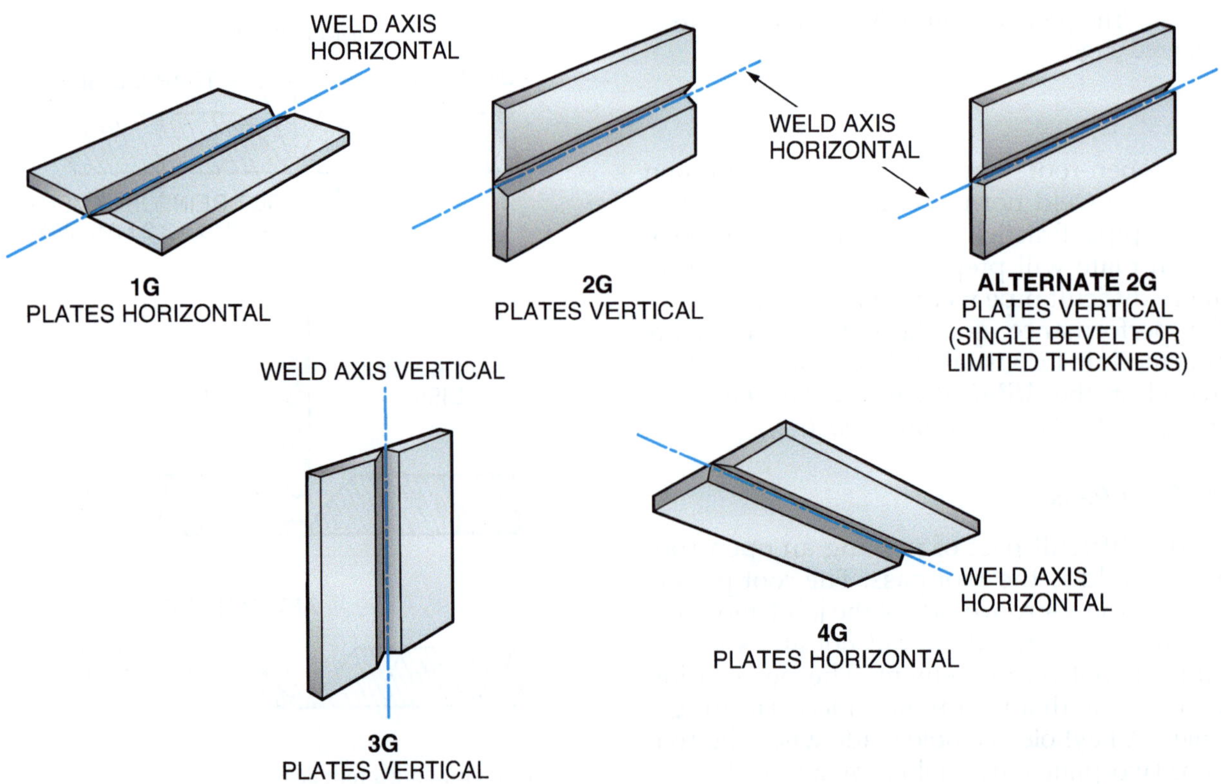

Figure 12 Open-root groove weld positions for plate.

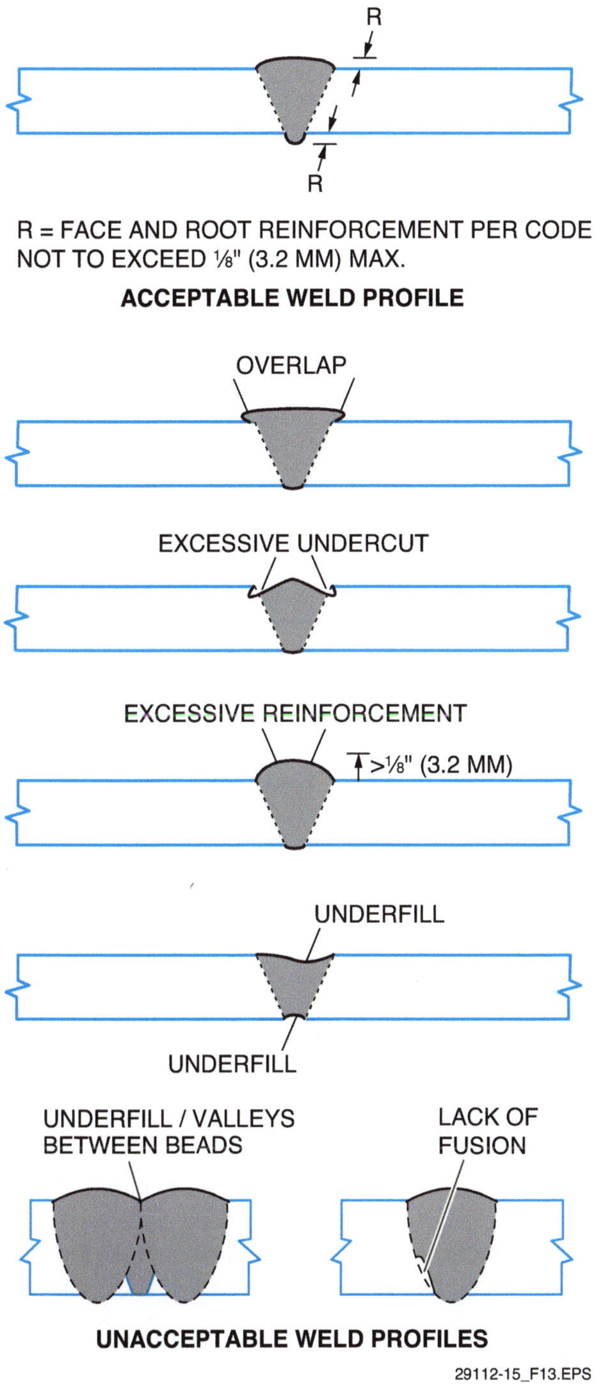

Figure 13 Acceptable and unacceptable groove weld profiles.

CAUTION

Quenching is only done on practice coupons. Never cool test coupons or functional welds with water. Cooling with water can cause weld cracks and affect the mechanical properties of the base metal.

Follow these steps to practice open-root V-groove welds in the flat position:

Step 1 Tack weld the practice coupon together as explained earlier.

Step 2 Clean and **feather** the tack welds with the edge of a grinding wheel. Feathering the ends of the tack welds with a grinder will help to fuse the tack welds into the root pass.

Step 3 Place the weld coupon in a flat position.

Step 4 Run the root pass using E6010 electrodes.

Step 5 Remove the electrode stub from the electrode holder and discard as soon as you stop welding to prevent arcing to nearby surfaces, and place the stub in a proper container.

Step 6 Chip, clean, and brush the weld between all passes. Grind if required.

Step 7 Run the fill passes, cleaning the weld after each pass, to complete the weld using E7018 electrodes.

2.1.2 Horizontal Welds (2G Position)

Horizontal welds can be made on a vertically positioned flat surface or on a horizontal open-root V-groove. This section explains welding horizontal beads and making horizontal open-root V-groove welds.

Before welding an open-root V-groove in the horizontal position, practice running horizontal stringer beads by building a pad in the horizontal position (*Figure 15*). Use E6010 electrodes. The electrode angle should be a 10- to 15-degree drag angle and be perpendicular to the plate. To control the weld puddle, the electrode can be dropped slightly, but no more than 10 degrees.

Follow the WPS and Site Procedures

Refer to your site's WPS for specific requirements on groove welds. The information in this module is provided as a general guideline only. The site WPS or quality specifications must be followed for all welds. Check with your supervisor if you are unsure of the specifications for your application.

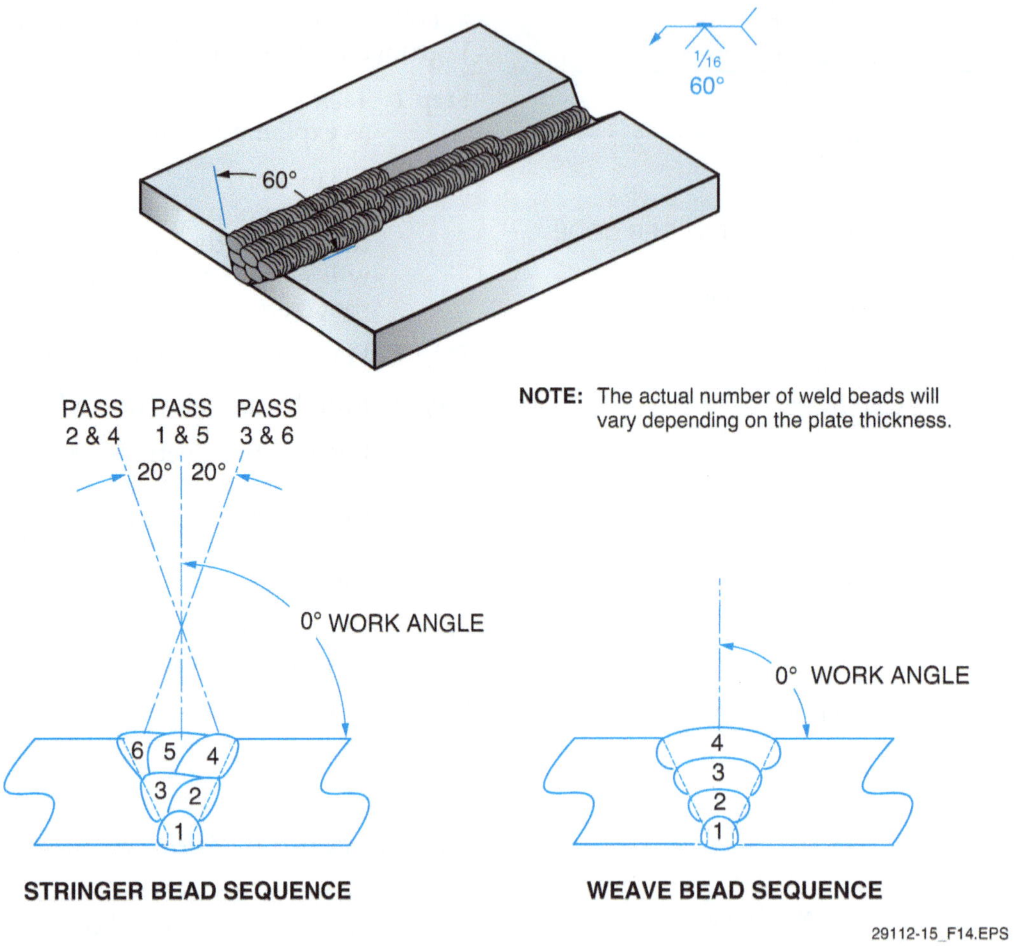

Figure 14 Multipass 1G weld sequences and work angles.

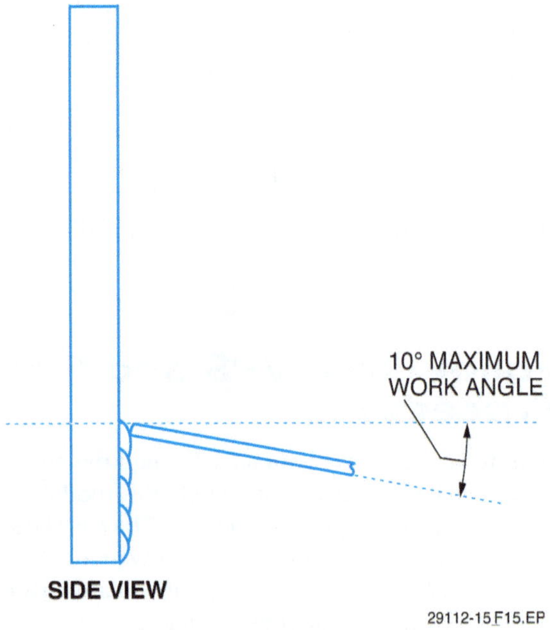

Figure 15 Work angle in the horizontal position.

Follow these steps to practice welding beads in the horizontal position:

Step 1 Tack weld the flat plate welding coupon into the horizontal position.

Step 2 Clean and feather the tack welds with the edge of a grinding wheel. Feathering the ends of the tack welds with a grinder will help to fuse the tack welds into the root pass.

Step 3 Run the first pass along the bottom of the coupon using E6010 electrodes.

Step 4 Remove the electrode stub from the electrode holder and discard as soon as you stop welding to prevent arcing to nearby surfaces, and place the stub in a proper container.

Step 5 Chip, clean, and brush the weld between all passes.

Step 6 Continue running beads to complete the pad.

A horizontal open-root V-groove weld is shown in *Figure 16*. Practice horizontal open-root V-groove welds using E6010 and E7018 electrodes and stringer beads. For the electrode angles, use a 10- to 15-degree drag angle and adjust the work angle as required. Pay particular attention at the termination of the weld to fill the crater. Note that the weave bead is not recommended for the 2G position.

Follow these steps to practice welding open-root V-groove welds in the horizontal position:

Step 1 Tack weld the practice coupon together as explained earlier. Use the standard or alternate horizontal weld coupon as directed by your instructor.

Step 2 Clean and feather tack welds with the edge of a grinding wheel.

Step 3 Tack weld the flat plate weld coupon into the horizontal position.

Step 4 Run the root pass using E6010 electrodes.

Step 5 Remove the electrode stub from the electrode holder and discard as soon as you stop welding to prevent arcing to nearby surfaces, and place the stub in a proper container.

Step 6 Chip, clean, and brush the weld between all passes.

Step 7 Run the fill and cover passes, cleaning the weld after each pass, to complete the weld. Use E7018 electrodes and stringer beads.

2.2.0 Open V-Groove Welding – 3G and 4G Positions

Now that the open-root V-groove techniques in the 1G and 2G positions have been practiced, the more challenging 3G and 4G positions can be attempted.

> **Can You Spare Some Change?**
>
> A nickel and a dime make excellent gauges for fitting open-root V-groove weld coupons. The thickness of the nickel can be used to check the root face, while the dime is used to set the root opening.

2.2.1 Vertical Welds (3G Position)

Vertical welds can be made on a vertically positioned flat surface or on a vertical open-root V-groove, as shown in *Figure 17*. This section explains welding vertical beads and making vertical open-root V-groove welds.

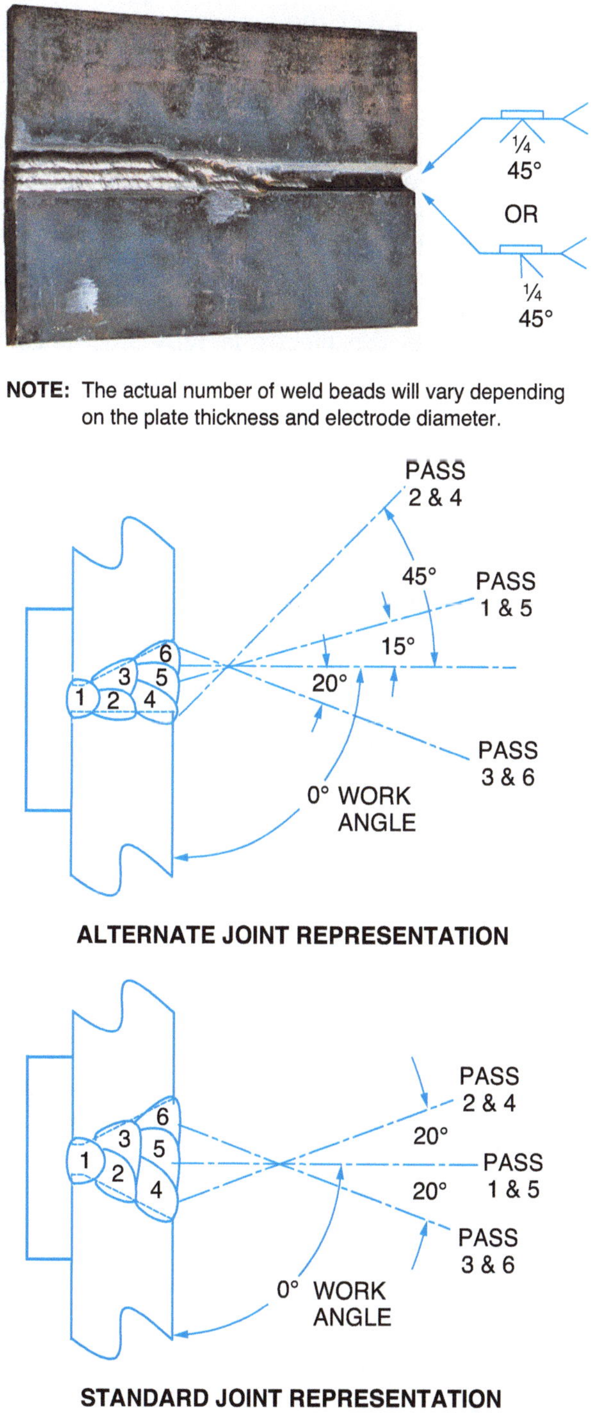

Figure 16 Multipass 2G weld sequences and work angles.

> **NOTE**
> Perform this practice if needed. Otherwise, move on to the next section.

Before welding a joint in the vertical position, practice running vertical weave beads by building a pad in the vertical position (see *Figure 17*). Use E6010 electrodes. Set the amperage in the lower part of the allowable range for better puddle control. The electrode travel should be upward with a 0- to 10-degree push angle and perpendicular to the plate (0-degree work angle) for the first pass. A 0- to 10-degree work angle toward the previous bead should be used for subsequent beads. Use a side-to-side motion with a slight pause at each weld toe to tie-in and to control the weld puddle. The width of the weave beads should be no more than five times the electrode wire (metal core) diameter. Follow these steps to practice welding beads in the vertical position:

Step 1 Tack weld the flat plate welding coupon into the vertical position.

Step 2 Using a whipping motion, run the first pass along the vertical edge of the coupon.

Step 3 Remove the electrode stub from the electrode holder and discard as soon as you stop welding to prevent arcing to nearby surfaces, and place the stub in a proper container.

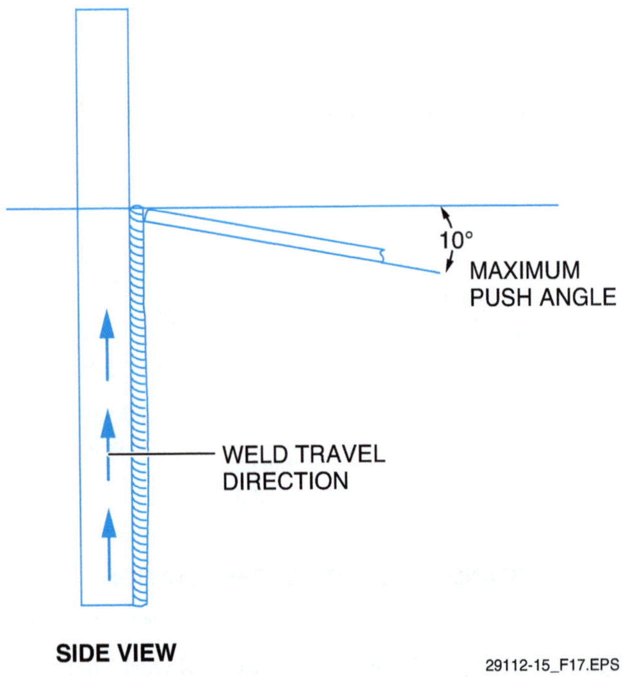

Figure 17 Work angle and direction of travel in the vertical position.

> **NOTE**
> Some codes do not allow weave beads. Always consult the WPS.

Step 4 Chip, clean, and brush the weld between all passes.

Step 5 Continue running beads and cleaning each pass to complete the pad.

An example of a vertical open-root V-groove weld is shown in *Figure 18*. Practice vertical open-root V-groove welds using E6010 for the root weld pass and E7018 electrodes and weave beads for the remaining passes. Set the amperage in the lower part of the range for better puddle control. Use an upward travel direction with a 0- to 10-degree push angle, but adjust the electrode work angle to tie-in the weld on one side or the other as needed. Pause slightly at each weld toe to penetrate and fill the crater to prevent undercut. Pay particular attention at the termination of the weld to fill the crater.

Figure 18 Multipass 3G weld sequences and work angles.

Follow these steps to practice open-root V-groove welds in the vertical position:

Step 1 Tack weld the practice coupon together as explained earlier.

Step 2 lean and feather the tack welds with the edge of a grinder wheel.

Step 3 Tack weld the weld coupon into the vertical position.

Step 4 Run the root pass using E6010 electrodes.

Step 5 Remove the electrode stub from the electrode holder and discard as soon as you stop welding to prevent arcing to nearby surfaces, and place the stub in a proper container.

Step 6 Chip, clean, and brush the weld between all passes.

Step 7 Run the fill and cover passes with E7018 electrodes to complete the weld. Clean the weld between each pass.

2.2.2 Overhead Welds (4G Position)

Overhead welds, as shown in *Figure 19*, can be made on an overhead flat surface or on an overhead open-root V-groove.

Before welding a joint in the overhead position, practice running overhead stringer beads by building a pad in the overhead position. Use E6010 electrodes. The electrode work angle should be 90 degrees to the plate with a 10- to 15-degree drag angle for the first pass. Then for the following beads, the work angle should be 10 degrees toward the previous beads. Follow these steps to practice welding beads in the overhead position:

Step 1 Tack weld the flat plate welding coupon into the overhead position.

Step 2 Run the first pass along the edge of the coupon.

Step 3 Remove the electrode stub from the electrode holder and discard as soon as you stop welding to prevent arcing to nearby surfaces, and place the stub in a proper container.

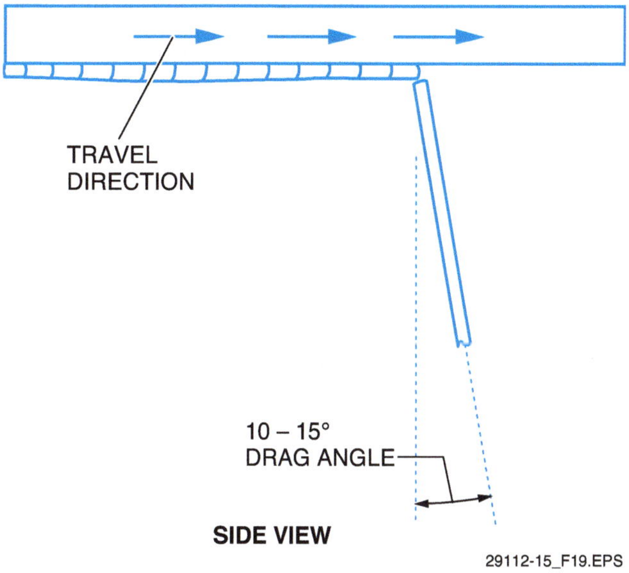

Figure 19 Building a pad in the overhead position.

Step 4 Chip, clean, and brush the weld between all passes.

Step 5 Continue running beads, cleaning the weld between passes, to complete the pad.

An overhead open-root V-groove weld is shown in *Figure 20*. Practice running overhead open-root V-groove welds using E6010 and E7018 electrodes and stringer beads. Use a 10- to 15-degree drag angle, but adjust the electrode work angle to tie-in the weld as needed. Pay particular attention at the termination of the weld to fill the crater. Follow these steps to practice welding open-root V-groove welds in the overhead position:

Step 1 Tack weld the practice coupon together as explained earlier.

Step 2 Clean and feather the tack welds with the edge of a grinder wheel.

Step 3 Tack weld the weld coupon into the overhead position.

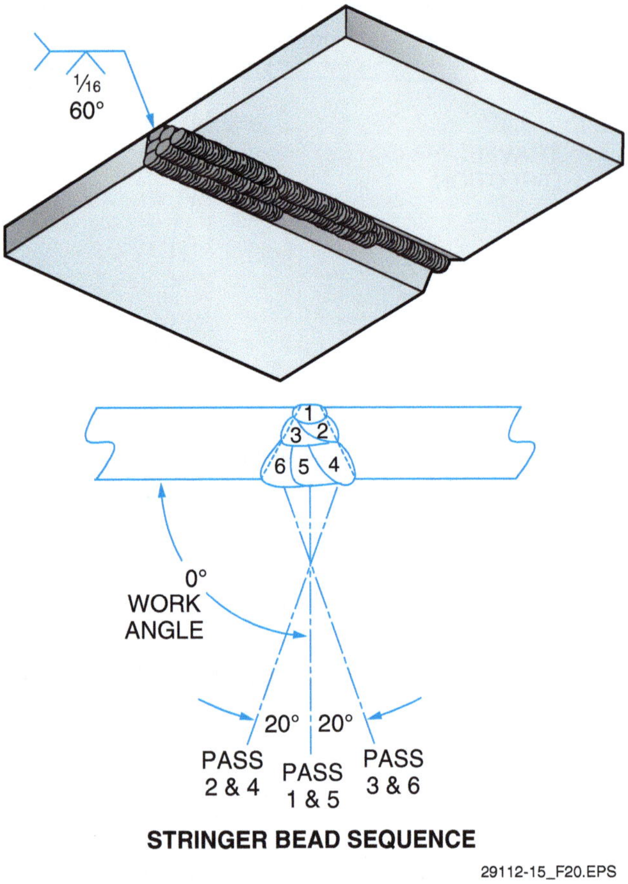

Figure 20 Multipass 4G weld sequences and work angles.

Step 4 Run the root pass using E6010 electrodes.

Step 5 Remove the electrode stub from the electrode holder and discard as soon as you stop welding to prevent arcing to nearby surfaces, and place the stub in a proper container.

Step 6 Chip, clean, and brush the weld between all passes.

Step 7 Run the fill and cover passes with E7018 electrodes to complete the weld. Clean the weld after each pass.

Hold It Open

Before beginning the root pass, tack the opposite end of the coupons (the end opposite the starting point) together with a bead about ½" (≈13 mm) long. This will prevent the root opening from closing up at the opposite end as you begin welding.

Using Your Bifocals and Trifocals

Many welders have used what are known as "cheater lenses" in welding hoods for years to help sharpen their vision when the hood is down. A cheater lens is simply a magnification lens. Although many welders wear prescription glasses with bifocal or trifocal lenses, it was impractical to use the glasses due to the design of the hood. Adjustment of the hood's position in front of the face was nonexistent, and the lens size restricted vision.

Newer welding hoods, like the Viking model shown here, allow welders to make use of their bifocal and trifocal prescription glasses. The hood distance from the front of the face can be adjusted in the headgear. This allows the hood to be tilted down farther, even to the chest. The welder can then tilt the head back and forth slightly, using the different focal lengths of the lenses, without losing sight of the task. The significantly larger lens also ensures that the task at hand remains in the welder's field of view.

Additional Resources

2014 Technical Training Guide. Current Edition. Cleveland, OH. USA: The Lincoln Electric Company. **www.lincolnelectric.com**

Welding Handbook. Current Edition. Miami, FL: The American Welding Society.

2.0.0 Section Review

1. Which of the following can be quenched without concern?
   a. Weld test coupons
   b. Functional welds
   c. Practice coupons
   d. Welds done in the 3G position

2. On an open-root V-groove weld in the 4G position, the electrode work angle should be _____.
   a. 30 degrees
   b. 45 degrees
   c. 90 degrees
   d. varied as needed to tie in

Summary

The ability to make open-root V-groove welds on plate is an essential welding skill. A welder must possess this skill to perform welding jobs and to progress to more difficult pipe welding procedures. Making open-root V-groove welds requires the welder to properly and safely set up and prepare the welding equipment and the coupons. V-groove welds can be made in the flat, horizontal, vertical, and overhead positions. Practice these welds until you can consistently produce acceptable welds in all positions.

Review Questions

1. Which of the following statements is true?
   a. Groove weld styles include single-, double-, and quadruple-groove styles.
   b. Groove welds can be combined with fillet welds to maintain strength.
   c. The root opening dimension of a groove weld is determined by the specified work angle.
   d. Weld symbols on drawings always show both metric and imperial dimensions.

2. Open-root weld coupons generally have a bevel angle of _____.
   a. 10 degrees
   b. 15 degrees
   c. 22½ degrees
   d. 30 degrees

3. Welding codes typically allow the root face on open-root groove welds to be within a range of _____.
   a. $\frac{1}{32}$" to $\frac{5}{32}$" (0.8 to 4.0 mm)
   b. $\frac{1}{32}$" to $\frac{3}{16}$" (0.8 to 4.8 mm)
   c. $\frac{1}{16}$" to ¼" (1.6 to 6.4 mm)
   d. 0" to ⅛" (0 to 3.2 mm)

4. When preparing to tack weld an open-root V-groove weld coupon, align the beveled strips on a flat surface with a _____.
   a. 0" to ⅛" (0 to 3.2 mm) root opening
   b. $\frac{3}{32}$" to $\frac{5}{32}$" (2.4 to 4.0 mm) root opening
   c. $\frac{1}{16}$" to ¼" (1.6 to 6.4 mm) root opening
   d. $\frac{1}{32}$" to $\frac{1}{16}$" (0.8 to 1.6 mm) root opening

5. When making an open-root V-groove weld, the hole made when the two root faces of two plate edges are melted away by the arc is called a(n) _____.
   a. bead
   b. overlap
   c. keyhole
   d. root pass

6. When the plates are vertical and the axis of the weld is also vertical, the code for this position of groove weld is _____.
   a. 1G
   b. 2G
   c. 3G
   d. 4G

7. When making weave beads on flat open-root V-groove welds, keep the electrode work angle at 90 degrees to the plate surface with a _____.
   a. 5- to 10-degree drag angle
   b. 10- to 15-degree drag angle
   c. 15- to 20-degree drag angle
   d. 20- to 22-degree drag angle

8. To control the weld puddle when welding an open-root V-groove in the horizontal position, the electrode work angle can be _____.
   a. dropped up to 10 degrees
   b. raised up to 15 degrees
   c. raised up to 20 degrees
   d. dropped up to 30 degrees

9. When practicing horizontal open-root V-groove welds, pay particular attention at the termination of the weld to fill the _____.
   a. pad
   b. root
   c. crater
   d. puddle

10. For better puddle control when welding open-root V-groove welds in the 3G position, _____.
    a. set the amperage to the lower part of the allowable range
    b. change to a downward direction of travel
    c. increase the work angle to 25 degrees
    d. change from a push to a drag angle

Trade Terms Quiz

Fill in the blank with the correct term that you learned from your study of this module.

1. The portion of the edge of a part to be joined by a groove weld that has not been beveled or grooved; previously referred to as the land is a _______.

2. Grinding a tack weld to a tapered edge is called a _______.

Trade Terms

Feather
Root face

Appendix

Performance Accreditation Tasks

The American Welding Society (AWS) School Excelling through National Skills Standards Education (SENSE) program is a comprehensive set of minimum Standards and Guidelines for Welding Education programs. The following Performance Accreditation Tasks (PATs) are aligned with and designed around the SENSE program. The PATs correspond to and support the learning objectives in *AWS EG2.0, Guide for the Training and Qualification of Welding Personnel: Entry-Level Welder*.

Note that in order to satisfy all learning objectives in *AWS EG2.0*, the instructor must also use the PATs contained in the second level of the NCCER Welding curriculum.

PATs 1 through 4 correspond to *AWS EG2.0, Module 4 – Shielded Metal Arc Welding*, Key Indicators: 3, 4, and 6

PATs provide specific acceptable criteria for performance and help to ensure a true competency-based welding program for students.

The following tasks are designed to evaluate your ability to run open-root V-groove welds with SMAW equipment in all positions using E6010 and E7018 electrodes. Perform each task when you are instructed to do so by your instructor. As you complete each task, take it to your instructor for evaluation. Do not proceed to the next task until instructed to do so.

OPEN V-GROOVE WITH E6010 AND E7018 ELECTRODES IN THE FLAT POSITION

Using 3⁄32" or 1⁄8" (2.4 or 3.2 mm) E6010 electrodes for the root pass and 3⁄32" or 1⁄8" (2.4 or 3.2 mm) E7018 electrodes for the fill and cover passes, make an open-root V-groove weld on carbon steel plate in the flat position as indicated.

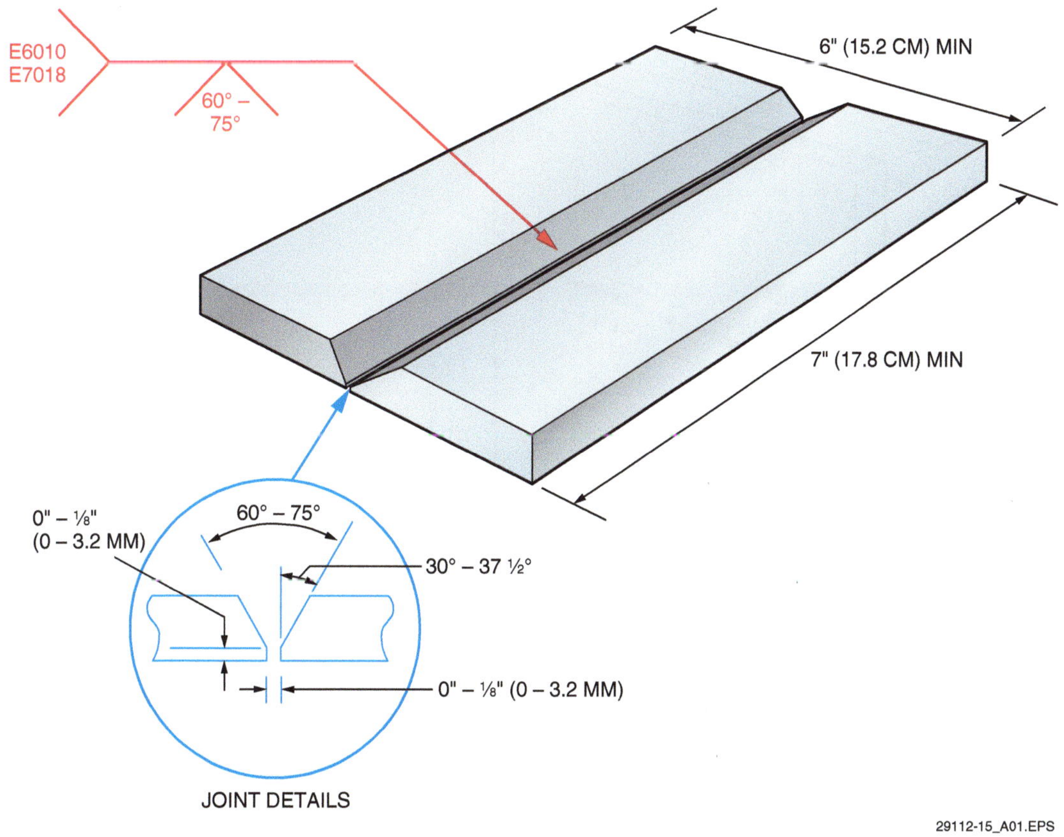

Criteria for Acceptance:

- No arc strikes outside the weld area ____________
- Uniform rippled appearance on the bead face ____________
- Craters and restarts filled to the full cross section of the weld ____________
- Uniform weld size ____________
- Acceptable weld profile in accordance with the *ASME Boiler and Pressure Vessel Code* ____________
- Smooth transition with complete fusion at the toes of the weld ____________
- Complete uniform root penetration at least flush with the base metal to a maximum buildup of 1⁄8" (3.2 mm) ____________
- No porosity ____________
- No excessive undercut ____________
- No inclusions ____________
- No cracks ____________
- Acceptable guided bend test results (instructor option) ____________

OPEN V-GROOVE WITH E6010 AND E7018 ELECTRODES IN THE HORIZONTAL POSITION

Using 3⁄32" or 1⁄8" (2.4 or 3.2 mm) E6010 electrodes for the root pass and 3⁄32" or 1⁄8" (2.4 or 3.2 mm) E7018 electrodes for the fill and cover passes, make an open-root V-groove weld on carbon steel plate in the horizontal position as shown.

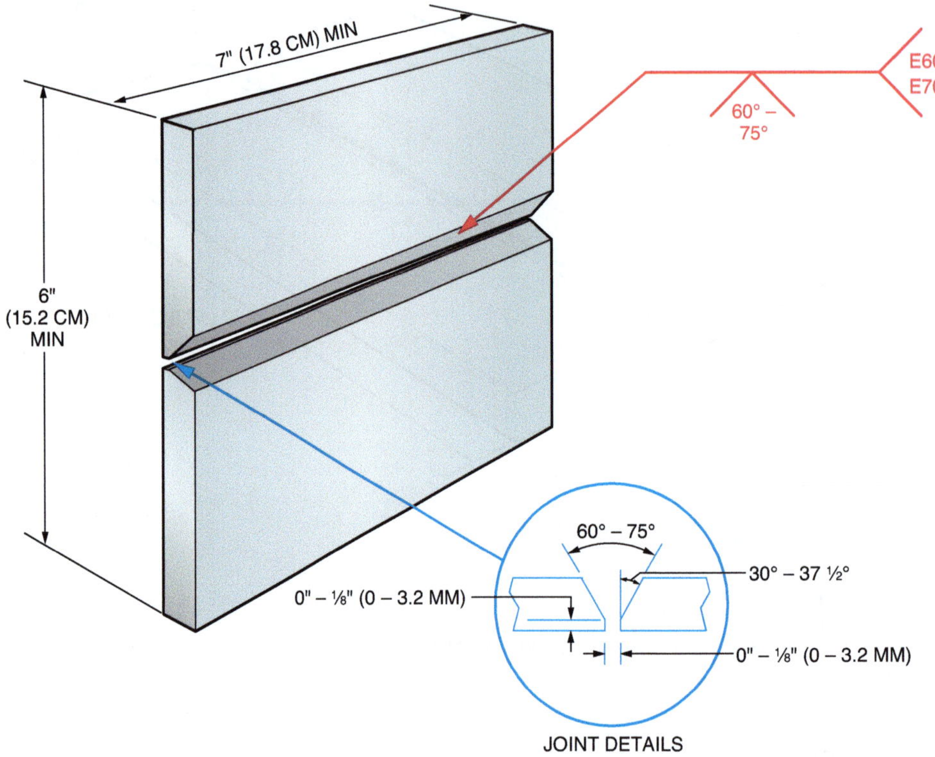

29112-15_A02.EPS

Criteria for Acceptance:

- No arc strikes outside the weld area ______
- Uniform rippled appearance on the bead face ______
- Craters and restarts filled to the full cross section of the weld ______
- Uniform weld size ______
- Acceptable weld profile in accordance with the *ASME Boiler and Pressure Vessel Code* ______
- Complete uniform root penetration at least flush with the base metal to a maximum buildup of 1⁄8" (3.2 mm) ______
- Smooth transition with complete fusion at the toes of the weld ______
- No porosity ______
- No excessive undercut ______
- No inclusions ______
- No cracks ______
- Acceptable guided bend test results (instructor option) ______

OPEN V-GROOVE WITH E6010 AND E7018 ELECTRODES IN THE VERTICAL POSITION

Using 3⁄32" to 1⁄8" (2.4 or 3.2 mm) E6010 electrodes for the root pass and 3⁄32" or 1⁄8" (2.4 or 3.2 mm) E7018 electrodes for the fill and cover passes, make an open-root V-groove weld on carbon steel plate in the vertical position as shown.

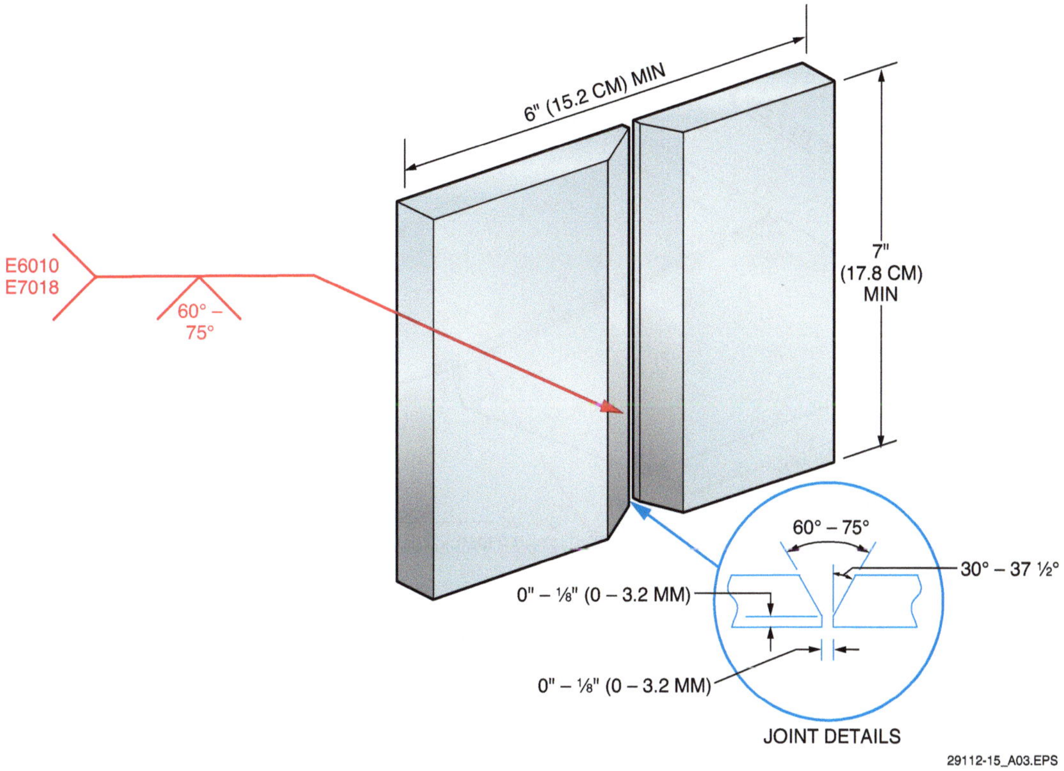

Criteria for Acceptance:

- No arc strikes outside the weld area __________
- Uniform rippled appearance on the bead face __________
- Craters and restarts filled to the full cross section of the weld __________
- Uniform weld size __________
- Acceptable weld profile in accordance with the *ASME Boiler and Pressure Vessel Code* __________
- Complete uniform root penetration at least flush with the base metal to a maximum buildup of 1⁄8" (3.2 mm) __________
- Smooth transition with complete fusion at the toes of the weld __________
- No porosity __________
- No excessive undercut __________
- No inclusions __________
- No cracks __________
- Acceptable guided bend test results (instructor option) __________

OPEN V-GROOVE WITH E6010 AND E7018 ELECTRODES IN THE OVERHEAD POSITION

Using ³⁄₃₂" to ⅛" (2.4 or 3.2 mm) E6010 electrodes for the root pass and ³⁄₃₂" or ⅛" (2.4 or 3.2 mm) E7018 electrodes for the fill and cover passes, make an open-root V-groove weld on carbon steel plate in the overhead position as indicated.

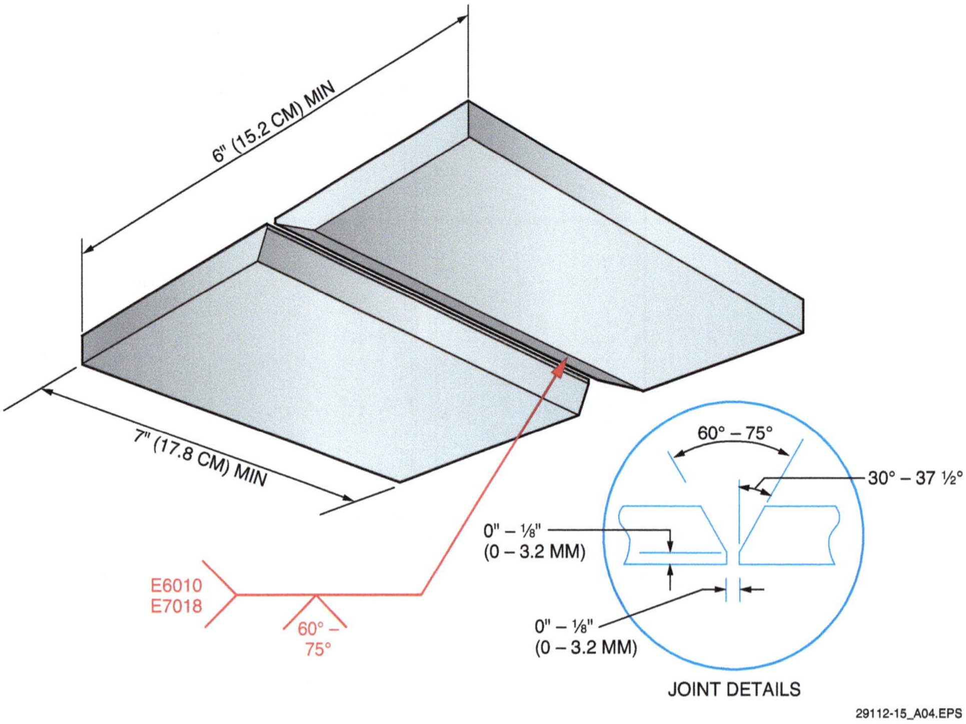

Criteria for Acceptance:

- No arc strikes outside the weld area ____________
- Uniform rippled appearance on the bead face ____________
- Craters and restarts filled to the full cross section of the weld ____________
- Uniform weld size ____________
- Acceptable weld profile in accordance with the *ASME Boiler and Pressure Vessel Code* ____________
- Complete uniform root penetration at least flush with the base metal to a maximum buildup of ⅛" (3.2 mm) ____________
- Smooth transition with complete fusion at the toes of the weld ____________
- No porosity ____________
- No excessive undercut ____________
- No inclusions ____________
- No cracks ____________
- Acceptable guided bend test results (instructor option) ____________

Trade Terms Introduced in This Module

Feather: The process of grinding a tack weld to a tapered edge.

Root face: The portion of the edge of a part to be joined by a groove weld that has not been beveled or grooved; previously referred to as the land.

Additional Resources

This module presents thorough resources for task training. The following resource material is suggested for further study.

2014 Technical Training Guide. Current Edition. Cleveland, OH. USA: The Lincoln Electric Company. **www.lincolnelectric.com**

Welding Handbook. Current Edition. Miami, FL: The American Welding Society.

Figure Credits

The Lincoln Electric Company, Cleveland, OH, USA, Module Opener, Figures 6, 7, 10, SA01

Topaz Publications, Inc., Figures 16 (photo), 18 (photo)

Section Review Answer Key

Answer	Section Reference	Objective
Section One		
1. a	1.1.3	1a
2. d	1.2.1	1b
Section Two		
1. c	2.1.1	2a
2. d	2.2.2	2b

NCCER CURRICULA — USER UPDATE

NCCER makes every effort to keep its textbooks up-to-date and free of technical errors. We appreciate your help in this process. If you find an error, a typographical mistake, or an inaccuracy in NCCER's curricula, please fill out this form (or a photocopy), or complete the online form at **www.nccer.org/olf**. Be sure to include the exact module ID number, page number, a detailed description, and your recommended correction. Your input will be brought to the attention of the Authoring Team. Thank you for your assistance.

Instructors – If you have an idea for improving this textbook, or have found that additional materials were necessary to teach this module effectively, please let us know so that we may present your suggestions to the Authoring Team.

NCCER Product Development and Revision

13614 Progress Blvd., Alachua, FL 32615

Email: curriculum@nccer.org
Online: www.nccer.org/olf

❑ Trainee Guide ❑ Lesson Plans ❑ Exam ❑ PowerPoints Other ______________

Craft / Level: Copyright Date:

Module ID Number / Title:

Section Number(s):

Description:

Recommended Correction:

Your Name:

Address:

Email: Phone:

GMAW and FCAW – Equipment and Filler Metals

Overview

Gas metal arc welding (GMAW) and flux-cored arc welding (FCAW) are two common welding processes. This module describes GMAW and FCAW processes and basic safety practices that are required for each type. It also provides an overview of GMAW and FCAW equipment and how it is set up for use. Various types of filler metals used in GMAW and FCAW are also examined.

Module 29205

Trainees with successful module completions may be eligible for credentialing through the NCCER Registry. To learn more, go to **www.nccer.org** or contact us at **1.888.622.3720**. Our website has information on the latest product releases and training, as well as online versions of our *Cornerstone* magazine and Pearson's product catalog.

Your feedback is welcome. You may email your comments to **curriculum@nccer.org**, send general comments and inquiries to **info@nccer.org**, or fill in the User Update form at the back of this module.

This information is general in nature and intended for training purposes only. Actual performance of activities described in this manual requires compliance with all applicable operating, service, maintenance, and safety procedures under the direction of qualified personnel. References in this manual to patented or proprietary devices do not constitute a recommendation of their use.

29205 V5

29205
GMAW and FCAW – Equipment and Filler Metals

Objectives

When you have completed this module, you will be able to do the following:

1. Describe basic GMAW/FCAW processes and related safety practices.
   a. Describe basic GMAW/FCAW processes.
   b. Identify GMAW/FCAW-related safety practices.
   c. Describe the various GMAW metal transfer modes.
   d. Describe the FCAW metal transfer process.
2. Describe GMAW and FCAW equipment and explain how to prepare for welding.
   a. Identify common GMAW/FCAW welding equipment.
   b. Describe power source control considerations.
   c. Identify and describe welding cables and terminations.
   d. Identify and describe external wire feeders and their controls.
   e. Identify and describe GMAW and FCAW guns, contact tips, and nozzles.
   f. Identify various shielding gases and their related equipment.
   g. Explain how to set up welding equipment for GMAW and FCAW welding.
3. Identify various GMAW and FCAW filler metals.
   a. Identify various GMAW filler metals.
   b. Identify various FCAW filler metals.

Performance Task

Under the supervision of your instructor, you should be able to do the following:

1. Set up GMAW and FCAW equipment with appropriate shielding gases and filler metals.

Trade Terms

Axially
Composite metal-core wire electrode
Globular transfer
Inductance
Pinch effect
Pulsed transfer
Short-circuiting transfer
Spray transfer
Stickout

Industry Recognized Credentials

If you are training through an NCCER-accredited sponsor, you may be eligible for credentials from NCCER's Registry. The ID number for this module is 29205. Note that this module may have been used in other NCCER curricula and may apply to other level completions. Contact NCCER's Registry at 888.622.3720 or go to **www.nccer.org** for more information.

Contents

Topics to be presented in this module include:

Figures and Tables

SECTION ONE

1.0.0 GMAW AND FCAW WELDING

Objective

Describe basic GMAW/FCAW processes and related safety practices.

a. Describe basic GMAW/FCAW processes.
b. Identify GMAW/FCAW-related safety practices.
c. Describe the various GMAW metal transfer modes.
d. Describe the FCAW metal transfer process.

Performance Task

1. Set up GMAW and FCAW equipment with appropriate shielding gases and filler metals.

Trade Terms

Axially: In a straight line along an axis.

Composite metal-cored wire electrode: A filler wire electrode with a hollow core containing powdered materials, primarily metals.

Globular transfer: A GMAW process in which the electrode wire does not touch the base metal, and the filler metal is transferred randomly in the arc as large, irregular globules.

Pulsed transfer: A GMAW welding process in which the welding power is cycled from a low level (background level) to a higher level (welding level), resulting in lower average voltage and current; this is a variation of the spray transfer process.

Short-circuiting transfer: A GMAW welding process in which the electrode wire shorts against the base metal. The welding current repeatedly and rapidly melts the wire and establishes an arc, which is extinguished when the wire again touches the base metal.

Spray transfer: A GMAW welding process in which the electrode wire does not touch the base metal, and the filler metal is transferred axially (in a straight line) in the arc as fine droplets.

In order to perform gas metal arc welding (GMAW) and flux-cored arc welding (FCAW) safely and effectively, a good understanding of how both processes work is required. It is also critical to understand the safety practices that are necessary for all welding operations.

1.1.0 GMAW and FCAW Processes

Gas metal arc welding (GMAW), shown in *Figure 1*, uses an electric arc to melt a consumable wire electrode and fuse it with the base metal. The arc generates an intense heat of 6,000°F to 10,000°F (3,316°C to 5,538°C). The arc is stabilized and the molten filler and base metals are protected from oxidation by a shielding gas dispensed from a GMAW gun nozzle. The wire is automatically fed into the weld puddle as it melts. The process usually operates with direct current electrode positive (DCEP).

Flux-cored arc welding (FCAW), shown in *Figure 2*, uses an electric arc to melt a flux-cored wire electrode and fuse it with the base metal to form a weld. The wire is tubular, allowing a flux material to be inserted inside the electrode. A wire feeder automatically feeds the flux-cored wire as it is being consumed in the arc. The equipment used for FCAW is basically the same as that used for GMAW. The main difference between the two processes is that GMAW uses solid or **composite metal-cored wire electrodes** with a shielding gas, while FCAW uses a tubular flux-cored wire electrode with or without a shielding gas.

In the two basic FCAW processes, one uses an external shielding gas while the other is self-shielding. Whether a shielding gas is used is determined by the type of flux-cored wire that is used. When the flux-cored wire does not require an external shielding gas, the process is called self-shielding FCAW, or FCAW-S. When an external shielding gas is required, the process is called gas-shielded FCAW, or FCAW-G. Equipment specifically designed for self-shielding FCAW does not have provisions for shielding gas (gas

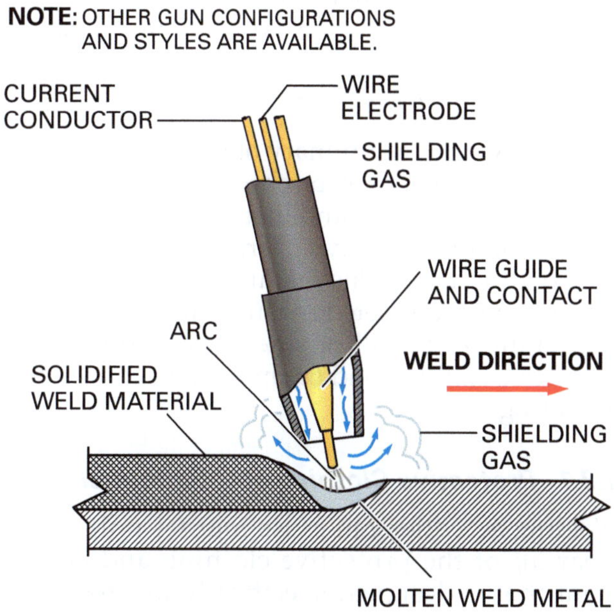

Figure 1 Basic GMAW process.

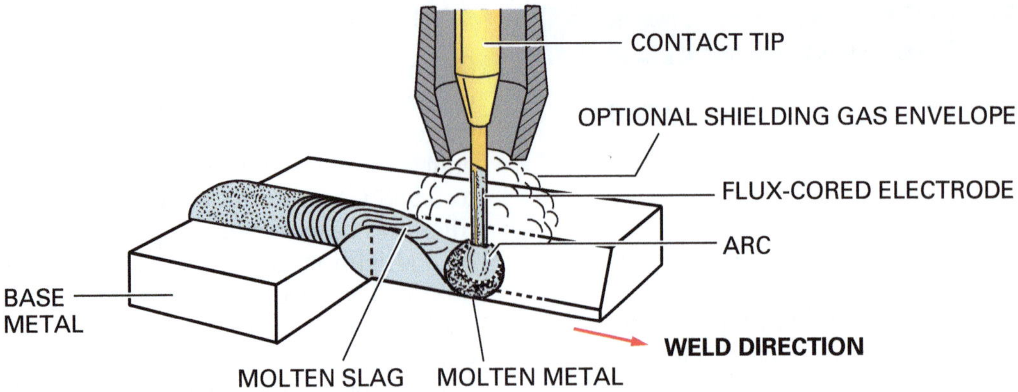

Figure 2 Basic FCAW process.

connections, solenoids, and gas preflow and postflow timers), and the guns do not have gas nozzles. Equipment used for gas-shielded FCAW welding is the same as that used for GMAW except that the wire feed drive assemblies have to be able to handle the cored wire, which is softer and generally larger in diameter than solid wire.

> **CAUTION**
>
> A shielding gas should not be applied to a self-shielding, flux-cored electrode. Doing so will destroy the mechanical properties of the weld. Even though the weld may actually run smoother, it will not be a good weld.

FCAW is presently limited to the welding of ferrous metals. These include low- and medium-carbon steels, some low-alloy steels, cast irons, and some stainless steels. When critical welds are required, the gas-shielded FCAW-G process is often used because the combination of flux and separate shielding gas produces welds of very high quality.

1.2.0 Safety Summary

The following is a summary of safety procedures and practices that must be observed while cutting or welding. Keep in mind that this is only a summary. Complete safety coverage is provided in the Welding Level One module, *Welding Safety*. If you have not completed that module, do so before continuing. Above all, be sure to wear appropriate protective clothing and equipment when welding or cutting.

1.2.1 Protective Clothing and Equipment

Welding activities can cause injuries unless you wear all of the protective clothing and equipment that is designed specifically for the welding industry. The following information includes important safety guidelines about protective clothing and equipment:

- Always use safety glasses with a full face shield or a welding helmet. The glasses, face shield, or helmet lens must have the proper light-reducing shade for the type of welding or cutting to be performed. Never directly or indirectly view an electric arc without using a properly tinted lens.
- Wear proper protective leather and/or flame-retardant clothing along with welding gloves that will protect you from flying sparks, molten metal, and heat (*Figure 3*).

Figure 3 A fully protected welder wearing leathers, gloves, and hood.

- Wear high-top safety shoes or boots. Make sure that the tongue and lace area of the footwear will be covered by a pant leg. If the tongue and lace area is exposed or the footwear must be protected from burn marks, wear leather spats under the pants or chaps over the front and top of the footwear.
- Wear a solid material hat, preferably 100-percent cotton, with or without a bill. Avoid hats with a fabric-covered metal button on the top. The bill can be pointed to the rear or over one shoulder to help keep sparks from going down inside the shirt when overhead cutting or welding is required. If overhead welding or welding in awkward positions is required, a full leather hood with a welding face plate and the correctly tinted lens is recommended. If a hard hat is required, use one that allows the attachment of rear deflector material and a face shield.
- Wear earmuffs or at least earplugs to protect ear canals from sparks.

1.2.2 Fire/Explosion Prevention

Welding activities usually involve the use of fire or extreme heat to melt metal. Whenever fire is used, it must be controlled and contained. Welding or cutting activities are often performed on vessels that may once have contained flammable or explosive materials. Residues from those materials can catch fire or explode when a welder begins work on such a vessel. The following are fire and explosion prevention guidelines associated with welding:

- Never carry matches or gas-filled lighters in your pockets. Sparks can cause the matches to ignite or the lighter to explode, causing serious injury.
- Never use oxygen to blow dust or dirt from clothing. The oxygen can remain trapped in the fabric for some time. If a spark hits the clothing during this time, the clothing can burn rapidly and violently out of control.
- Make sure that any flammable material in the work area is moved or shielded by a fire-resistant covering. Approved fire extinguishers must be available before attempting any heating, welding, or cutting operations. If a hot-work permit and a fire watch are required, be sure those items are in place before beginning and ensure that all site requirements have been observed.
- Never release a large amount of oxygen or use oxygen in place of compressed air. Its presence around flammable materials or sparks can cause rapid and uncontrolled combustion. Keep oxygen away from oil, grease, and other petroleum products.
- Never release a large amount of fuel gas, especially acetylene. Methane and propane tend to concentrate in and along low areas and can ignite at a considerable distance from the release point. Acetylene is lighter than air, but is even more dangerous than propane. When mixed with air or oxygen, acetylene will explode at much lower concentrations than any other fuel.
- To prevent fires, maintain a neat and clean work area, and make sure that any metal scrap or slag is cold before disposal.
- Before cutting containers such as tanks or barrels, check to see if they have contained any explosive, hazardous, or flammable materials, including petroleum products, citrus products, or chemicals that decompose into toxic fumes when heated. As a standard practice, always clean and then fill any tanks or barrels with water, or purge them with a flow of inert gas such as nitrogen to displace any oxygen. Containers must be cleaned by steam cleaning, flushing with water, or washing with detergent until all traces of the material have been removed.

WARNING!

Welding or cutting must never be performed on drums, barrels, tanks, vessels, or other containers until they have been emptied and cleaned thoroughly, eliminating all flammable materials and all substances (such as detergents, solvents, greases, tars, or acids) that might produce flammable, toxic, or explosive vapors when heated. Clean containers only in well-ventilated areas. Vapors can accumulate during cleaning, causing explosions or injury.

Proper procedures for cutting or welding hazardous containers are described in the *American Welding Society (AWS) F4.1, Safe Practices for the Preparation of Containers and Piping for Welding and Cutting.*

1.2.3 Work Area Ventilation

Vapors and fumes tend to rise in the air from their sources. Welders often have to work above the welding area where the fumes are being created. Welding fumes can cause personal injuries. Good work area ventilation helps to remove the vapors and to protect the welder. The following is a list of work area ventilation guidelines to consider before and during welding activities:

- Make sure confined space procedures are followed before conducting any welding or cutting in the confined space.

- Never use oxygen for ventilation in confined spaces.
- Always perform cutting or welding operations in a well-ventilated area. Cutting or welding operations involving zinc or cadmium materials or coatings result in toxic fumes. For long-term cutting or welding of such materials, always wear an approved full-face, supplied-air respirator that uses breathing air supplied from outside the work area. For occasional, very short-term exposure, a high-efficiency, particulate arresting (HEPA)-rated or metal-fume filter may be used on a standard respirator.
- Make sure that confined spaces are properly ventilated for cutting or welding purposes. Use powered extraction systems when available. *Figure 4* shows a welder working with a portable extraction (vacuum) system designed to remove and filter welding fumes and vapors.

1.3.0 GMAW Metal Transfer Modes

GMAW can use any of several methods of metal transfer to conduct the filler metal to the molten weld pool. The principal modes include the following:

- Spray transfer
- Globular transfer
- Short-circuiting transfer
- Pulsed transfer

Figure 4 Welder working with a portable extraction system.

1.3.1 Spray Transfer

In the spray transfer mode (*Figure 5*), the welding voltage is set in the higher ranges to keep the end of the wire electrode away from the work and to maintain a continuous arc. The high heat of the arc melts the end of the wire electrode, and fine droplets are carried by the arc axially to the molten weld puddle on the base metal. The spray transfer deposits filler metal at a high rate and produces very little spatter. However, the weld puddle is relatively large, and the mode works best in the flat or horizontal fillet positions. Also, only certain shielding gases or gas combinations can be used with spray transfer. Shielding gases are covered later in this module.

1.3.2 Globular Transfer

In the globular transfer mode (*Figure 6*), the welding voltage is set lower than it is for spray transfer. The end of the wire electrode remains a short distance from the work to maintain a continuous arc. Because of its weight, the electrode wire tip periodically melts off in a large glob, which the

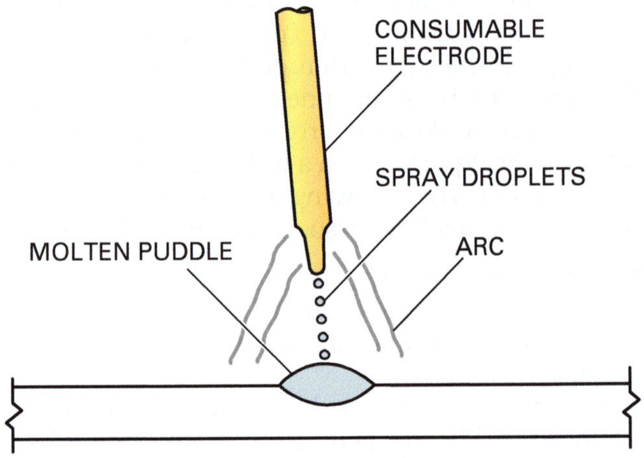

Figure 5 Spray transfer.

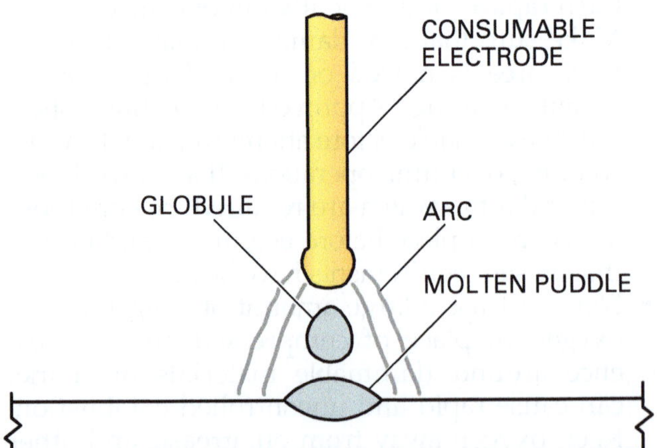

Figure 6 Globular transfer.

arc then carries to the workpiece, where it mixes with the weld puddle. Globular transfer can be used to deposit filler metal at high rates, and it is capable of producing high quality welds in all positions. However, bead appearance can be rough, and spatter may be heavy in some applications.

1.3.3 Short-Circuiting Transfer

Short-circuiting transfer (GMAW-S) uses the lowest voltage settings of the three modes (*Figure 7*). It is the best mode for welding light-gauge sheet metal. The wire electrode is fed against the workpiece. When contact is made, current flow heats the wire electrode until it melts enough for a droplet to separate and mix with the weld puddle. When the droplet separates, a gap is formed, and an arc is initiated. The continuously feeding wire closes the gap and contacts the workpiece, shorting out the arc. The current again heats the wire until another droplet separates, and the arc is again established. This process repeats many times each second. The short-circuiting process can be used to weld thin materials, and it is suitable for all positions. However, it does not penetrate deeply and may produce incomplete fusion on thicker materials because of the low heat input.

A number of manufacturers have introduced proprietary versions of short-circuiting transfer that rely on intelligent inverter power sources. These adjust the voltage and current moment-by-moment throughout each stage of the short-circuit transfer process so the droplets form and then transfer to the weld puddle optimally. The result is minimal spatter, excellent fusion, and higher-quality welds. These new technologies are especially popular in settings involving stainless steel pipe welds.

Adaptive Control Welding

Some newer inverter power sources are capable of using adaptive control welding during pulse transfer. These units sense and alter the frequency of the pulse, its width, the peak current, and wire speed automatically to maintain optimum welding characteristics during welding.

1.3.4 Pulsed Spray Transfer

Pulsed-transfer (GMAW-P) welding is a variation of spray transfer welding (*Figure 8*). The process transfers molten metal in droplets while maintaining an arc with the base metal. The pulsed GMAW process also allows the transfer action to occur while lower current levels are used. The advantages of the lower current levels are less spatter and better penetration without melting through the base metal. Pulsed transfer is especially good for aluminum welding because it results in a smaller heat-affected zone (HAZ).

In pulsed-transfer welding, the welding power is electronically controlled by a special power source that pulses the output power to the consumable electrode. Although different GMAW machines may have different types of power sources, the results are the same for each. The welding current is generated in pulses of high current at specified intervals of time. The frequency of these pulses varies from one power source to the next, but some power sources are capable of producing pulses over a wide time range. During the period of time between each pulse, the welding current is reduced, but it is still maintained at a high enough level to keep the

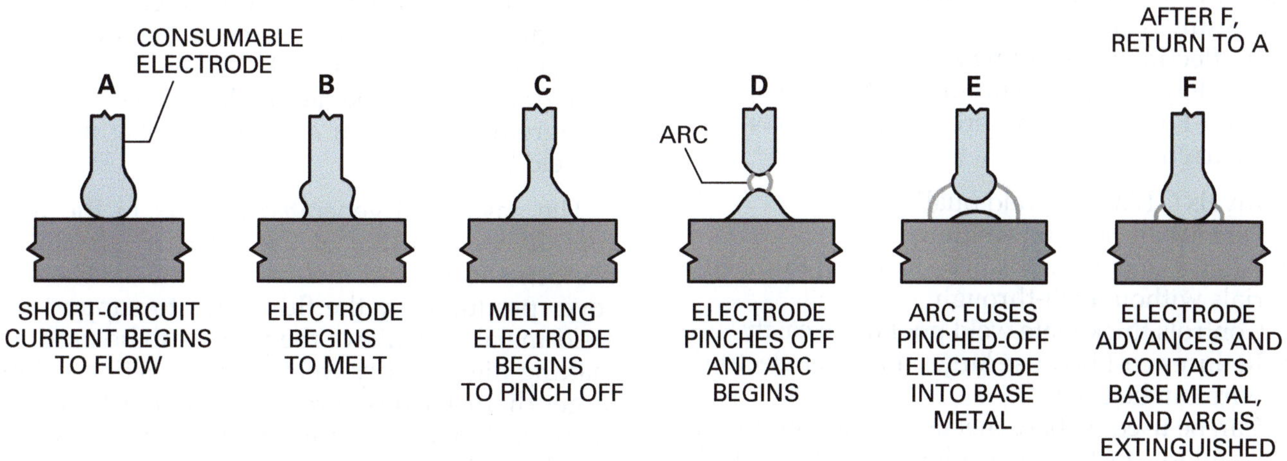

Figure 7 Short-circuiting transfer.

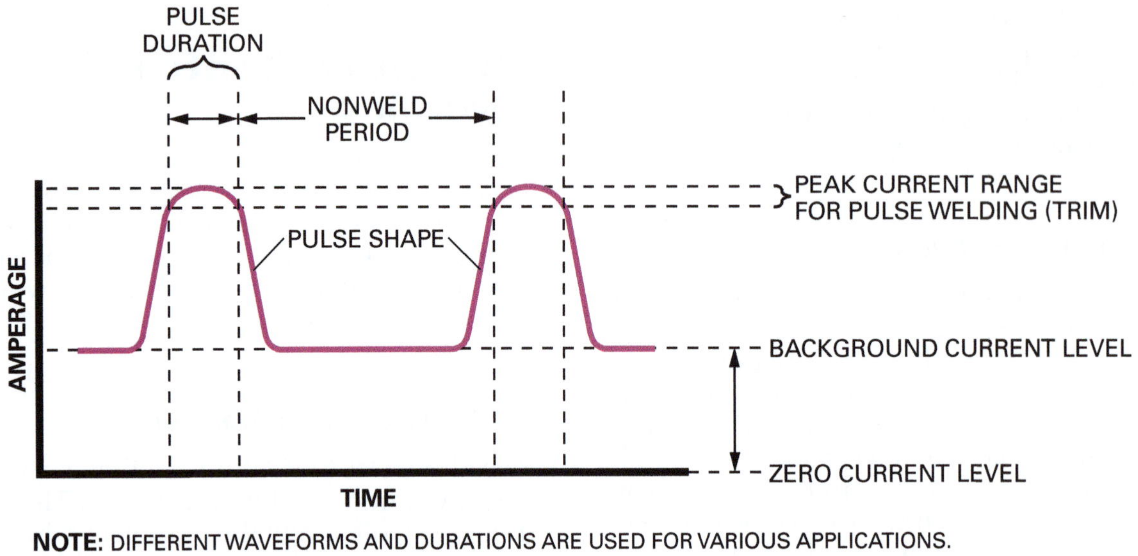

Figure 8 Output pulses of high welding current occurring over time.

wire heated and an arc established. This lowered current level is called the background current.

After an arc has been established and maintained between the consumable electrode and the base metal, the welding current to the electrode is momentarily increased to form a pulse, and then it returns to its lower background level. Each pulse of welding current must peak strongly enough and have enough of a duration to make a single drop of molten metal separate from the wire that is maintaining an arc with the base metal.

Figure 9 shows the GMAW-P process relative to one of the pulse shapes. Points A through E on the diagram correspond to the electrode arc illustrations, labeled A through E, above the diagram.

When the single drop of molten metal separates from the electrode, it is transferred across the arc to form the molten puddle on the base metal. This transfer process creates less spatter and reduces the chance of melt-through. When these pulses are repeated many times per second, the result is a high-quality weld with minimal heat affecting the welding zone.

The following is a summary of advantages of GMAW-P over welding processes using continuous currents:

- Rivals GTAW for weld quality
- Permits all-position welding
- Allows smooth welds to be made on thin materials without melt-through
- Controls thick plate weld pool more easily
- Reduces cost because larger diameter electrode wire can be used
- Can be used on base metals such as aluminum, stainless steel, copper alloy, and nickel steels, as well as carbon steels
- Reduces weld contamination from surface oxides because less surface area is exposed during the weld
- Creates a smaller heat-affected zone (HAZ), which favors its use with aluminum

1.4.0 FCAW Metal Transfer Process

The FCAW process (*Figure 10*) uses an electric arc to melt the filler and base metals and to transfer fine droplets of filler metal to the weld puddle. Globular transfer is commonly used in FCAW processes, but axial spray can be achieved if the correct electrode and shielding gas are used.

The electromagnetic force that propels the metal spray is strong, but the process produces very little spatter. The flux inside the core contains various ingredients that perform the following functions:

- Ionizers to stabilize the arc
- Deoxidizers to purge the weld of gases
- Additional metals and elements to enhance the quality of the weld metal
- Ingredients to generate shielding gas
- Ingredients to form a protective slag over the weld bead

The gas shield generated by flux in the wire core, as well as that of the external gas shield, protects the arc and molten weld metal. The slag cover produced by the flux helps to support the weld metal and allows the process to be used for making high-quality welds in all positions. With larger diameter flux-cored wires, deposition rates of about 25 pounds per hour (≈11 kg per hour) can be achieved.

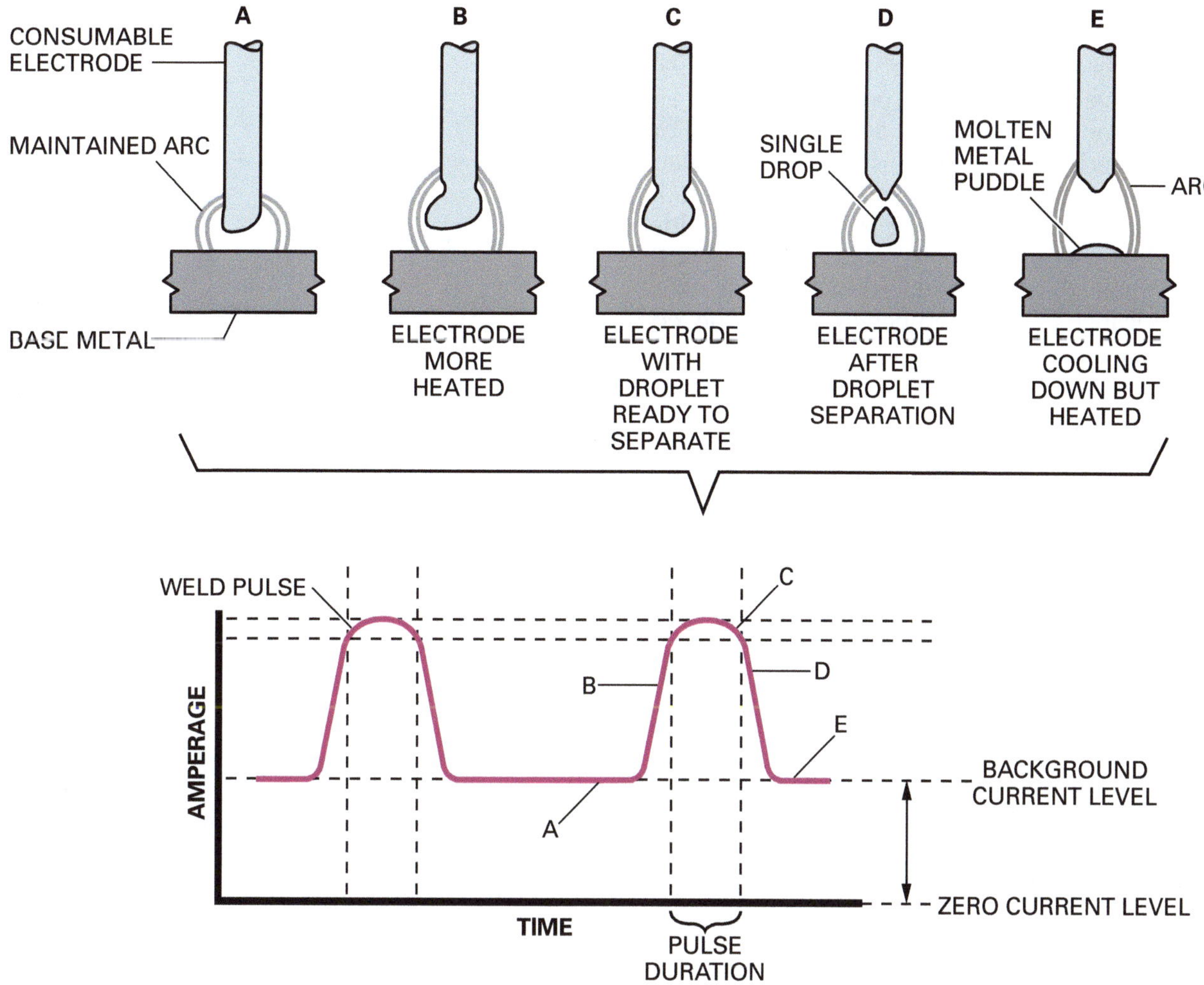

Figure 9 Typical pulsed welding process associated with one pulse shape.

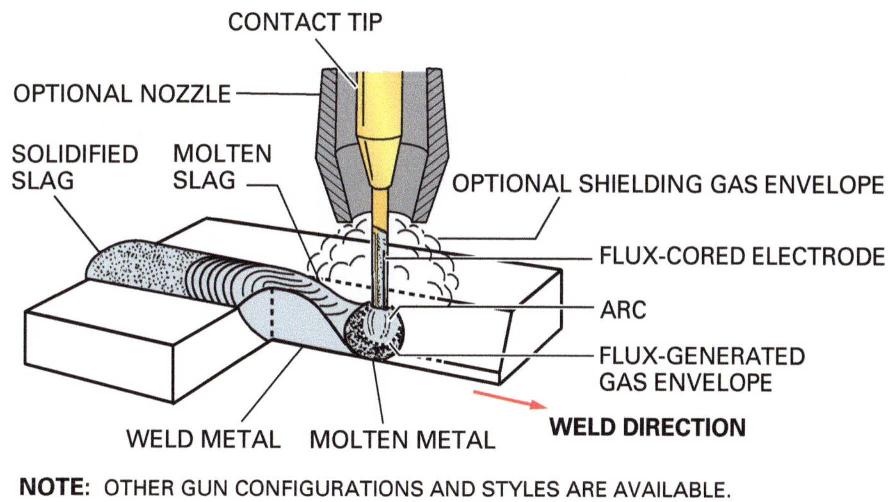

Figure 10 FCAW process.

1.4.1 *FCAW Weld Penetration*

FCAW welds made with carbon dioxide (CO_2) shielding have a much deeper penetration than either SMAW or FCAW without a shielding gas. For this reason, edge preparation is often not needed for double-welded butt joints on plate up to ⅜-inch (≈10 mm) thick. When making FCAW fillet welds using CO_2 shielding, a smaller size fillet weld will have as much strength as a larger fillet made with SMAW because of the deep penetration achieved with FCAW and the CO_2 shielding.

1.4.2 *FCAW Joint Design*

FCAW uses the same basic joint designs (*Figure 11*) as SMAW. However, when making groove welds with FCAW using a CO_2 shielding gas, the square groove (square butt) joint can be used for plate thicknesses up to ⅜ of an inch (10 mm). Above this thickness, beveled joints should be used.

Because FCAW electrode wire is much thinner than comparable SMAW electrodes and FCAW penetration with CO_2 is superior to SMAW, the included angles of V-joints can be reduced to half the angle required for SMAW. For example, an included angle of 60 degrees can be reduced to 30 degrees when using FCAW with CO_2 shielding. This can save approximately 50 percent on filler metal and a considerable amount of welding time when compared with SMAW.

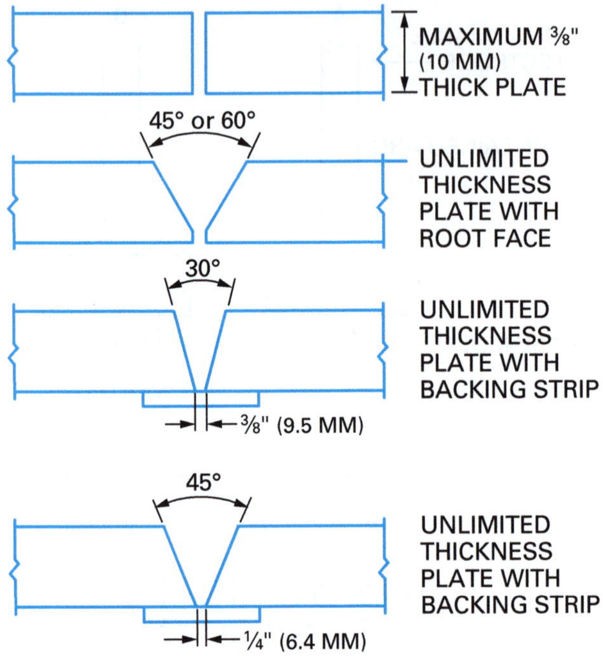

Figure 11 Joint designs for FCAW with CO_2 shielding.

CAUTION

Always refer to the Welding Procedure Specification (WPS) or site quality standards for specific information on joint requirements. Information in this module is provided as a general guideline only.

Additional Resources

AWS B1.10M/B1.10 Guide for the Nondestructive Examination of Welds. Miami, FL: American Welding Society.

AWS B1.11 Guide for the Visual Examination of Welds. Miami, FL: American Welding Society.

Modern Welding Technology. Howard B. Cary. Englewood Cliffs, NJ: Prentice Hall, Inc.

OSHA 1926.351, Arc Welding and Cutting. Current Edition. Washington, DC: Occupational Safety and Health Administration (OSHA).

Welding Handbook. Volume 1. Welding Science & Technology. Miami, FL: American Welding Society, 2001.

Welding Handbook. Volume 2. Part 1: Welding Processes. Miami, FL: American Welding Society, 2004.

Lincoln Electric website: **http://www.lincolnelectric.com** offers sources for products and training.

1.0.0 Section Review

1. Flux-cored arc welding (FCAW) uses a tubular flux-cored wire electrode with or without a(n) ____.
   a. oxidizer tip
   b. flux insulator
   c. shielding gas
   d. pulse generator

2. One good way to minimize fire hazards and vent a confined space before a cutting or welding procedure is to ____.
   a. spray the entire space with water
   b. use a powered extraction system
   c. blow compressed oxygen into the area
   d. purge the area with methane

3. The GMAW metal transfer method that uses the lowest voltage settings is the ____.
   a. globular-transfer method
   b. pulsed-transfer method
   c. spray-transfer method
   d. short-circuiting transfer method

4. In an FCAW metal transfer process, the flux inside the core of the electrode contains ionizers that are used to ____.
   a. purge the weld of gases
   b. stabilize the electric arc
   c. generate a shielding gas
   d. form a protective slag

Section Two

2.0.0 GMAW and FCAW Equipment

Objective

Describe GMAW and FCAW equipment and explain how to prepare for welding.

a. Identify common GMAW/FCAW welding equipment.
b. Describe power source control considerations.
c. Identify and describe welding cables and terminations.
d. Identify and describe external wire feeders and their controls.
e. Identify and describe GMAW and FCAW guns, contact tips, and nozzles.
f. Identify various shielding gases and their related equipment.
g. Explain how to set up welding equipment for GMAW and FCAW welding.

Performance Task

1. Set up GMAW and FCAW equipment with appropriate shielding gases and filler metals.

Trade Terms

Inductance: An electrical circuit property, adjustable on some FCAW power supplies, that influences the pinch effect.

Pinch effect: The tendency of the end of the wire electrode to melt into a droplet during short-circuit GMAW welding.

Stickout: The length of electrode that projects from the contact tube, measured from the arc tip to the contact tube.

Typical GMAW and FCAW systems (*Figure 12*) are made of four interconnected pieces of equipment, including the following:

- DC power source (welding machine)
- Wire feeder
- Welding gun
- Shielding gas supply (if required)

In some systems, equipment functions may be integrated or combined. For example, the wire feeder can be built into the same cabinet as the power source, or it may be incorporated into the welding gun. Welding guns are shown in a later section of this module.

Most systems have a separate wire feeder (push system) that pushes the wire from the feeder through the gun cable and gun. Other systems may incorporate the wire feeder into the welding gun itself (pull system) or do both (push-pull system). There are even systems that have synchronized wire feeders located at intervals along very long feeder conduits. Various types of wire feeders are covered in a later section. Heavy duty, water-cooled guns may also be equipped with a closed-loop water-cooling system. Systems that are used for FCAW-S do not require gas cylinders, since the shielding gas comes from the flux-core electrode.

2.1.0 Types of Welding Power Sources

Direct current (DC) power sources, or welding machines, are usually designed to produce either constant-current (CC) or constant-voltage (CV) DC power. A constant-current machine produces a constant-current output over a wide voltage range; this is typical of a SMAW welding machine. A constant-voltage machine maintains a constant voltage as the output current varies; this is the typical power source used in GMAW or FCAW welding machines. The output voltages and output currents of a welding machine can be plotted on a graph to form a curve. These curves show how the output voltage relates to the output current as either changes (*Figure 13*).

Different types of welding machines (power sources) are available for GMAW, including the following:

- Transformer-rectifier power sources
- Engine-driven generator power sources
- Inverter power sources

Constant Current/ Constant Voltage Power Sources

Transformer-rectifiers can be designed to produce constant-current or constant-voltage DC. Some will produce either variable or constant voltage by setting a switch. These machines are sometimes referred to as multiprocess power sources, since they can be used for any welding process, including GMAW and FCAW.

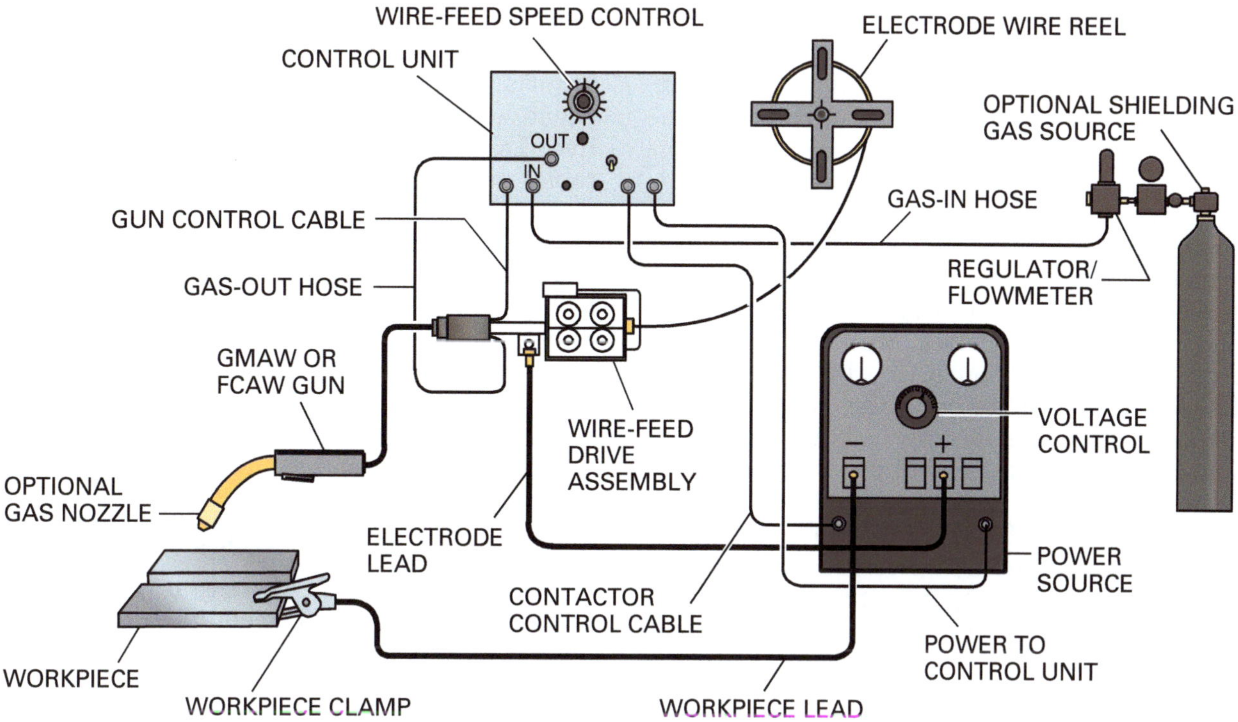

NOTE: The polarity of the gun and workpiece leads is determined by the type of filler metal and application.

Figure 12 Simplified GMAW/FCAW system diagram.

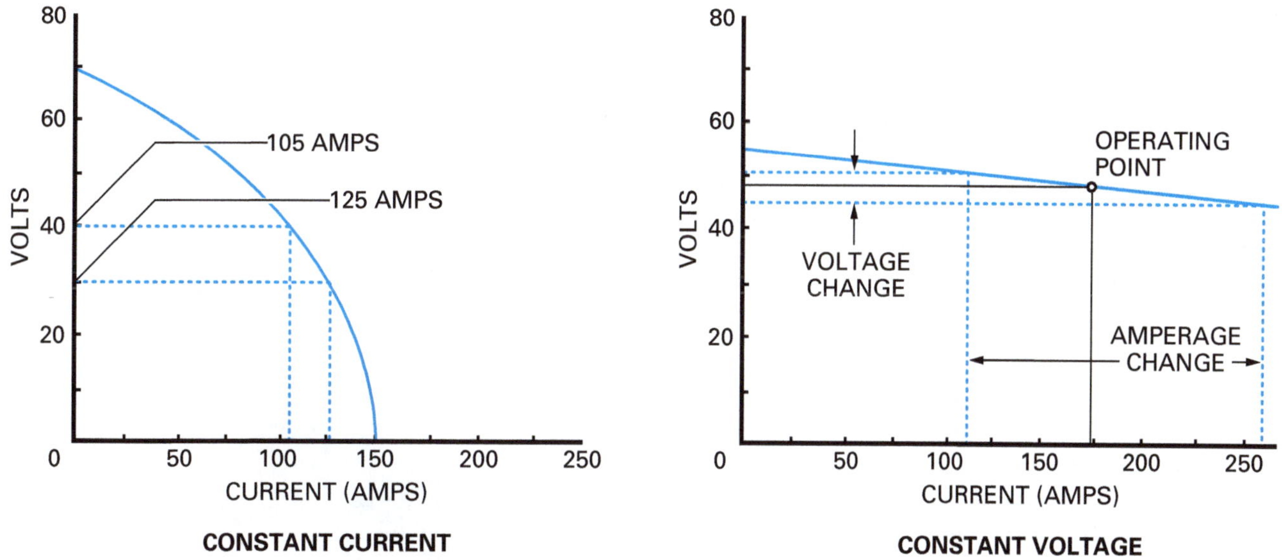

Figure 13 Constant-current and constant-voltage output curves.

2.1.1 Characteristics of Welding Current

The current produced by a welding machine has different characteristics from the current flowing through utility power lines. Welding current has low voltage and high amperage, while the power line current has high voltage and low amperage.

Voltage is the measure of the electromotive force or pressure that causes current to flow in a circuit. The specific unit of measure is the volt.

There are two types of voltage associated with welding current: open-circuit voltage and operating voltage. Open-circuit voltage is the voltage present when the machine is on but no welding is being done. There are usually ranges of open-circuit voltages that can be selected, up to about 40 volts. Operating voltage, or arc voltage, is the voltage after the arc has been struck.

Arc voltage is generally slightly lower than the open-circuit voltage. The arc voltage is typically 2 to 3 volts lower than the open-circuit voltage for each 100 amperes of current, but it depends on the range selected.

Amperage is the measure for electric current flow in a circuit. The specific unit of measure for amperage is the ampere, which is commonly abbreviated to amp.

In welding, the current flows in a closed loop through two welding cables; the workpiece lead connects the power source to the base metal, and the other cable connects the power source to the wire electrode. During welding, an arc is established between the end of the wire electrode and the work. The arc generates intense heat at 6,000°F to 10,000°F (3,316°C to 5,538°C), melting the base metal and the wire electrode to form the weld. The amount of amperage produced by the welding machine determines the intensity of the arc and the amount of heat available to melt the work and the electrode.

2.1.2 Transformer-Rectifier Power Sources

The transformer-rectifier power source uses a transformer to convert the primary current to welding current and a rectifier to change the current from AC to DC. Transformer-rectifier power sources can be designed to produce either AC and DC, or DC only. Transformer-rectifiers that produce both AC and DC are usually lighter duty than those that produce DC only. Transformer-rectifier power sources that produce DC only are sometimes called rectifier power sources or just rectifiers. Depending on the size, transformer-rectifier power sources may require 120/240-volt single-phase input power, 208/230/240-volt three-phase input power, or 480-volt three-phase input power.

WARNING!

Coming into contact with the primary voltage of a welding machine can cause major shock or electrocution. Make sure that all welding machines are properly grounded to prevent injury.

Transformer-rectifiers used for GMAW and FCAW have an On/Off switch and a voltage control. If the machine has a CC/CV or mode switch, it also has an amperage control for SMAW or GTAW. When the switch is set to CV for either GMAW or FCAW, the amperage control is disabled or becomes the voltage control.

The welding cables (electrode cable and workpiece cable) are connected to terminals marked Electrode and Ground, or Positive (+) and Negative (–). Welding machines often have selector switches to select direct current electrode negative (DCEN) or direct current electrode positive (DCEP). If there is no selector switch, the cables must be manually changed on the machine terminals to select the type of current desired.

Figure 14 shows a typical industrial transformer-rectifier CC/CV power source with a wire feeder. These units usually have a 100 percent duty cycle at their rated output and a 60 percent duty cycle at their maximum output. They require a wire feeder and a GMAW/FCAW gun or spool gun. These devices are described in a later section.

2.1.3 Engine-Driven Welding Machines

Welding machines can also be powered by gasoline or diesel engines. The engine is connected to a generator that produces the power needed for the welding machine. The size and type of engine used depends on the requirements of the welding machine.

To produce sufficient electrical power, the engine generator must turn at a required number of revolutions per minute (rpm). The engines of power generators have governors to control the engine speed. Most governors have a welding speed switch. The switch can be set to idle the engine when no welding is taking place. When the electrode is touched to the base metal, the governor automatically increases the speed of the engine to the required speed for welding. If no

Figure 14 Industrial transformer-rectifier CC/CV power source with wire feeder.

welding takes place for about 15 seconds, the engine will automatically return to idle. The switch can also be set for the engine to run continuously at the welding speed.

Figure 15 shows a diesel engine-driven welding machine. This particular unit can produce CC/AC or DC and CV/DC outputs up to 400 amps at a 100 percent duty cycle for SMAW, GTAW, GMAW, and FCAW. For GMAW and FCAW, it is available with a portable wire feeder that provides a constant wire-feed speed for use with CV power sources. When used with CV power sources, the wire feeder provides wire as needed when it senses that an arc is struck. The wire feeder is shown with an FCAW-S gun, but other guns can also be used.

Engine-driven units often have an auxiliary power unit that produces 120 volts alternating current (VAC) for lighting, power tools, and other electrical equipment. When 120 VAC is required, the engine-driven generator or alternator must run continuously at its normal welding speed.

Engine-driven units have both engine controls and welding machine controls. The engine controls vary with the type and size of the unit but normally include the following:

- Starter
- Voltage meter
- Temperature gauge
- Fuel gauge
- Hour meter

Engine-driven welding machines have a voltage/amperage control. They may have a polarity or mode selector switch, or the welding cables at the welding current terminals might have to be changed to change the polarity. If the machine has a CC/CV or mode selector switch, it also has an amperage control. When the switch is set to CV or GMAW/FCAW mode, the amperage control is disabled and becomes a voltage control.

4-WHEEL TRAILER WITH MULTI-PROCESS WELDER

PORTABLE INDUSTRIAL WIRE FEEDER

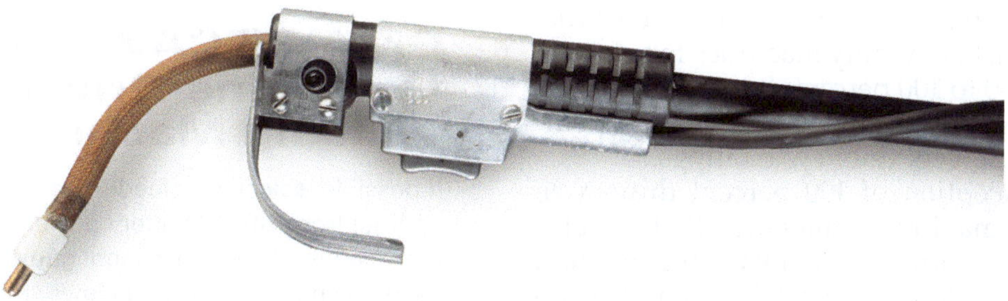

SEMIAUTOMATIC FLUX-CORED, SELF-SHIELDING (FCAW-S) GUN

Figure 15 Diesel engine-driven welding machine with wire feeder and FCAW-S gun.

The advantage of engine-driven generators is that they are portable and can be used in the field where electricity is not available for other types of welding machines. The disadvantage is that engine-driven generators are more costly to purchase, operate, and maintain.

2.1.4 Inverter Power Sources

Inverter welding machines (*Figure 16*) increase the frequency of the incoming primary power. This provides a smaller, lighter power source with a faster response time and more control for pulse welding. Inverter welding machines are used where space is limited and portability is important. The controls on these welding machines vary according to size and application. Smaller inverter machines are often mounted on lightweight carts for greater portability.

Typical industrial advanced inverter welding machines designed specifically for GMAW or FCAW operations usually have controls similar to those shown in *Figure 17*. For specific control functions for the power supply you are using, refer to the operator's manual.

A typical program sequence for one of the internal welding programs available in an advanced inverter power source is shown in *Figure 18*.

Figure 16 Inverter welding machine.

2.1.5 Welding Machine Ratings

The rating of a welding machine is determined by the amperage output of the machine at a given duty cycle. The duty cycle of a welding machine is based on a 10-minute period and specified by a percentage. It is the percentage of 10 minutes during which the machine can continuously produce its rated amperage without overheating. For example, a machine with a rated output of 300 amps at a 60-percent duty cycle can deliver 300 amps of welding current for 6 minutes out of every 10 (60 percent) without overheating.

The duty cycle of a welding machine will be 10, 20, 30, 40, 60, or 100 percent. A welding machine with a duty cycle of 10 to 40 percent is considered a light- to medium-duty machine. Most industrial, heavy-duty machines for manual welding have 60 to 100 percent duty-cycle ratings. Machines designed for automatic welding operations have 100 percent duty-cycle ratings.

With the exception of 100 percent duty-cycle machines, the maximum amperage that a welding machine can produce is always higher than its rated capacity. A welding machine rated at 300 amps and a 60-percent duty cycle will generally put out a maximum of 375 to 400 amps. But since the duty cycle is a function of its rated capacity, the duty cycle will decrease as the amperage is raised over 300 amps. Welding at 375 amps with a welding machine rated 300 amps at 60-percent duty cycle will lower the duty cycle to about 30 percent. If welding continues for more than 3 out of 10 minutes, the machine will overheat. Most welding machines have a heat-activated circuit breaker that shuts off the machine automatically when it overheats. Leave the machine plugged in and turned on to allow the cooling fans to continue running when this occurs.

On the other hand, if the amperage is set below the rated amperage, the duty cycle increases. Setting the amperage at 200 amps for a welding machine rated at 300 amps and a 60-percent duty cycle will increase the duty cycle to 100 percent. *Figure 19* shows the relationship between amperage and duty cycle.

2.2.0 Power Source Control

As described in earlier sections, GMAW and FCAW are best performed with a constant-voltage

> **GMAW/FCAW Feeders for Engine-Driven Welding Machines**
>
> If possible, select wire feeders or spool guns that operate from 24VDC battery power and arc voltage. If AC-powered feeder or spool gun controls are used, the power source will have to operate in the continuous run mode to supply AC power to operate the feeder or spool gun.

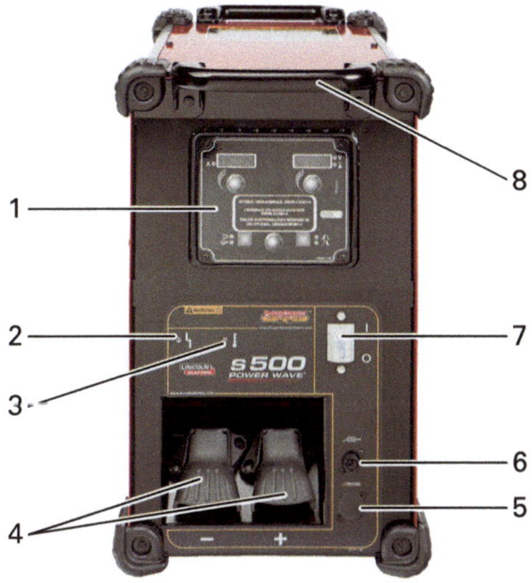

FRONT

1. OPTIONAL S-SERIES USER INTERFACE KIT FOR STICK, TIG AND CV MIG WITH VOLTAGE SENSING FEEDER
2. STATUS LIGHT
3. THERMAL FAULT INDICATOR LIGHT
4. OUTPUT STUDS
5. OUTPUT CONTROL RECEPTACLE KNOCKOUT PLATE – 12-PIN
6. WORK SENSE LEAD RECEPTACLE
7. MAIN POWER SWITCH
8. REVERSIBLE HANDLES

Figure 17 Typical industrial inverter unit controls.

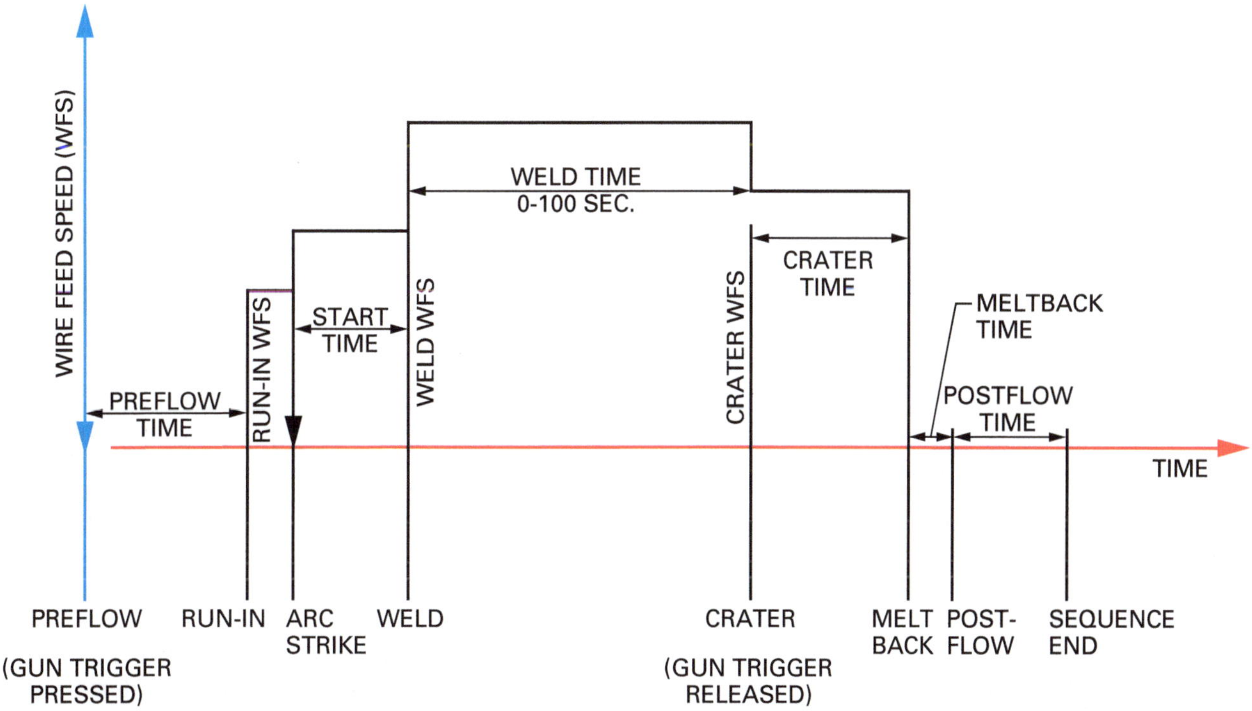

Figure 18 Typical welding program sequence.

or constant-voltage/constant-current (CV/CC) DC welding machine. In constant-voltage welding machines, the open-circuit voltage and the welding voltage are nearly the same. Constant-current variable-voltage DC welding machines, the type used for SMAW in which the output voltage varies between high open-circuit voltage and lower arc voltage, are sometimes used. However, they require specially designed voltage-sensing wire feeders. The use of CC machines for GMAW and FCAW is prohibited by some codes. CV/CC welding machines are required for GMAW-P operations.

Welding machines usually contain controls for adjusting arc voltage, slope, and inductance, and they have meters to monitor amperage and voltage. The simplest machines are power sources with built-in wire feeders. A simple combination welder and wire feeder may have only a wire-feed speed control and a voltage control.

2.2.1 Slope

The constant-voltage power sources used for GMAW and FCAW are not precisely constant in their voltage output, because the voltage always drops a little as the current increases. This voltage-to-amperage relationship forms a slight curve when plotted as a graph. The general angle

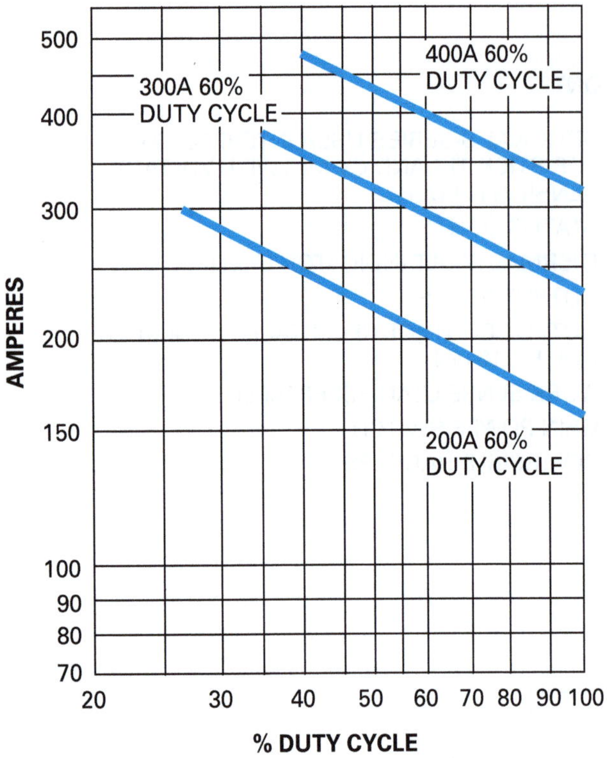

Figure 19 Relationship between amperage and duty cycle.

of the volt-ampere curve is known as the slope. This slope is adjustable on some units. A flatter (more horizontal) slope is better for spray-transfer welding, while a steeper slope is better for short-circuiting welding (*Figure 20*).

2.2.2 Pinch Effect and Inductance

When using short-circuiting welding in GMAW, the current passing through the wire electrode heats and then melts the wire. This is sometimes called the **pinch effect** because the molten wire appears to be pinched off, forming a droplet of molten metal (*Figure 21*).

An electrical circuit property called inductance influences the pinch effect. Low inductance

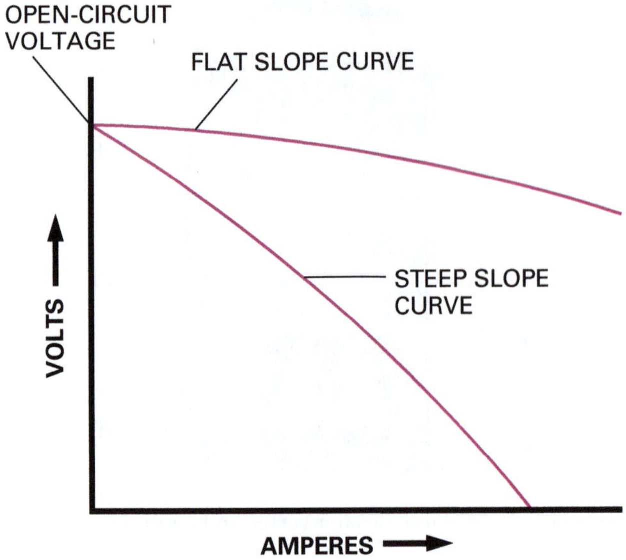

Figure 20 Volt-ampere curve slope profiles for GMAW.

produces high pinch effect. Conversely, high inductance produces low pinch effect. With low circuit inductance, the current rise is very fast, and the high pinch effect can cause the wire droplet to explode or spatter. When the circuit inductance is high, the current rises more slowly, decreasing the number of short circuits per second and extending the arc duration. The pinch effect forms the droplet more gently, which results in a more fluid puddle, smoother weld, and less spatter.

Constant-voltage power sources are self-regulating and automatically produce current as it is required. Inductance controls the rate at which the current output rises. Some inductance is desirable in FCAW because it prevents explosive arc starts by slowing down the rate of current rise. Many FCAW power sources have a control to adjust the inductance. In general, lower thermal conductivity materials use a higher inductance, and higher thermal conductivity materials use a lower inductance.

All-in-One Welding Machines

Portable, light industrial GMAW/FCAW welding machines are also available. The MIG unit shown is equipped with an internal wire feeder and has dual voltage inputs of 120 or 240VAC. With a 120VAC input, it is rated at 100 amps with a 40 percent duty cycle. With a 240V input, it is rated at 200 amps with a 25 percent duty cycle. This machine can also be used to perform TIG welding.

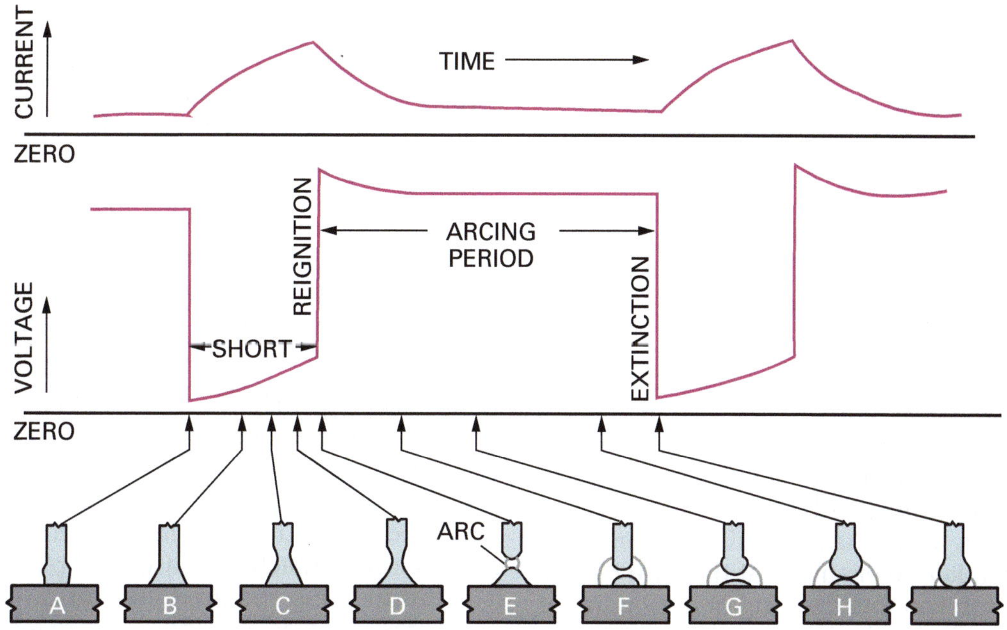

Figure 21 Pinch effect.

2.2.3 Arc Blow

When operating in GMAW spray-transfer mode or in FCAW, arc blow can sometimes be a problem. Arc blow is the deflection of the arc from its normal course because of the arc magnetic field's attraction to or repulsion from the weld current's magnetic field in the base metal. The amount and direction of arc deflection depends upon the relative position, direction, and density of the base metal's magnetic field. Arc blow can result in excess spatter and weld defects. It can be minimized by relocating the workpiece clamp or changing the weld angle of the gun.

2.3.0 Welding Cable

Cables used to carry welding current are designed for maximum strength and flexibility. The conductors inside the cable are made of fine strands of copper wire. The copper strands are covered with layers of rubber reinforced with nylon or Dacron® cord (*Figure 22*).

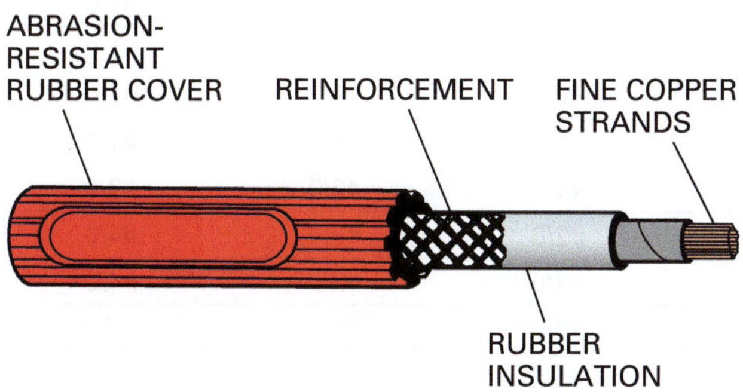

Figure 22 Cut-away section of welding cable.

The size of a welding cable is based on the number of copper strands it contains. Large-diameter cable has more copper strands and can carry more welding current. Typical welding cable sizes range from #4 to #3/0 (spoken as "3 aught"), as shown in *Table 1*.

When selecting the welding cable size, the current flow (amperage) as well as the distance the current will travel must be considered. The longer the distance the current has to travel, the larger the cable must be. Cable that is too small will become hot, and the voltage provided under load will drop due to the cable's electrical resistance. Excessive voltage drop may affect the performance of the welding machine. When selecting welding cable, use the rated capacity of the welding machine for the cable amperage requirement. For the distance, measure both the electrode and the workpiece leads, and add the two lengths together. To identify the welding cable size required, refer to a recommended welding cable size table, furnished by most welding cable manufacturers.

2.3.1 Welding Cable End Connections

Welding cables must be equipped with the proper end connections or terminals.

Lugs are used at the end of the welding cable to connect the cable to the welding machine terminals. The lugs come in various sizes to match the welding cable size, and are mechanically secured to the welding cable.

> **CAUTION**
>
> If the end connection is not tightly secured to the cable, the connection will overheat and oxidize. An overheated connection will cause variations in the welding current and permanent damage to the connector and/or cable. Check connections for tightness and repair loose or overheated connections. If quick-disconnects are used, they must be properly sized to the cable or overheating may occur.

Quick-disconnects are also mechanically connected to the cable ends. They are insulated and serve as cable extensions for splicing two lengths of cable together. Quick-disconnects are connected or disconnected with a half twist or similar motion. When using quick-disconnects, care must be taken to ensure that they are tightly connected to prevent overheating or arcing in the connector. *Figure 23* shows some typical quick-disconnects and lugs used as welding cable connectors.

The workpiece clamp (*Figure 24*) provides the connection between the end of the workpiece lead and the workpiece. Workpiece clamps are mechanically connected to the welding cable and come in a variety of shapes and sizes. The size of a workpiece clamp is the rated amperage that it can carry without overheating. Workpiece clamps must at least have the same rating as the rated capacity of the power source on which they will be used.

2.4.0 External Wire Feeders

An external wire feeder pulls the electrode wire from a spool and pushes it through the gun cable and the gun. It consists of a wire spool holder with a drag brake, an electric motor that drives either one or two sets of opposing slotted rollers, and various controls. The wire feeder also usually contains the shielding gas connections, the gas control solenoid, and (optionally) the cooling-water flow-control solenoid. Some of the controls that may be located on a wire feeder include the following:

- *Spool drag* – This is a manual adjustment used to adjust the drag on the wire spool to prevent the uncontrolled unwinding of the wire.
- *Wire feed speed control* – This is a variable control used to set the wire feed speed. Since GMAW welding machines are self-regulating (they produce the required current on demand), increasing or decreasing the wire feed speed will cause the welding machine to automatically increase or decrease its current output.
- *Voltage control* – This control is used to set the welding voltage.
- *Wire jog* – The manual wire jog switch is used to start and stop the wire feeder without energizing the welding current.
- *Wire meltback timer (burnback)* – This is an adjustable time delay that causes the welding current to continue after the trigger has been released. This consumes the length of wire that continues to feed after the trigger has been released so that the proper **stickout** is maintained to start the next weld.

Table 1 Welding Cable Sizes

Machine Size in Amperes	Duty Cycle (%)	Copper Cable Sizes for Combined Lengths of Electrodes Plus Workpiece Lead				
		Up to 50 feet (15 m)	50–100 feet (15–30 m)	100–150 feet (30–46 m)	150–200 feet (46–61 m)	200–250 feet (61–76 m)
100	20	#8	#4	#3	#2	#1
180	20	#5	#4	#3	#2	#1
180	30	#4	#4	#3	#2	#1
200	50	#3	#3	#2	#1	#1/0
200	60	#2	#2	#2	#1	#1/0
225	20	#4	#3	#2	#1	#1/0
250	30	#3	#3	#2	#1	#1/0
300	60	#1/0	#1/0	#1/0	#2/0	#3/0
400	60	#2/0	#2/0	#2/0	#3/0	#4/0
500	60	#2/0	#2/0	#3/0	#3/0	#4/0
600	60	#3/0	#3/0	#3/0	#4/0	* * *
650	60	#3/0	#3/0	#4/0	* *	* * *

** Use double strand of #2/0
*** Use double strand of #3/0

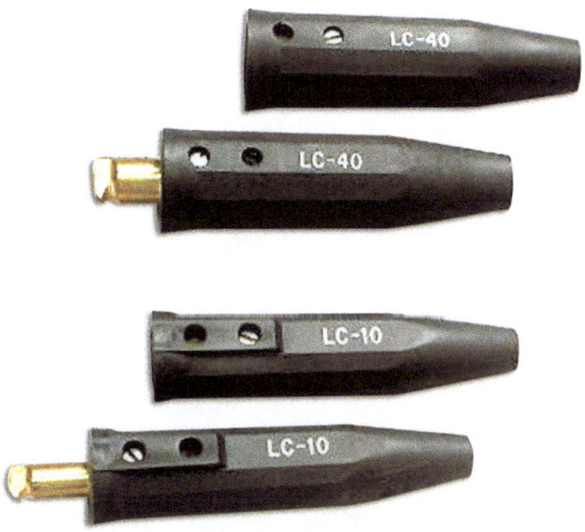

(A) QUICK-DISCONNECTS

(B) LUGS

(C) DINSE CONNECTORS

Figure 23 Quick-disconnects, lugs, and DINSE connectors.

- *Gas purge* – The manual gas purge switch is used to operate the gas solenoid in order to purge the shielding gas system prior to actual welding. It is also used to adjust the flowmeter.
- *Trigger hold* – This switch enables or disables a trigger hold function activated by the gun trigger. It enables welding without holding the trigger.
- *Gas preflow timer* – This is an adjustable time delay that prevents the welding current contactor from closing until the shielding gas is flowing.
- *Gas postflow timer* – This is an adjustable time delay that is used to keep shielding gas flowing for a short period after the gun trigger has been released.

Figure 25 shows some basic wire feeder units, while *Figure 26* shows other types of wire feeders.

Two critical properties of electrode wire must be considered before selecting the wire to be used with a wire feeder. Those two properties are cast and helix (*Figure 27*). Cast and helix are determined by cutting a strand of the wire several feet long and laying it on a flat surface. The cast of the wire is the diameter of the circle formed by the strand of wire. The helix is the height, or vertical distance, of the end of the wire from the flat surface. The cast and helix of electrode wire primarily affect the wire feeder's ability to move the wire through the gun cable and gun. The cast and helix can also cause excessive wear to the contact tip and misguide the wire, causing it to be diverted away from the joint. These factors can affect the quality of the weld.

During operation, a wire feeder grips the wire between the grooves of opposing grooved rollers. There can be one or two sets of opposed rollers (*Figure 28*). The grooves can be V-shaped or U-shaped in their cross section.

V-grooves are generally used for hard wire. One roller slot in each V-groove set may be serrated for better gripping. U-groove rollers are used for soft wire or flux-cored wire, as they crush more easily. For extremely soft wire or flux-cored wire, cogged U-groove rollers are used.

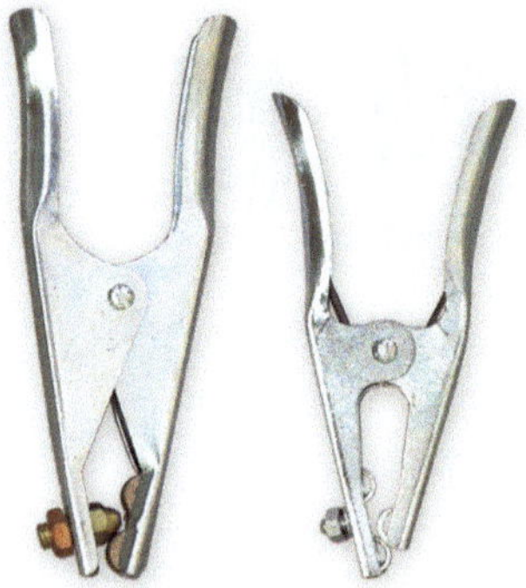

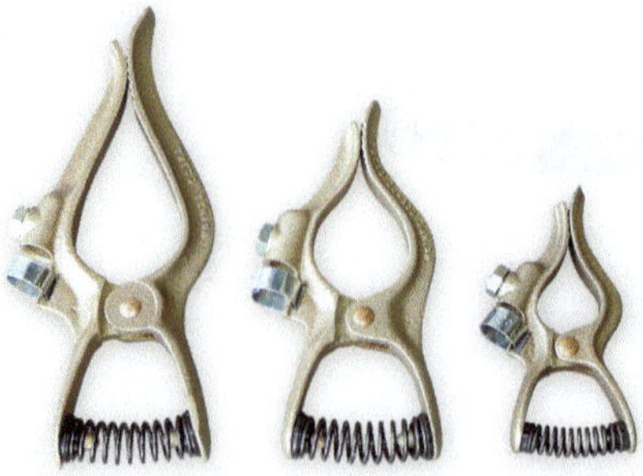

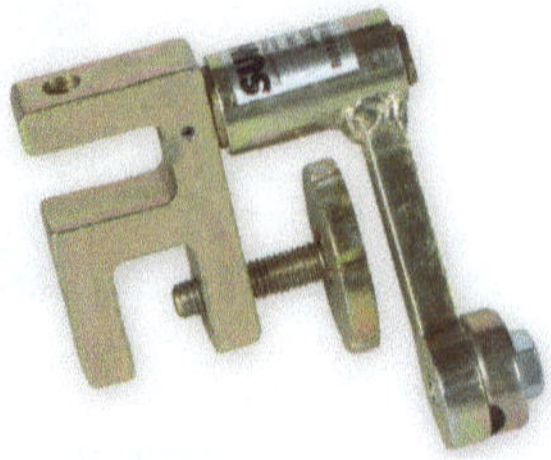

Figure 24 Workpiece clamps.

Some rollers are made in two pieces with different-size half-grooves formed on the outside edges of each half-roller. The half-rollers are assembled to fit two different ranges of wire size. *Figure 29* shows several different groove styles and combination rollers.

When dust or dirt is a problem, a felt wiper can be added to the wire between the spool and the inlet guide to the feed rollers. The felt wipers are treated with a silicone lubricant that cleans and lubricates the wire. Special liquid silicone cleaner/lubricants can occasionally be added to the felt wiper.

CAUTION

Use only special cleaner/lubricants made specifically for wire feed wipers. Use of other cleaners or lubricants may contaminate and cause porosity in the weld.

2.5.0 GMAW/FCAW-G Guns

The GMAW/FCAW-G gun (*Figure 30*) supports and guides the wire electrode and provides the electrical contact between the wire and the welding machine electrode lead. The gun also contains a remote switch to start and stop the welding current and the shielding gas flow. The tip of the gun has a nozzle to direct the shielding gas around the arc. Within the nozzle, the wire passes through a contact tip that provides the actual connection between the welding current and the electrode wire. Nozzles and contact tips are subject to wear and electrical erosion, so both components are replaceable.

The duty cycle of GMAW guns is determined and rated with CO_2. The rating is reduced from 10 to 50 percent when mixed gases are used, depending on gas mixture and welding parameters. The chosen mode of transfer (short-circuit,

Figure 25 Basic GMAW/FCAW wire feeder units.

spray, or pulsed) may further derate their duty cycle.

Some types of guns contain motor-driven wire pullers within the gun handle (*Figure 31*). These work in conjunction with a push wire feeder located with the power source. This system was developed for feeding soft wire, such as aluminum, that tends to bend easily. With the wire being both pushed and pulled, neither feeder mechanism needs to grip the wire quite as tightly.

In addition, there are spool guns designed for aluminum wire. These guns contain both a wire feeder and a small spool of wire. In some applications, an auxiliary gun control unit is required for proper operation. *Figure 32* shows one style of spool gun.

Light-duty guns are gas-cooled by the shielding gas flow. Heavy-duty guns are water-cooled. The cooling-water flow may be controlled by a solenoid in the wire feeder or by a separate dedicated cooling system. Cooling water is piped to and from the gun through flexible tubing integrated into the gun cable. Some cooling systems use domestic water to cool the gun. Others use a closed-loop system (*Figure 33*) that circulates demineralized water or special cooling fluid that will not corrode internal gun surfaces or plug the cooling passages with mineral scale.

(A) DOUBLE REEL BENCH
GMAW/FCAW WIRE FEEDER

(B) SINGLE REEL PROGRAMMABLE
BENCH WIRE FEEDER FOR GMAW-P

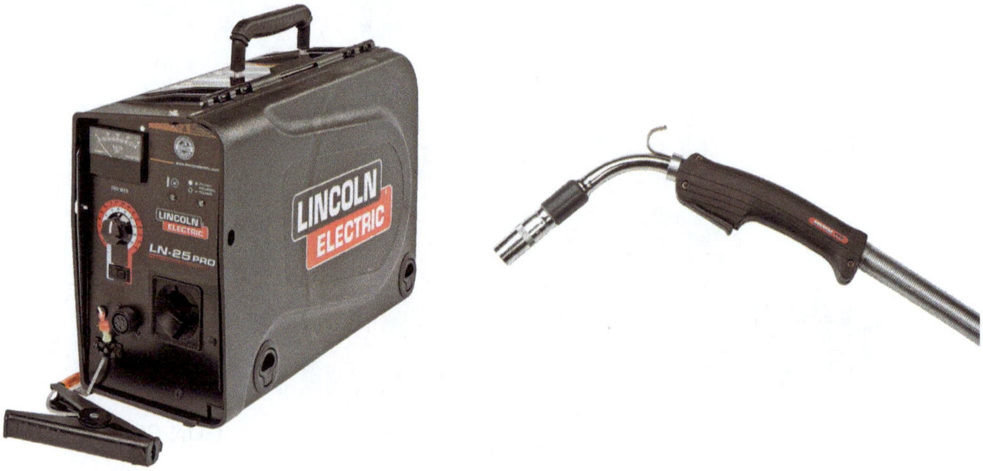

(C) WEATHERPROOF PORTABLE GMAW/FCAW FEEDER
WITH FCAW-S GUN

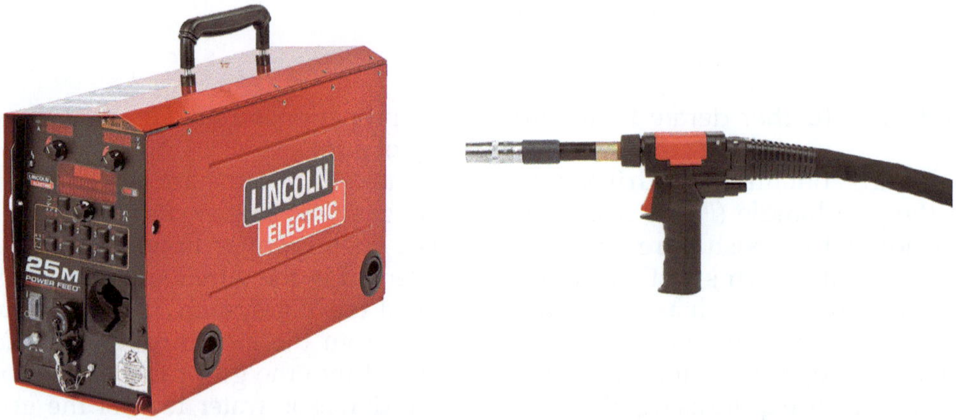

(D) PROGRAMMABLE PORTABLE PUSH-PULL WIRE FEEDER
AND PUSH-PULL GUN

Figure 26 Various types of wire feeders.

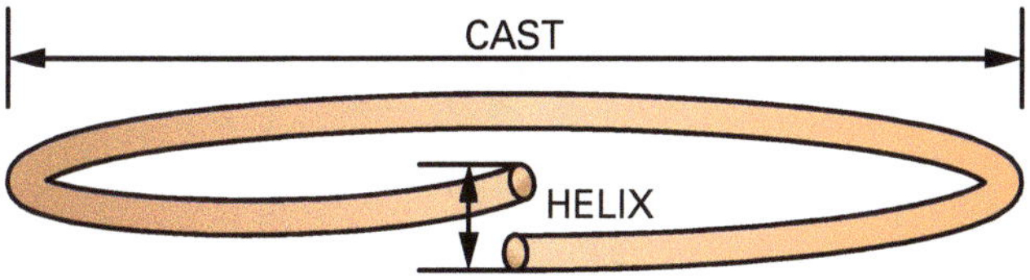

Figure 27 Cast and helix of a wire electrode.

2.5.1 FCAW-S Guns

FCAW-S guns are similar to GMAW/FCAW-G guns, but they are often designed to handle larger-diameter wire and they do not provide shielding gas flow. The FCAW gun supports and guides the wire electrode. It also contains a remote switch for starting and stopping the welding current. The wire electrode passes through the contact tip at the end of the gun, which provides the electrical connection from the welding machine electrode lead to the electrode wire.

Guns designed for self-shielded FCAW do not have provisions for delivering shielding gas. This makes the gun nozzles smaller in diameter. To protect the contact tip, these guns have a fiber insulator that threads onto the end of the gun nozzle.

FCAW-S generates smoke, which can be removed at the gun tip with a special smoke extraction nozzle. The nozzle is connected to a vacuum pump or blower by a flexible tube that draws off smoke, which improves visibility.

2.5.2 Contact Tips and Nozzles

Contact tips and nozzles are a vital part of effective welding. The contact tip provides the connection between the welding current and the electrode wire. The nozzle directs shielding gas around the arc. Both of these components are

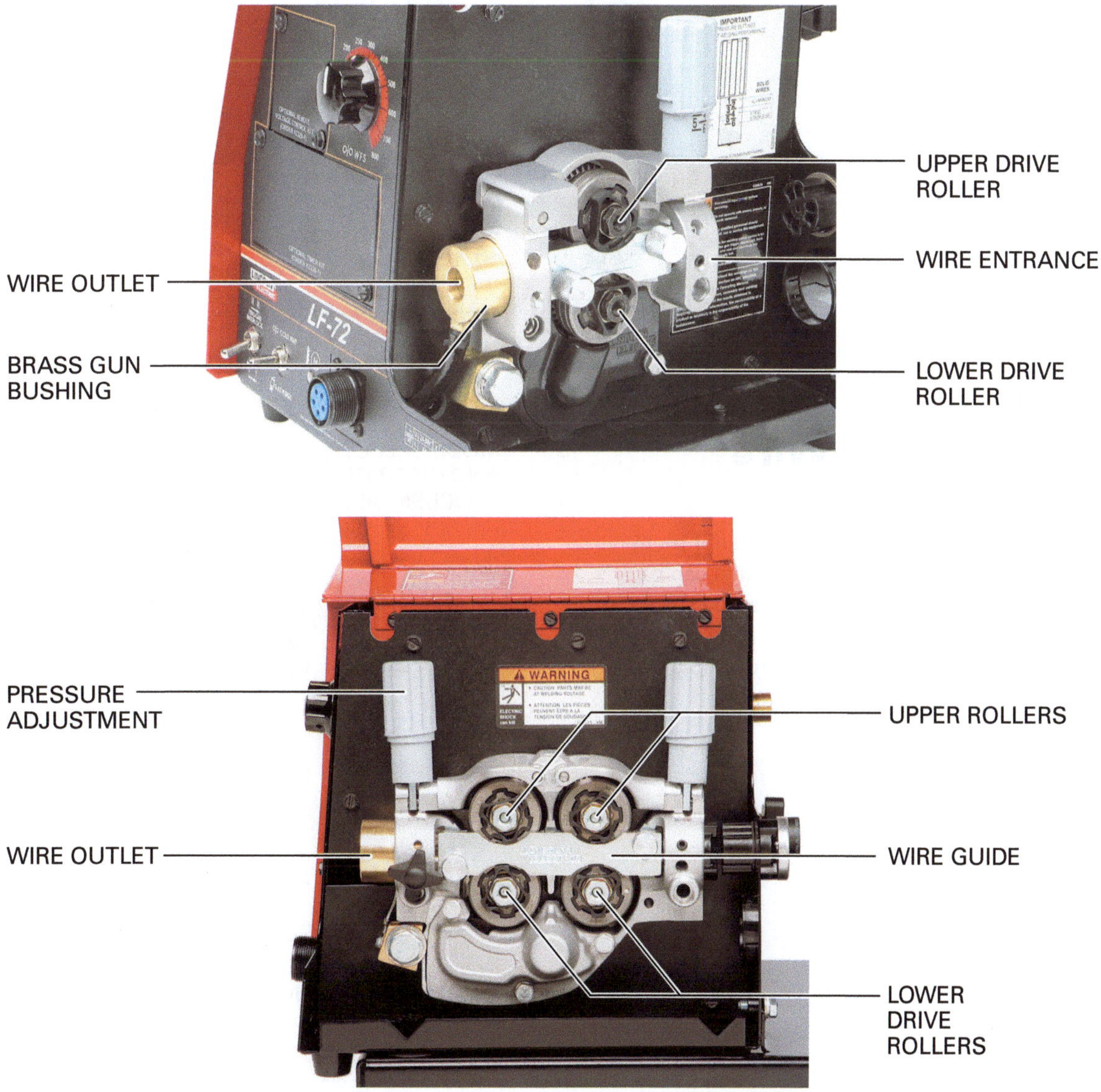

Figure 28 Wire feeder drive roll systems.

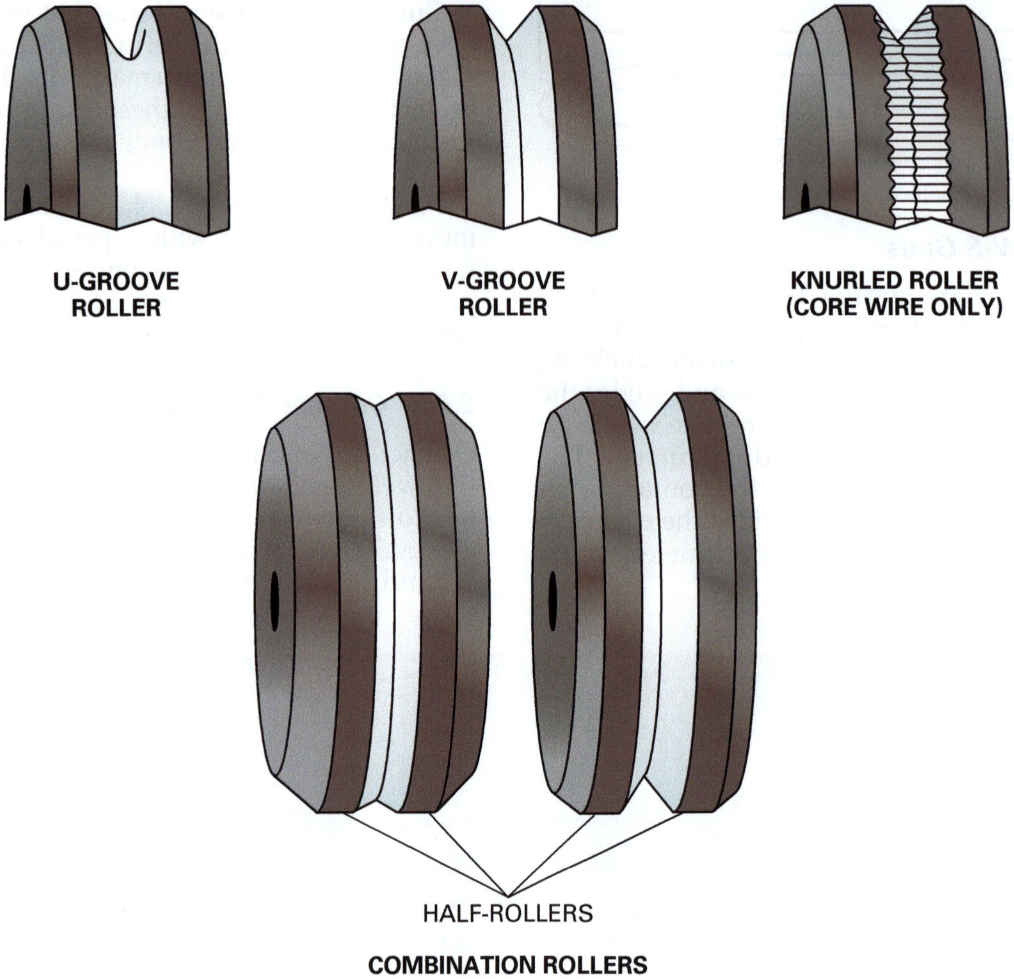

Figure 29 Groove styles and combination rollers.

Programmable Wire Feeders

Programmable wire feeders are available for many power sources. Custom welding programs can be created, or preset welding programs can be stored in the feeder or in the power source. The program can be activated by the various controls and displays on the feeder.

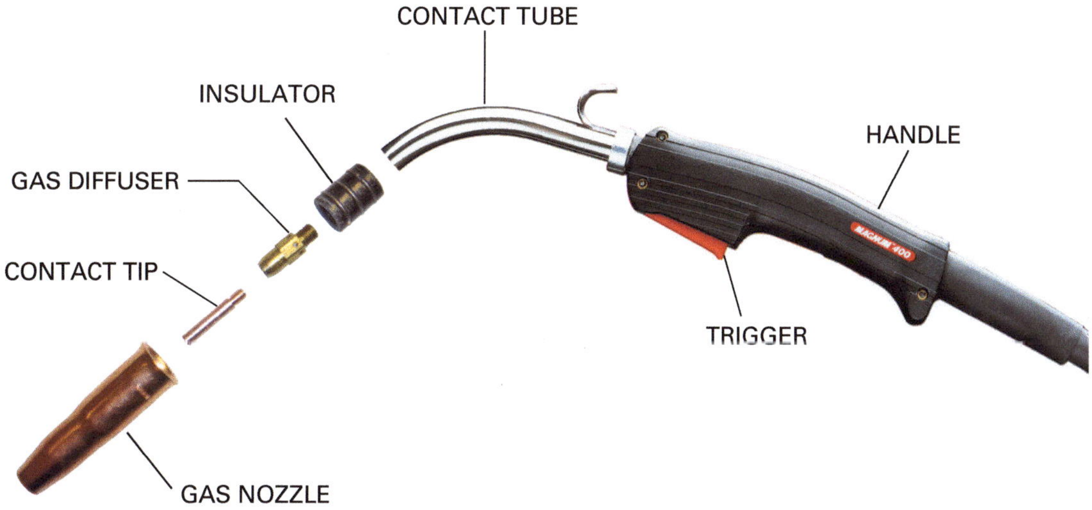

Figure 30 Construction of a typical GMAW/FCAW-G gun.

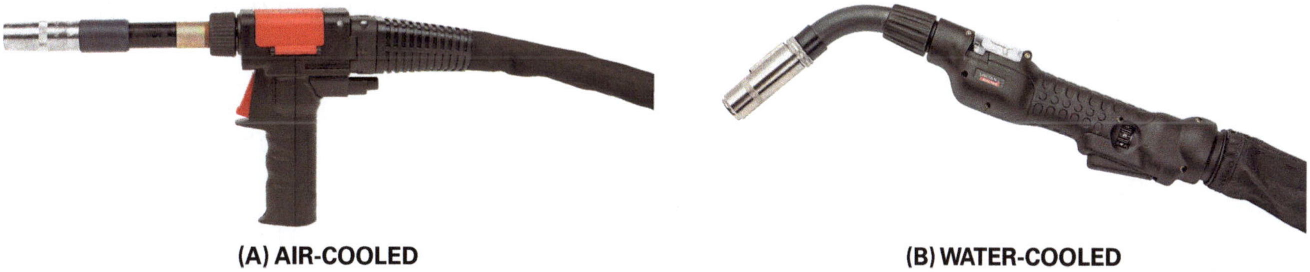

Figure 31 Pull guns for push-pull wire feeders.

Figure 32 Spool gun.

Figure 33 Typical closed-loop cooling unit.

available in many different shapes and sizes (*Figure 34*), depending on the application requirements. For that reason, a number of factors must be taken into consideration when selecting contact tips and nozzles. The chosen tip and nozzle must be compatible with the equipment in use.

A contact tip should always correspond to the wire electrode size. For example, if a wire electrode with a diameter of 0.125 inches (3.2 mm) is being used, the contact tip should also be 0.125 inches (3.2 mm). If a contact tip is too small for the wire being used, the wire will jam. A contact tip that is larger than necessary may work, but it is not the ideal setup. If the contact tip is too large, poor electrical contact will cause damage from arcing between the tip and the wire.

Other factors to consider when selecting a contact tip include the following:

- *Tapered vs. non-tapered* – Tapered tips are usually longer to allow for welding in areas with restricted access. Non-tapered tips have more mass and tend to last longer than tapered ones.
- *Threaded vs. non-threaded* – Threaded tips are screwed into the gas diffuser. These are the most common contact tips. Non-threaded tips are dropped into the gas diffuser.

Figure 34 Various contact tips and nozzles.

- *Standard vs. heavy duty* – Contact tips can be small standard-duty types or larger heavy-duty types. Most low-amperage applications can be handled with a standard contact tip. High-amperage or extensive welding applications are best handled with a heavy-duty contact tip.

Like contact tips, there are a wide variety of nozzles. They can be tapered or non-tapered, threaded or non-threaded, have small or large inner diameters, and come in numerous shapes and sizes. Ideally, it is best to choose the largest nozzle possible that still allows access to the weld joint. This way the greatest amount of shielding gas coverage is provided. If the welding joint has limited access, a smaller tapered nozzle will need to be used.

Nozzles are available in copper or brass. Copper nozzles are able to withstand extreme heat, so they are best suited for heavy-duty applications. Brass nozzles are subject to melting and burning if they are used in high-temperature applications, so they are more suited for light-duty applications. However, brass nozzles resist welding spatter better than copper nozzles.

2.6.0 Shielding Gases

Shielding gases are used to displace atmosphere from the weld zone and to prevent contamination of the weld puddle by oxygen, nitrogen, and moisture. The three principal shielding gases are argon, helium, and CO_2. These gases are often mixed in various proportions for specific applications.

The GMAW/FCAW-G process always uses shielding gas to protect the electrode and the weld from contamination. The shielding gas is directed around the end of the electrode and the weld zone by the nozzle of the gun.

Some FCAW core wires require the use of a shielding gas. The shielding gas is directed around the electrode end and the weld zone by the nozzle of the gun. The shielding gas used depends on the filler wire and base metal being welded. The most common shielding gases for FCAW carbon and low-alloy steels are CO_2 or an argon-CO_2 mixture. For FCAW high-alloy or stainless steels, argon or an argon-CO_2 mixture is generally used. These gases are available separately or premixed in pressurized cylinders. They are also available separately in bulk liquid tanks, so they can be mixed at the point of use.

CAUTION

Check your WPS or site quality standards for the type of shielding gas to use. If these are not available, refer to the filler metal supplier's specifications.

Each shielding gas has distinctive performance characteristics. Each gas performs differently in the various transfer modes and with different base metals. Mixtures of gases often have the best features of the individual constituents. The following sections explain some of the features of the principal gases and their primary uses.

Gas flow rate from the gun tip is important because it affects the quality and the cost of a weld. Too low a flow rate will not shield the weld zone adequately, which will result in a poor-quality weld. Excessive gas flow wastes expensive gases and can cause turbulence at the weld. The turbulence can pull atmosphere into the weld zone, causing porosity in the weld. Welding specifications indicate appropriate shielding gas flow rates.

2.6.1 Argon

Argon (Ar), is the least expensive and the most commonly used inert shielding gas. Argon can be used alone or in combination with other shielding gases because of its current transfer and ionization properties. It forms a tight arc column with high current density, which concentrates the arc in a small area and produces deep penetration.

Pure argon is used with aluminum, nickel-based alloys, copper alloys, and the reactive metals zirconium, titanium, and tantalum. It provides excellent arc stability, penetration, and bead shape. For ferrous metals, argon is usually mixed with oxygen, helium, hydrogen, carbon dioxide, and/or nitrogen.

2.6.2 Helium

Helium is likely the most expensive of the inert gases and supplies are diminishing. Because of thermal conductivity (heat transfer), helium (He) is used when deeper penetration and higher travel speed are required.

Arc stability with helium is not as good as it is with argon. The helium arc column is wider than the argon arc column, resulting in reduced current density. Helium is better for welding aluminum, magnesium, and copper alloys because its use results in better puddle fluidity and bead wetting. Some mixtures of helium and argon have the beneficial characteristics of both gases. Helium mixtures of 90 percent helium, 7.5 percent argon, and 2.5 percent CO_2 are widely used for short-circuit GMAW of stainless steel in all positions.

2.6.3 Carbon Dioxide

CO_2 is commonly used to weld mild steel because of its weld performance and low cost. In the welding arc, it is not an inert gas because it breaks down into carbon monoxide (CO) and free oxygen. The oxygen combines with elements in the weld puddle to form slag. CO_2 does not support spray transfer, so short-circuiting transfer and globular transfer must be used. CO_2 provides good penetration. However, in GMAW the weld surface is heavily oxidized and has high weld spatter. This is not true in FCAW. Normally, argon is mixed with no more than 25 percent CO_2 to improve the performance characteristics. *Figure 35* shows the effects of shielding gas on the welding arc in GMAW and FCAW.

2.6.4 Gas Mixtures

The principal shielding gases are often mixed to improve their overall welding characteristics. For steel, small amounts of oxygen added to argon stabilize the arc, increase filler metal deposition rate, and improve wetting and bead shape. Small percentages of oxygen added to CO_2 improve arc char-

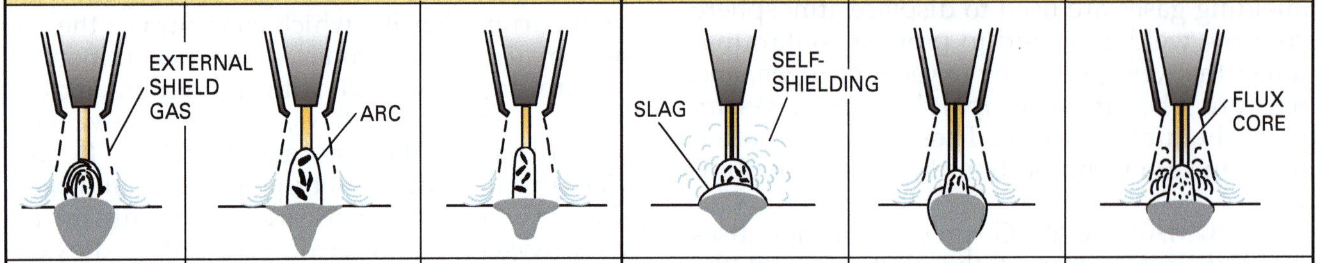

GMAW			FCAW		
CO_2	$Ar\text{-}O_2$	$Ar\text{-}CO_2$	SELF-SHIELDING	DUAL-SHIELDING with CO_2	DUAL-SHIELDING with $Ar\text{-}CO_2$
Shorter, Colder Arc	Hotter Arc Narrow Penetration	Eliminates Under Cutting	Better Arc Ionization Provides Covering (Slag)	Increases Penetration	Increases Arc Length, Improves Spray, and Decreases Penetration

Figure 35 Effects of shield gas on the welding arc.

acteristics. Helium is added to argon to increase the arc voltage and heat for welding nonferrous metals. Small amounts of nitrogen and oxygen are added to argon for welding stainless steel when the austenitic characteristics should be preserved. Helium and CO_2 are added to argon to improve arc stability, wetting, and bead profile. Some examples and uses of gas mixtures are shown in *Table 2*. Many other typical gas mixtures are available.

2.6.5 Shielding Gas Selection

Selecting the best shielding gas for a particular welding task involves considering many factors. The shielding gas selected will affect arc shape, arc density, arc temperature, arc stability, rate of filler metal transfer, degree of spatter, weld penetration, weld bead shape, weld bead appearance, weld porosity, weld chemistry, and weld quality.

Shielding gases affect the welding process in three ways:

- Gas thermal conductivity affects arc voltage and heat delivered to the weld. The thermal conductivity of both helium and CO_2 is much higher than that of argon; both helium and CO_2 supply more heat to the weld. As a result, they also require more voltage and power to maintain a stable arc.
- Some gases can react with the filler and base metals and can also affect arc stability. CO_2 and most oxygen-bearing gases cannot be used with aluminum because they form oxides. However, they provide better fusion and arc stability when used with steels.
- Gases affect the mode of metal transfer and penetration depth. Mixtures containing more than 15 percent CO_2 do not allow true spray transfer.

Common short-circuiting shielding gases and their applications in short-circuiting transfer welding are shown in *Table 3*. Common shielding gases and their applications in spray transfer welding are shown in *Table 4*.

2.6.6 Gas Regulators/Flowmeters

A gas pressure regulator and a flowmeter are required to meter the shielding gas to the gun at the proper flow rate and pressure (*Figure 36*). The two devices can be matched to each other in a single assembly, or they may be separate devices. A typical cylinder pressure regulator includes two pressure gauges—one for cylinder pressure and another for the outlet pressure. The flowmeter may also have a gauge. The flow rate indicator is a clear cylinder with a ball that floats freely inside the housing to indicate the actual flow rate. Increased flow rates cause the ball to rise higher in the housing. The metering, or flow, valve on the flowmeter is used to adjust the gas flow rate to the gun nozzle. The flow indicator shows the gas flow rate in cubic feet per hour (cfh) or liters per minute (lpm).

Gas flow to the gun is started and stopped by an electric solenoid valve that is usually located inside the control unit or the wire feeder unit. The solenoid is controlled by the operator with the trigger switch on the gun. The same trigger switch simultaneously starts and stops the wire feeder and welding current.

2.6.7 Cylinder Safety

Cylinders must always be handled with great care because of the extremely high-pressure gases they contain. When a valve is broken off a high-pressure cylinder, the gas escapes with explosive

Table 2 Examples and Uses of Gas Mixtures

Gas Mixture	Use
Argon with 1% oxygen	Spray transfer on stainless steels
Argon with 2% oxygen	Spray transfer on carbon steels, low-alloy steels, and stainless steels
Argon with 5% oxygen	General carbon steel welding using highly deoxidized wires
Argon with 8% to 12% oxygen	Single-pass spray transfer welds with high puddle fluidity
Argon with 3% to 10% carbon dioxide	Spray transfer and short-circuiting transfer on carbon steels
Argon with 11% to 20% carbon dioxide	High-speed and out-of-position welds on carbon steels and low-alloy steels
Argon with 21% to 25% carbon dioxide	Short-circuiting transfer on mild steels
Argon with 25% helium	Sometimes used for aluminum where bead appearance is important
Argon with 75% helium	Welding heavy aluminum
Argon with 90% helium	Welding aluminum over 3" thick and copper over 1" thick
Argon with 5% to 10% carbon dioxide and 1% to 3% oxygen	Welding carbon steel, low-alloy steel, and stainless steel with all transfer modes
Argon with 10% to 30% helium and 5% to 15% carbon dioxide	Pulse spray transfer on heavy out-of-position carbon and low-alloy steel
60% to 70% helium with 20% to 35% argon and 4% to 5% carbon dioxide	Short-circuiting transfer of high-strength steels in all positions
90% helium with 7.5% argon and 2.5% carbon dioxide	Short-circuiting transfer of stainless steels in all positions
Argon, helium, carbon dioxide, and oxygen	High-rate deposition of low-alloy, high-tensile metals, and mild steels

Table 3 Short-Circuiting Shielding Gases and Their Applications

Gas Mixture	Shielding Gas	Thickness	Advantages
Carbon steel	75% Argon + 25% Carbon Dioxide	Less than ⅛" (3.2 mm)	High welding speeds without burn-through; minimum distortion and spatter
	75% Argon + 25% Carbon Dioxide	More than ⅛" (3.2 mm)	Minimum spatter; clean weld appearance; good puddle control in vertical and overhead positions
	50% Argon + 50% Carbon Dioxide	—	Deeper penetration; faster welding speeds
Stainless steel	90% Helium + 7.5% Argon + 2.5% Carbon Dioxide	—	No effect on corrosion resistance; small heat-affected zone; no undercutting; minimum distortion
Low-alloy steel	60 to 70% Helium + 25 to 35% Argon + 4.5% Carbon Dioxide	—	Minimum reactivity; excellent toughness; excellent arc stability, wetting characteristics, and bead contour; little spatter
	75% Argon + 25% Carbon Dioxide	—	Fair toughness; excellent arc stability, wetting characteristics, and bead contour; little spatter
Aluminum, copper, magnesium, nickel, and their alloys	Argon and Argon + Helium mixtures	Over ⅛" (3.2 mm)	Argon satisfactory on sheet metal; argon-helium preferred base material

Table 4 Spray Transfer Shielding Gases and Their Applications

Gas Mixture	Shielding Gas	Thickness	Advantages
Aluminum	100% Argon	0 to 1" (0 to 25 mm)	Best metal transfer and arc stability, least splatter
	65% Argon + 35% Helium	1 to 3" (27 to 76 mm)	Higher heat input than straight argon; improved fusion characteristics with 5XXX series Al-Mg alloys
	75% Helium + 25% Argon	Over 3" (76 mm)	Highest heat input; minimizes porosity
Magnesium	100% Argon	–	Excellent cleaning action
Carbon steel	Argon + 3 to 5% Oxygen	–	Improves arc stability; produces a more fluid and controllable weld puddle; good coalescence and bead contour; minimizes undercutting; permits higher speeds than pure argon
	Argon + 8 to 10% Carbon Dioxide	–	High-speed mechanized welding; low-cost manual welding
Low-alloy steel	98% Argon + 2% Oxygen	–	Minimizes undercutting; provides good toughness
Stainless steel	99% Argon + 1% Oxygen	–	Improves arc stability; produces a more fluid and controllable weld puddle; good coalescence and bead contour; minimizes undercutting on heavier stainless steels
	98% Argon + 2% Oxygen	–	Provides better arc stability, coalescence, and welding speed than 1% oxygen mixture for thinner stainless steel materials
Nickel, copper, and their alloys	100% Argon	Up to ⅛" (3.2 mm)	Provides good wetting; decreases fluidity of weld material
	Argon + Helium mixtures		Higher heat inputs of 50 and 75% helium mixtures offset high heat dissipation of heavier gauges
Titanium	100% Argon		Good arc stability; minimum weld contamination; inert gas backing is required to prevent air contamination on back of weld area

force and can severely injure personnel from the blast and blown debris. The cylinder itself can become a powerful, uncontrolled projectile. When transporting and handling cylinders, always observe the following rules:

WARNING!

Do not remove the protective valve cap unless the cylinder is secured. If a cylinder falls over and the valve breaks off, the cylinder may rocket away uncontrollably, causing severe injury or death to anyone in its way.

- The safety cap should always be installed over the valve except when the cylinder is connected for use.
- When in use, the cylinder must always be secured to prevent it from falling. It should be chained or clamped to the welding machine or to a post, beam, or pipe.
- A cylinder cart should be used to transport a cylinder. Cylinders should be secured in the cart.
- Cylinders should never be hoisted by slings or magnets. They can slip out of the sling or fall from the magnets. Do not route a sling through the cap openings. Always use a hoisting basket or similar device.
- Always open the cylinder valve slowly. Once pressure has been applied to the system, the valve should be completely opened to its back-seated position to prevent leakage from around the valve stem.

2.7.0 Welding Equipment Setup

To weld safely and efficiently, the welding equipment must be properly set up (*Figure 37*). The following sections explain the steps for setting up welding equipment.

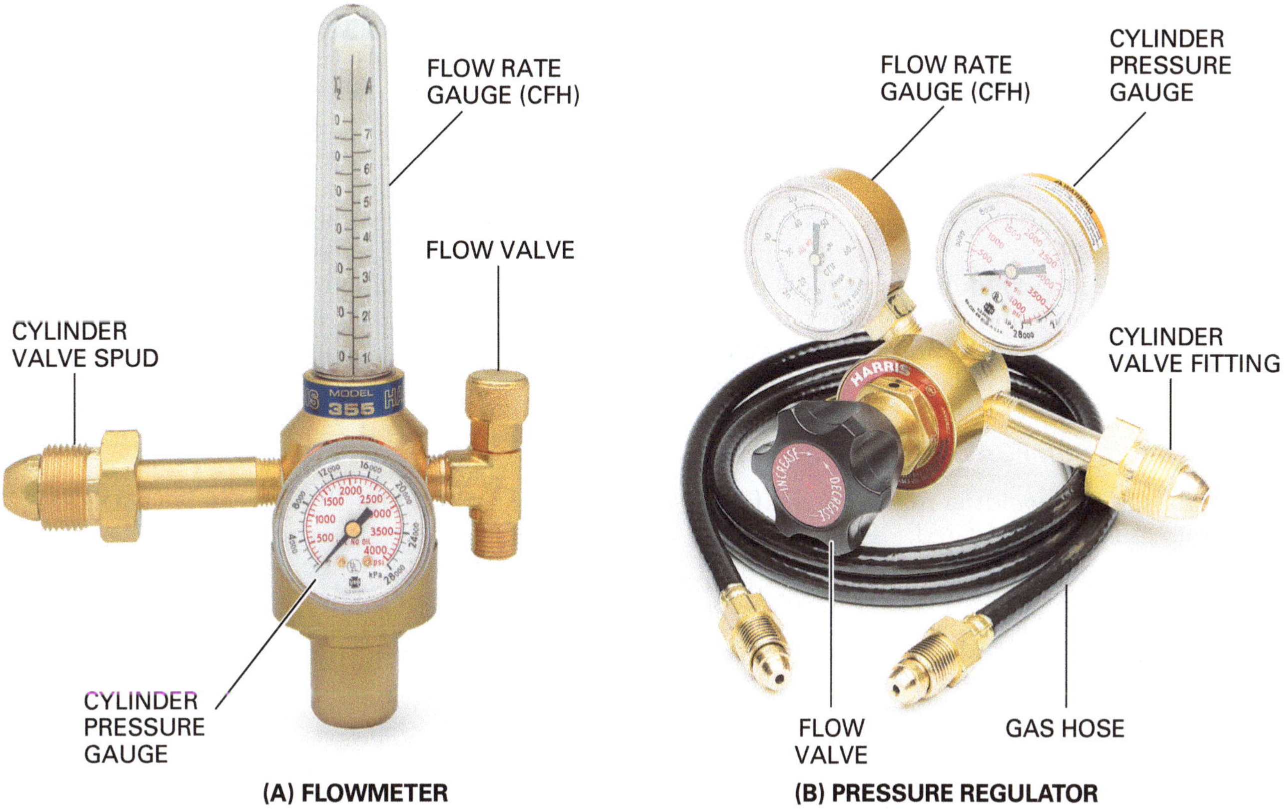

Figure 36 Shielding-gas flowmeter and pressure regulator.

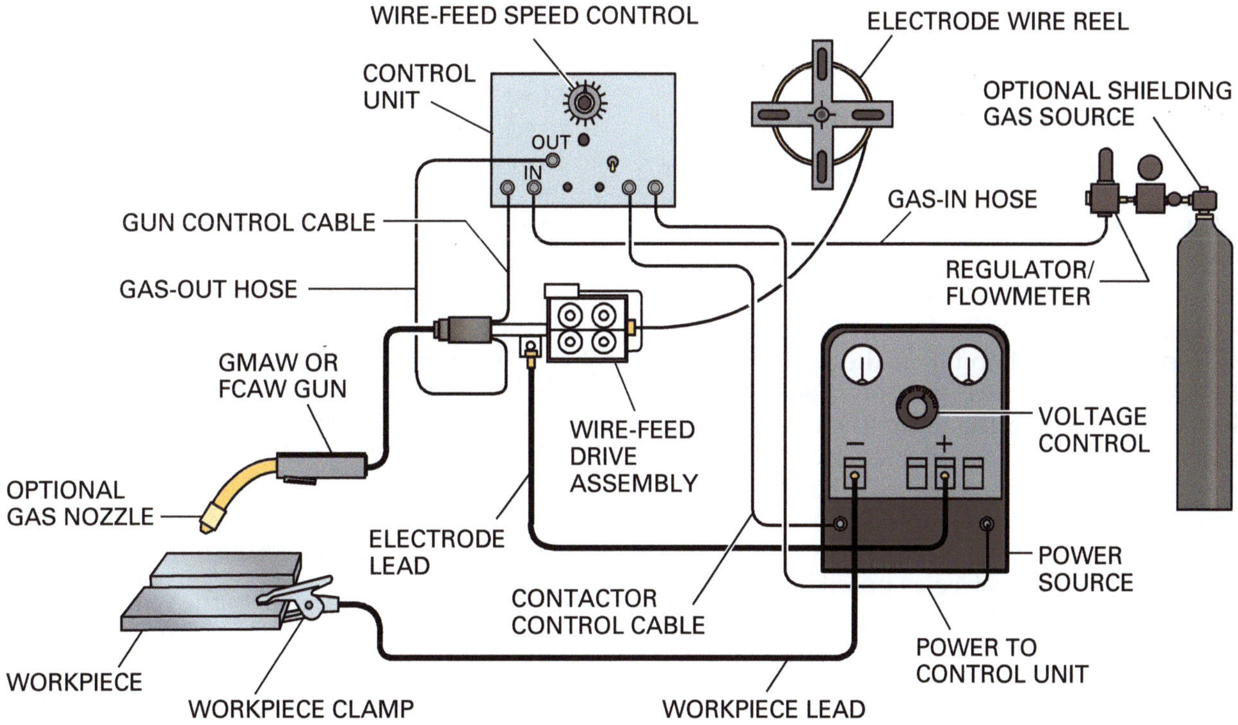

NOTE: The polarity of the gun and workpiece leads is determined by the type of filler metal and application.

Figure 37 Diagram of a GMAW/FCAW system

2.7.1 Selecting a GMAW Welding Machine

To select a GMAW welding machine, the following factors must be considered:

- GMAW requires a constant-voltage or constant-voltage/constant-current welding machine.
- A DCEP-type welding current is required.
- The maximum amperage required for the task must be available.
- The primary power requirements must be met. Ensure that the proper voltage and circuit current capacity is available, and that the welding machine plug and receptacle are compatible.

2.7.2 Selecting an FCAW Welding Machine

To select an FCAW welding machine, the following factors must be considered:

- FCAW requires a constant-voltage power source.
- The welding current required for FCAW is DC.
- The required operating amperage ranges from 150 amps with a 0.045" (1.2 mm) electrode to 650 amps or more with a ⅛" (3.2 mm) electrode; larger electrodes require even higher amperage.
- The primary power requirements must be met. Ensure that the proper voltage and circuit current capacity is available, and that the welding machine plug and receptacle are compatible.

2.7.3 Positioning the Equipment

Because the gun cables are short, the wire feeder must be positioned close to the work to be performed. The power source must also be located near the wire feeder because of the limited length of the contactor/control cable that runs from the power source to the wire feeder. Normally, the wire feeder is located on top of the power source, on the floor, or a cart near the power source.

Select a site where the equipment will not be in the way but will be protected from welding, cutting, or grinding sparks. There should be good air circulation to keep the machine cool. The environment should be free from explosive or corrosive fumes and as free as possible from dust and dirt. Welding machines have internal cooling fans that will pull these materials into the welding machine if they are present. Also, dust and dirt can collect on the wire, causing weld contamination and clogging of the liner in the gun cable. The site should be free of standing water or water leaks. If an engine-driven unit is used, position it so that it can be easily refueled and serviced.

There should be easy access to the site so that the equipment can be started, stopped, or adjusted as needed. If the machine will be plugged into an outlet, be sure that the outlet has been properly installed by a qualified electrician to ensure that it is grounded. Also, be sure to identify the location of the electrical disconnect before plugging the welding machine into the outlet.

2.7.4 Moving Welding Power Sources

Large engine-driven generators are mounted on a trailer frame, and they can easily be moved by a pickup truck or tractor with a trailer hitch. Other types of welding machines may have a skid base or be mounted on steel or rubber wheels. Be careful when moving wheel-mounted welding machines by hand. Some machines are top-heavy and could fall over in a tight turn or if the floor or ground is uneven or soft.

WARNING!

Secure or remove the wire feeder and gas cylinder before attempting to move a welding machine. If a welding machine starts to fall over, do not attempt to hold it. Welding machines are very heavy, and severe crushing injuries can occur if a welding machine falls on you.

Most welding machines have a lifting eye that is used to move machines mounted on skids or to lift any machine. Before lifting a welding machine, check the equipment specifications for the weight. Be sure the lifting device and tackle can handle the weight of the machine. Always use a sling and/or a shackle. Never attempt to lift a machine by placing the lifting hook directly in the machine's lifting eye. This is because the safety latch on the hook cannot be closed. Also, before lifting or moving a welding machine, be sure the welding cables are secure. *Figure 38* shows the proper way to lift a welding machine.

2.7.5 Connecting the Shielding Gas

The shielding gas connects to the wire feeder unit or to the power source if the wire feeder is integrated with the power source.

Follow these steps to connect the shielding gas:

Step 1 Identify the required shielding gas by referring to the WPS or site quality standards.

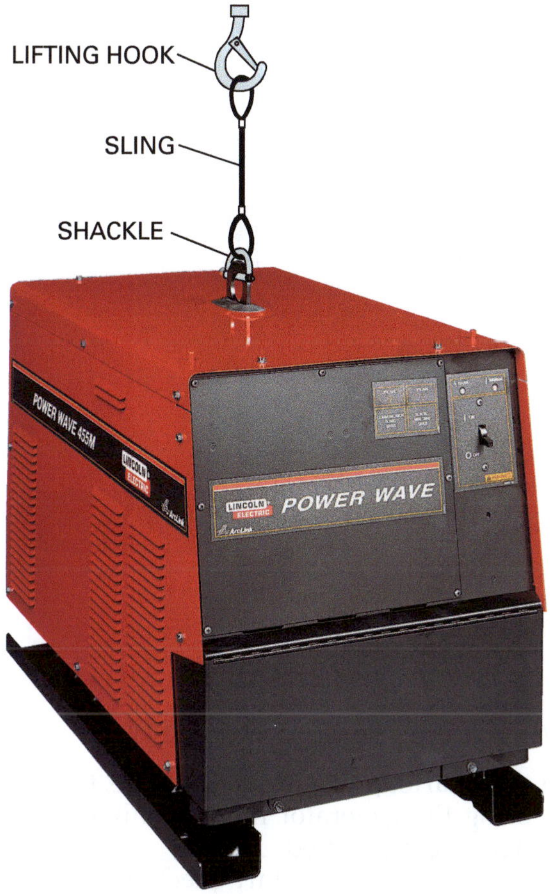

Figure 38 Lifting a welding machine.

Step 2 Locate a cylinder of the correct gas or mixture, and secure it near the wire feeder. Be sure to secure the cylinder so that it cannot fall over.

Step 3 Remove the cylinder's protective cap, momentarily crack the cylinder valve open to blow out any dirt, and then close it.

WARNING!

Even though they are available, do not use an adapter to connect a regulator/flowmeter equipped with one type of Compressed Gas Association (CGA) connection to a gas cylinder that has a different CGA connection. The CGA connections are specific to the types of gas and cylinder pressures permitted for the connection.

Step 4 Using a regulator/flowmeter with the correct CGA connection for the cylinder, install the pressure regulator/flowmeter on the cylinder.

Step 5 Connect the gas hose to the flowmeter and to the gas connection on the wire feeder (or welding machine).

Step 6 Slowly crack open the cylinder valve, and then open it completely.

WARNING!

Do not open the valve quickly or the flowmeter gauges may break. If a gauge breaks, the glass may shatter and the gas may escape, potentially injuring the operator.

Step 7 Adjust the gas flow to the specified flow rate using the flowmeter valve.

Some flowmeters are equipped with several scales of different calibrations around the same sight tube, for monitoring the flows of different gases. These scales differ due to the different densities of shielding gases. Be sure to rotate the scales or read the correct side for the type of gas being used.

2.7.6 Selecting and Installing Filler Wire

Follow these steps to select and install the filler wire:

Step 1 Identify the filler wire required by referring to the WPS or site quality standard.

Step 2 Locate a spool of the wire, and mount it on the wire feeder spool holder. Lock the holder, and set the drag brake for a slight drag. The brake control is usually located on the end of the spool axle.

Step 3 Check (and change if necessary) the wire feed drive wheels, liner, and gun contact tube to make sure that they are the correct size for the wire selected. Adjust the wheel tension if necessary. For detailed instructions on changing or adjusting the wire feeder drive wheels or changing the gun's contact tube or liner, refer to the specific manufacturer's instructions.

Step 4 Feed the wire into the feeder wheels, and then use the JOG control to feed it through the gun cable and the gun contact tube.

2.7.7 Placing the Workpiece Clamp

The workpiece clamp must be properly placed to prevent damage to surrounding equipment. If the electrical welding current travels through a bearing, seal, valve, or contacting surface, it could cause severe damage from heat and arcing, which would require that these items be replaced. Carefully check the area to be welded, and position the workpiece clamp so the welding current will not pass through any contacting surface. When in doubt, ask your supervisor for assistance before proceeding.

> **CAUTION**
>
> Welding current passing through electrical or electronic equipment will cause severe damage to the affected equipment. Also, before welding on any type of equipment with a battery, the ground lead at the battery must be disconnected to protect the electrical system.

> **WARNING!**
>
> If welding is to be done near a battery, the battery must be removed. Batteries produce hydrogen gas, which is extremely explosive. A welding spark could cause the battery to explode, showering the area with battery acid.

Workpiece clamps must never be connected to pipes carrying flammable or corrosive materials. The welding current could cause overheating or sparks, resulting in an explosion or fire.

The workpiece clamp must make a good electrical contact when it is connected. Dirt and paint will inhibit the connection and cause arcing, resulting in overheating of the workpiece clamp. Dirt and paint also make the welding current unstable and can cause defects in the weld. Clean the surface before connecting the workpiece clamp. If the workpiece clamp is damaged or does not close securely onto the surface, replace it.

2.7.8 Energizing the Power Source

Electrically powered welding machines are energized by plugging them into an electrical outlet. The electrical requirements will be located on the equipment specification tag that is prominently displayed on the machine. Most machines requiring single-phase, 240VAC power have a three-pronged plug. Machines requiring three-phase, 208 or 460VAC power have a four-pronged plug. Both types of plugs are shown in *Figure 39*.

If a welding machine does not have a power plug, an electrician must connect it. The electrician will add a plug or hard-wire the machine directly into an electrical box.

> **WARNING!**
>
> Never use a welding machine until you identify the location of the electrical disconnect switch for that specific machine or receptacle. In the event of an emergency, you must be able to turn off the power quickly to the welding machine at the disconnect switch.

2.7.9 Starting Engine-Driven Generators

Before welding can take place with an engine-driven generator, the engine must be checked and then started. As with a car engine, the engine powering the generator must also have routine maintenance performed.

Figure 39 Typical three- and four-pronged plugs.

Many sites have prestart checklists that must be completed and signed prior to starting or operating an engine-driven generator. Check with your supervisor. If your site has such a checklist, complete and sign it. If your site does not have a prestart checklist, perform the following checks before starting the engine:

- Check the oil, using the engine oil dipstick. If the oil is low, add the appropriate grade of oil for the conditions, according to the manufacturer's maintenance instructions.
- Check the coolant level in the radiator if the engine is liquid-cooled. If the coolant level is low, add the proper coolant.

> **CAUTION**
>
> Do not add plain water to radiators that contain antifreeze. Antifreeze not only protects radiators from freezing in cold weather, it also has rust inhibitors and additives to protect heat-transfer surfaces. If the antifreeze is diluted, it will not provide the proper protection.

- Check the fuel. The unit may have a fuel gauge or a dipstick. If the fuel is low, add the correct fuel (diesel or gasoline) to the fuel tank. The type of fuel required should be marked on the fuel tank. If it is not marked, contact your supervisor to verify the fuel required, and have the tank marked.

CAUTION

Adding gasoline to a diesel engine or diesel to a gasoline engine will cause severe engine problems, and may create a fire hazard. Always be sure to add the correct fuel to the fuel tank.

- Check the battery water level unless the battery is sealed. Add demineralized water if the battery water level is low.
- Check the electrode holder to be sure it is not grounded. If the electrode holder is grounded, it will arc and overheat the welding system when the welding machine is started. This may cause damage to the equipment.
- Open the fuel shut-off valve if the equipment has one. If there is a fuel shut-off valve, it should be located in the fuel line between the fuel tank and the carburetor.
- Record the hours from the hour meter. An hour meter records the total number of hours that the engine runs. This information is used to determine when the engine needs to be serviced. The hours will be displayed on a gauge similar to an odometer.
- Clean the unit. Use a compressed air hose to blow off the engine and generator or alternator. Use a rag to remove heavier deposits that cannot be removed with the compressed air.

WARNING!

Always wear eye protection when using compressed air to blow dirt and debris from surfaces. Never point the compressed air nozzle at yourself or anyone else.

Most engines have an On/Off ignition switch and a starter. They may be combined into a key switch similar to the ignition on a car. To start the engine, turn on the ignition switch and press the starter. Release the starter when the engine starts. The engine speed is controlled by the governor. If the governor switch is set for idle, the engine will slow to an idle after a few seconds. If the governor is set to welding speed, the engine will continue to run at the welding speed.

Small engine-driven generators may have an On/Off switch and a pull cord. These are started by turning on the ignition switch and pulling the cord, in much the same way as a lawn mower is started. Engine-driven generators should be started about 5 to 10 minutes before they are needed for welding. This will allow the engine to warm up before a welding load is applied.

If no welding is required for 30 minutes or more, stop the engine by turning off the ignition switch. If you are finished with the welding machine for the day, close the fuel valve if there is one.

Engine-driven generators require regular preventive maintenance to keep the equipment operating properly. Most sites have a preventive maintenance schedule based on the hours that the engine operates. In severe conditions, such as in very dusty environments or cold weather, maintenance may have to be performed more frequently.

CAUTION

To prevent damage to the equipment, perform preventive maintenance as recommended by the site procedures or by the manufacturer's maintenance schedule in the equipment manual.

The responsibility for performing preventive maintenance varies by site. Check with your supervisor to determine who is responsible for performing preventive maintenance. Even though the welder may not be responsible for conducting preventive or periodic maintenance, all welders share an obligation to monitor and protect the equipment, and to report problems. As is true for most crafts, a welder cannot produce welds without functional and reliable equipment.

Adjusting Flowmeters

In most cases, the top of the floating flowmeter ball is used as the reference point for setting the gas flow. However, always check the manufacturer's instructions to determine the correct adjustment method. Use the lowest flow rate as often as possible to conserve gas and protect the environment.

Gas Flow Rates

More gas flow is not always better. Too much flow can deflect off the workpiece. The resultant swirling gas can create turbulence that draws in air, contaminating the weld and creating pinholes. If black flecks occur in the weld puddle or pinholes occur in the weld, try reducing the gas flow.

When performing preventive maintenance, follow the manufacturer's guidelines in the equipment manual. Typical procedures to be performed as a part of preventive maintenance include the following:

- Changing the oil
- Changing the gas filter
- Changing the air filter
- Checking/changing the antifreeze
- Greasing the undercarriage
- Repacking the wheel bearings

Dust Removal

Blow out the welding machine and drive unit with compressed air occasionally to remove dust. A significant buildup of dust inside the machine can provide an alternate path for the electricity and could result in some of the components shorting out.

Additional Resources

AWS B1.10M/B1.10 Guide for the Nondestructive Examination of Welds. Miami, FL: American Welding Society.

AWS B1.11 Guide for the Visual Examination of Welds. Miami, FL: American Welding Society.

Lincoln Electric website: **http://www.lincolnelectric.com** offers sources for products and training.

Modern Welding Technology. Howard B. Cary. Englewood Cliffs, NJ: Prentice Hall, Inc.

OSHA 1926.351, Arc Welding and Cutting. Current Edition. Washington, DC: Occupational Safety and Health Administration (OSHA).

Welding Handbook. Volume 1. Welding Science & Technology. Miami, FL: American Welding Society, 2001.

Welding Handbook. Volume 2. Part 1: Welding Processes. Miami, FL: American Welding Society, 2004.

2.0.0 Section Review

1. The two types of voltage associated with welding current are ____.
   a. open-circuit voltage and operating voltage
   b. operating voltage and arc voltage
   c. closed-circuit voltage and welding voltage
   d. operating voltage and closed-circuit voltage

2. The deflection of an arc caused by the attraction or repulsion of the arc's magnetic field with the weld current's magnetic field in the base metal is known as ____.
   a. pinch effect
   b. slope curvature
   c. arc blow
   d. inductance bending

3. Quick disconnects used on welding cable are connected and disconnected with a ____.
   a. terminal lug
   b. screwing action
   c. pronged plug
   d. half twist

4. A device that can be used to start and stop the wire feeder on a welder without energizing the welding current is a ____.
   a. spool drag brake
   b. wire jog switch
   c. feed speed control
   d. wire meltback timer

5. The part of a GMAW/FCAW gun that provides the electrical connection from the welding machine electrode to the electrode wire is the ____.
   a. gun handle
   b. fiber insulator
   c. contact tip
   d. extraction nozzle

6. The most commonly used shielding gas for GMAW/FCAW-G processes is ____.
   a. argon (Ar)
   b. CO_2
   c. helium (He)
   d. carbon monoxide (CO)

7. What rigging device should never be placed into a welding machine's lifting eye when the machine is being lifted or moved?
   a. A synthetic sling
   b. The lifting hook
   c. A wire rope sling
   d. Any type of shackle

SECTION THREE

3.0.0 FILLER METALS

Objective

Identify various GMAW and FCAW filler metals.

a. Identify various GMAW filler metals.
b. Identify various FCAW filler metals.

Performance Task

1. Set up GMAW and FCAW equipment with appropriate shielding gases and filler metals.

GMAW and FCAW filler metals are explained in this section. The WPS specifies the filler metals to be used. The improper selection of filler metal can cause the weld to fail.

Regardless of the type of filler metal being used, there is often a need to cut and manipulate the wire. *Figure 40* shows an example of a special tool that is used for this purpose. Commonly called MIG welding pliers or wire feed pliers, this tool is designed for drawing out wire, cutting wire, removing and installing contact tips and nozzles, removing welding spatter from nozzles, and light hammering.

3.1.0 GMAW Filler Metals

GMAW uses a continuous solid wire electrode for filler material. This filler wire is drawn from high-grade pure alloys compounded for different applications. During manufacture, the wire is drawn, cleaned and inspected for defects several times. Standard wire diameters are 0.023" (0.6 mm), 0.030" (0.7 mm), 0.035" (0.9 mm), 0.045" (1.2 mm), 0.052" (1.4 mm), and 0.062" (1.5 mm). Note

Figure 40 Wire feed pliers.

Filler Wire and Filler Metal

The terms *filler wire* and *filler metal* are both used to describe the electrode used in GMAW and FCAW. Filler metal is also referred to as filler rod when used for GTAW.

that the metric equivalents used here represent actual metric wire sizes, and are not direct equivalents. The finished wire is wound on spools 4" (102 mm), 8" (204 mm), 12" (30 cm), or 14" (36 cm) in diameter. The most common spool is 12" (30 cm) in diameter and weighs 30 pounds (13.6 kg). Reels or drums up to 30" (76 cm) in diameter and weighing up to 1,000 pounds (454 kg) are also available for stationary production welding.

Several industry organizations and the United States government publish specification standards for welding wire. The most common are the AWS specifications. The purpose of the AWS specifications is to set standards that all manufacturers must follow when manufacturing welding consumables. This ensures consistency for the user regardless of who manufactured the product. The specifications set standards for the following:

- Classification system, identification, and marking
- Chemical composition of the deposited weld metal
- Mechanical properties of the deposited weld metal

The following AWS specifications refer to gas metal arc welding wires and rods:

- *A5.7: Specification for Copper and Copper Alloy Bare Welding Rods and Electrodes*
- *A5.9/A5.9M: Specification for Bare Stainless Steel Welding Electrodes and Rods*
- *A5.10/A.510M: Specification for Bare Aluminum and Aluminum Alloy Welding Electrodes and Rods*
- *A5.14/A5.14M: Specification for Nickel and Nickel Alloy Bare Welding Electrodes and Rods*
- *A5.16/A5.16M: Specification for Titanium and Titanium Alloy Welding Electrodes and Rods*
- *A5.18/A5.18M: Specification for Carbon Steel Electrodes and Rods for Gas Shielded Arc Welding*
- *A5.19: Specification for Magnesium Alloy Welding Electrodes and Rods*
- *A5.28/A5.28M: Specification for Low-Alloy Steel Electrodes and Rods for Gas Shielded Metal Arc Welding*

NOTE

An AWS specification number is generally followed by a four-digit number that indicates the year that the specification was last revised.

Filler wire is graded for three major areas of use:

- *General use* – Wire meets specifications. No record of chemical composition, strength, or similar factors is supplied to the user with the wire purchase.
- *Rigid control fabrication* – A Certificate of Conformance is supplied with the wire at purchase. The stock is identified by code numbers located on the roll package.
- *Critical use* – A Certified Chemical Analysis report is supplied, and records are kept by the fabricator of welds and processes for later reference on aircraft, nuclear reactors, and pressure vessels.

The following factors affect the selection of a GMAW filler wire:

- Base metal chemical composition
- Base metal mechanical properties
- Weld joint design
- Service or specification requirements
- Shielding gas used

Commonly used filler wire types include the following:

- Carbon steel
- Stainless steel
- Aluminum and aluminum alloy
- Copper and copper alloy
- Low-alloy steel

Settings for electrode wire of the same classification can vary widely depending upon the manufacturer. It is important to check the manufacturer's recommended settings for each specific wire used. This information can typically be found on the wire manufacturer's website or catalog.

Rod and Electrode Designations

The letter R that follows the E in an AWS classification number indicates that the classification is used as a metal filler rod as well as an electrode. GMAW and GTAW filler metals are both labeled the same way when used for gas tungsten arc welding (GTAW).

3.1.1 Carbon Steel Filler Metals

Carbon steel filler metals are identified by specification *AWS A5.18*. The filler metal is available in wire reels and rod form. The wire classification number is located on a label on the side of the reel or container (*Figure 41*).

All carbon steel filler metals contain alloys such as silicon, manganese, aluminum, and carbon. Other alloys such as nickel, chromium, and molybdenum are also often added. The purpose of the alloys is as follows:

- *Silicon (Si)* – Concentrations of 0.40 to 1.00 percent are employed to deoxidize the puddle and to strengthen the weld. Silicon above 1 percent may make the welds crack-sensitive. Silicon also lowers the melting temperature of the wire and promotes wetting.
- *Manganese (Mn)* – Concentrations of 1 to 2 percent are also employed as a deoxidizer and to strengthen the weld. Manganese also decreases hot-crack sensitivity.
- *Aluminum (Al), titanium (Ti) and zirconium (Zr)* – One or more of these may be added in very small amounts for deoxidizing. Strength may also be increased.
- *Carbon (C)* – Concentrations of 0.05 to 0.12 percent are used to add strength without adversely affecting ductility, porosity, or toughness.
- *Nickel (Ni), chromium (Cr), and molybdenum (Mo)* – May be added in small amounts to improve corrosion resistance, strength, and toughness.

In many cases, carbon steel filler metal wire is copper coated. The copper coating protects the filler metal wire from corrosion and makes electrical contact easier as the wire passes through the copper contact tube at the end of a gun.

AWS carbon steel filler metal wire classifications and their uses are listed in *Table 5*.

3.1.2 Low-Alloy Steel Filler Metals

Low-alloy steel filler metals are identified by *AWS Specification A5.28*. The filler metal is available in wire reel and rod form. The wire classification number is located on a label on the side of the reel or container. The classification is similar to the carbon steel filler metal classification except that there is a letter in the suffix at the end of the classification (*Figure 42*). The letter suffix indicates that the filler metal contains alloys (nickel, manganese, molybdenum, etc.) that give the deposited weld metal special properties. Also, the tensile strength of low-alloy filler metals is often higher than that of carbon steel filler metals. For specific

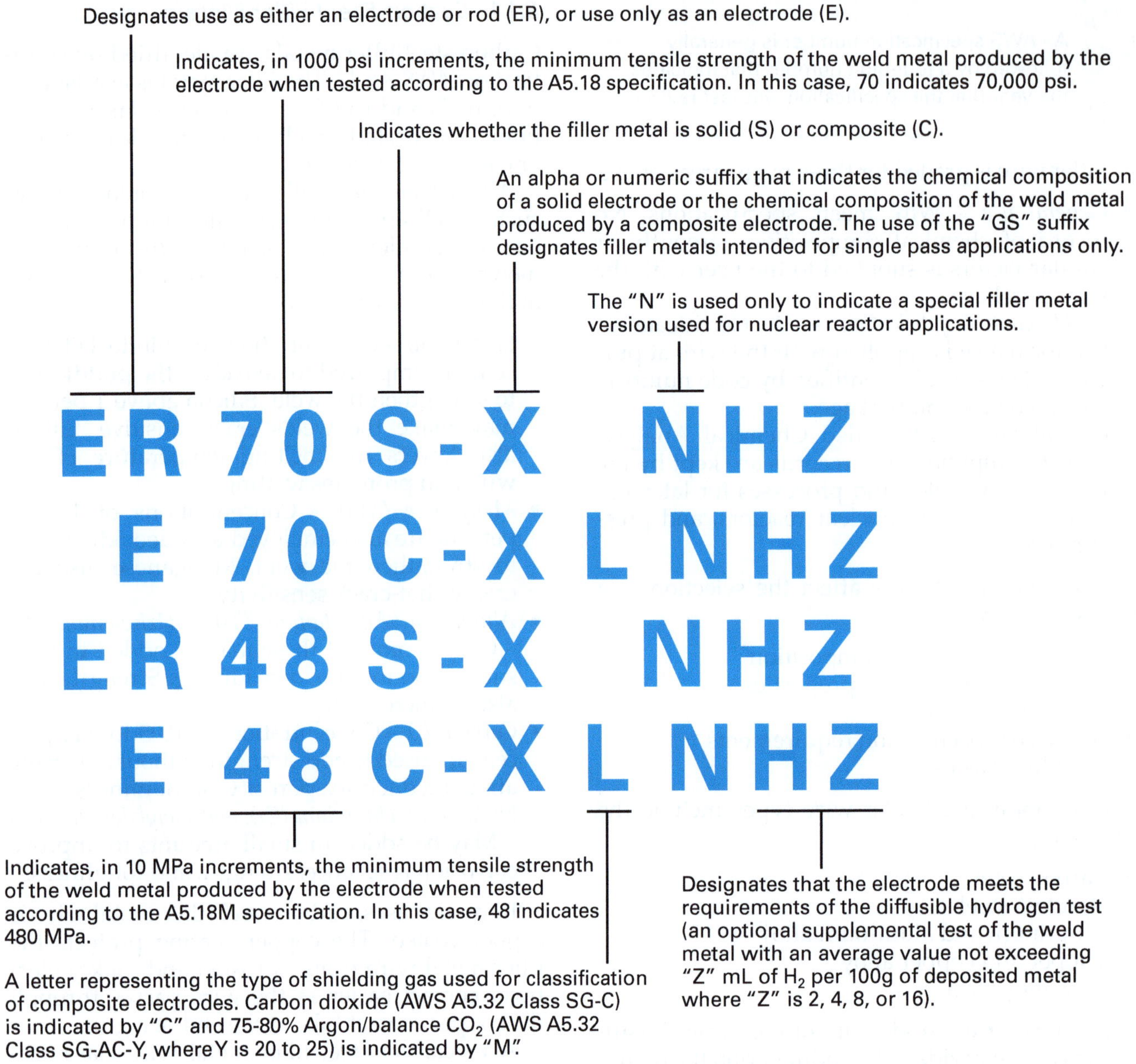

Figure 41 AWS classification for carbon steel filler metal.

information on low-alloy filler metal composition and intended use, refer to the manufacturer's specifications.

3.1.3 Stainless Steel Filler Metals

Stainless steel filler metals are identified by specification *AWS A5.9*. The filler metal is available in wire reel, flat strip, and rod form (*Figure 43*).

Stainless steel filler metal should be selected to closely match the alloy composition of the base metal. Stainless steel filler metals also require specific shielding gases or gas mixtures.

3.1.4 Aluminum and Aluminum Alloy Filler Metals

Aluminum filler metals are covered by *AWS Specification A5.10*. Aluminum filler metals usually contain magnesium, manganese, zinc, silicon, or copper for increased strength. Corrosion resistance and weldability are also considerations. Aluminum filler metals are designed to weld specific types of aluminum and should be selected for compatibility. The most widely used aluminum filler metals are ER4043 (contains silicon) and ER5356 (contains magnesium). The filler metal classification number is located on a label on the side of the reel or container (*Figure 44*).

Table 5 AWS Carbon Steel Wire Classifications and Uses

Wire Classification	Use
ER70S-2 ER48S-2	Welds all grades of carbon steel, including killed (highly deoxidized by adding aluminum powder or ferrosilicon), semi-killed, and rimmed (slightly deoxidized by adding a small amount of aluminum powder). Also welds carbon steel with rusted surfaces. Acceptable shielding gases include Ar-O_2, Ar-CO_2, and CO_2. The viscous weld puddle is well suited to out-of-position welding. Wetting is helped by keeping the oxygen or CO_2 percentage relatively high.
ER70S-3 ER48S-3	Welds killed and semi-killed steels (used in appliances, automobiles, and farm equipment) with CO_2, Ar-O_2, or Ar-CO_2 shielding gases.
ER70S-4 ER48S-4	Welds semikilled and rimmed steels (used in boilers and pressure vessels, piping, ship steels, structural steels, and A515 grades 55-70 steels) with CO_2, Ar-O_2, or Ar-CO_2 shielding gases.
ER70S-6 ER48S-6	Welds many carbon steels with CO_2 and high welding current. High-speed welding can be done with Ar-O_2 with 5% or more O_2. Either spray transfer or short-circuiting transfer can be used.
ER70S-7 ER48S-7	Welds a wide variety of steels with superior weld quality and appearance. Porosity and weld defects caused by mill scale or rust are minimized.
ER70S-G ER48S-G E70C-G E48C-G	Filler metals not included in the preceding classes that only have certain mechanical properties specified. Absence of a C or M suffix means that the manufacturer should be consulted for the recommended shielding gas and intended applications.
E70C-GS E48C-GS	Composite stranded or metal cored electrodes intended for only single pass applications. These electrodes may have higher alloy content and the manufacturer should be consulted for specific applications.
E70C-3 E48C-3 E70C-6 E48C-6	Composite stranded or metal cored electrodes intended for both single and multipass applications on steel. They are characterized by a spray arc and excellent bead wash characteristics.

3.1.5 Copper and Copper Alloy Filler Metals

Copper filler metals are covered by *AWS Specification A5.7*. Most copper filler metals contain other elements that increase strength, deoxidize the weld metal, and match the base metal composition. *Figure 45* shows the AWS classification for copper filler metal.

3.1.6 Nickel and Nickel Alloy Filler Metals

Nickel-based filler metals (*Figure 46*) are covered by *AWS Specification A5.14*. These filler metals contain other elements to match base metal applications and to increase the strength and quality of the weld metal. For GTAW, DCEN is used with high-purity argon, helium, or both as a shielding gas. For GMAW, DCEP is used. Argon shielding gas is often used, as are mixtures of argon and helium.

3.1.7 Magnesium Alloy Filler Metals

Magnesium alloy filler metals (*Figure 47*) are covered by *AWS Specification A5.19*. These filler metals are usually used with GTAW, GMAW, and plasma arc welding (PAW) processes. Oxyfuel welding should be used only for temporary repair work. For GTAW welding, the techniques and equipment are similar to those for aluminum. Argon, helium, or both are used for shielding. Alternating current is preferred for arc cleaning and penetration. Direct current is also employed: DCEP is used for thin materials, and DCEN is used for mechanized welding with helium for deep penetration. GTAW is recommended for the defect repair of clean magnesium castings. GMAW of magnesium alloys is the same as for other metals. Argon is generally used for shielding; however, mixtures of argon and helium are occasionally used. Pulsed and short-circuit GMAW are used for magnesium alloys, and spray transfer without pulsing is also used; however, globular transfer is not suitable.

3.1.8 Titanium and Titanium Alloy Filler Metals

Titanium and titanium alloy filler metals (*Figure 48*) are covered by *AWS Specification A5.16*. These filler metals are generally used with GTAW, GMAW, submerged arc welding (SAW), and PAW

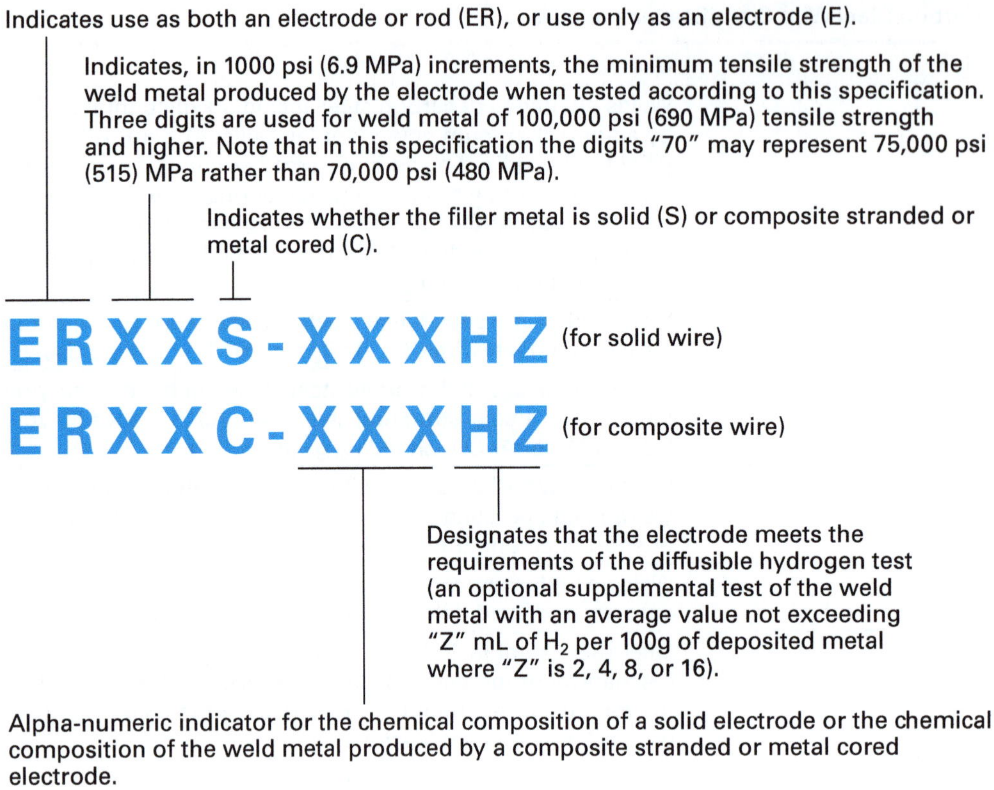

Figure 42 AWS classification for low-alloy steel filler metal.

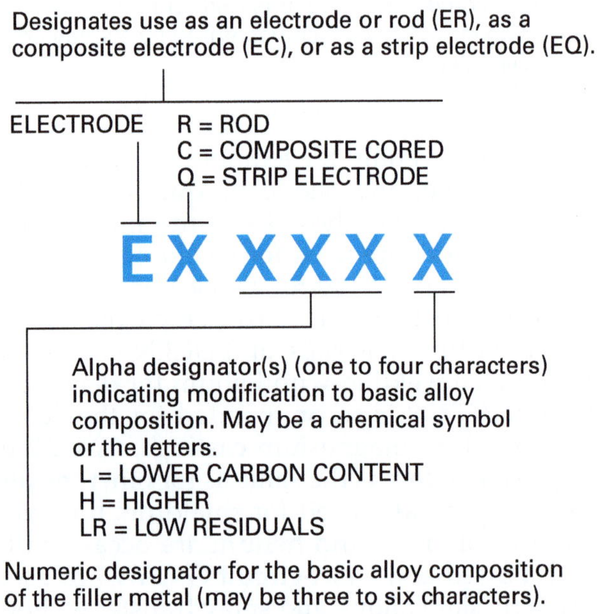

Figure 43 AWS classification for stainless steel filler metal.

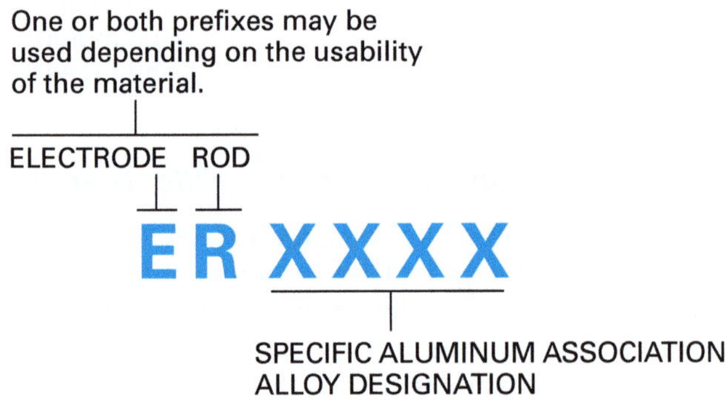

Figure 44 AWS classification for aluminum filler metal.

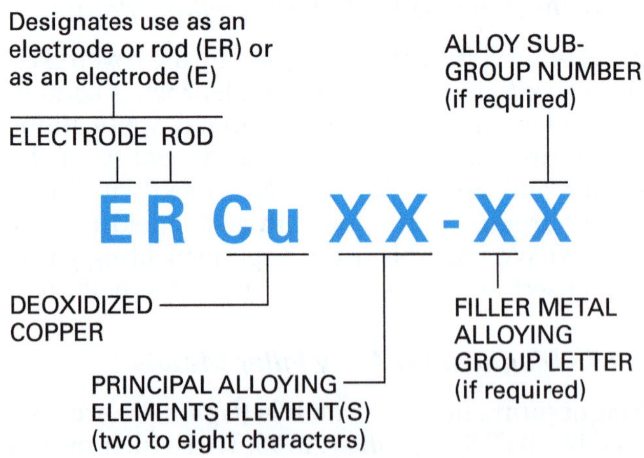

Figure 45 AWS classification for copper filler metal.

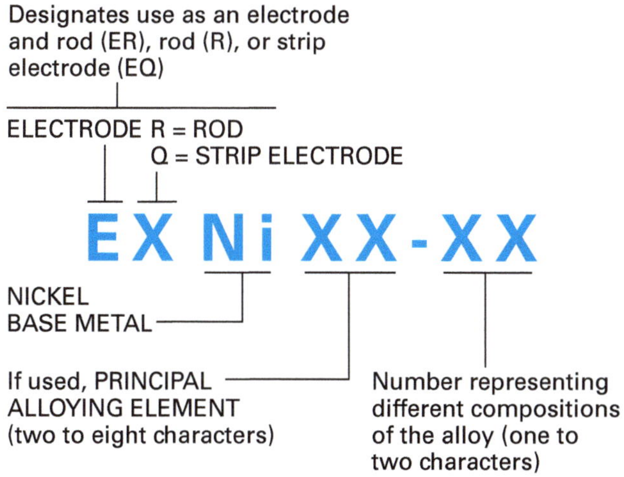

Figure 46 AWS classification for nickel filler metal.

processes. Titanium is sensitive to embrittlement by oxygen, nitrogen, and hydrogen at temperatures above 500°F (260°C). Like aluminum, titanium requires weld cleaning and high-purity gas shielding, and an especially adequate post-flow shielding to avoid embrittlement. Titanium can be successfully fusion welded to zirconium, tantalum, niobium, and vanadium. Titanium should not be welded to copper, iron, nickel, and aluminum. As with magnesium, titanium fines are also flammable, so don't allow them to accumulate.

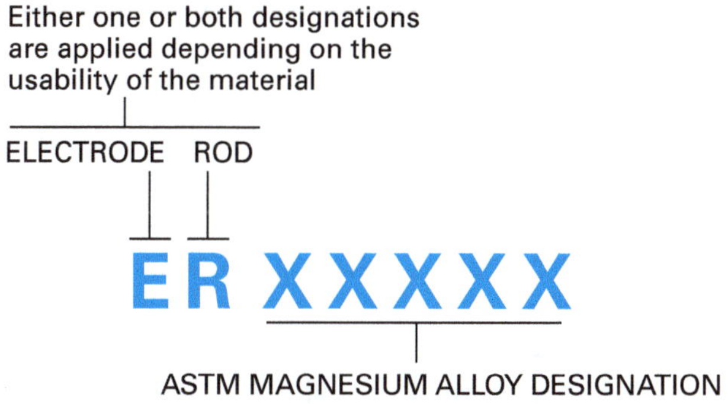

Figure 47 AWS classification for magnesium filler metal.

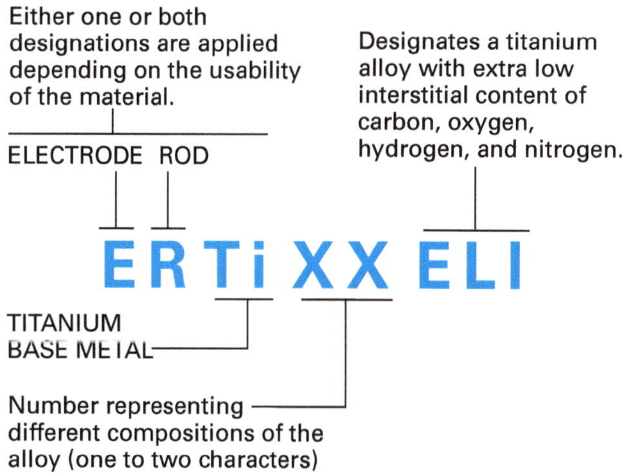

Figure 48 AWS classification for titanium filler metal.

3.2.0 FCAW Filler Metals

FCAW is primarily used for the welding of ferrous metals. These include low- and medium-carbon steels, some low-alloy steels, cast irons, and some stainless steels.

The FCAW electrode is a continuous flux-cored wire electrode. The wire is made by forming a thin filler metal strip into a U-shape, filling the U with the flux material, squeezing the strip shut around the flux, and then drawing the tube through dies to size the wire and compact the flux. During manufacture, the wire is carefully inspected for defects. Flux-cored wires are manufactured in a range of sizes beginning at 0.023" (0.6 mm). The finished flux-cored wire is coiled on all the standard spool sizes and forms.

Keep in mind that all flux-cored wires are considered to be low hydrogen. For this reason, proper electrode control procedures must be followed to keep the wire dry.

By varying the power settings on the welding machine, many smaller sizes of electrode wire can be used both for thin metals and for multiple-pass welds on metals of unlimited thickness in all positions.

Magnesium Fires

While the ignition of magnesium is a very remote possibility when welding, the ignition will cease when the heat source is removed. Ignition of the weld pool is also prevented by the gas shielding used in GTAW, GMAW, and PAW processes. Most magnesium fires occur with accumulations of fines (powder-sized granules) from grinding or filing, or chips from machining. Accumulations of fines on clothing should be avoided. Graphite-based or salt-based powders recommended for extinguishing magnesium fires should be stored in the work area. If large amounts of fines are produced, they should be collected in a water-wash dust collector designed for use with magnesium. Special precautions for the handling and final disposal of wet magnesium fines must then followed.

There have been several AWS standards related to flux-cored wire over the years. However, a new standard has recently been introduced by AWS. *AWS A5.36/A5.36M* is presently in effect and incorporates the information historically provided in several other standards. The new standard covers carbon and low-alloy carbon steel flux-cored products. Manufacturers may continue to label packaging in reference to the previous standards until January 1, 2016. After this date, labeling per the new standard will be required. As a result, both the old and new standards will be seen until all of the products manufactured and labeled before the deadline have been distributed and consumed.

Common AWS specifications for flux-cored electrodes that will be discussed briefly in this section include the following:

- *AWS A5.20/A5.20M: Specification for Carbon Steel Electrodes for Flux Cored Arc Welding*
- *AWS A5.22: Specification for Stainless Steel Electrodes for Flux Cored Arc Welding and Stainless Steel Flux Cored Rods for Gas Tungsten Arc Welding*
- *AWS A5.29/A5.29M: Specification for Low-Alloy Steel Electrodes for Flux Cored Arc Welding*

Note that these standards are discussed below, even though the *AWS A5.36/A5.36M* standard is already in effect. However, a great deal of product awaiting consumption is labeled according to the previous standards. Standards *AWS A5.20/A5.20M* and *AWS A5.29/A5.29M* will be archived and their contents are now reflected in the new standard. The standard related to flux-cored stainless steel electrodes, *AWS A5.22*, will remain in effect.

Each standard defines specifications for the following:

- Classification, identification, and marking of the electrode wire
- Chemical composition of the deposited weld metal
- Chemical composition of the flux filling
- Mechanical properties of the deposited weld metal

FCAW flux-cored electrodes are now classified by the AWS in the new standard using either a fixed or an open classification system. The fixed classification system has been carried over from an earlier system that provided limited descriptors for the welding characteristics of the electrode. As an example, *Figure 49* shows the AWS fixed classification for mild steel and low-alloy steel FCAW electrodes under the newer *AWS A5.36* standard.

The open classification system allows more designators to be defined, which provides more information to personnel for selecting electrodes based on process, welding characteristics, strengths, and weld positions. *Figure 50* shows the AWS open classification for mild steel and low-alloy steel FCAW electrodes.

3.2.1 Carbon Steel Flux-Cored Electrodes

Figure 51 shows the AWS A5.20 specification format for carbon steel flux-cored electrodes.

The carbon steel flux-cored electrode welding characteristics as indicated by the AWS classification are shown in *Table 6*. Carbon steel flux-cored electrode classes and uses are shown in *Table 7*.

3.2.2 Low-Alloy Steel Flux-Cored Electrodes

Figure 52 shows the *AWS A5.29* specification format for low-alloy steel flux-cored electrodes.

3.2.3 Stainless Steel Flux-Cored Electrodes and Rods

Figure 53 shows the *AWS A5.22* specification format for stainless steel flux-cored electrodes and rods.

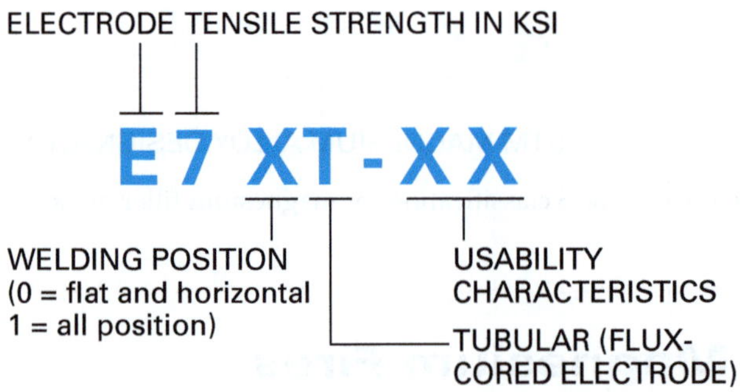

Figure 49 AWS fixed classification for mild steel and low-alloy steel FCAW electrodes.

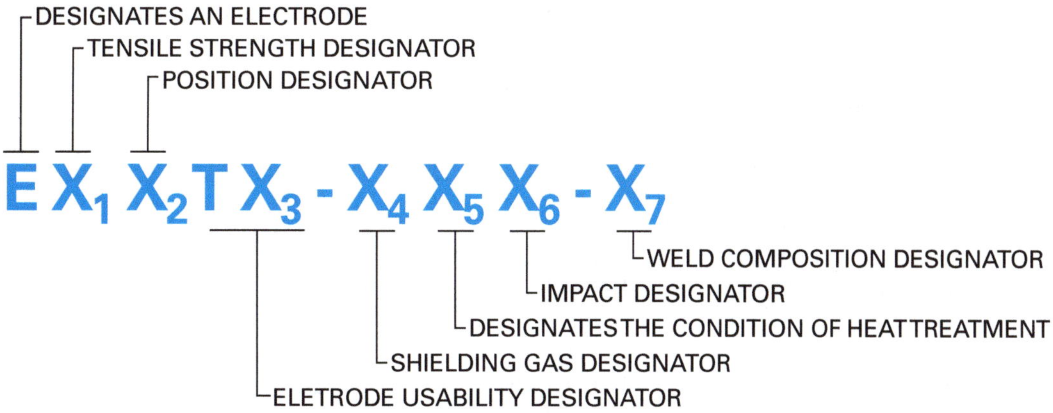

Figure 50 AWS open classification for mild steel and low-alloy steel FCAW electrodes.

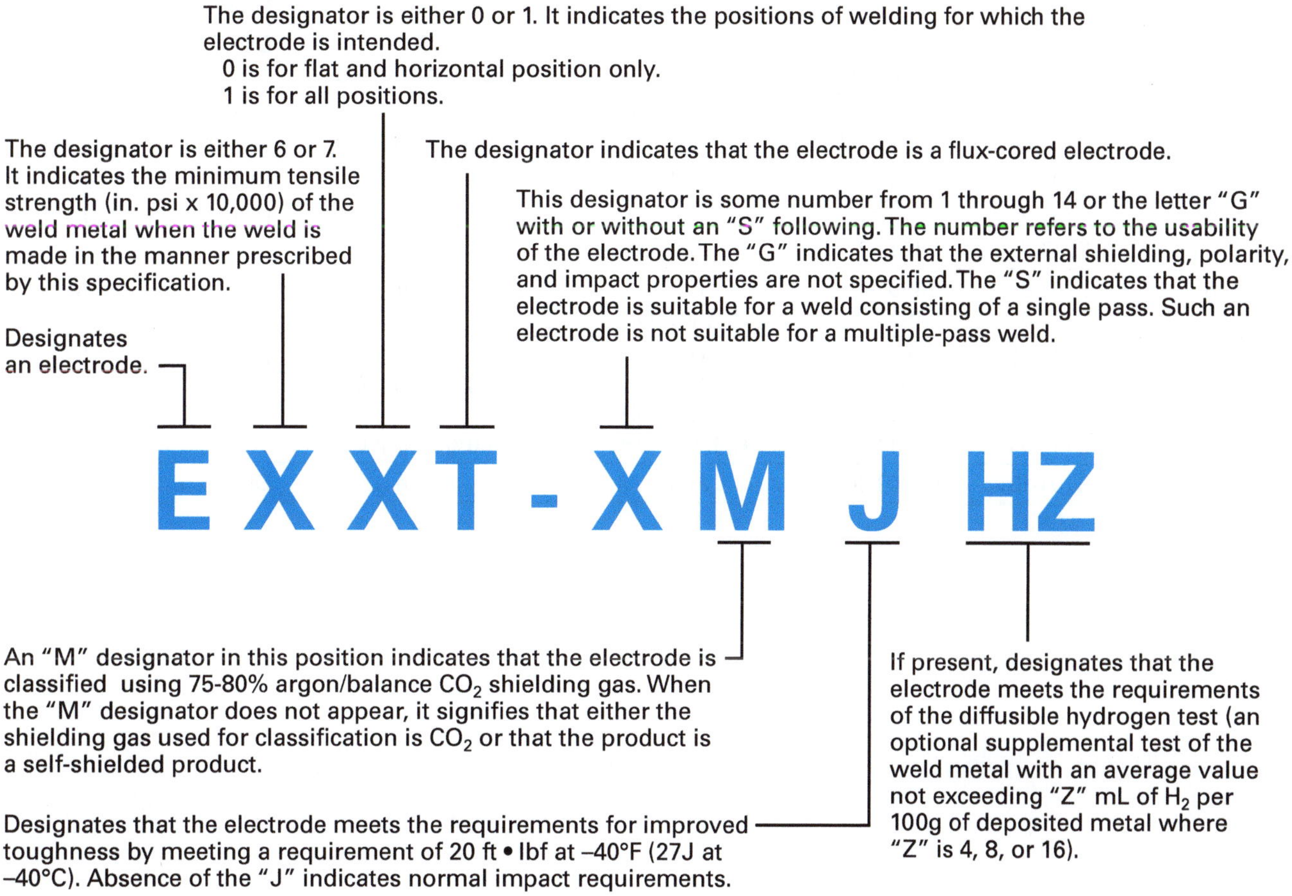

Figure 51 AWS classification for carbon steel flux-cored electrodes.

Table 6A Carbon Steel Flux-Cored Electrode Welding Position, Shielding, Polarity, and Application Requirements (Sheet 1 of 2)

Usability Designator	AWS Classification		Position of Welding[a,b]	External Shielding[c]	Polarity[d]	Application[e]
	A5.20	A5.20M				
1	E70T-1C E70T-1M E71T-1C E71T-1M	E490T-1C E490T-1M E491T-1C E491T-1M	H, F H, F H, F, VU, OH H, F, VU, OH	CO_2 75-80% Ar/bal CO_2 CO_2 75-80% Ar/bal CO_2	DCEP	M
2	E70T-2C E70T-2M E71T-2C E71T-2M	E490T-2C E490T-2M E491T-2C E491T-2M	H, F H, F H, F, VU, OH H, F, VU, OH	CO_2 75-80% Ar/bal CO_2 CO_2 75-80% Ar/bal CO_2	DCEP	S
3	E70T-3	E490T-3	H, F	None	DCEP	S
4	E70T-4	E490T-4	H, F	None	DCEP	M
5	E70T-5C E70T-5M	E490T-5C E490T-5M	H, F H, F	CO_2 75-80% Ar/bal CO_2	DCEP	M
	E70T-5C E70T-5M	E491T-5C E491T-5M	H, F, VU, OH H, F, VU, OH	CO_2 75-80% Ar/bal CO_2	DCEP or DCEN[f]	
6	E70T-6	E490T-6	H, F	None	DCEP	M
7	E70T-7 E71T-7	E490T-7 E491T-7	H, F H, F, VU, OH	None	DCEN	M
8	E70T-8 E71T-8	E490T-8 E491T-8	H, F H, F, VU, OH	None	DCEN	M
9	E70T-9C E70T-9M E71T-9C E71T-9M	E490T-9C E490T-9M E491T-9C E491T-9M	H, F H, F H, F, VU, OH H, F, VU, OH	CO_2 75-80% Ar/bal CO_2 CO_2 75-80% Ar/bal CO_2	DCEP	M
10	E70T-10	E490T-10	H, F	None	DCEN	S

Table 6B Carbon Steel Flux-Cored Electrode Welding Position, Shielding, Polarity, and Application Requirements (Sheet 2 of 2)

Usability Designator	AWS Classification		Position of Welding[a,b]	External Shielding[c]	Polarity[d]	Application[e]
	A5.20	A5.20M				
11	E70T-11 E71T-11	E490T-11 E491T-11	H, F H, F, VU, OH	None	DCEN	M
12	E70T-12C E70T-12M E71T-12C E71T-12M	E490T-12C E490T-12M E491T-12C E491T-12M	H, F H, F H, F, VU, OH H, F, VU, OH	CO_2 75-80% Ar/bal CO_2 CO_2 75-80% Ar/bal CO_2	DCEP	M
13	E61T-13 E71T-13	E431T-13 E491T-13	H, F, VU, OH	None	DCEN	S
14	E71T-14	E491T-14	H, F, VU, OH	None	DCEN	S
	E60T-G E70T-G	E430T-G E490T-G	H, F	Not Specified	Not Specified	M
	E61T-G E71T-G	E431T-G E491T-G	H, F, VD or VU, OH	Not Specified	Not Specified	M
G	E60T-GS E70T-GS	E430T-GS E490T-GS	H, F	Not Specified	Not Specified	S
	E61T-GS E71T-GS	E431T-GS E491T-GS	H, F, VD or VU, OH	Not Specified	Not Specified	S
	EX0T-Gg EX1T-Gg EX0T-GSg EX1T-GSg	EX0T-Gg EX1T-Gg EX0T-GSg EX1T-GSg	H, F H, F, VD or VU, OH H, F H, F, VD or VU, OH	Not Specified Not Specified Not Specified Not Specified	Not Specified Not Specified Not Specified Not Specified	S

Notes:

[a] H = horizontal position; F = flat position; OH = overhead position; VD = vertical position with downward progression; VU = vertical position with upward progression.

[b] Electrode sizes suitable for out-of-position welding, i.e., welding position other than flat or horizontal, usually are those sizes that are smaller than the ³⁄₃₂" (2.4 mm) size. For that reason, electrodes meeting the requirements for the groove weld tests and the fillet weld tests may be classified as EX1T-X or EX1T-XM (where X represents the tensile strength and usability designators) regardless of their size.

[c] Properties of weld metal from electrodes that are used with external gas shielding (EXXT-1, EXXT-1M, EXXT-2, EXXT-2M, EXXT-5, EXXT-5M, EXXT-9, EXXT-9M, EXXT-12, and EXXT-12M) vary according to the shielding gas employed. Electrodes classified with the specified shielding gas should not be used with other shielding gases without first consulting the manufacturer of the electrode.

[d] The term "DCEP" refers to direct current electrode positive (DC, reverse polarity). The term "DCEN" refers to direct current electrode negative (DC, straight polarity).

[e] M = single or multiple pass; S = single pass only. (Single-pass only classes are highly oxidizing and are used on metals that are rusty or have mill scale.)

[f] Some E71T-5 and E71T-5M electrodes may be recommended for use on DCEN for improved out-of-position welding. Consult the manufacturer for the recommended polarity.

[g] The letter "X" can be replaced with either a "6" or "7" to designate the weld metal's tensile strength.

Table 7A Carbon Steel Flux-Cored Electrode Classes and Uses (1 of 2)

AWS Classification	Intended Uses
EXXT-1 EXXT-1M	Electrodes designated as –1 use CO_2 shielding. However, other gas mixtures can be used to improve arc stability if recommended by the manufacturer. Electrodes designated as –1M use an argon-CO_2 mixture to increase manganese and silicon in the weld metal to increase yield and tensile strengths. Reducing argon in a gas mixture or using CO_2 shielding only with –1M electrodes may cause deterioration of arc and out-of-position welding characteristics and loss of strength. Both electrodes have spray transfer characteristics and low spatter. Most electrodes of this class have rutile-base slag and produce high disposition rates.
EXXT-2	These electrodes are similar to –1 and –1M types with higher manganese or silicon, or both. The higher levels of deoxidizers allow single-pass welding of heavily oxidized or rimmed steel. The –2M electrodes have an even higher manganese content and can be used on metals with heavier mill scale, rust, or foreign matter that cannot be tolerated by some –1 or –1M electrodes.
EXXT-3	These self-shielding electrodes are designed for single-pass, very high-speed welding and use spray transfer. Besides the listed positions, they can be used on 20-degree inclined downhill welds of sheet metal. They are not recommended for T or lap joints in materials thicker than 3⁄16" or for butt, edge, or corner joints in materials thicker than 1⁄4".
EXXT-4	These self-shielding electrodes use globular-type transfer and are designed for very high deposition rates. They produce low-sulfur welds that resist hot cracking. They are also designed for low penetration beyond the root of the weld because of poor joint fitup.
EXXT-5 EXXT-5M	These electrodes have a globular transfer and a lime-fluoride base slag that produces welds that have better impact strength and hot and/or cold cracking resistance than other similar classes with rutile-based slags. The –5 electrodes are used with a CO_2 shielding gas, however, argon-CO_2 mixes can be used to reduce spatter if recommended by the manufacturer. Reducing argon in a gas mixture or using CO_2 shielding only with –5M electrodes may cause deterioration of arc an out-of-position welding characteristics and loss of strength. Operator appeal for both types of electrodes is not as good as other similar classes with rutile slag.
EXXT-6	These self-shielding electrodes use spray-type transfer. The slag system is designed for good penetration into the root of the weld and excellent slag removal, even in deep grooves.
EXXT-7	These self-shielding electrodes use a droplet/spray-type transfer and produce a low-sulfur weld that is very resistant to cracking. The slag system is designed to produce high deposition rates for larger size electrodes in the flat or horizontal positions.
EXXT-8	These self-shielded electrodes have small droplet/spray-type transfer. They are suitable for all positions. The weld metal has good toughness, low temperature performance, and crack resistance.
EXXT-9 EXXT-9M	These electrodes are essentially the same as the –1 and –M electrodes except the deposit weld metal with improved impact properties. Some electrodes in this class require joints that are relatively clean of oil, excess oxide, and scale.
EXXT-10	These self-shielding electrodes have small droplet transfer. They are used for high speed, single-pass welds on any thickness material in the flat, horizontal, and 20-degree inclined vertical positions.
EXXT-11	These self-shielding electrodes have a smooth spray-type transfer and are general-purpose electrodes for all positions. Their use on material thicknesses greater than 3⁄4" is not generally recommended unless preheat and interpass temperatures are maintained. Consult the manufacturer for specific recommendations.

Table 7B Carbon Steel Flux-Cored Electrode Classes and Uses (2 of 2)

AWS Classification	Intended Uses
EXXT-12 EXXT-12M	These electrodes are essentially the same as –1 and –1M electrodes except they are modified to improve impact toughness and to lower manganese requirements of the A-1 Analysis Group in the *ASME Boiler and Pressure Code Section IX*. As a result, tensile strength and hardness will be decreased from –1 and –1 electrodes. Users should check hardness when a required hardness level is specified or required.
EXXT-13	These self-shielding electrodes are usually welded with a short-circuiting transfer. The slag system allows an all-position root pass on pipe welds. They are recommended only for the root pass and are not generally recommended for multiple passes.
EXXT-14	These self-shielding electrodes have a smooth spray-type transfer. The slag system allows high speed welds in all positions. They are used to make welds on sheet metal up to 3/16" thick and are often designed especially for galvanized, aluminized or other coated steels. They are generally not recommended for T or lap joints in materials thicker than 3/16" or butt, edge, or corner joints in materials thicker than 1/4". The manufacturer should be consulted for specific recommendations.
EXXT-G	These are multipass electrodes not covered by other classifications. Except for requirements to assure carbon steel deposits and tensile strengths, other requirements are agreed upon by the purchaser and manufacturer.
EXXT-GS	These are single-pass electrodes not covered by other classifications. Except for tensile strength, other requirements are agreed upon by the purchaser and the manufacturer.

FCAW of Structural Steel

For large structural welding projects such as bridges and skyscrapers, many construction companies prefer carbon or low-alloy steel FCAW to meet the toughness requirements for new construction using the newer self-shielding E71T-8 or gas-shielded E71T-1 electrodes. Many ironworkers who use SMAW are reluctant to use FCAW, and those who have used FCAW predominantly relied upon E71T-11 (a limited-thickness electrode) for structural welds because it was easy to use in all positions. However, welds with –11 wire sometimes had cracking problems and had to be rewelded. When welders are properly retrained for FCAW using the newer electrodes, the cracking problems are eliminated and the deposition rates are higher because the newer electrodes run hotter and faster than the –11 electrodes or SMAW electrodes.

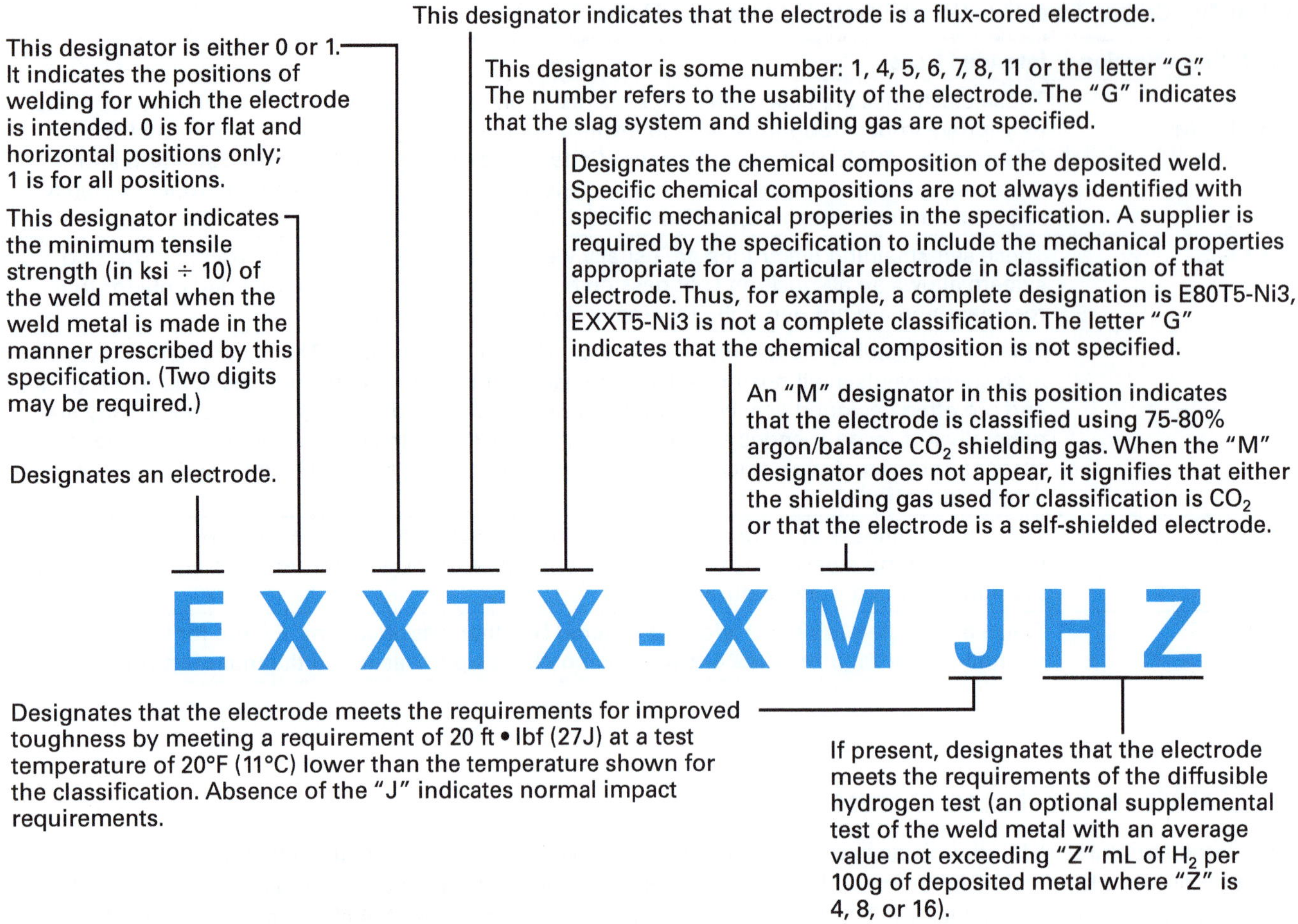

Figure 52 AWS classification for low-alloy steel flux-cored electrodes.

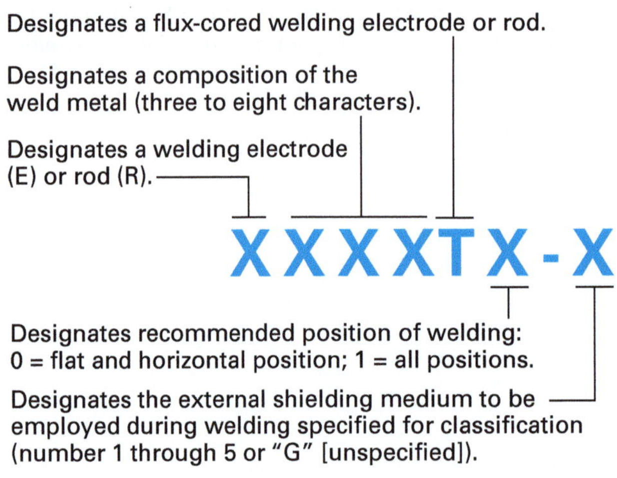

Figure 53 AWS specification for stainless steel flux-cored electrodes and rods.

Additional Resources

AWS B1.10M/B1.10 Guide for the Nondestructive Examination of Welds. Miami, FL: American Welding Society.

AWS B1.11 Guide for the Visual Examination of Welds. Miami, FL: American Welding Society.

Lincoln Electric website: **http://www.lincolnelectric.com** offers sources for products and training.

Modern Welding Technology. Howard B. Cary. Englewood Cliffs, NJ: Prentice Hall, Inc.

OSHA 1926.351, Arc Welding and Cutting. Current Edition. Washington, DC: Occupational Safety and Health Administration (OSHA).

Welding Handbook. Volume 1. Welding Science & Technology. Miami, FL: American Welding Society, 2001.

Welding Handbook. Volume 2. Part 1: Welding Processes. Miami, FL: American Welding Society, 2004.

3.0.0 Section Review

1. Which of these filler metals is sensitive to embrittlement by oxygen, nitrogen, and hydrogen at temperatures above 500°F (260°C)?
   a. Nickel
   b. Magnesium
   c. Titanium
   d. Aluminum

2. In FCAW operations, it is possible to use a smaller size of electrode wire for thin metals and multiple-pass welds by varying the welding machine's ____.
   a. pinch
   b. polarity cycling
   c. feed drag
   d. power settings

Summary

Performing gas metal arc welding and flux-cored arc welding safely and effectively requires a good understanding of the welding equipment, the filler metals that are used, and the proper safety precautions that should be followed. There are many different types of GMAW and FCAW equipment and components. Before any GMAW or FCAW can begin, workers must first select the appropriate safety and welding equipment, as well as the proper filler metals. They must also know how to safely move and set up the equipment. This module has provided enough information about these basic responsibilities to enable workers to use GMAW and FCAW equipment safely and efficiently.

Review Questions

1. In GMAW, the arc is stabilized and the molten filler and base metals are protected from oxidation by the use of ____.
   a. differential pressure
   b. self-fluxing alloy
   c. intense heat
   d. shielding gas

2. Of the two basic FCAW processes, one is self-shielding while the other one ____.
   a. requires no shielding gas
   b. uses ferrous core vapors
   c. uses external shielding gas
   d. utilizes an oxygen generator

3. Areas where oil, grease, and other petroleum products are present, should be kept clear of accumulated ____.
   a. oxygen
   b. helium
   c. argon
   d. nitrogen

4. Which metal transfer mode uses a high welding voltage that ultimately causes fine droplets of electrode wire to be carried axially to the molten weld puddle?
   a. Globular transfer
   b. Inverter transfer
   c. Short-circuiting transfer
   d. Spray transfer

5. Short-circuiting metal transfer is the best mode to use for welding ____.
   a. cast iron blocks
   b. light-gauge sheet metal
   c. heavy steel beams
   d. aluminum plates

6. FCAW welds made with CO_2 shielding gas produce ____.
   a. deep penetration
   b. a slag-free weld
   c. shallow penetration
   d. a lot of spatter

7. Depending on the range selected, for every 100 amperes of current the arc voltage of a welding machine is typically lower than the open-circuit voltage by ____.
   a. 0 to 1 volts
   b. 1 to 2 volts
   c. 2 to 3 volts
   d. 4 to 5 volts

8. The welding arc generates intense heat that ranges from ____.
   a. 1,000°F to 5,000°F (530°C to 2,760°C)
   b. 6,000°F to 10,000°F (3,316°C to 5,538°C)
   c. 5,000°F to 12,500°F (2,760°C to 6,927°C)
   d. 8,500°F to 12,500°F (4,704°C to 6,927°C)

9. One advantage of using engine-driven generators as a way to power welders is that the equipment is ____.
   a. less expensive than single-phase welders
   b. easier to operate than all other welders
   c. free of most maintenance requirements
   d. portable so it can be used in the field

10. Which welding power source increases the frequency of the incoming primary power and is well suited for applications where space is limited?
    a. Inverter
    b. Engine-driven generator
    c. Axial
    d. Transformer-rectifier

11. When plotted on a graph, the general angle showing the voltage-to-amperage relationship for GMAW is known as the ____.
    a. pinch
    b. arc blow
    c. slope
    d. hypotenuse

12. Two factors that must be considered when selecting the size of a welding cable are the amperage load and the ____.
    a. type of copper strands in the cable
    b. distance the current will travel
    c. speed of the wire electrode feed
    d. pinch effect imparted on the cable

13. The duty cycle of a GMAW gun is determined and rated with ____.
    a. carbon dioxide (CO_2)
    b. helium (He)
    c. argon (Ar)
    d. oxygen (O)

14. Light-duty GMAW guns are cooled by ____.
    a. domestic water flow
    b. special cooling fluid
    c. shielding gas flow
    d. demineralized water

15. Because of its current transfer and ionization properties, the most common shielding gas that can be used alone or in combination with other gases is ____.
    a. carbon dioxide (CO_2)
    b. helium (He)
    c. nitrogen (N)
    d. argon (Ar)

16. What is required to meter shielding gas to a welding gun at the proper flow rate?
    a. A gas mixer/stabilizer
    b. A variable vacuum tip
    c. A gas regulator/flowmeter
    d. An auxiliary gun controller

17. After checking the equipment specifications to determine the weight of a welding machine, what is the proper way to lift and move the machine?
    a. Use a wire rope basket hitch under the welder.
    b. Attach a sling and/or shackle to the welder lift eye.
    c. Use numerous synthetic slings attached to the corners of the welder.
    d. Attach the lifting hook from the lifting device to the welder lift eye.

18. A tool designed for cutting and manipulating wire, removing and installing contact tips and nozzles, and cleaning welding spatter from nozzles is a pair of ____.
    a. cast and helix straightener bars
    b. knurled combination rollers
    c. push-pull wire straighteners
    d. MIG welding, or wire feed, pliers

19. Many types of GMAW carbon steel filler wire are protected from corrosion by a(n) ____.
    a. copper coating
    b. aluminum core
    c. mercury alloy
    d. nonferrous skin

20. The primary use of FCAW is for the welding of ____.
    a. nonferrous metals
    b. aluminum
    c. ferrous metals
    d. titanium

Trade Terms Introduced in This Module

Axially: In a straight line along an axis.

Composite metal-core wire electrode: A filler wire electrode with a hollow core containing powdered materials, primarily metals.

Globular transfer: A GMAW process in which the electrode wire does not touch the base metal, and the filler metal is transferred randomly in the arc as large, irregular globules.

Inductance: An electrical circuit property, adjustable on some FCAW power supplies, that influences the pinch effect.

Pinch effect: The tendency of the end of the wire electrode to melt into a droplet during short-circuit GMAW welding.

Pulsed transfer: A GMAW welding process in which the welding power is cycled from a low level (background level) to a higher level (welding level), resulting in lower average voltage and current; this is a variation of the spray transfer process.

Short-circuiting transfer: A GMAW welding process in which the electrode wire shorts against the base metal. The welding current repeatedly and rapidly melts the wire and establishes an arc, which is extinguished when the wire again touches the base metal.

Spray transfer: A GMAW welding process in which the electrode wire does not touch the base metal, and the filler metal is transferred axially (in a straight line) in the arc as fine droplets.

Stickout: The length of electrode that projects from the contact tube, measured from the arc tip to the contact tube.

Additional Resources

This module is intended as a thorough resource for task training. The following reference works are suggested for further study.

AWS B1.10M/B1.10 Guide for the Nondestructive Examination of Welds. Miami, FL: American Welding Society.

AWS B1.11 Guide for the Visual Examination of Welds. Miami, FL: American Welding Society.

Lincoln Electric website: **http://www.lincolnelectric.com** offers sources for products and training.

Modern Welding Technology. Howard B. Cary. Englewood Cliffs, NJ: Prentice Hall, Inc.

OSHA 1926.351, Arc Welding and Cutting. Current Edition. Washington, DC: Occupational Safety and Health Administration (OSHA).

Welding Handbook. Volume 1. Welding Science & Technology. Miami, FL: American Welding Society, 2001.

Welding Handbook. Miami, FL: American Welding Society, 2004.

Figure Credits

The Lincoln Electric Company, Cleveland, OH, USA, Module opener, Figures 3, 4, 14–17, SA01, 23C, 25, 26, 28, 30, SA02, 31–34, 36, 38 (photo)

Miller Electric Mfg. Co., Figure 18

AWS C5.6:1989, Figure 6, Reproduced with permission from the American Welding Society (AWS), Miami, FL, USA, Figure 21

Topaz Publications, Inc., Figures 23A and B, 24A and B, 39

Courtesy of Sumner Manufacturing Co., Figure 24C

Magswitch Technology Pty Ltd, Figure 24D

Courtesy of Irwin Tools, Figure 40

AWS A5.18/A5.18M:2005, Figure A.1, Reproduced with permission from the American Welding Society (AWS), Miami, FL, USA, Figure 41

AWS PHB-4:2000, Page 9, Reproduced with permission from the American Welding Society (AWS), Miami, FL, USA, Tables 3, 4

AWS A5.20/A5.20M:2005, Figure A.1, Reproduced with permission from the American Welding Society (AWS), Miami, FL, USA, Figure 51

AWS A5.20/A5.20M:2005, Table 2, Reproduced with permission from the American Welding Society (AWS), Miami, FL, USA, Table 6

AWS A5.29:2005, Figure A.1, Reproduced with permission from the American Welding Society (AWS), Miami, FL, USA, Figure 52

Section Review Answer Key

Answer	Section Reference	Objective
Section One		
1. c	1.1.0	1a
2. b	1.2.3	1b
3. d	1.3.3	1c
4. b	1.4.0	1d
Section Two		
1. a	2.1.1	2a
2. c	2.2.3	2b
3. d	2.3.1	2c
4. b	2.4.0	2d
5. c	2.5.2	2e
6. a	2.6.1	2f
7. b	2.7.4	2g
Section Three		
1. c	3.1.8	3a
2. d	3.2.0	3b

NCCER CURRICULA — USER UPDATE

NCCER makes every effort to keep its textbooks up-to-date and free of technical errors. We appreciate your help in this process. If you find an error, a typographical mistake, or an inaccuracy in NCCER's curricula, please fill out this form (or a photocopy), or complete the online form at **www.nccer.org/olf**. Be sure to include the exact module ID number, page number, a detailed description, and your recommended correction. Your input will be brought to the attention of the Authoring Team. Thank you for your assistance.

Instructors – If you have an idea for improving this textbook, or have found that additional materials were necessary to teach this module effectively, please let us know so that we may present your suggestions to the Authoring Team.

NCCER Product Development and Revision

13614 Progress Blvd., Alachua, FL 32615

Email: curriculum@nccer.org
Online: www.nccer.org/olf

❑ Trainee Guide ❑ Lesson Plans ❑ Exam ❑ PowerPoints Other ______________________

Craft / Level: Copyright Date:

Module ID Number / Title:

Section Number(s):

Description:

Recommended Correction:

Your Name:

Address:

Email: Phone:

GMAW – Plate

Overview

Gas metal arc welding (GMAW) is one of the most common welding processes. GMAW equipment is available in many different sizes and varieties, however, the basic operating principles of GMAW apply to all makes and models of equipment. This module describes how GMAW equipment is used, identifies GMAW-related safety practices, and examines different techniques used to produce various types of basic weld beads.

Module 29209

Trainees with successful module completions may be eligible for credentialing through the NCCER Registry. To learn more, go to **www.nccer.org** or contact us at **1.888.622.3720**. Our website has information on the latest product releases and training, as well as online versions of our *Cornerstone* magazine and Pearson's product catalog.

Your feedback is welcome. You may email your comments to **curriculum@nccer.org**, send general comments and inquiries to **info@nccer.org**, or fill in the User Update form at the back of this module.

This information is general in nature and intended for training purposes only. Actual performance of activities described in this manual requires compliance with all applicable operating, service, maintenance, and safety procedures under the direction of qualified personnel. References in this manual to patented or proprietary devices do not constitute a recommendation of their use.

29209 V5

29209
GMAW – Plate

Objectives

When you have completed this module, you will be able to do the following:

1. Identify GMAW-related safety practices and explain how to set up for welding.
   a. Describe basic GMAW processes.
   b. Identify GMAW-related safety practices.
   c. Explain how to safely set up the equipment and work area for welding.
2. Describe equipment control and welding procedures for GMAW and explain how to produce basic weld beads.
   a. Describe equipment control and welding techniques related to GMAW.
   b. Explain how to produce basic GMAW weld beads.
3. Describe the welding procedures needed to produce proper fillet and V-groove welds using GMAW welding techniques.
   a. Describe the welding procedures needed to produce proper fillet welds using GMAW welding techniques.
   b. Describe the welding procedures needed to produce proper V-groove welds using GMAW welding techniques.

Performance Tasks

1. Make multiple-pass GMAW-S (short-circuit) fillet welds on carbon steel plate coupons in all four 1F through 4F positions, using solid or composite electrode.
2. Make multiple-pass GMAW-S (short-circuit) V-groove welds on carbon steel plate coupons in all four 1G through 4G positions, with or without backing, using solid or composite electrode.
3. Make multiple-pass GMAW spray-transfer fillet welds on carbon steel plate coupons in both the 1F and 2F positions, using solid or composite electrode.
4. Make multiple-pass GMAW spray-transfer V-groove welds on carbon steel plate coupons in the 1G position, with backing, using solid or composite electrode.

Trade Terms

Axis
Synergic
Trim

Industry Recognized Credentials

If you are training through an NCCER-accredited sponsor, you may be eligible for credentials from NCCER's Registry. The ID number for this module is 29209. Note that this module may have been used in other NCCER curricula and may apply to other level completions. Contact NCCER's Registry at 888.622.3720 or go to **www.nccer.org** for more information.

Contents

Topics to be presented in this module include:

Figures

SECTION ONE

1.0.0 GMAW WELDING

Objective

Identify GMAW-related safety practices and explain how to prepare for welding.

a. Describe basic GMAW processes.
b. Identify GMAW-related safety practices.
c. Explain how to safely set up the equipment and work area for welding.

Performance Tasks

1. Make multiple-pass GMAW-S (short-circuit) fillet welds on carbon steel plate coupons in all four 1F through 4F positions, using solid or composite electrode.
2. Make multiple-pass GMAW-S (short-circuit) V-groove welds on carbon steel plate coupons in all four 1G through 4G positions, with or without backing, using solid or composite electrode.
3. Make multiple-pass GMAW spray-transfer fillet welds on carbon steel plate coupons in both the 1F and 2F positions, using solid or composite electrode.
4. Make multiple-pass GMAW spray-transfer V-groove welds on carbon steel plate coupons in the 1G position, with backing, using solid or composite electrode.

GMAW is often called MIG welding. MIG is an acronym for metal inert gas. As the names imply, GMAW joins together two metals using a filler metal electrode and an inert shielding gas. This section describes GMAW equipment and how it works, identifies safety practices related to GMAW, and examines how to prepare for GMAW.

1.1.0 Welding Processes

Arc welding processes use the intense heat produced by an electrical arc between a filler metal electrode and a workpiece to melt the metals and cause them to merge, or coalesce. Gas metal arc welding, or GMAW, is one common form of arc welding used extensively in industry. Performing GMAW safely and effectively requires a good understanding of how the process and the equipment work. It is also critical to understand the safety practices that are necessary during a GMAW operation.

1.1.1 The GMAW Process

In the GMAW process, a continuous, consumable, solid or composite-cored electrode is used for the filler metal while shielding gas is used to protect the weld zone. Composite-cored electrodes are a subset of GMAW solid electrode carbon steel electrodes as defined in *AWS Specification A5.18.* Keep in mind that the proper term for the consumable material used for filler metal is electrode, although it is commonly referred to as wire.

GMAW is a fast and effective method of making high-quality welds (*Figure 1*). The GMAW process is commonly used to make welds on carbon steel plate as well as on aluminum and other metals.

1.1.2 GMAW Equipment

The equipment used for gas-shielded GMAW includes guns equipped with a gas nozzle, gas hoses, a gas control solenoid valve, a gas regulator/flowmeter, and a source of shielding gas. *Figure 2* is an example of typical GMAW equipment in operation.

1.2.0 Safety Summary

The following is a summary of safety procedures and practices that must be observed while cutting or welding. Keep in mind that this is only a summary. Complete safety coverage is provided in the Welding Level One module, *Welding Safety*. That module should be completed before continuing with this one. Above all, be sure to wear appropriate protective clothing and equipment when welding or cutting.

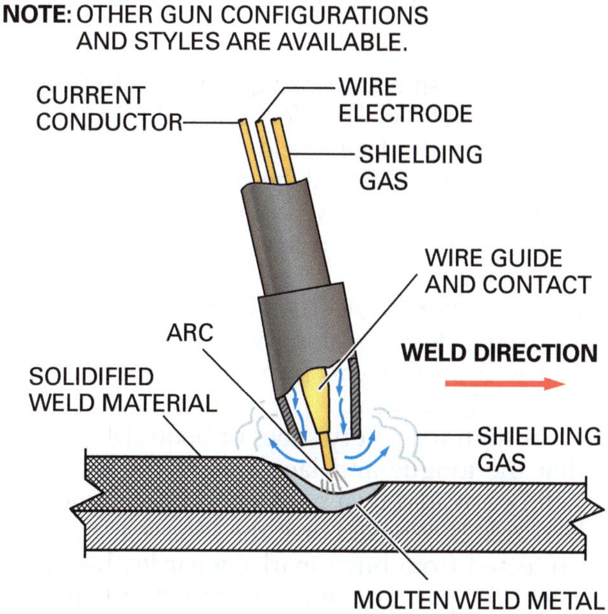

Figure 1 GMAW process.

Figure 2 Gas-shielded GMAW equipment.

1.2.1 Protective Clothing and Equipment

Welding activities can cause injuries unless you wear all of the protective clothing and equipment that is designed specifically for the welding industry. The following information includes important safety guidelines about protective clothing and equipment:

- Wear a face shield over snug-fitting cutting goggles or safety glasses for gas welding or cutting. Either the face shield or the lenses of the welding goggles must be an approved shade for the application. A welding hood equipped with a properly tinted lens is best for all forms of welding.
- Wear proper protective leather and flame-retardant clothing along with welding gloves that will protect you from flying sparks, molten metal, and heat.
- Wear high-top safety shoes or boots. Make sure that the tongue and lace area of the footwear will be covered by a pant leg. If the tongue and lace area is exposed or if the footwear must be protected from burn marks, wear leather spats under the pants or chaps over the front and top of the footwear.
- Wear a 100 percent cotton cap with no mesh material included in its construction. The bill of the cap points to the rear. If a hard hat is required for the environment, use one that allows the attachment of rear deflector material and a face shield. A hard hat with a rear deflector is generally preferred when working overhead, and may be required by some employers and job sites.
- Wear earmuffs or at least earplugs to protect ear canals from sparks.

1.2.2 Fire/Explosion Prevention

Welding activities usually involve the use of fire or extreme heat to melt metal. Whenever fire is used, it must be controlled and contained. Welding or cutting activities are often performed on vessels that may once have contained flammable or explosive materials. Residues from those materials can catch fire or explode when a welder begins work on such a vessel. The following are fire and explosion prevention guidelines associated with welding:

- Never carry matches or gas-filled lighters in pockets. Sparks can cause the matches to ignite or the lighter to explode, causing serious injury.
- Never use oxygen to blow dust or dirt from clothing. The oxygen can remain trapped in the fabric for a time. If a spark hits the clothing during this time, the clothing can burn rapidly and violently out of control.
- Make sure that any flammable material in the work area is moved or shielded by a fire-resistant covering. Approved fire extinguishers must be available before attempting any heating, welding, or cutting operations. If a hot-work permit and a fire watch are required, be sure those items are in place before beginning and ensure that all site requirements have been observed.
- Never release a large amount of oxygen or use oxygen as compressed air. Its presence around flammable materials or sparks can cause rapid and uncontrolled combustion. Keep oxygen away from oil, grease, and other petroleum products.
- Never release a large amount of fuel gas, especially acetylene. Propane tends to concentrate in and along low areas and can ignite at a considerable distance from the release point. Acetylene is lighter than air, but is even more dangerous than propane. When mixed with air or oxygen, acetylene will explode at much lower concentrations than any other fuel.
- To prevent fires, maintain a neat and clean work area, and make sure that any metal scrap or slag is cold before disposal.

- Before cutting containers such as tanks or barrels, check to see if they have contained any explosive, hazardous, or flammable materials, including petroleum products, citrus products, or chemicals that decompose into toxic fumes when heated. As a standard practice, always clean and then fill any tanks or barrels with water, or purge them with a flow of inert gas such as nitrogen to displace any oxygen. Containers must be cleaned by steam cleaning, flushing with water, or washing with detergent until all traces of the material have been removed.

WARNING!

Welding or cutting must never be performed on drums, barrels, tanks, vessels, or other containers until they have been emptied and cleaned thoroughly, eliminating all flammable materials and all substances (such as detergents, solvents, greases, tars, or acids) that might produce flammable, toxic, or explosive vapors when heated. Clean containers only in well-ventilated areas. Vapors can accumulate during cleaning, causing explosions or injury.

Proper procedures for cutting or welding hazardous containers are described in the *American Welding Society (AWS) F4.1, Safe Practices for the Preparation of Containers and Piping for Welding and Cutting.*

1.2.3 Work Area Ventilation

Vapors and fumes tend to rise in the air from their sources. Welders often have to work above the welding area where the fumes are being created. Welding fumes can cause personal injuries. Good work area ventilation helps to remove the vapors and to protect the welder. The following is a list of work area ventilation guidelines to consider before and during welding activities:

- Make sure to follow confined space procedures before conducting any welding or cutting in the confined space.
- Never use oxygen to provide ventilation in confined spaces.
- Always perform cutting or welding operations in a well-ventilated area. Cutting or welding operations involving zinc or cadmium materials or coatings result in toxic fumes. For long-term cutting or welding of such materials, always wear an approved full-face, supplied-air respirator that uses breathing air supplied from outside the work area. For occasional, very short-term exposure, a HEPA-rated or metal-fume filter may be used on a standard respirator.

Machine Bevels and Root Faces for GMAW

For V-groove joints made in thick materials, machined bevels and root faces provide the cleanest and most uniform way to prepare materials for satisfactory GMAW welds. While SMAW, FCAW, and GTAW can tolerate some poor fitup for root passes, controlling penetration of root passes with poor fitup using GMAW is difficult.

- Make sure confined spaces are properly ventilated for cutting or welding purposes. Use powered extraction systems when available. *Figure 3* shows mobile and portable extraction equipment used to remove cutting or welding fumes and vapors from work areas.

1.3.0 Welding Equipment Setup

Before welding can take place, the area must be made ready, the welding equipment must be set up, and the metal to be welded must be prepared. The following sections explain how to set up equipment for welding.

1.3.1 Preparing the Welding Area

To practice welding, a welding table, bench, or stand is needed. The welding surface must be steel, and provisions must be made for placing weld coupons out of position (*Figure 4*).

Follow these steps to set up the area for welding:

Step 1 Make sure that the area is properly ventilated. Make use of doors, windows, and fans.

Step 2 Check the area for fire hazards. Remove any flammable materials before proceeding.

Step 3 Know the location of the nearest fire extinguisher. Know how to use the fire extinguisher and do not proceed unless the extinguisher is charged.

Step 4 Set up welding curtains around the welding area.

1.3.2 Preparing the Practice Weld Coupons

If possible, the practice weld coupons should be ⅜" (10 mm plate) thick carbon steel. If this size is not readily available, ¼" to ¾" (6 mm to 20 mm plate) thick steel can be used to practice welding. Use a wire brush or grinder to remove heavy mill scale or corrosion.

MOBILE UNIT

PORTABLE UNIT

Figure 3 Mobile and portable extraction equipment.

Prepare weld coupons for practicing the following:

- *Running beads* – The coupons can be any size or shape that can be easily handled.
- *Overlapping beads* – The coupons can be any size or shape that can be easily handled.
- *Fillet welds* – Cut the metal into 2" × 6" (51 mm × 152 mm) rectangles for the base and 1" × 6" (25 mm × 152 mm) rectangles for the web. *Figure 5* shows the weld coupons for fillet welding.
- *Open V-groove welds* – As shown in *Figure 6*, cut the metal into 3" × 7" (76 mm × 178 mm) rectangles with one (or both) of the lengths beveled at 30 to 37½ degrees for carbon steel and low-alloy plate. Note that the bevel angle is 22½

Figure 4 Welding station.

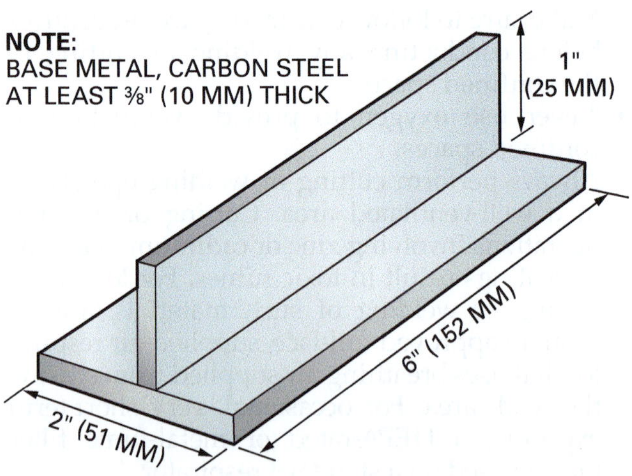

Figure 5 Fillet weld coupons.

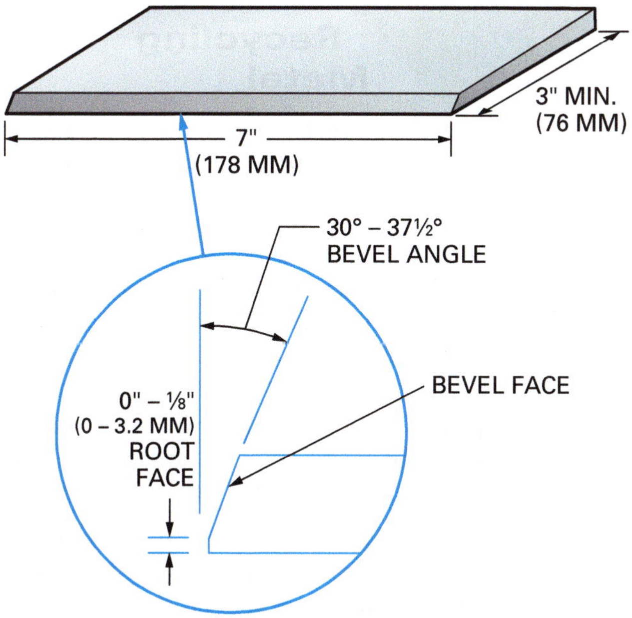

Figure 6 Metal cut for open V-groove weld coupons.

degrees for stainless steel coupons. Grind up to a ⅛-inch (3.2 mm) root face on the bevel as chosen and directed by the instructor.

Follow these steps to prepare each V-groove weld coupon:

Step 1 Check the bevel face of each coupon. There should be no dross and a 0" to ⅛" (0 to 3.2 mm) root face. The bevel angle should be 30 to 37½ degrees. An example of this is shown in *Figure 7*.

Step 2 Align the beveled coupons on a flat surface with a root opening determined by the instructor. Place tack welds on the back side of the joint. Use from three to four ¼" (6.4 mm) tack welds.

1.3.3 The Welding Machine

Identify the proper welding machine for GMAW use, and follow these steps to set it up for welding:

Step 1 Verify that the welding machine can be used for GMAW.

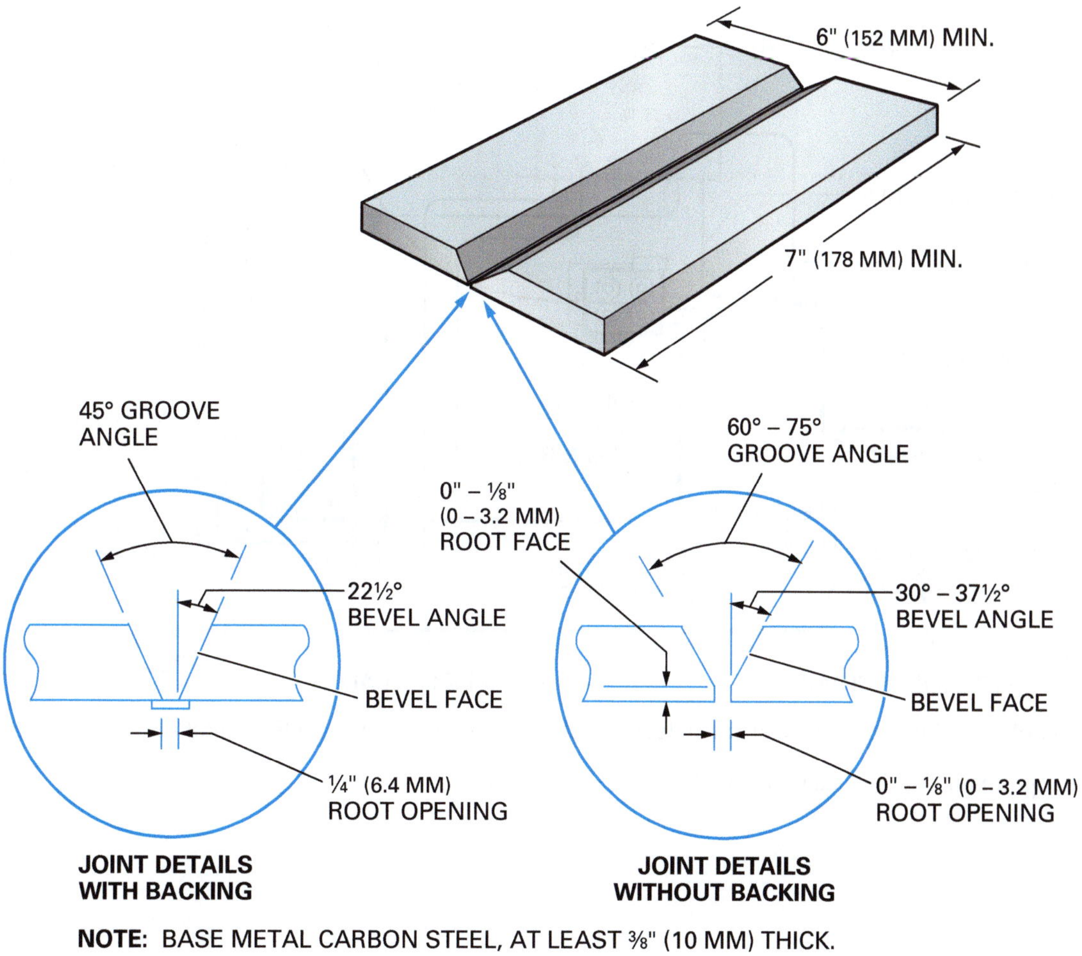

Figure 7 Open V-groove weld coupon.

Step 2 Verify the location of the primary disconnect.

Step 3 Configure the welding machine for GMAW as directed by the instructor (*Figure 8*). Configure the gun polarity, and equip the gun with the correct liner material and contact tip for the diameter of the electrode being used and the correct nozzle for the application.

Step 4 In accordance with the manufacturer's instructions, load the wire feeder and gun with a solid or composite electrode of the proper diameter as directed by the instructor.

Step 5 Connect the proper shielding gas for the application as specified by the electrode manufacturer, the WPS, the site quality standards, or the instructor.

Step 6 Connect the clamp of the workpiece lead to the workpiece.

GOING GREEN

Recycling Metal

When metals are no longer usable in the welding shop or on the job, they should be collected and sold to a recycling company rather than sent to a landfill. Selling the scrap metal can make money for the company or welding school, and it protects the environment. Steel is the most recycled material in the world.

Step 7 Turn on the welding machine and, if necessary, purge the gun as directed by the gun manufacturer's instructions.

Step 8 Set the initial welding voltage and wire feed speed for the type and size of electrode as recommended by the manufacturer, directed by the instructor, or provided in the WPS.

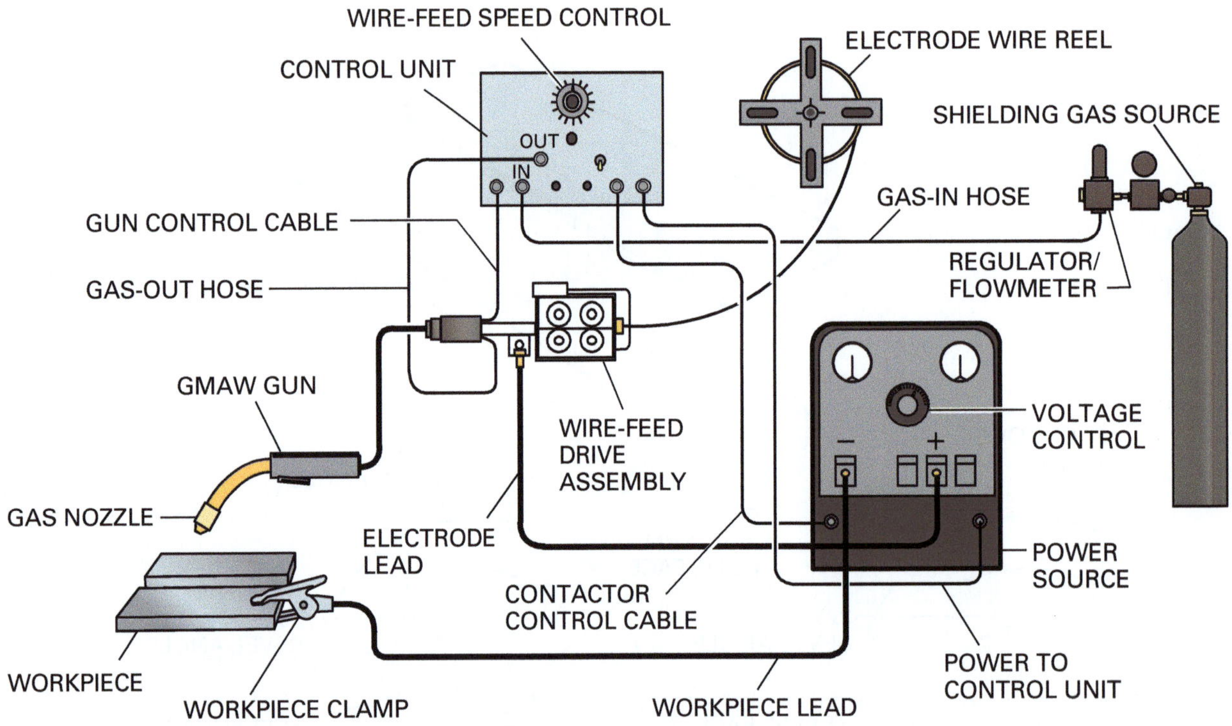

NOTE: The polarity of the gun and workpiece lead is determined by the type of filler metal and application.

Figure 8 Configuration diagram of a typical GMAW welding machine.

Conserving Material

Steel for practice welding is expensive and difficult to obtain. Every effort should be made to conserve the material that is available. Reuse weld coupons until all surfaces and joints have been used for welding, then cut the weld coupon apart and reuse the pieces. Use material that cannot be cut into weld coupons to practice running beads.

Additional Resources

AWS B1.10M/B1.10 Guide for the Nondestructive Examination of Welds. Miami, FL: American Welding Society.

AWS B1.11 Guide for the Visual Examination of Welds. Miami, FL: American Welding Society.

AWS C5.6-89R Recommended Practices for Gas Metal Arc Welding. Miami, FL: American Welding Society.

AWS D3.5-93R Guide for Steel Hull Welding. Miami, FL: American Welding Society.

GMAW Welding Guide. Current Edition. Cleveland, OH: The Lincoln Electric Company.

Lincoln Electric website: **http://www.lincolnelectric.com** offers sources for products and training.

OSHA 1910.269, Appendix *C, Protection from Step and Touch Potentials.* Current Edition. Washington, DC: Occupational Safety & Health Administration (OSHA).

OSHA 1926.351, Arc Welding and Cutting. Current Edition. Washington, DC: Occupational Safety and Health Administration (OSHA).

The Procedure Handbook of Arc Welding. 14th Edition. Cleveland, OH: The James F. Lincoln Arc Welding Foundation, 2000.

Welding Handbook. Volume 1. Welding Science & Technology. Miami, FL: American Welding Society, 2001.

Welding Handbook. Volume 2. Part 1: Welding Processes. Miami, FL: American Welding Society, 2004.

1.0.0 Section Review

1. In the GMAW process, the shielding gas that is used to protect the weld zone comes from a ____.
   a. flux-cored electrode
   b. shielding gas cylinder or other gas source
   c. solid-cored wire
   d. gaseous filler metal

2. The best alternative for safely cutting into a previously used vessel is to first fill the container with ____.
   a. sand
   b. air
   c. oxygen
   d. water

3. When setting up GMAW equipment for practice welding, the table, bench, or stand that will be used must have a ____.
   a. steel welding surface
   b. secondary disconnect
   c. shielding gas liner
   d. dry, wooden pedestal

Section Two

2.0.0 Welding Procedures and Basic Beads

Objective

Describe equipment control and welding procedures for GMAW and explain how to produce basic weld beads.

a. Describe equipment control and welding techniques related to GMAW.
b. Explain how to produce basic GMAW weld beads.

Performance Tasks

1. Make multiple-pass GMAW-S (short-circuit) fillet welds on carbon steel plate coupons in all four 1F through 4F positions, using solid or composite electrode.
2. Make multiple-pass GMAW-S (short-circuit) V-groove welds on carbon steel plate coupons in all four 1G through 4G positions, with or without backing, using solid or composite electrode.
3. Make multiple-pass GMAW spray-transfer fillet welds on carbon steel plate coupons in both the 1F and 2F positions, using solid or composite electrode.
4. Make multiple-pass GMAW spray-transfer V-groove welds on carbon steel plate coupons in the 1G position, with backing, using solid or composite electrode.

Trade Terms

Axis: The center line of the weld or prepared base metal.

Synergic: Welding technology that controls voltage, current, and wire feed using a program optimized for each welding situation.

Trim: A control found on synergic welding machines that is similar to the voltage control on regular machines and acts as a kind of override to the program.

Effective welding requires the welder to properly control the equipment being used and employ various welding techniques to produce strong and acceptable weld beads. This section examines factors that affect GMAW and how these factors are controlled. It also describes how to produce basic GMAW weld beads.

2.1.0 Welding Procedures

Welding can be affected by numerous factors, including the welding voltage, the welding amperage, the position and travel speed of the gun, and other aspects of the welding equipment and process.

2.1.1 Welding Voltage

Arc length is determined by voltage, which is set at the power source. Arc length is the distance from the electrode tip to the base metal or molten pool at the base metal. If voltage is set too high, the arc will be too long, which could cause the electrode to fuse to the tip. Too high a voltage also causes porosity and excessive spatter. Voltage must be increased or decreased as wire feed speed is increased or decreased. Set the voltage so that the arc length is just above the surface of the base metal (*Figure 9*).

Synergic welding machines control voltage, current, and wire feed moment by moment using a program that's customized for each welding situation. These machines will have a **trim** control that works similarly to the voltage control on other machines. Essentially, it overrides the program.

2.1.2 Welding Amperage

With a standard welding machine, the wire feed speed control on the power source controls the welding amperage when the welder is making the initial recommended setting. The welding machine provides the amperage necessary to melt the electrode while maintaining the selected welding voltage. Within limits, when the wire feed speed is increased, the welding amperage and deposition rate also increase. This results in higher welding heat, deeper penetration, and higher beads. When the wire feed speed is

Electrode Manufacturer's Recommendations

Always obtain and follow the manufacturer's recommendations or WPS for shielding gas and for initially setting the welding voltage, current (if variable), and wire feed speed parameters. These balanced parameters are critical and are based on the welding position, size, and composition of the solid or composite electrode being used.

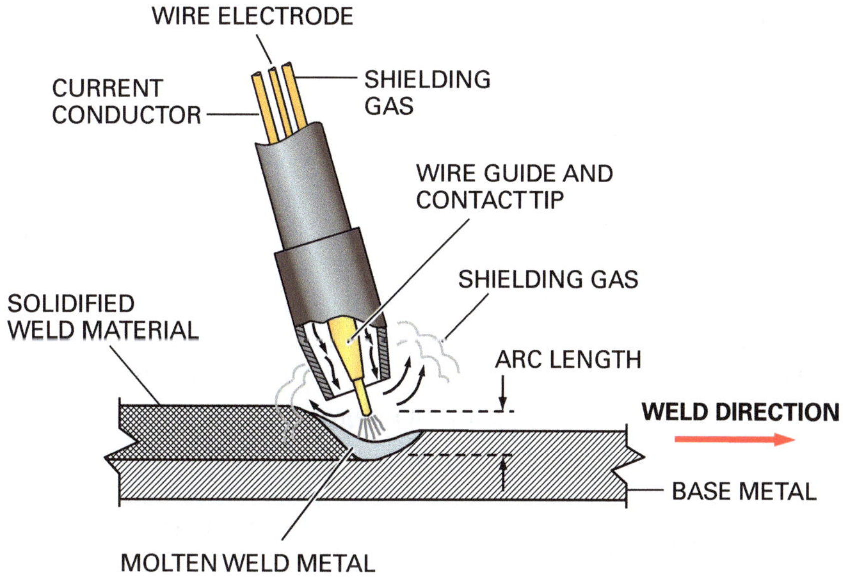

Figure 9 Arc length.

decreased, the welding amperage automatically decreases. With lower welding amperage and less heat, the deposition rate drops and the weld beads are smaller and have less penetration.

Note that some constant-voltage welding machines used for GMAW provide varying degrees of current modification, such as slope or induction adjustments. Power sources with a slope adjustment allow the amount of amperage change to vary in relation to voltage range of the unit. This allows the correct short-circuiting current to be established for certain welding applications and electrode requirements. Standard constant-voltage GMAW units have current slope that is fixed by the manufacturer for general welding applications and conditions. Other units allow adjustment of the response time for short-circuit current by using an induction control.

In an instantaneous short-circuit current draw mode of operation, the machine draws a high level of current as soon as the electrode touches the workpiece. In the adjustable induction mode of operation, the machine gradually increases the welding current after the initial contact of the electrode to the workpiece. This smoothes out the arc characteristics during short-circuiting transfer.

Pulse-transfer welding machines allow the adjustment of the peak current for the pulse and the background current between pulses to match specific welding applications and electrode requirements. In other cases, where a voltage-sensing wire feeder is used with a constant-current machines, the current is adjustable and the wire feeder varies the wire feed speed to maintain the proper arc voltage at the electrode tip.

Arc Voltage

A minimum arc voltage is needed to maintain spray transfer. However, penetration is not directly related to voltage. Penetration will increase with voltage for a time, but beyond an optimum voltage, it will actually decrease with further voltage increases.

2.1.3 Welding Travel Speed

Weld travel speed is the speed of the electrode tip across the base metal in the direction of the weld, measured in inches per minute (ipm) or a metric equivalent such as centimeters per minute. Travel speed has a great effect on penetration and bead size. Slower travel speeds build higher beads with deeper penetration. Faster travel speeds build smaller beads with less penetration. Ideally, the welding parameters should be adjusted so that the electrode tip is positioned at the leading edge of the weld puddle during travel.

Inexperienced welders are tempted to turn down the wire feed speed if they experience difficulty controlling the weld puddle. Electrodes must be run at certain balanced parameters that cannot be changed individually. Voltage, wire feed speed, and travel speed are adjusted and balanced together to control the weld puddle.

2.1.4 Gun Position

Gun position for GMAW has an effect on weld penetration along with voltage and travel speed. The gun work angle (*Figure 10*) is an angle less

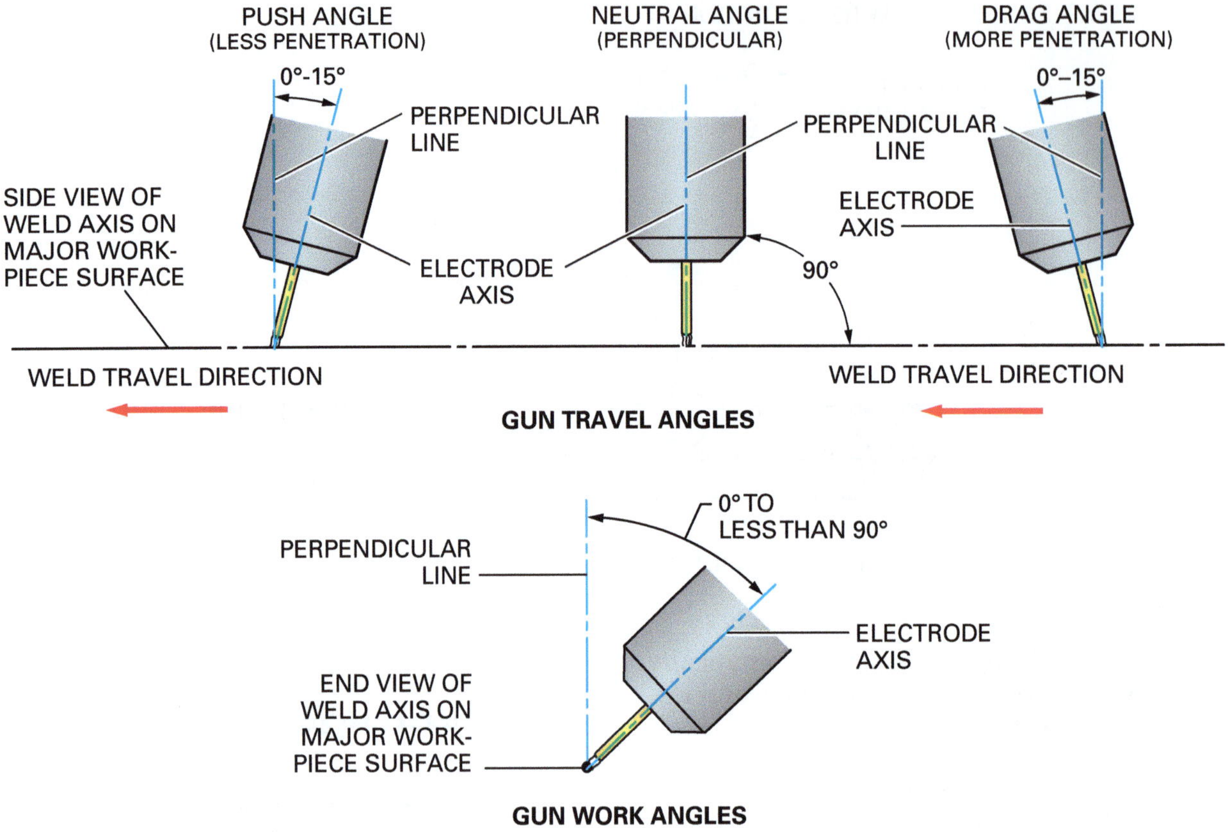

Figure 10 Gun travel and work angles.

than 90 degrees between a line perpendicular to the major workpiece surface at the point of electrode contact, and a plane determined by the electrode axis and the weld axis. For a T-joint or corner joint, the line is perpendicular to the nonbutting member. For pipe, the plane is determined by the electrode axis and a line tangent to the pipe surface at the same point.

The gun travel angle as shown in the figure is defined as an angle less than 90 degrees between the electrode axis and a line perpendicular to the weld axis at the point of electrode contact in a plane determined by the electrode axis and the weld axis. For pipe, the plane is determined by the electrode axis and a line tangent to the pipe surface at the same point.

- *Neutral angle* – The term *neutral angle* describes the travel angle when the center line of the gun electrode is exactly perpendicular (90 degrees) to the welding surface. The term is nonstandard for a 0-degree travel angle. Weld penetration is moderate when the gun electrode is used in this position.
- *Push angle* – A push angle is generally used for welding thin materials or when shallow penetration is required. A push angle is the travel angle

GMAW of Stainless Steel

For stainless steel, a drag angle is normally used for fillet welds and a push angle for square or grooved butt joints. Stainless steel is usually welded in the flat and horizontal positions with spray transfer. Vertical and overhead welds usually employ short-circuiting transfer for at least the root passes, if not the fill and cover passes as well. Pulsed-spray transfer can also be used for fill and cover passes. Remember that all open joints in stainless steel materials must be backed with copper, stainless steel, or an inert gas to prevent air from reaching the weld puddle until the puddle solidifies. When practicing welding using stainless steel electrodes and carbon steel coupons, use the travel angles and transfer methods outlined, along with carbon steel backing strips or a backing fixture for open or butt joints, when required. When welding plate that is thicker than 1/4" (6.4 mm), the gun should be moved back and forth in the travel direction, with a slight side-to-side motion. On thinner materials, only the back and forth motion is necessary.

created when the welding gun is tilted back so that the electrode tip precedes the gun in the direction of the weld. In this position, the electrode tip and shielding gas are being directed ahead of the weld bead. Welding with this angle produces a smaller weld bead with less penetration. Push angles greater than 15 degrees are not recommended for manual welding.

- *Drag angle* – A drag angle is the travel angle created when the gun is tilted forward so that the electrode tip trails the gun in the direction of the weld. In this position, the electrode tip and shielding gas are being directed back toward the weld bead. Welding with this angle produces a higher and wider weld bead with more penetration. Maximum penetration is achieved with a drag angle of approximately 25 degrees.

2.1.5 Electrode Extension, Stickout, and Standoff Distance

Electrode extension is the length of the electrode that extends beyond the contact tip of the welding gun. Changing the extension influences the welding current for low-conductivity metal electrodes because it changes the preheating of the electrode caused by the resistance of the electrode to current flow. As the extension increases, preheating of the electrode increases. Since the welding power source is self-regulating, it does not have to furnish as much welding current to melt the electrode, so the current output automatically decreases. This results in less penetration and increased deposition rates. Increasing extension is useful for bridging gaps and compensating for mismatch, but it can cause overlap or lack of fusion and a ropy bead appearance. When the extension is decreased, the power source is forced to increase its current output to burn off the electrode. Too little extension can cause the electrode to weld to the contact tip or develop porosity in the weld. For metal electrodes which are highly conductive, the preheating effect of electrode resistance is minimal, and electrode speed and voltage settings have a more direct effect.

GMAW extensions vary between ¼" to 1" (6.4 mm to 25.4 mm), depending on the transfer mode. *Figure 11* shows typical electrode extensions for GMAW gun configurations.

Additional gun nomenclature is defined in *Figure 11*. Stickout is the distance from the gas nozzle or insulating nozzle to the end of the electrode. Standoff distance is the distance from the gas nozzle or insulating nozzle to the workpiece. Contact tip extension or setback is usually dependent on the transfer mode for the GMAW application.

2.1.6 Gas Nozzle Cleaning

As a GMAW welding machine is being used, weld spatter accumulates on the gas nozzle and the contact tip. If the gas nozzle is not properly cleaned, it will restrict the shielding gas flow and cause porosity in the weld.

Clean the gas nozzle with a reamer or round file or with the tang of a standard file. After cleaning, the nozzle can be sprayed or dipped in a special anti-spatter compound. The anti-spatter compound helps prevent the spatter from sticking to the nozzle. Ensure that any anti-spatter product used is compatible with the welding application and equipment in use.

Arc Blow

When current flows, magnetic fields are created. The magnetic fields tend to concentrate in corners, deep grooves, or the ends of the base metal. When the arc approaches these concentrated magnetic fields, it is deflected from its intended path. Arc blow can cause defects such as excessive weld spatter and porosity. If arc blow occurs, try one or more of these methods to control it:

- Change the position of the workpiece clamp. This will change the flow of welding current, affecting the way the magnetic fields are created.
- Change the angle of the gun. Reducing or even reversing the angle of the gun can compensate for the arc blow.

Anti-Spatter Materials

Use only anti-spatter material specifically designed for GMAW gas nozzles. Water-based products may cause hydrogen embrittlement in hydrogen-sensitive steel. Always refer to the SDS/MSDS for specific safety considerations.

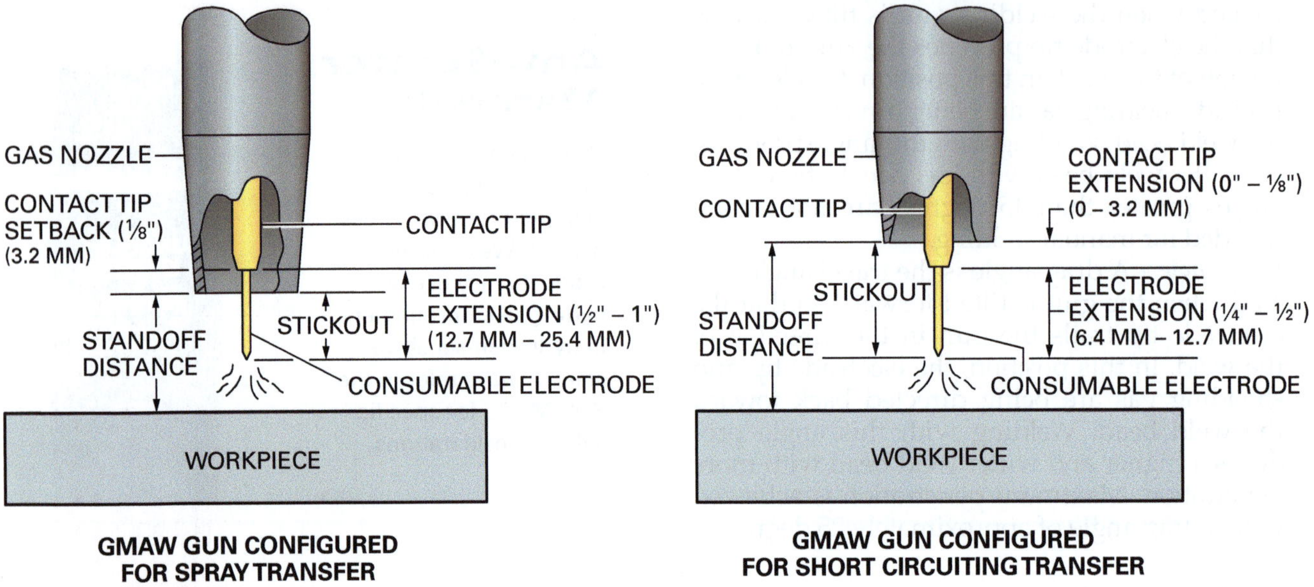

Figure 11 GMAW gun configurations.

2.2.0 Bead Types

Different types of weld beads can be created using GMAW equipment. The two basic bead types are stringer beads and weave beads.

2.2.1 Stringer Beads

Stringer beads are made with little or no side-to-side motion of the electrode.

> **CAUTION**
>
> The width of stringer beads is specified in the welding code or WPS being used at a site. Do not exceed the widths specified for the site.

Practice running stringer beads in the flat position. Experiment with different voltages, travel angles, and stickouts. Change the wire feed speed and observe the bead size. Try different travel angles and observe the bead shape and size. Use solid- or composite-core electrodes and the appropriate shielding gas as directed by the instructor.

Follow these steps to run stringer beads:

Step 1 Hold the gun at the desired travel angle (drag or push), with the electrode tip directly over the point where the weld is to begin, and pull the gun trigger.

Step 2 Hold the arc in place until the weld puddle begins to form.

Step 3 Slowly advance the arc while maintaining the gun angle. Stay in the leading edge of the puddle to prevent incomplete fusion.

Step 4 Continue to weld until a bead about 2" to 3" (50 mm to 76 mm) long is formed.

Step 5 Stop, raise the gun to a neutral (0-degree) travel angle, pause until the crater is filled, and then release the trigger.

Step 6 Inspect the bead for the following:

- Straightness of the bead
- Uniform appearance of the bead face
- Smooth, flat transition with complete fusion at the toes of the weld
- No porosity
- No undercut
- No overlap
- No lack of fusion
- Complete penetration
- No pinholes (fisheyes)
- No inclusions
- Crater filled
- No cracks

Step 7 Continue practicing stringer beads until acceptable welds are made every time.

2.2.2 Weave Beads

Weave beads are made with wide, side-to-side motions of the electrode. The width of a weave bead is determined by the amount of side-to-side motion. *Figure 12* shows weave beads made with 0.035" (0.9 mm) diameter electrode.

When making a weave bead, use care at the toes to be sure there is proper tie-in to the base metal. To ensure proper tie-in at the toes, slow down or pause slightly at the edges. The pause at the edges will also flatten out the weld, giving it the proper profile.

Figure 12 Close-up view of GMAW weave beads.

Practice running weave beads about ⅝" (16 mm) wide in the flat position. Experiment with different weave motions, travel angles, and stickouts. Change the wire feed speed and observe the bead size. Try different gun angles and observe the bead shape and size. Use solid- or composite-core electrodes and appropriate shielding gas as directed by the instructor.

Follow these steps to run weave beads:

Step 1 Hold the gun at the desired travel angle with the electrode tip directly over the point where the weld is to begin.

Step 2 Pull the gun trigger.

Step 3 Hold the arc in place until the weld puddle begins to form.

Step 4 Slowly advance the arc in a weaving motion while maintaining the gun angle (*Figure 13*).

Step 5 Continue to weld until a bead about 2" to 3" (50 mm to 76 mm) long is formed.

Step 6 Stop, raise the gun to a neutral (0-degree) travel angle, pause until the crater is filled, and then release the trigger.

Step 7 Inspect the bead for the following:

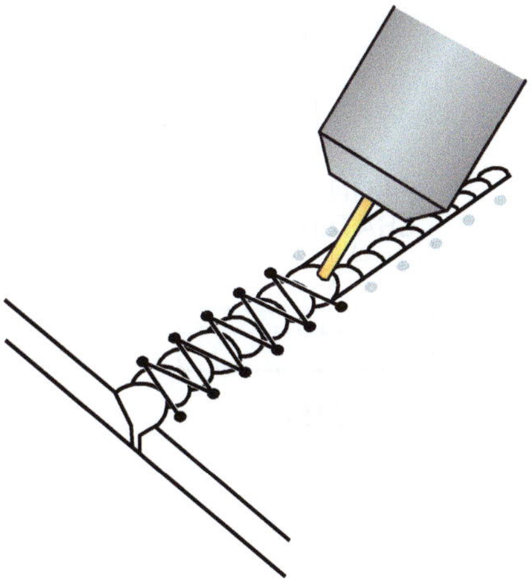

Figure 13 Weave motion.

- Straightness of the bead
- Uniform appearance of the bead face
- Smooth, flat transition with complete fusion at the toes of the weld
- No porosity
- No undercut
- No overlap
- No lack of fusion
- Complete penetration
- No pinholes (fisheyes)
- No inclusions
- Crater filled
- No cracks

Step 8 Continue practicing weave beads until able to make acceptable welds every time.

2.2.3 Weld Restarts

A restart is the junction where a new weld connects to and continues the bead of a previous weld. Restarts are important because an improperly made restart creates a weld discontinuity. One advantage of the semi-automatic GMAW process is that restarts can be minimized. Plan the weld to eliminate restarts, if possible. If a restart has to be made, it must be made so that it blends smoothly with the previous weld and does not stand out. The technique for making a restart is the same for both stringer and weave beads. Practice weld restarts using GMAW as directed by the instructor.

Follow these steps to make a restart:

Step 1 With the weld clean, hold the torch at the proper travel angle while restarting the arc directly over the center of the crater. Remember that welding codes do not allow arc strikes outside the area of permanent welds.

Step 2 Pull the gun trigger.

Step 3 Move the electrode tip over the crater in a small circular motion to fill the crater with a molten puddle.

Weave Beads and Stainless Steel

Weave beads are not normally used with stainless steel because of gas shielding problems and excessive heat input to the base metal. However, some fully automatic processes can make reasonable weave beads using pulse-transfer techniques.

Step 4 Move to the leading edge of the puddle; as soon as the puddle fills the crater, continue the stringer or weave bead pattern being used.

Step 5 Inspect the restart. A properly made restart will blend into the bead, making it hard to detect. If the restart has undercut, not enough time was spent in the crater to fill it. If the undercut is on one side or the other, use more of a side-to-side motion while moving back into the crater. If the restart has a lump, it was overfilled; too much time was spent in the crater before resuming the forward motion.

Step 6 Continue to practice restarts until they are correct.

2.2.4 Weld Terminations

A weld termination is made at the end of a weld. A termination normally leaves a crater. When making a termination, the welding codes require that the crater must be filled to the full cross section of the weld. This can be difficult, since most terminations are at the edge of a plate, where welding heat tends to build up, making filling the crater more difficult. *Figure 14* shows the proper technique for terminating a weld.

Follow these steps to make a weld termination:

Step 1 Start to bring the gun up to a 0-degree travel angle, and slow the forward travel as the end of the weld is approached.

Step 2 Stop the forward movement about ⅛" (3.2 mm) from the end of the plate.

Step 3 With the gun at a 0-degree travel angle, allow the crater to fill, and then move about ⅛" (3.2 mm) toward the start of the weld.

Step 4 Release the trigger when the crater is completely filled. Do not remove the gun from the weld until the puddle has solidified. The gas postflow that continues after the welding stops protects the molten metal. If the gun and shielding gas are removed before the crater has solidified, crater porosity or cracks can occur.

Step 5 Inspect the termination. The crater should be filled to the full cross section of the weld.

2.2.5 Overlapping Beads

Overlapping beads are made by depositing connective weld beads parallel to one another. The parallel beads overlap, forming a flat surface. This is also called padding. Overlapping beads are used to build up a surface and to make multiple-pass welds. Both stringer and weave beads can be overlapped. When viewed from the end, properly overlapped beads will have a relatively flat surface. *Figure 15* shows proper and improper overlapping string and weave beads.

Follow these steps to weld overlapping stringer or weave beads, using a specific diameter solid or composite electrode and the proper shielding gas as directed by the instructor.

Step 1 Mark out a square on a piece of steel.

Step 2 Weld a stringer or weave bead along one edge at a 0-degree work angle.

Step 3 Clean the weld.

Step 4 Position the gun at a work angle of 10 to 15 degrees toward the side of the previous bead to get proper tie-in (*Figure 16*), and pull the trigger.

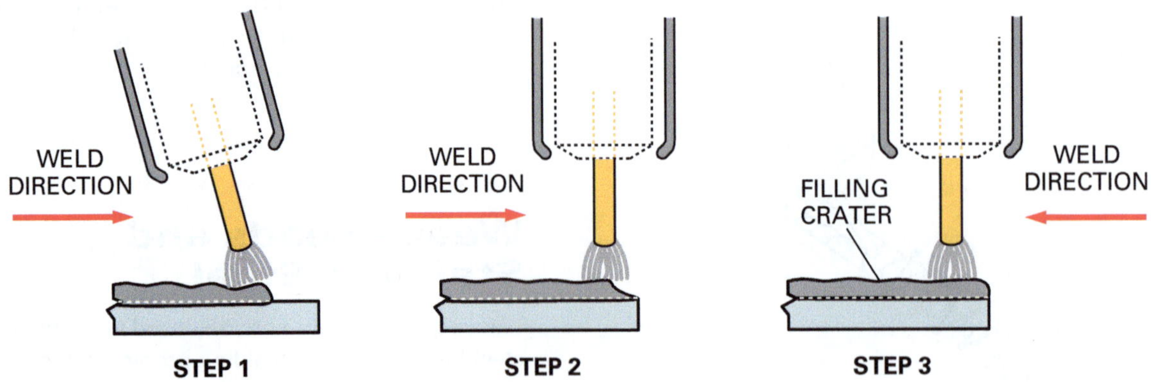

Figure 14 Terminating a weld.

Step 5 Continue running overlapping stringer or weave beads until the square is covered. Clean each bead before running the new bead.

Step 6 Continue building layers of stringer or weave beads, one on top of the other, until perfecting the technique.

WARNING! The base metal will get very hot as it is being built up with multiple layers.

Build a pad using stringer beads (*Figure 17*). Repeat building the pad using weave beads.

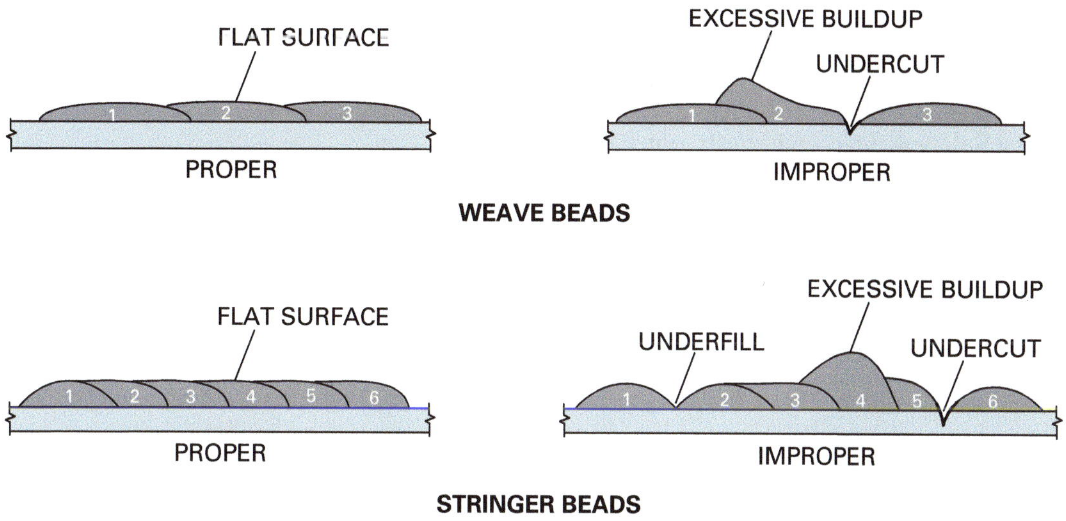

Figure 15 Proper and improper overlapping beads.

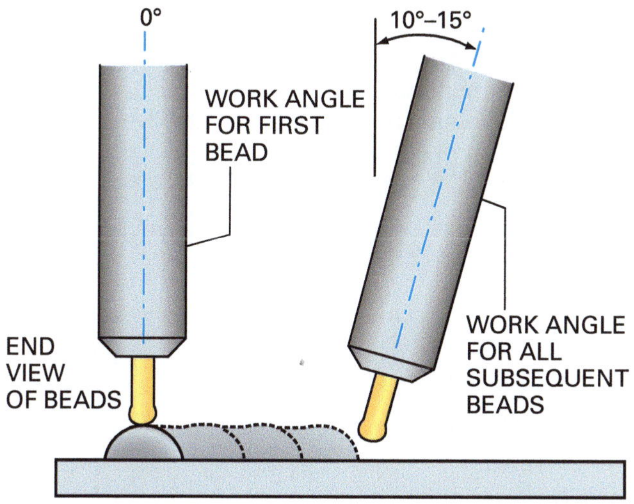

Figure 16 Work angles for overlapping beads.

Figure 17 Pad built with stringer beads.

A Few Suggestions

To improve your technique, practice with both hands in opposing directions when welding in the 1G, 2G, and 4G positions. Also, the first weld bead can be done up the center of the pad, with the remaining beads being worked from the center out. This tactic gives you practice overlapping beads on the pad as you might do in a groove weld.

Additional Resources

AWS B1.10M/B1.10 Guide for the Nondestructive Examination of Welds. Miami, FL: American Welding Society.

AWS B1.11 Guide for the Visual Examination of Welds. Miami, FL: American Welding Society.

AWS C5.6-89R Recommended Practices for Gas Metal Arc Welding. Miami, FL: American Welding Society.

AWS D3.5-93R Guide for Steel Hull Welding. Miami, FL: American Welding Society.

GMAW Welding Guide. Current Edition. Cleveland, OH: The Lincoln Electric Company.

Lincoln Electric website: **http://www.lincolnelectric.com** offers sources for products and training.

OSHA 1910.269, Appendix *C, Protection from Step and Touch Potentials.* Current Edition. Washington, DC: Occupational Safety & Health Administration (OSHA).

OSHA 1926.351, Arc Welding and Cutting. Current Edition. Washington, DC: Occupational Safety and Health Administration (OSHA).

The Procedure Handbook of Arc Welding. 14th Edition. Cleveland, OH: The James F. Lincoln Arc Welding Foundation, 2000.

Welding Handbook. Volume 1. Welding Science & Technology. Miami, FL: American Welding Society, 2001.

Welding Handbook. Volume 2. Part 1: Welding Processes. Miami, FL: American Welding Society, 2004.

2.0.0 Section Review

1. Depending on the transfer mode, GMAW electrode extensions vary between ____.
   a. ¹⁄₁₆" to ¼" (1.6 to 6.4 mm)
   b. ⅛" to ½" (3.2 to 12.7 mm)
   c. ¼" to 1" (6.4 to 25.4 mm)
   d. ½" to 1½" (12.7 to 32 mm)

2. The width of a weave bead is determined by the amount of ____.
   a. forward and backward gun movement
   b. stringer bead padding overlap
   c. filler material in the flux core
   d. side-to-side motion of the electrode

SECTION THREE

3.0.0 GMAW FILLET AND V-GROOVE WELDS

Objective

Describe the welding procedures needed to produce proper fillet and V-groove welds using GMAW welding techniques.

a. Describe the welding procedures needed to produce proper fillet welds using GMAW welding techniques.
b. Describe the welding procedures needed to produce proper V-groove welds using GMAW welding techniques.

Performance Tasks

1. Make multiple-pass GMAW-S (short-circuit) fillet welds on carbon steel plate coupons in all four 1F through 4F positions, using solid or composite electrode.
2. Make multiple-pass GMAW-S (short-circuit) V-groove welds on carbon steel plate coupons in all four 1G through 4G positions, with or without backing, using solid or composite electrode.
3. Make multiple-pass GMAW spray-transfer fillet welds on carbon steel plate coupons in both the 1F and 2F positions, using solid or composite electrode.
4. Make multiple-pass GMAW spray-transfer V-groove welds on carbon steel plate coupons in the 1G position, with backing, using solid or composite electrode.

Two common types of welds produced with GMAW are fillet welds and V-groove welds. This section examines the procedures used to make each of these welds in four different positions: flat, horizontal, vertical, and overhead.

3.1.0 Fillet Welds

Fillet welds require little base metal preparation, except for the cleaning of the weld area and the removal of any excess material from cut surfaces. Any dross from cutting will cause porosity in the weld. For this reason, the codes require that this material is entirely removed prior to welding.

The most common fillet welds are made in lap and T-joints. The weld position for plate is determined by the weld axis and the orientation of the workpiece. The positions for fillet welding on plate are flat, or 1F (the letter F stands for fillet); horizontal, or 2F; vertical, or 3F; and overhead, or 4F. In the 1F and 2F positions, the weld axis can be inclined up to 15 degrees. Any weld axis inclination for the other positions varies with the rotational position of the weld face as specified in AWS standards. *Figure 18* shows the four fillet welding positions for plate.

Fillet welds can be concave or convex, depending on the WPS or site quality standards. The welding codes require a fillet weld to have a uniform concave or convex face, although a slightly

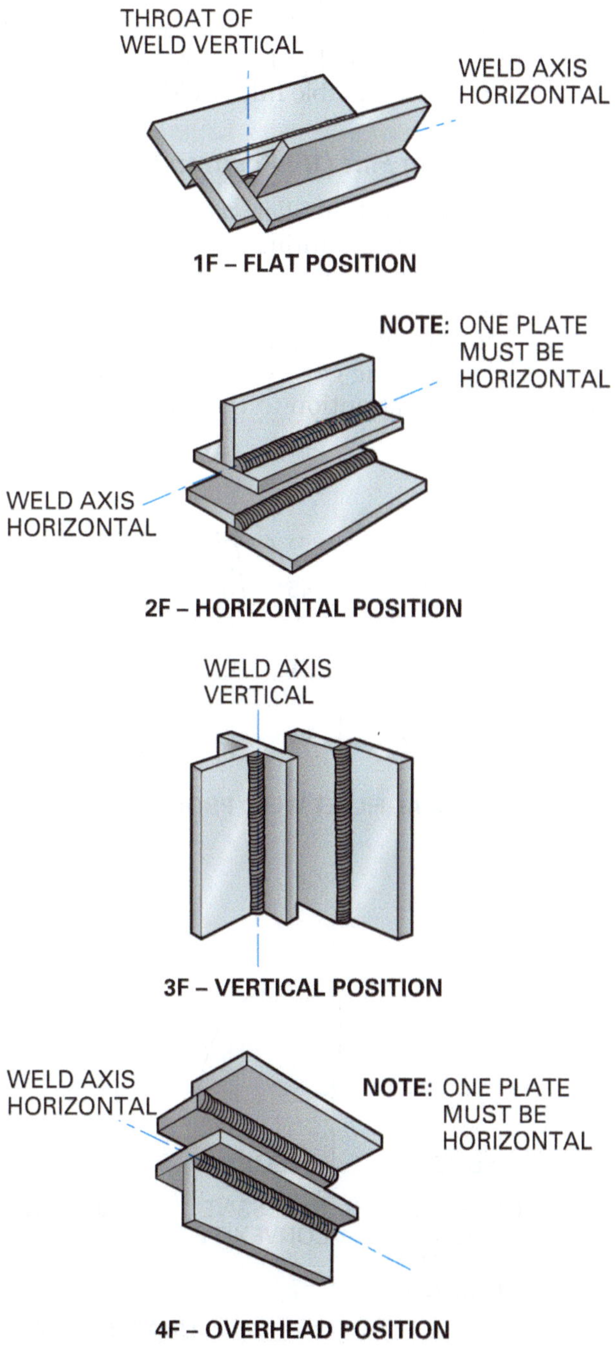

Figure 18 Fillet weld positions.

non-uniform face is acceptable. Convexity is the distance the weld extends above a line drawn between the toes of the weld. The convexity of a fillet weld or individual surface bead must not exceed that permitted by the applicable code or standard. A fillet weld must be replaced if the profile has defects.

A fillet weld is unacceptable and must be replaced if the profile has insufficient throat, excessive convexity, undercut, overlap, insufficient leg, or incomplete fusion. In single-pass, weave-bead fillet welds where two workpieces are being joined at an angle (not lap joints), flat or slightly convex faces are usually preferred because weld stresses are more uniformly distributed through the fillet weld and workpieces. *Figure 19* shows ideal, acceptable, and unacceptable fillet weld profiles.

3.1.1 Practicing Flat (1F) Position Fillet Welds

Practice flat (1F) position fillet welds using GMAW spray and/or short-circuiting transfer as directed by the instructor. Weld multiple-pass (six-pass) convex fillet welds in a T-joint. Use an instructor-specified size of solid or composite electrode with appropriate shielding gas. When making flat fillet welds, pay close attention to the work angle and travel speed. For the first bead, the work angle is vertical (45 degrees to both plate surfaces). The angle is adjusted for all subsequent beads. Increase or decrease the travel speed to control the amount of weld metal buildup. *Figure 20* shows the work angles for 1F position fillet welds.

WARNING!

Cooling with water (quenching) is done only on practice coupons. Never cool test coupons or on-the-job welds with water. Cooling with water can cause weld cracks and affect the mechanical properties of the base metal.

WARNING!

Use pliers to handle the hot practice coupons. Wear gloves when placing the practice coupon in water. Steam will rise off the coupon and can burn or scald unprotected hands and arms.

Follow these steps to make a flat fillet weld. Be sure to clean all weld beads thoroughly before beginning the next bead.

Step 1 Tack two plates together to form a T-joint for the fillet weld coupon (*Figure 21*). Clean the tack welds.

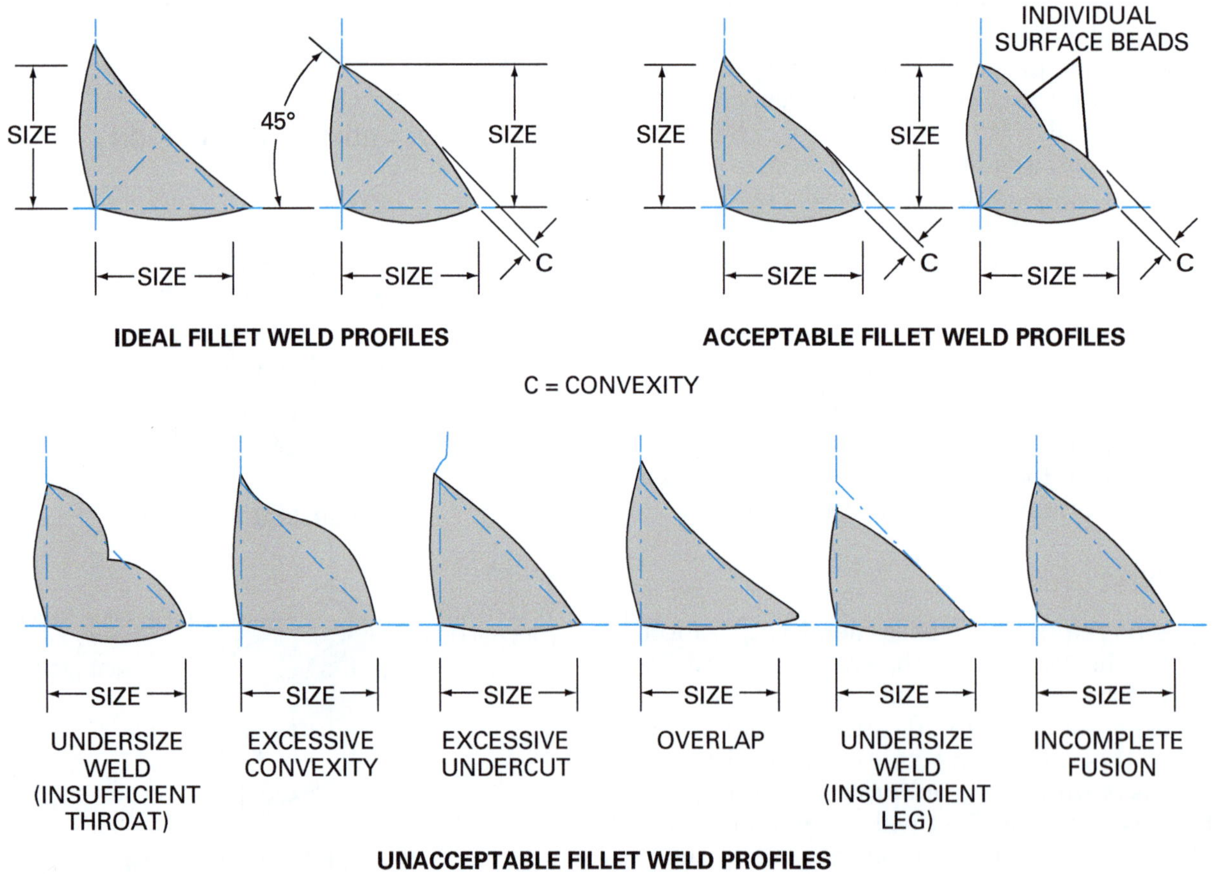

Figure 19 Ideal, acceptable, and unacceptable fillet weld profiles.

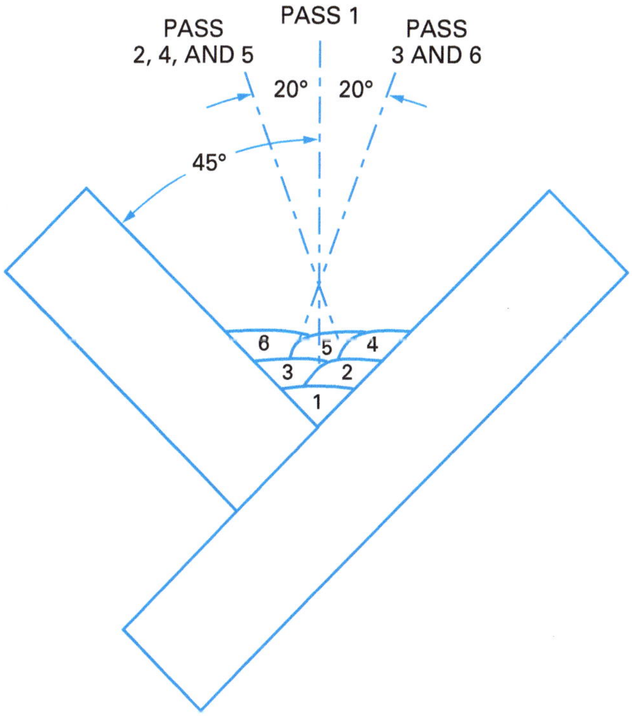

Figure 20 Multiple-pass 1F bead sequence and work angles.

Figure 21 Fillet weld coupon.

Step 2 Clamp or tack weld the coupon in the flat position.

Step 3 Run the first bead along the root of the joint, using a work angle of 45 degrees and a 5- to 10-degree drag angle. Use a slight side-to-side motion (oscillation).

Step 4 Run the second bead along a toe of the first weld, overlapping about 75 percent of the first bead. Alter the work angle as shown in *Figure 20*, and use a 5- to 10-degree drag angle with a slight oscillation.

Step 5 Run the third bead along the other toe of the first weld, filling the groove created when the second bead was run. Use the work angle shown in *Figure 20* and a 5- to 10-degree drag angle with a slight oscillation.

Step 6 Run the fourth bead along the outside toe of the second weld, overlapping about half the second bead. Use the work angle shown in *Figure 20* and a 5- to 10-degree drag angle with a slight oscillation.

Step 7 Run the fifth bead along the inside toe of the fourth weld, overlapping about half the fourth bead that was run. Use the work angle shown in *Figure 20* and a 5- to 10-degree drag angle with a slight oscillation.

Step 8 Run the sixth bead along the toe of the fifth weld, filling the groove created when the fifth bead was run. Use the work angle shown in *Figure 20* and a 5- to 10-degree drag angle with a slight oscillation.

Step 9 Inspect the weld. The weld is acceptable if it has the following features:

- Uniform appearance on the bead face
- Craters and restarts filled to the full cross section of the weld
- Uniform weld size ±1⁄16" (1.6 mm)
- Acceptable weld profile in accordance with the applicable code or standard
- Smooth transition with complete fusion at the toes of the weld
- No porosity
- No undercut
- No overlap
- No inclusions
- No cracks

T-Joint Heat Dissipation

In T-joints, the welding heat dissipates more rapidly in the thicker or non-butting member. On various bead passes, the arc may have to be concentrated slightly more on the thicker or the non-butting member to compensate for the heat loss.

T-Joint Vertical Plate Undercut

The most common defect for T-joints is an undercut on the vertical plate of the joint. If this problem occurs, angling the arc slightly more toward the vertical plate and the bead at the top of the weave will force more metal into the bead at the top edge of the weld.

3.1.2 *Practicing Horizontal (2F) Position Fillet Welds*

Practice horizontal (2F) fillet welding using GMAW spray and/or short-circuiting transfer as directed by the instructor. Place multiple-pass fillet welds in a T-joint using solid or composite-core electrode sizes and the appropriate shielding gas as directed by the instructor. When making horizontal fillet welds, pay close attention to the gun angles and travel speed. For the first bead, the electrode work angle is 45 degrees. The work angle is adjusted for all other welds. Increase or decrease the travel speed to control the amount of weld metal buildup.

Follow these steps to make a horizontal fillet weld:

Step 1 Tack two plates together to form a T-joint for the fillet weld coupon. Clean the tack welds.

Step 2 Clamp or tack weld the coupon in the horizontal position.

Step 3 Run the first bead along the root of the joint using a work angle of approximately 45 degrees with a 5- to 10-degree drag angle and a slight side-to-side oscillation (*Figure* 22).

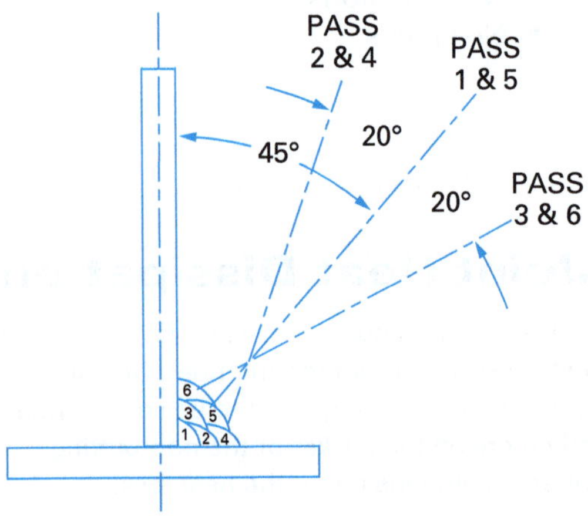

***Figure* 22** Multiple-pass 2F bead sequence and work angles.

Step 4 Clean the weld.

Step 5 Run the remaining passes at the appropriate work angles using a 5- to 10-degree drag angle and a slight oscillation at the appropriate work angles as shown in *Figure* 22. Overlap each previous pass and clean the weld after each pass.

Step 6 Have the instructor inspect the weld. The weld is acceptable if it has the following features:

- Uniform appearance on the bead face
- Craters and restarts filled to the full cross section of the weld
- Uniform weld size ±1⁄16" (1.6 mm)
- Acceptable weld profile in accordance with the applicable code or standard
- Smooth transition with complete fusion at the toes of the weld
- No porosity
- No undercut
- No overlap
- No inclusions
- No cracks

3.1.3 *Practicing Vertical (3F) Position Fillet Welds*

Practice vertical (3F) fillet welding using GMAW as directed by the instructor. Place multiple-pass fillet welds in a T-joint using solid- or composite-core electrode sizes and appropriate shielding gas as directed by the instructor. Normally, vertical welds are made welding uphill (also called vertical-up welding) from the bottom to the top, using a gun push angle (up-angle). However, downhill welding (also called vertical-down) can also be accomplished using a slight drag angle. When vertical welding, either stringer or weave beads can usually be used. Check with the instructor to see if stringer or weave beads, or both techniques, should be practiced. However, weave beads are avoided on stainless steel or in downhill welding. On the job, the site WPS or quality standard will specify which technique to use.

When making vertical fillet welds, pay close attention to the gun angles and travel speed. For the first bead, the work angle is approximately 45 degrees. The work angle is adjusted for all other welds. Increase or decrease the travel speed to control the amount of weld metal buildup.

Follow these steps to make an uphill fillet weld:

Step 1 Tack two plates together to form a T-joint for the fillet weld coupon.

Vertical Weave Bead

To help control undercut on a root pass, an alternate pattern, such as a small triangular weave, can be used in vertical welds. By pausing the rod at the edges, the previous undercut can be filled. This action also creates undercut at the existing weld pool, but that will be filled in by the next weave. This technique should only be applied to the root pass.

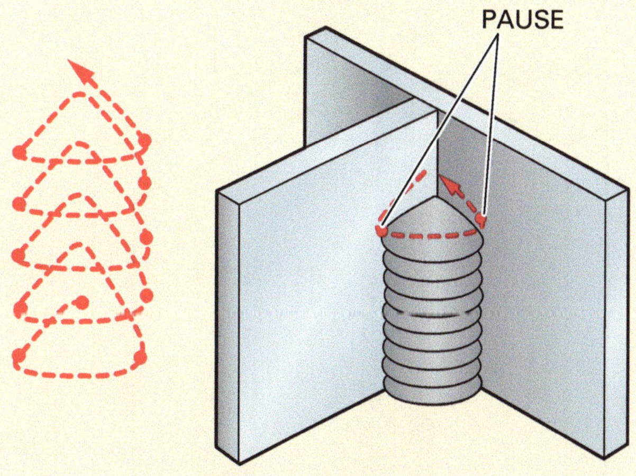

TRIANGULAR WEAVE PATTERN

Step 2 Clamp or tack weld the coupon in the vertical position.

Step 3 Starting at the bottom, run the first bead along the root of the joint using a work angle of approximately 45 degrees and a 10- to 15-degree push angle. Pause in the weld puddle to fill the crater.

Step 4 Clean the weld.

Step 5 Run the remaining passes using a 5- to 10-degree push angle and a side-to-side weave technique with a 45-degree work angle (*Figure 23*). Use a slow motion across the face of the weld, pausing at each toe for penetration and to fill the crater. Adjust the travel speed across the face of the weld to control the buildup. Overlap each previous pass as required, and clean the weld after each pass.

Step 6 Have the instructor inspect the weld. The weld is acceptable if it has the following features:

- Uniform appearance on the bead face
- Craters and restarts filled to the full cross section of the weld
- Uniform weld size ±1⁄16" (1.6 mm)
- Acceptable weld profile in accordance with the applicable code or standard
- Smooth transition with complete fusion at the toes of the weld
- No porosity
- No undercut
- No overlap
- No inclusions
- No cracks

Repeat vertical fillet (3F) welding using stringer beads. Use a slight oscillation and side-to-side motion, pausing slightly at each toe to prevent undercut. For stringer beads, use a 0- to 10-degree push angle and the work angles shown in *Figure 23* as required.

3.1.4 Practicing Overhead (4F) Position Fillet Welds

Practice overhead (4F) fillet welding using GMAW as directed by the instructor. Place multiple-pass fillet welds in a T-joint using solid or composite electrode sizes and appropriate shielding gas as directed by the instructor. When making overhead fillet welds, pay close attention to the gun angles and travel speed. For the first bead, the work angle is approximately 45 degrees. The work angle is adjusted for all other welds. Increase or decrease the travel speed to control the amount of weld metal buildup.

Follow these steps to make an overhead fillet weld:

Step 1 Tack two plates together to form a T-joint for the fillet weld coupon.

Step 2 Clamp or tack weld the coupon so it is in the overhead position.

Step 3 Run the first bead along the root of the joint, using a work angle of approximately 45 degrees with a 5- to 10-degree drag angle. Use a slight oscillation to tie-in the weld at the toes.

Step 4 Clean the weld.

Step 5 Using a slight oscillation, run the remaining passes using a 5- to 10-degree

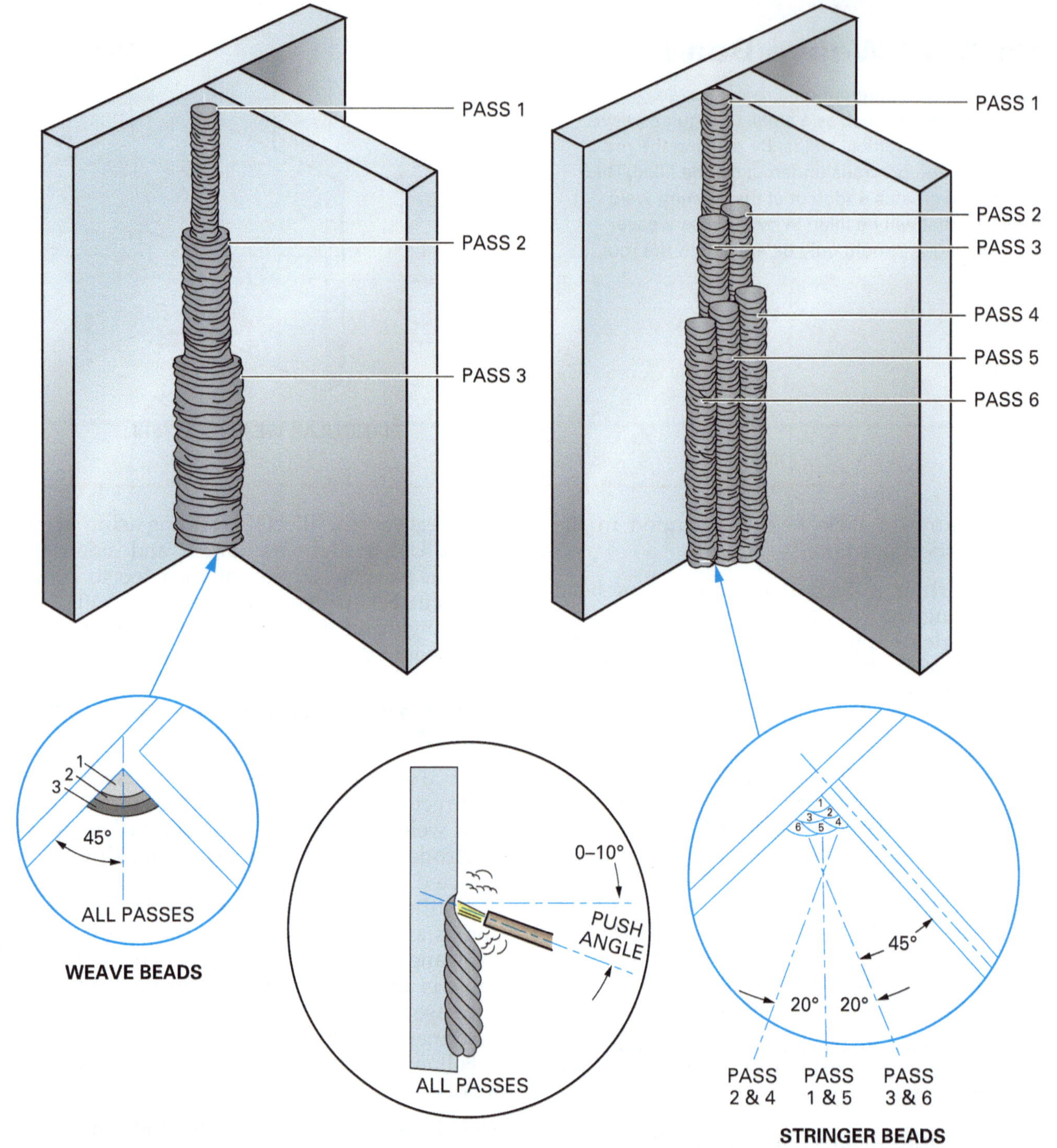

Figure 23 Multiple-pass 3F bead sequences and work angles for weave and stringer beads.

drag angle and the work angles shown in *Figure 24*. Overlap each previous pass and clean the weld after each pass.

Step 6 Have the instructor inspect the weld. The weld is acceptable if it has the following features:

- Uniform appearance on the bead face
- Craters and restarts filled to the full cross section of the weld
- Uniform weld size ±1/16" (1.6 mm)
- Acceptable weld profile in accordance with the applicable code or standard
- Smooth transition with complete fusion at the toes of the weld
- No porosity
- No undercut
- No overlap
- No inclusions
- No cracks

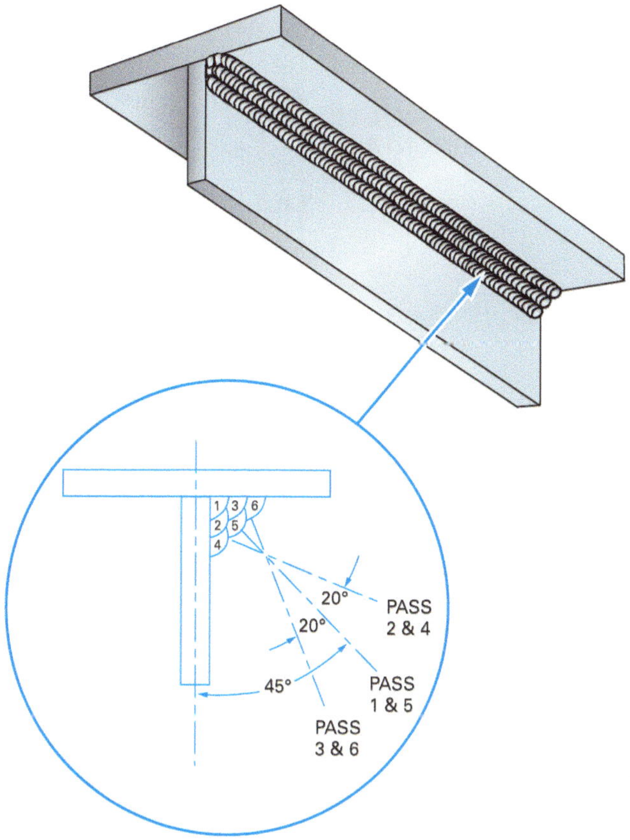

Figure 24 Multiple-pass 4F bead sequence and work angles.

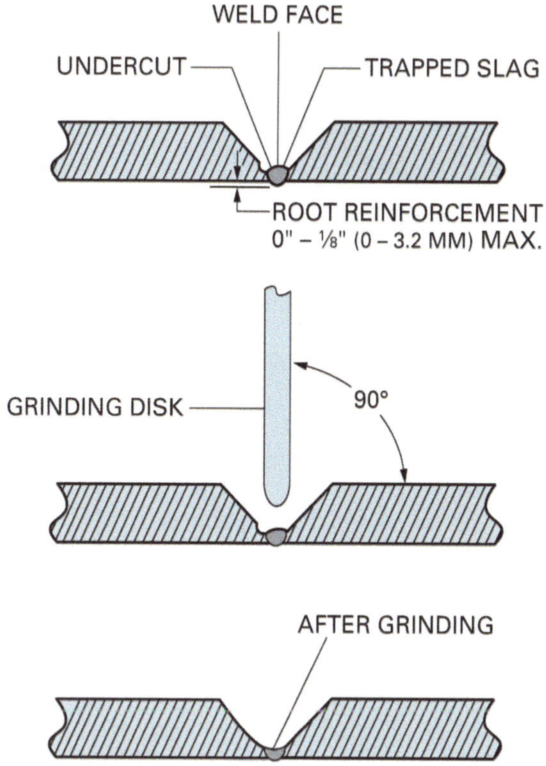

Figure 25 Grinding a root pass.

3.2.0 Open V-Groove Welds

The open V-groove weld is a common groove weld normally made on plate and pipe. Practicing the open V-groove weld on plate will help with making the more difficult pipe welds covered in the next level of the Welding curriculum.

3.2.1 Root Pass

The most difficult part of making an open V-groove weld is the root pass. The root pass is made from the V-groove side of the joint and must have complete penetration, but not an excessive amount of root reinforcement. The penetration is controlled with a side-to-side motion or oscillation.

After the root pass has been run, it should be cleaned and inspected. Inspect the face of the weld for excess buildup or undercut (*Figure 25*). Remove excess buildup or undercut with a hand grinder by grinding the face of the root pass with the edge of a grinding disk. Use care not to grind through the root pass or widen the groove.

3.2.2 Groove Weld Positions

As shown in *Figure 26*, groove welds can be made in all positions. The weld position for plate is determined by the axis of the weld and the orientation of the workpiece. Groove weld positions for plate are flat, or 1G (groove); horizontal, or 2G; vertical, or 3G; and overhead, or 4G.

In the 1G and 2G positions, the weld axis can be inclined up to 15 degrees. Any weld axis inclination for the other positions varies with the rotational position of the weld face as specified in the WPS.

3.2.3 Acceptable and Unacceptable Groove Weld Profiles

Groove welds should be made with slight reinforcement and a gradual transition to the base metal at each toe. Groove welds must not have excessive reinforcement or undercut, underfill, or overlap (*Figure 27*). If a groove weld has any of these defects, it must be replaced.

3.2.4 Practicing Flat (1G) Position Open V-Groove Welds

Practice flat (1G) open V-groove welds, as shown in *Figure 28* using GMAW (spray and short-circuiting transfer) as directed by the instructor. Use solid- or composite-core electrode sizes and appropriate shielding gas as directed by the instructor. Except for a root pass, use weave beads and keep the gun angle at 90 degrees to the plate surface (0-degree work angle) with a 5- to 10-

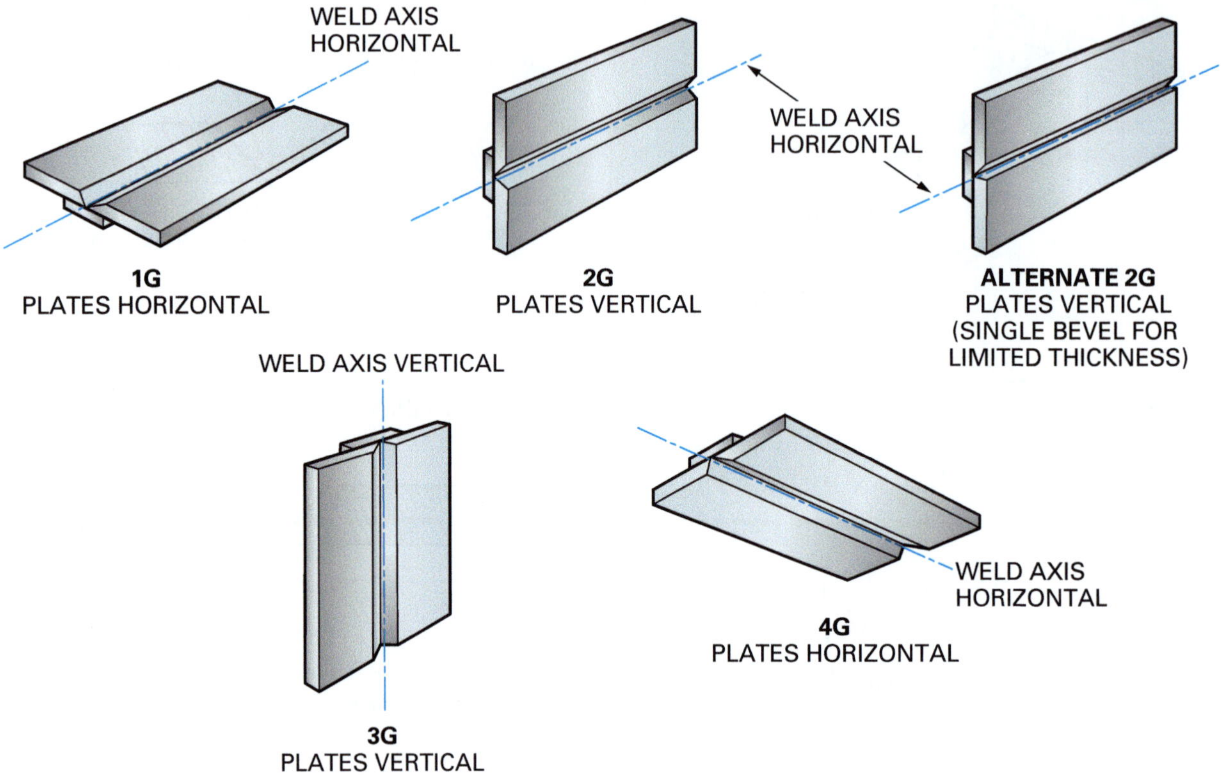

Figure 26 Groove weld positions for plate.

degree drag angle. When using stringer beads, use a 5- to 10-degree drag angle (except for the root pass), but adjust the work angle to tie in and fill the bead. Pay special attention at the termination of the weld to fill the crater. If the practice coupon gets too hot between passes, it may have to be cooled in water.

Follow these steps to practice open V-groove welds in the flat position:

> **NOTE**
>
> Avoid using weave beads for stainless steel.

Step 1 Tack weld the practice coupon together as explained earlier.

Step 2 Clamp or tack weld the weld coupon in the flat position above the welding table surface.

Step 3 Run the root pass using a 5- to 10-degree drag angle and a slight side-to-side motion to control penetration.

Step 4 Clean the root pass and grind if required.

Step 5 Run the remaining passes using a 5- to 10-degree drag angle and applicable work angles. Overlap each previous pass and clean the weld after each pass.

Step 6 Inspect the weld. The weld is acceptable if it has the following:

- Uniform appearance on the bead face
- Craters and restarts filled to the full cross section of the weld
- Uniform weld width ±1⁄16" (1.6 mm)
- Acceptable weld profile in accordance with the applicable code or standard
- Complete uniform root reinforcement at least flush with the base metal to a maximum buildup of 1⁄8" (3.2 mm)
- Smooth, flat transition with complete fusion at the toes of the weld
- No porosity
- No undercut
- No inclusions
- No cracks

Root Pass Welding

In practice, root passes are usually made on various metals using GTAW or SMAW methods; then the fill and cover passes are made using FCAW or GMAW.

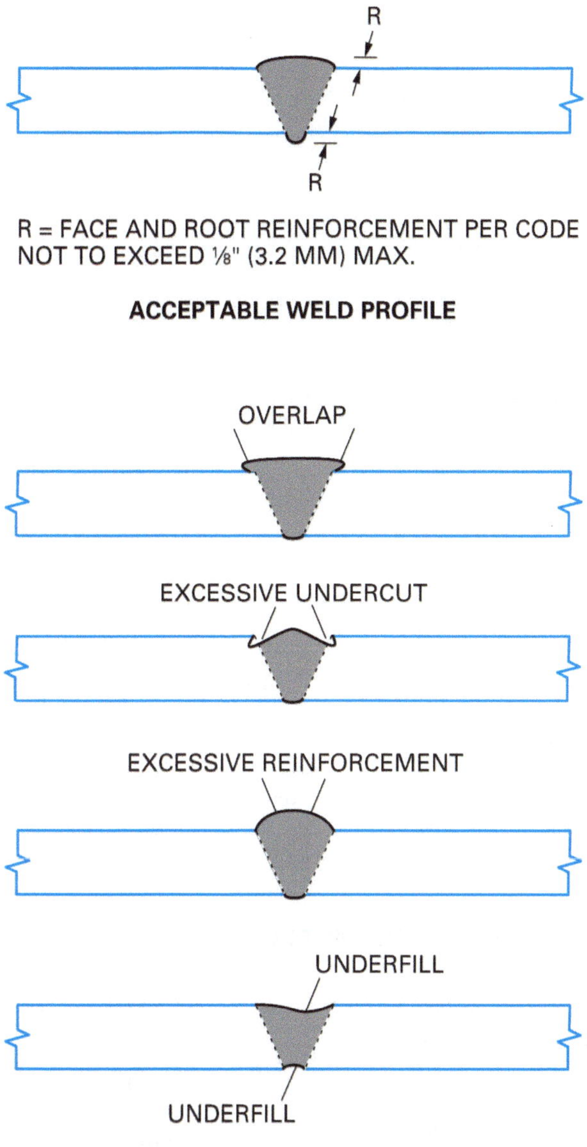

Figure 27 Acceptable and unacceptable groove weld profiles.

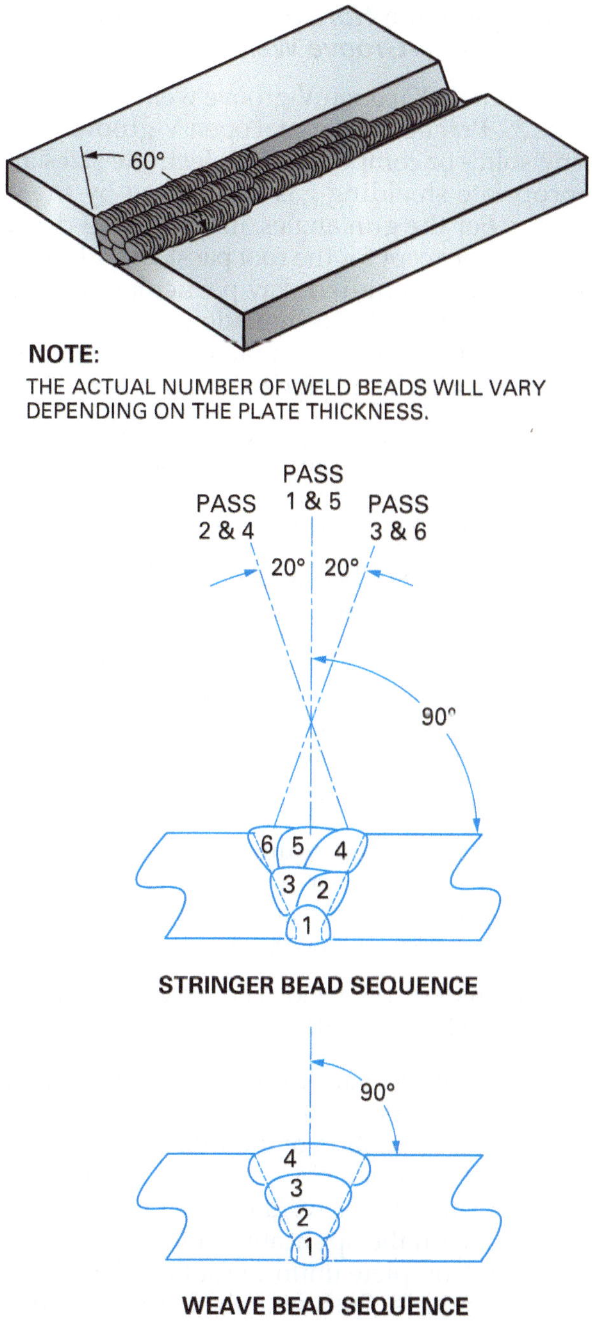

Figure 28 Multiple-pass 1G bead sequences and work angles.

Whiskers

When making a root pass, make sure that the electrode does not get ahead of the leading edge of the weld puddle. If it does, a portion of the electrode may penetrate through the joint and be melted off and fused into the back side of the weld. This will cause pieces of electrode, sometimes called whiskers, to protrude from the back of the weld.

3.2.5 *Practicing Horizontal (2G) Position Open V-Groove Welds*

A horizontal (2G) open V-groove weld is shown in *Figure 29*. Practice horizontal open V-groove welds using solid- or composite-core electrode sizes and appropriate shielding gas as directed by the instructor. For the gun angles, use a 5- to 10-degree drag angle (except for the root pass) and adjust the work angle as required. Pay particular attention to fill the crater at the termination of the weld.

Follow these steps to practice welding open V-groove welds in the horizontal position:

Step 1 Tack weld the practice coupon together as explained earlier. Use the standard or alternate horizontal weld coupon as directed by the instructor.

Step 2 Tack weld the flat plate weld coupon into the horizontal position.

Step 3 Run the root pass using a 5- to 15-degree drag angle and the appropriate work angle. Use a slight side-to-side motion to control penetration.

Step 4 Clean the weld.

Step 5 Run the remaining passes using a 5- to 10-degree drag angle and appropriate work angles. Overlap each previous pass and clean the weld after each pass.

Step 6 Inspect the weld. The weld is acceptable if it has the following:

- Uniform appearance on the bead face
- Craters and restarts filled to the full cross section of the weld
- Uniform weld width ±1⁄16" (1.6 mm)
- Acceptable weld profile in accordance with the applicable code or standard
- Complete uniform root reinforcement at least flush with the base metal to a maximum buildup of 1⁄8" (3.2 mm)
- Smooth, flat transition with complete fusion at the toes of the weld
- No porosity
- No undercut
- No inclusions
- No cracks

3.2.6 *Practicing Vertical (3G) Position Open V-Groove Welds*

Practice vertical (3G) multiple-pass groove welds using stringer beads or weave beads. Set the welding parameters at the lower ends of the suggested ranges. Out-of-position welds are made at lower welding voltages and currents than flat or horizontal welds to prevent the molten weld metal from sagging or running.

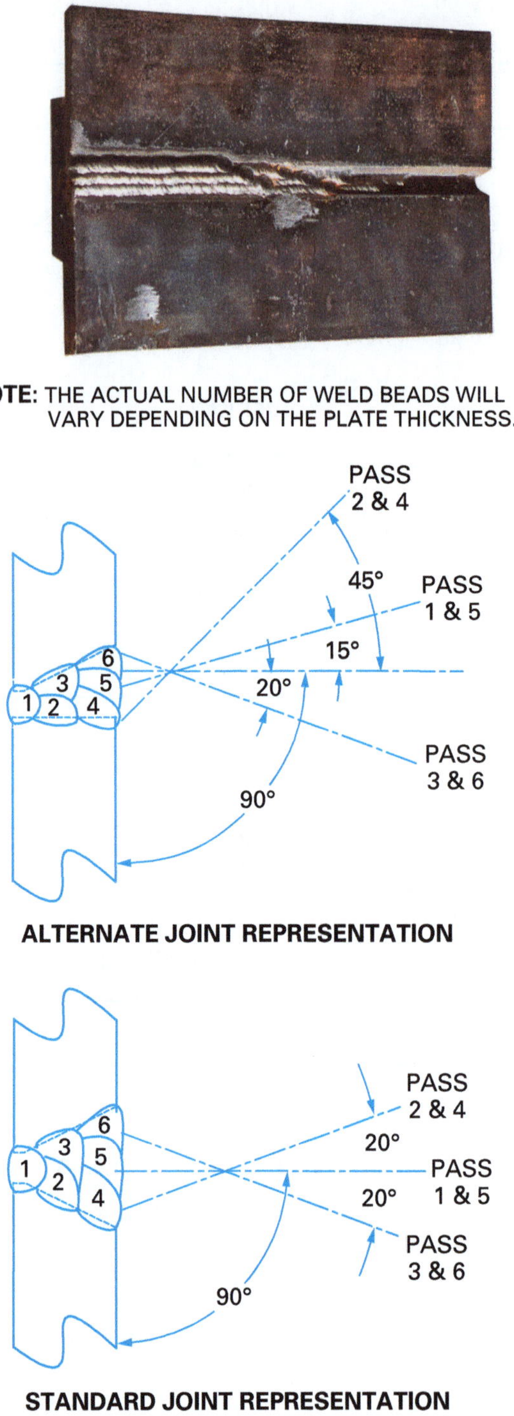

Figure 29 Multiple-pass 2G bead sequences and work angles.

When welding vertically, either stringer or weave beads can be used. However, weave beads are not normally used on stainless steel. On the job, the WPS or site quality standards will specify which technique to use. Check with the instructor to see if stringer or weave beads or both techniques should be practiced. The root bead can be run either uphill or downhill. The fill and cover beads can be run uphill or downhill as well. However, downhill weave beads are not normally used. Use the direction preferred at the site or as specified in the WPS or site quality standards.

Run the root bead, using a work angle of 0 degrees (nozzle center line bisects the V-groove) and a 5- to 15-degree push angle. The work angle is adjusted for all other welds.

Practice 3G open V-groove welds using GMAW as directed by the instructor. Use solid or composite electrode sizes and appropriate shielding gas as directed by the instructor.

Follow these steps to make a vertical groove weld with stringer beads:

Step 1 Tack weld the practice coupon together as explained earlier.

Step 2 Clamp or tack weld the coupon in the vertical position.

Step 3 Run the root pass uphill with a work angle of 0 degrees (nozzle center line bisects the V-groove) and a 5- to 15-degree push angle.

Step 4 Clean the weld.

Step 5 Run the remaining passes at the work angles shown in *Figure 30*, using a 5- to 15-degree push angle. Overlap each previous pass and use a slight side-to-side oscillation to tie-in the welds at the toes. Clean the weld after each pass.

Step 6 Inspect the weld. The weld is acceptable if it has the following features:

- Uniform appearance on the bead face
- Craters and restarts filled to the full cross section of the weld
- Uniform weld width ±1/16" (1.6 mm)
- Acceptable weld profile in accordance with the applicable code or standard
- Complete uniform root reinforcement at least flush with the base metal to a maximum buildup of 1/8" (3.2 mm)
- Smooth, flat transition with complete fusion at the toes of the weld
- No porosity
- No undercut
- No inclusions
- No cracks

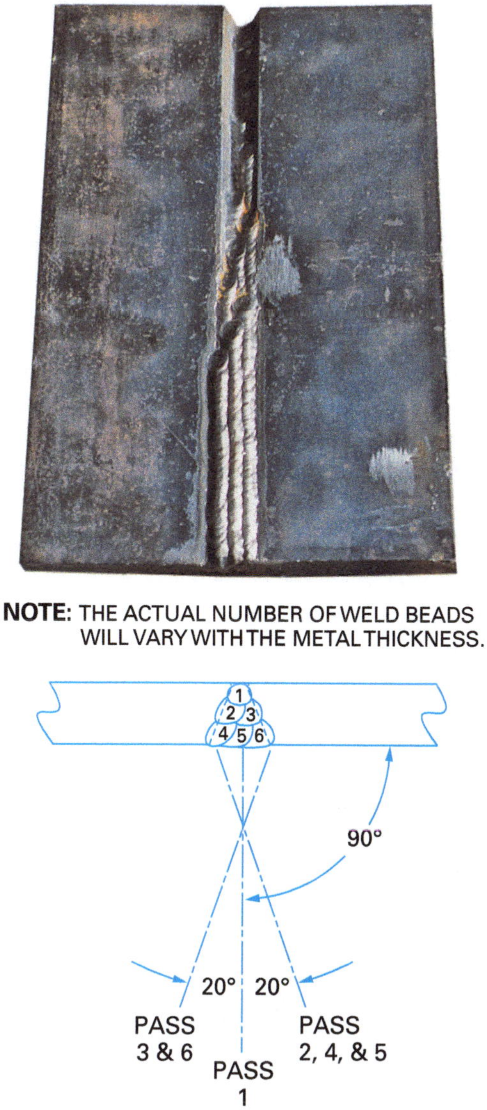

Figure 30 Multiple-pass 3G bead sequence and work angles for stringer beads.

Follow these steps to make a vertical (3G) groove weld with weave beads:

> **NOTE**
> Avoid using weave beads for stainless steel.

Step 1 Tack weld the practice coupon together as explained earlier.

Step 2 Clamp or tack weld the coupon in the vertical position.

Step 3 Run the root pass uphill with a work angle of 0 degrees (90 degrees from the plate) and a 5- to 15-degree push angle. Use a slight side-to-side oscillation to control the penetration.

Step 4 Clean the weld.

Step 5 Run the remaining passes as uphill weave beads across the root bead at a 0-degree work angle. Clean the weld after each pass. Use a zigzag weave pattern across the root bead and pause at each toe for good fusion. Use a push angle of 5 to 15 degrees. Increase or decrease the travel speed to control the amount of metal buildup. *Figure 31* shows the zigzag pattern and bead sequence for weave beads.

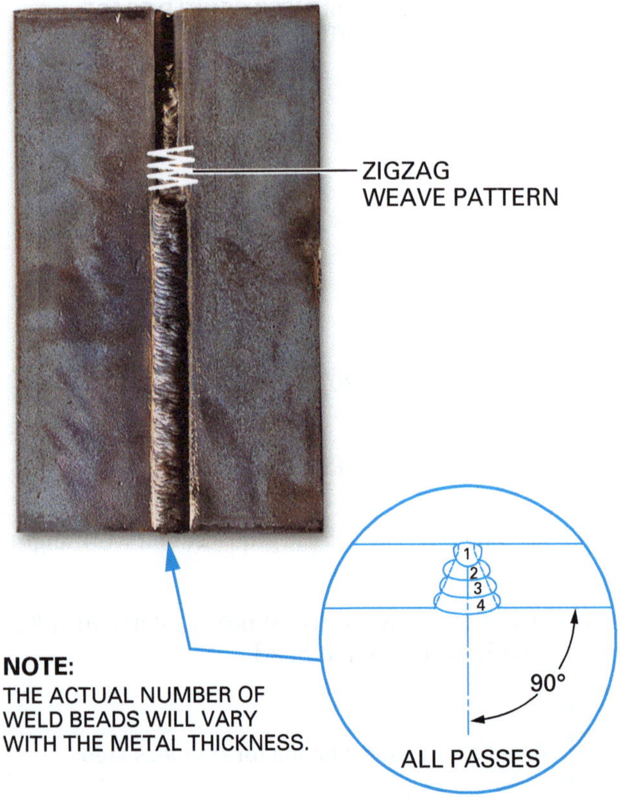

Figure 31 Multiple-pass 3G bead sequence and work angle for weave beads.

Step 6 Inspect the weld. The weld is acceptable if it has the following features:

- Uniform appearance on the bead face
- Craters and restarts filled to the full cross section of the weld
- Uniform weld width ±⅛" (3.2 mm)
- Acceptable weld profile in accordance with the applicable code or standard
- Complete uniform root reinforcement at least flush with the base metal to a maximum buildup of ⅛" (3.2 mm)
- Smooth, flat transition with complete fusion at the toes of the weld
- No porosity
- No undercut
- No inclusions
- No cracks

3.2.7 Practicing Overhead (4G) Position Open V-Groove Welds

Practice overhead (4G) open V-groove welds using GMAW as directed by the instructor. Use solid or composite electrodes and appropriate shielding gas as directed by the instructor.

> **NOTE**
> To reduce heat and prevent the molten weld metal from sagging or running, GMAW overhead welds are made with lower welding voltages and currents than flat or horizontal welds.

Use a work angle of 0 degrees (90 degrees to plate) and a 5- to 15-degree drag angle for the root bead.

Follow these steps to make an overhead groove weld with stringer beads:

Step 1 Tack weld the practice coupon together as explained earlier.

Step 2 Clamp or tack weld the coupon in the overhead position.

Step 3 Run the root pass, using a work angle of 0 degrees and a 5- to 15-degree drag angle. Use a slight side-to-side oscillation to control the penetration.

Step 4 Clean the weld.

Step 5 Run the remaining passes at the work angles shown in *Figure 32* using a 5- to 10-degree drag angle. Overlap each previous pass, and use a slight side-to-side motion to tie in the welds at the toes. Each bead should be cleaned before running the next bead.

Step 6 Inspect the weld. The weld is acceptable if it has the following features:

- Uniform appearance on the bead face
- Craters and restarts filled to the full cross section of the weld
- Uniform weld width ±1⁄16" (1.6 mm)
- Acceptable weld profile in accordance with the applicable code or standard
- Complete uniform root reinforcement at least flush with the base metal to a maximum buildup of 1⁄8" (3.2 mm)
- Smooth, flat transition with complete fusion at the toes of the weld
- No porosity
- No undercut
- No inclusions
- No cracks

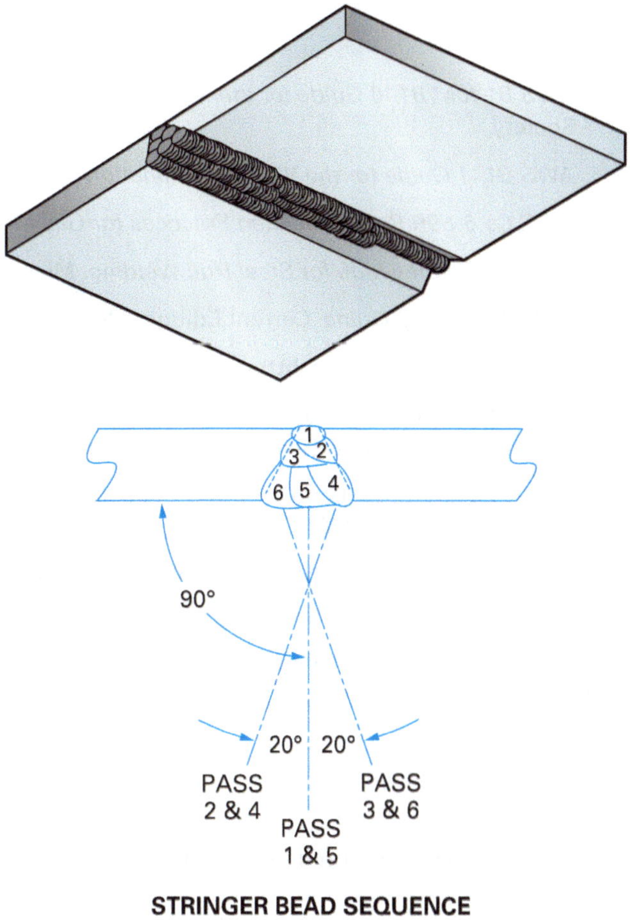

Figure 32 Multiple-pass 4G bead sequence and work angles.

Additional Resources

AWS B1.10M/B1.10 Guide for the Nondestructive Examination of Welds. Miami, FL: American Welding Society.

AWS B1.11 Guide for the Visual Examination of Welds. Miami, FL: American Welding Society.

AWS C5.6-89R Recommended Practices for Gas Metal Arc Welding. Miami, FL: American Welding Society.

AWS D3.5-93R Guide for Steel Hull Welding. Miami, FL: American Welding Society.

GMAW Welding Guide. Current Edition. Cleveland, OH: The Lincoln Electric Company.

Lincoln Electric website: **http://www.lincolnelectric.com** offers sources for products and training.

OSHA 1910.269, Appendix *C, Protection from Step and Touch Potentials.* Current Edition. Washington, DC: Occupational Safety & Health Administration (OSHA).

OSHA 1926.351, Arc Welding and Cutting. Current Edition. Washington, DC: Occupational Safety and Health Administration (OSHA).

The Procedure Handbook of Arc Welding. 14th Edition. Cleveland, OH: The James F. Lincoln Arc Welding Foundation, 2000.

Welding Handbook. Volume 1. Welding Science & Technology. Miami, FL: American Welding Society, 2001.

Welding Handbook. Volume 2. Part 1: Welding Processes. Miami, FL: American Welding Society, 2004.

3.0.0 Section Review

1. The most common fillet welds are made in ____.
   a. horizontal piping
   b. lap and T-joints
   c. root passes
   d. vertical piping

2. The position of a 3G groove weld is ____.
   a. vertical
   b. overhead
   c. flat
   d. horizontal

Summary

A welder must possess a number of essential skills in order to perform basic welding jobs and to progress to more difficult welding procedures. These skills include setting up GMAW equipment, preparing the welding work area, running stringer and weave beads, and making acceptable fillet welds and open V-groove welds on plate in all positions. The ability to consistently produce acceptable welds requires an understanding of the necessary welding techniques as well as dedicated practice.

Review Questions

1. The filler metal in a GMAW process is provided by ____.
   a. individual rods, or fillets, of steel
   b. a solid- or composite-cored electrode
   c. the expansion of molten base metal
   d. the material in the flux-cored wiring
2. If a hard hat is required during a welding operation, the hat should allow for the attachment of ____.
   a. shielding gas tubing
   b. a portable air vacuum
   c. rear deflector material
   d. leather side spats
3. A gas that is commonly used to purge areas or containers prior to GMAW operations is ____.
   a. nitrogen
   b. propane
   c. oxygen
   d. acetylene
4. If possible, practice weld coupons for GMAW should be ____.
   a. ¼" (6.4 mm) thick stainless steel
   b. ⅜" (10 mm) thick carbon steel
   c. ¾" (19.1 mm) thick low-alloy plate
   d. ⅞" (22.2 mm) thick composite steel
5. Setting up a GMAW machine requires loading the wire feeder and gun with the proper electrode as well as ____.
   a. locking out and tagging out the primary disconnect
   b. removing any liner material from the feeder and gun
   c. attaching the ground lead to the gun handle
   d. connecting the proper shielding gas for the application
6. In GMAW, the distance from the electrode tip to the base metal or the molten pool at the base metal is known as ____.
   a. arc length
   b. strikeout distance
   c. stickout length
   d. slope reach
7. When weave beads are made, the motion of the electrode is ____.
   a. quickly forward and backward
   b. slightly side-to-side
   c. widely side-to-side
   d. slowly circular and forward
8. If a weld restart has undercut, it is because ____.
   a. there was a previous weld termination
   b. not enough time was spent filling in the crater
   c. there was too little fillet in the electrode
   d. too much time was spent filling in the crater
9. Unless the proper technique is used, a weld termination normally leaves ____.
   a. a crack
   b. porosity
   c. spatter
   d. a crater
10. The process of depositing connective weld beads parallel to one another is known as ____.
    a. padding
    b. stringing
    c. pooling
    d. weaving
11. When flat (1F) fillet welds are being made, the travel speed can be increased or decreased to control the amount of ____.
    a. base metal fusion
    b. root penetration
    c. weld metal buildup
    d. flux-core shielding
12. What work angle should be used for the first bead in a horizontal (2F) fillet weld?
    a. 25 degrees
    b. 45 degrees
    c. 60 degrees
    d. 90 degrees
13. With GMAW, groove welds can be made in ____.
    a. only the 1G and 2G positions
    b. only the 2G and 3G positions
    c. only the 1G, 2G, and 3G positions
    d. all positions

14. When making a 3G open V-groove weld with weave beads, use a zigzag pattern across the root bead and pause at each toe ____.
    a. for complete penetration
    b. to prevent excessive spatter
    c. for good fusion
    d. to fill in the crater

15. One of the criteria for an acceptable overhead open V-groove weld with stringer beads is a uniform weld width of ____.
    a. ±1⁄16" (1.6 mm)
    b. ±1⁄8" (3.2 mm)
    c. ±3⁄8" (9.5 mm)
    d. ±1⁄2" (12.7 mm)

Trade Terms Introduced in This Module

Axis: The center line of the weld or prepared base metal.

Synergic: Welding technology that controls voltage, current, and wire feed using a program optimized for each welding situation.

Trim: A control found on synergic welding machines that is similar to the voltage control on regular machines and acts as a kind of override to the program.

Performance Accreditation Tasks

The Performance Accreditation Tasks (PATs) correspond to and support the learning objectives in *AWS EG2.0, Curriculum Guide for the Training of Welding Personnel: Level 1 – Entry Welder.*

PATs provide specific acceptable criteria for performance and help to ensure a true competency-based welding program for students.

The following tasks are designed to evaluate your ability to run fillet and groove welds with GMAW equipment. Perform each task when you are instructed to do so by the instructor. As you complete each task, show it to the instructor for evaluation. Do not proceed to the next task until instructed to do so by the instructor.

Have trainees complete PAT 1 through PAT 8, according to the acceptance criteria.

PAT 1, Make a Fillet Weld in the (1F) Flat Position. This task corresponds to *AWS EG2.0, Module 5, Gas Metal Arc Welding (GMAW-S, GMAW),* Key Indicators 5 and 7.

PAT 2, Make a Fillet Weld in the (2F) Horizontal Position. This task corresponds to *AWS EG2.0, Module 5, Gas Metal Arc Welding (GMAW-S, GMAW),* Key Indicators 5 and 7.

PAT 3, Make a Fillet Weld in the (3F) Vertical Position. This task corresponds to *AWS EG2.0, Module 5, Gas Metal Arc Welding (GMAW-S, GMAW),* Key Indicators 5 and 7.

PAT 4, Make a Fillet Weld in the (4F) Overhead Position. This task corresponds to *AWS EG2.0, Module 5, Gas Metal Arc Welding (GMAW-S, GMAW),* Key Indicators 5 and 7.

PAT 5, Make a Groove Weld, With or Without Backing, in the (1G) Flat Position. This task corresponds to *AWS EG2.0, Module 5, Gas Metal Arc Welding (GMAW-S, GMAW),* Key Indicators 6 and 11.

PAT 6, Make a Groove Weld, With or Without Backing, in the (2G) Horizontal Position. This task corresponds to *AWS EG2.0, Module 5, Gas Metal Arc Welding (GMAW-S, GMAW),* Key Indicators 6 and 11.

PAT 7, Make a Groove Weld, With or Without Backing, in the (3G) Vertical Position. This task corresponds to *AWS EG2.0, Module 5, Gas Metal Arc Welding (GMAW-S, GMAW),* Key Indicators 6 and 11.

PAT 8, Make a Groove Weld, With or Without Backing, in the (4G) Overhead Position. This task corresponds to *AWS EG2.0, Module 5, Gas Metal Arc Welding (GMAW-S, GMAW),* Key Indicators 6 and 11.

MAKE A FILLET WELD IN THE (1F) FLAT POSITION

As directed by the instructor, use GMAW with carbon steel solid electrodes and appropriate shielding gas to make a six-pass fillet weld using stringer beads on carbon steel plate, as shown.

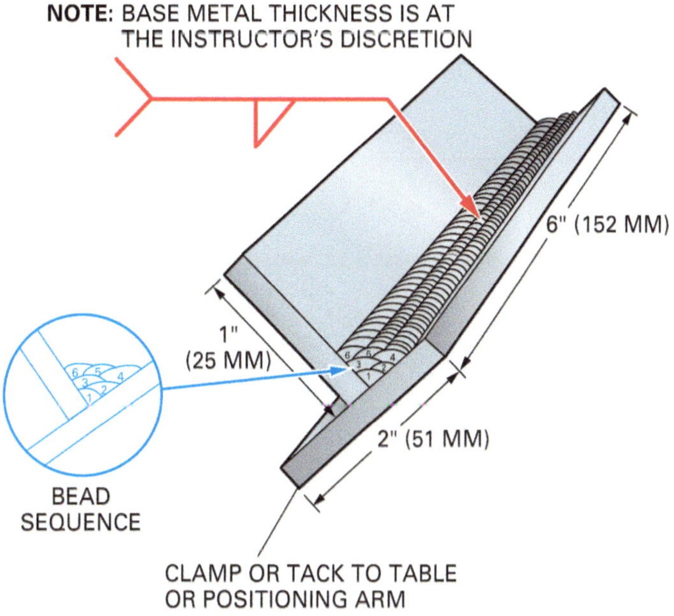

Criteria for Acceptance:

- Uniform rippled appearance on the bead face ____________
- Craters and restarts filled to the full cross section of the weld ____________
- Uniform weld width ±1⁄16" (1.6 mm) ____________
- Smooth flat transition with complete fusion at the toes of the welds ____________
- Acceptable weld profile in accordance with the applicable code or standard ____________
- No porosity ____________
- No undercut ____________
- No overlap ____________
- No inclusions ____________

MAKE A FILLET WELD IN THE (2F) HORIZONTAL POSITION

As directed by the instructor, use GMAW with carbon steel solid electrodes and appropriate shielding gas to make a six-pass fillet weld using stringer beads on carbon steel plate, as shown.

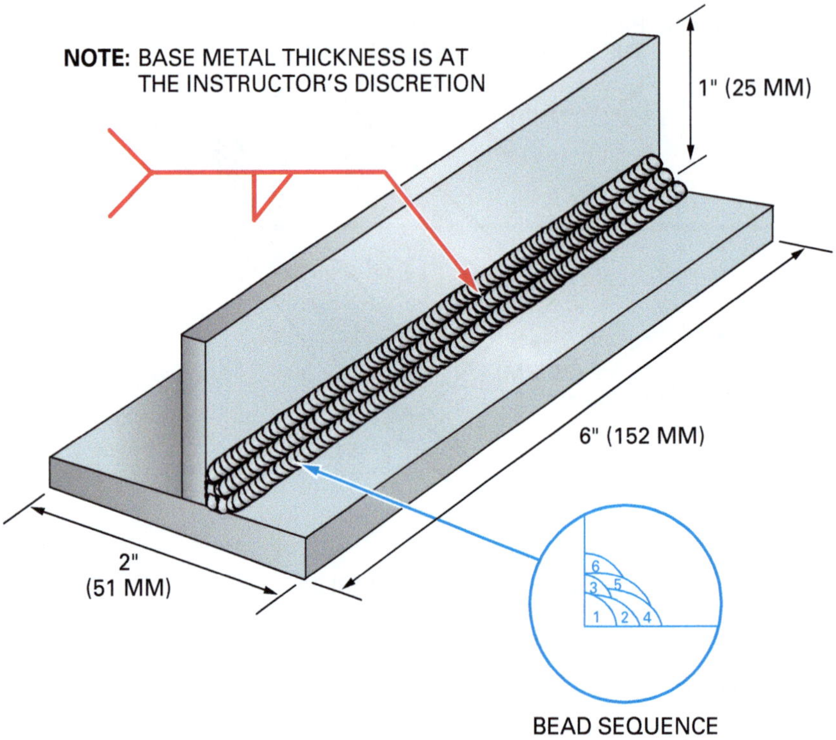

Criteria for Acceptance:

- Uniform rippled appearance on the bead face ______
- Craters and restarts filled to the full cross section of the weld ______
- Uniform weld width ±1⁄16" (1.6 mm) ______
- Smooth flat transition with complete fusion at the toes of the welds ______
- Acceptable weld profile in accordance with the applicable code or standard ______
- No porosity ______
- No undercut ______
- No overlap ______
- No inclusions ______

MAKE A FILLET WELD IN THE (3F) VERTICAL POSITION

As directed by the instructor, use GMAW with carbon steel solid electrodes and appropriate shielding gas to make a vertical fillet weld on carbon steel plate, as shown.

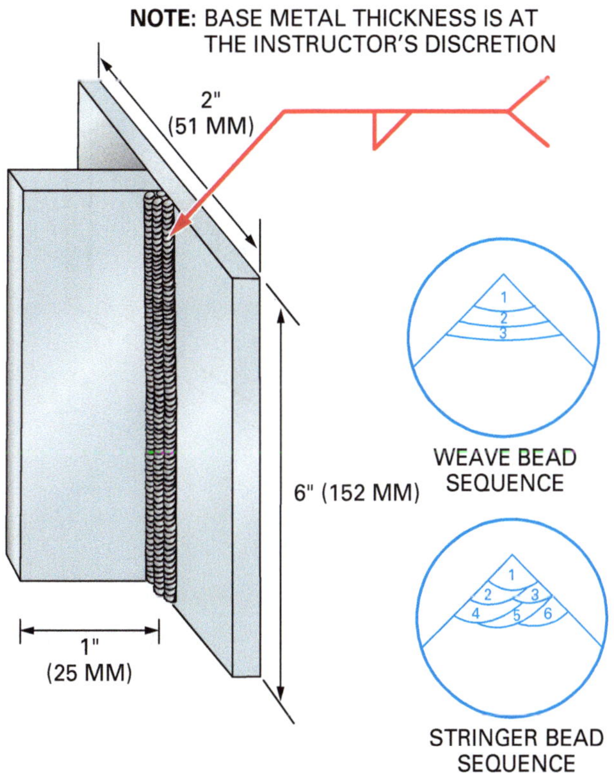

Criteria for Acceptance:

- Uniform rippled appearance on the bead face ____________
- Craters and restarts filled to the full cross section of the weld ____________
- Uniform weld width ±1⁄16" (1.6 mm) ____________
- Acceptable weld profile in accordance with the applicable code or standard ____________
- Smooth flat transition with complete fusion at the toes of the weld ____________
- No porosity ____________
- No undercut ____________
- No inclusions ____________
- No cracks ____________

MAKE A FILLET WELD IN THE (4F) OVERHEAD POSITION

As directed by the instructor, use GMAW with carbon steel solid electrodes and appropriate shielding gas to make a six-pass fillet weld using stringer beads on carbon steel plate, as shown.

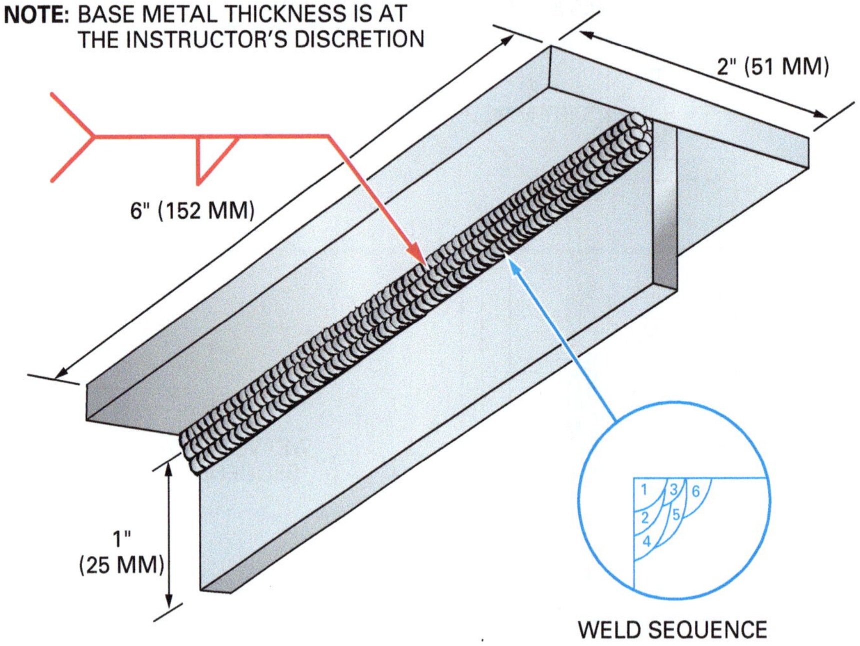

Criteria for Acceptance:

- Uniform rippled appearance on the bead face ________
- Craters and restarts filled to the full cross section of the weld ________
- Uniform weld width ±1⁄16" (1.6 mm) ________
- Acceptable weld profile in accordance with the applicable code or standard ________
- Smooth flat transition with complete fusion at the toes of the weld ________
- No porosity ________
- No undercut ________
- No inclusions ________
- No cracks ________

MAKE A GROOVE WELD, WITH OR WITHOUT BACKING, IN THE (1G) FLAT POSITION

As directed by the instructor, use GMAW with carbon steel solid electrodes and appropriate shielding gas to make a multiple-pass groove weld using stringer beads on carbon steel plate, with or without backing, as shown.

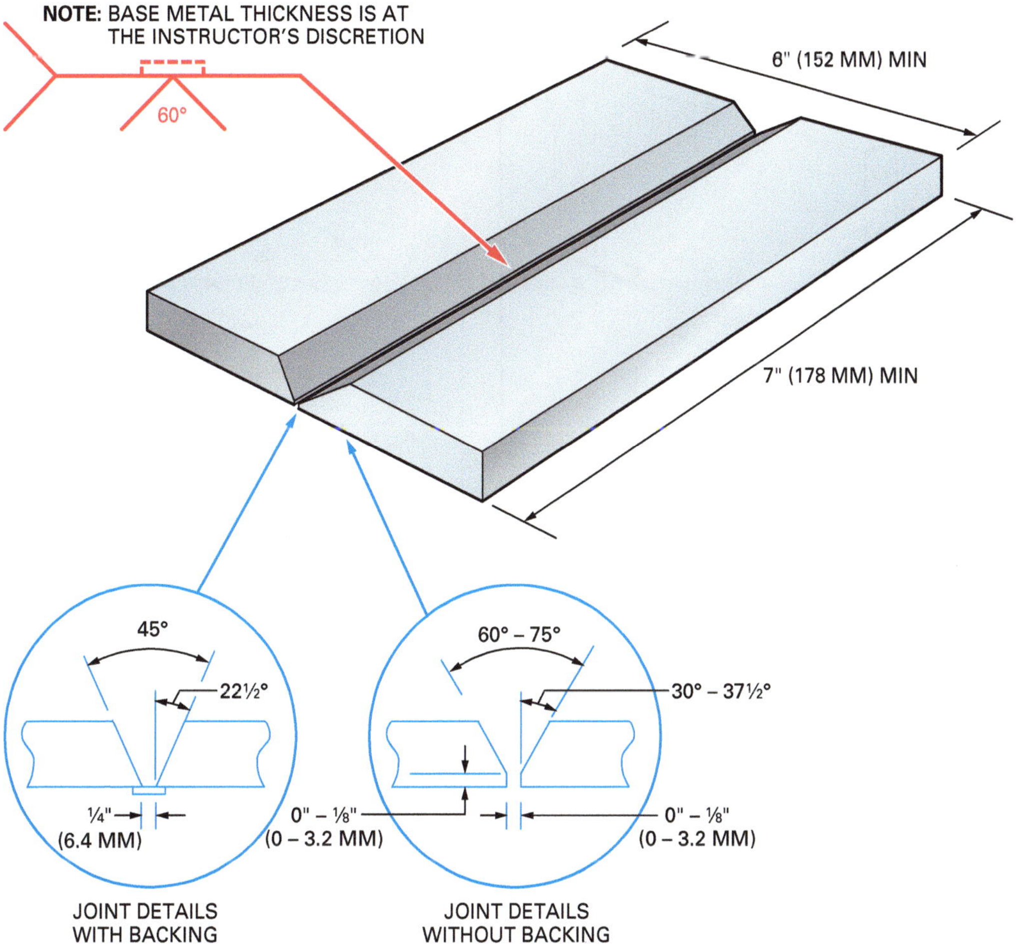

Criteria for Acceptance:

- Uniform rippled appearance on the bead face ________
- Craters and restarts filled to the full cross section of the weld ________
- Uniform weld width ±1/16" (1.6 mm) ________
- Acceptable weld profile in accordance with the applicable code or standard ________
- Smooth transition with complete fusion at the toes of the weld ________
- Complete joint penetration with uniform root and face reinforcement at least flush with the base metal to a maximum buildup of ⅛" (3.2 mm), if applicable ________
- No porosity ________
- No undercut ________
- No inclusions ________
- No cracks ________

MAKE A GROOVE WELD, WITH OR WITHOUT BACKING, IN THE (2G) HORIZONTAL POSITION

As directed by the instructor, use GMAW with carbon steel solid electrodes and appropriate shielding gas to make a multiple-pass groove weld on carbon steel plate, with or without backing, as shown.

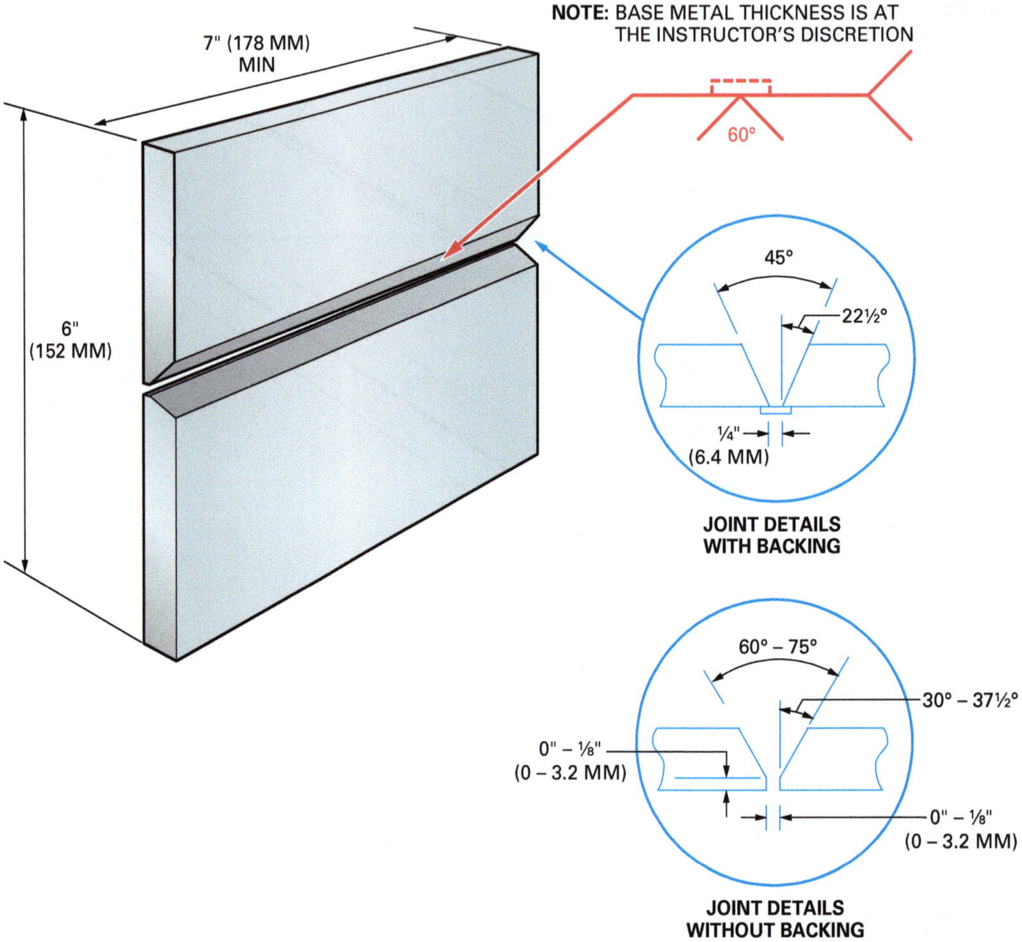

Criteria for Acceptance:

- Uniform rippled appearance on the bead face ________
- Craters and restarts filled to the full cross section of the weld ________
- Uniform weld width ±1⁄16" (1.6 mm) ________
- Acceptable weld profile in accordance with the applicable code or standard ________
- Complete joint penetration with uniform root and face reinforcement at least flush with the base metal to a maximum buildup of ⅛" (3.2 mm), if applicable ________
- Smooth transition with complete fusion at the toes of the weld ________
- No porosity ________
- No undercut ________
- No inclusions ________
- No cracks ________

MAKE A GROOVE WELD, WITH OR WITHOUT BACKING, IN THE (3G) VERTICAL POSITION

As directed by the instructor, use GMAW with carbon steel solid electrodes and appropriate shielding gas to make a multiple-pass groove weld on carbon steel plate, with or without backing, as shown.

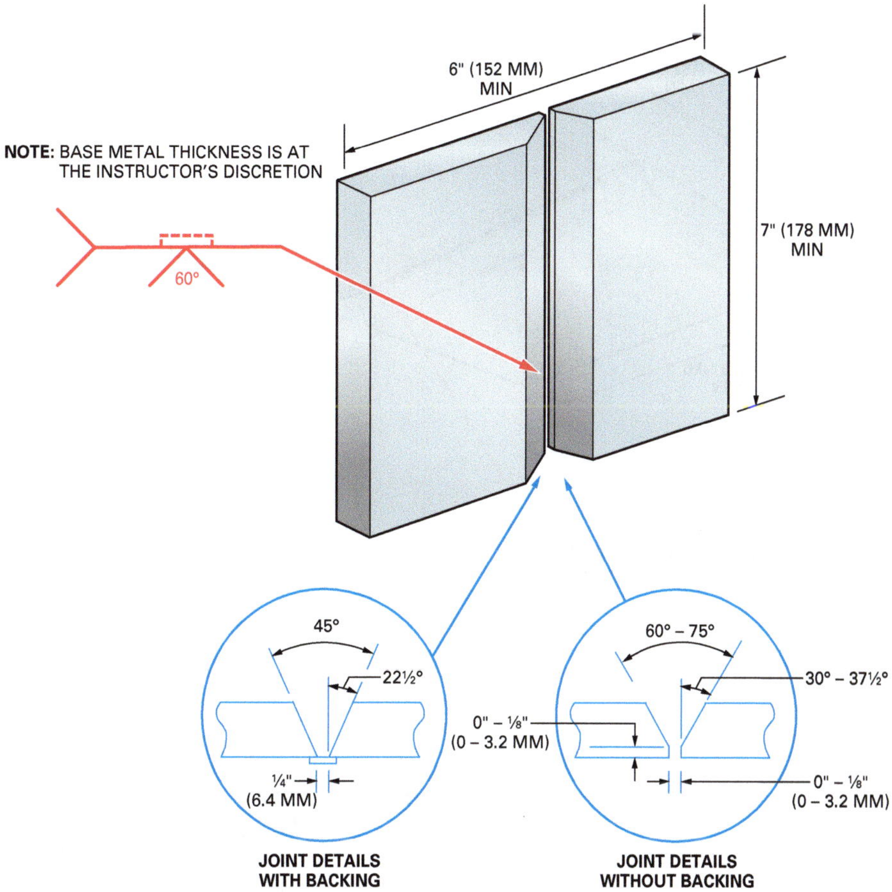

Criteria for Acceptance:

- Uniform rippled appearance on the bead face ______
- Craters and restarts filled to the full cross section of the weld ______
- Uniform weld width ±1⁄16" (1.6 mm) ______
- Acceptable weld profile in accordance with the applicable code or standard ______
- Complete joint penetration with uniform root and face reinforcement at least flush with the base metal to a maximum buildup of ⅛" (3.2 mm), if applicable ______
- Smooth transition with complete fusion at the toes of the weld ______
- No porosity ______
- No undercut ______
- No inclusions ______
- No cracks ______

MAKE A GROOVE WELD, WITH OR WITHOUT BACKING, IN THE (4G) OVERHEAD POSITION

As directed by the instructor, use GMAW with carbon steel solid electrodes and appropriate shielding gas to make a multiple-pass groove weld on carbon steel plate, with or without backing, as shown.

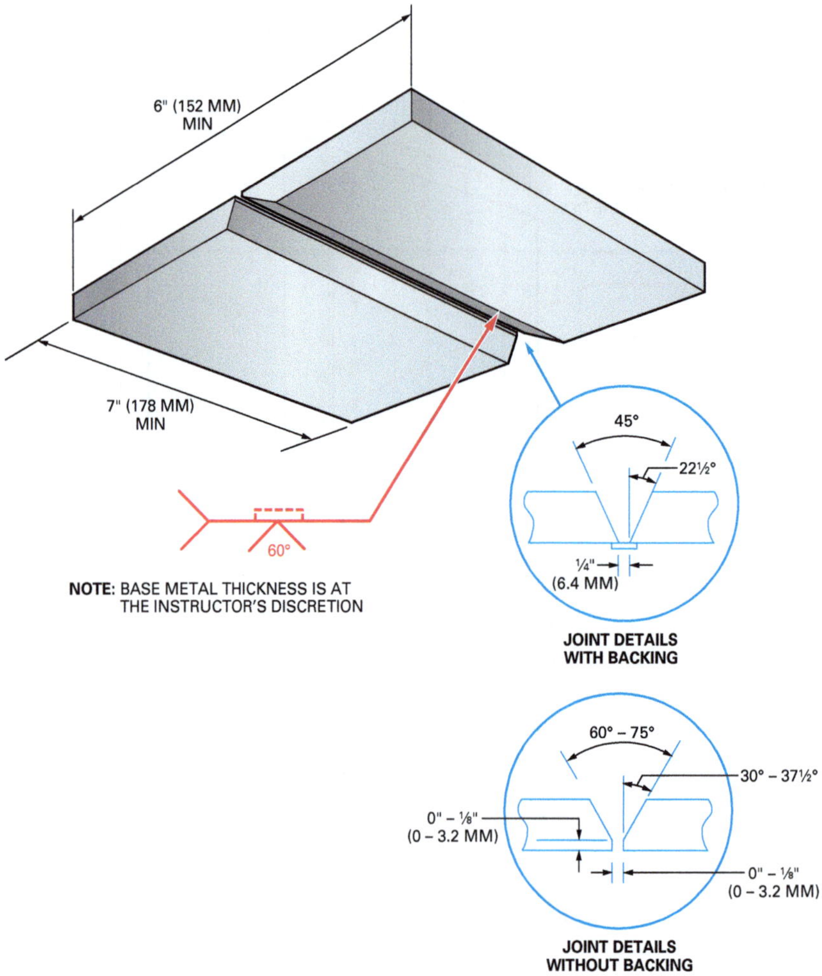

Criteria for Acceptance:

- Uniform rippled appearance on the bead face ________
- Craters and restarts filled to the full cross section of the weld ________
- Uniform weld width ±1⁄16" (1.6 mm) ________
- Acceptable weld profile in accordance with the applicable code or standard ________
- Complete joint penetration with uniform root and face reinforcement at least flush with the base metal to a maximum buildup of ⅛" (3.2 mm), if applicable ________
- Smooth transition with complete fusion at the toes of the weld ________
- No porosity ________
- No undercut ________
- No inclusions ________
- No cracks ________

Additional Resources

This module is intended as a thorough resource for task training. The following reference works are suggested for further study.

AWS B1.10M/B1.10 Guide for the Nondestructive Examination of Welds. Miami, FL: American Welding Society.

AWS B1.11 Guide for the Visual Examination of Welds. Miami, FL: American Welding Society.

AWS C5.6-89R Recommended Practices for Gas Metal Arc Welding. Miami, FL: American Welding Society.

AWS D3.5-93R Guide for Steel Hull Welding. Miami, FL: American Welding Society.

GMAW Welding Guide. Current Edition. Cleveland, OH: The Lincoln Electric Company.

Lincoln Electric website: **http://www.lincolnelectric.com** offers sources for products and training.

OSHA 1910.269, Appendix C, *Protection from Step and Touch Potentials*. Current Edition. Washington, DC: Occupational Safety & Health Administration (OSHA).

OSHA 1926.351, Arc Welding and Cutting. Current Edition. Washington, DC: Occupational Safety and Health Administration (OSHA).

The Procedure Handbook of Arc Welding. 14th Edition. Cleveland, OH: The James F. Lincoln Arc Welding Foundation, 2000.

Welding Handbook. Volume 1. Welding Science & Technology. Miami, FL: American Welding Society, 2001.

Welding Handbook. Volume 2. Part 1: Welding Processes. Miami, FL: American Welding Society, 2004.

Figure Credits

The Lincoln Electric Company, Cleveland, OH, USA, Module opener, Figures 2, 3, SA01, 17

Dan Sterry, TIC, Figure 12

Holley Thomas, SME, Figure 21

Terry Lowe, Figures 4, 29-31 (photos)

Section Review Answer Key

Answer	Section Reference	Objective
Section One		
1. b	1.1.2	1a
2. d	1.2.2	1b
3. a	1.3.1	1c
Section 2.0.0		
1. c	2.1.5	2a
2. d	2.2.2	2b
Section 3.0.0		
1. b	3.1.0	3a
2. a	3.2.2	3b

NCCER CURRICULA — USER UPDATE

NCCER makes every effort to keep its textbooks up-to-date and free of technical errors. We appreciate your help in this process. If you find an error, a typographical mistake, or an inaccuracy in NCCER's curricula, please fill out this form (or a photocopy), or complete the online form at **www.nccer.org/olf**. Be sure to include the exact module ID number, page number, a detailed description, and your recommended correction. Your input will be brought to the attention of the Authoring Team. Thank you for your assistance.

Instructors – If you have an idea for improving this textbook, or have found that additional materials were necessary to teach this module effectively, please let us know so that we may present your suggestions to the Authoring Team.

NCCER Product Development and Revision

13614 Progress Blvd., Alachua, FL 32615

Email: curriculum@nccer.org
Online: www.nccer.org/olf

❑ Trainee Guide ❑ Lesson Plans ❑ Exam ❑ PowerPoints Other ______________

Craft / Level: Copyright Date:

Module ID Number / Title:

Section Number(s):

Description:

Recommended Correction:

Your Name:

Address:

Email: Phone:

FCAW – Plate

Overview

Flux-cored arc welding (FCAW) is a common welding process that is used in industries throughout the world. FCAW equipment is available in different sizes and varieties, however, the basic operating principles of FCAW should apply to all makes and models. This module describes how FCAW equipment is used, identifies FCAW-related safety practices, and examines different techniques used to produce various types of weld beads.

Module 29210

Trainees with successful module completions may be eligible for credentialing through the NCCER Registry. To learn more, go to **www.nccer.org** or contact us at **1.888.622.3720**. Our website has information on the latest product releases and training, as well as online versions of our *Cornerstone* magazine and Pearson's product catalog.

Your feedback is welcome. You may email your comments to **curriculum@nccer.org**, send general comments and inquiries to **info@nccer.org**, or fill in the User Update form at the back of this module.

This information is general in nature and intended for training purposes only. Actual performance of activities described in this manual requires compliance with all applicable operating, service, maintenance, and safety procedures under the direction of qualified personnel. References in this manual to patented or proprietary devices do not constitute a recommendation of their use.

29210 V5

29210
FCAW – Plate

Objectives

When you have completed this module, you will be able to do the following:

1. Identify FCAW-related safety practices and explain how to prepare for welding.
   a. Describe basic FCAW processes.
   b. Identify FCAW-related safety practices.
   c. Explain how to safely set up the equipment and work area for welding.
2. Describe equipment control and welding procedures for FCAW and explain how to produce basic weld beads.
   a. Describe equipment control and welding techniques related to FCAW.
   b. Explain how to produce basic FCAW weld beads.
3. Describe the welding procedures needed to produce proper fillet and V-groove welds using FCAW welding techniques.
   a. Describe the welding procedures needed to produce proper fillet welds using FCAW welding techniques.
   b. Describe the welding procedures needed to produce proper V-groove welds using FCAW welding techniques.

Performance Tasks

Under the supervision of your instructor, you should be able to do the following:

1. Make multiple-pass FCAW-G/GM (gas-shielded) and/or FCAW-S (self-shielded) fillet welds on carbon steel plate coupons in all four 1F through 4F positions.
2. Make multiple-pass FCAW-G/GM (gas-shielded) and/or FCAW-S (self-shielded) V-groove welds on carbon steel plate coupons in all four 1G through 4G positions, with or without backing.

Trade Terms

Degassing
Synergic
Trim

Industry Recognized Credentials

If you are training through an NCCER-accredited sponsor, you may be eligible for credentials from NCCER's Registry. The ID number for this module is 29210. Note that this module may have been used in other NCCER curricula and may apply to other level completions. Contact NCCER's Registry at 888.622.3720 or go to **www.nccer.org** for more information.

Contents

Topics to be presented in this module include:

Figures

Figures

SECTION ONE

1.0.0 FCAW WELDING

Objective

Identify FCAW-related safety practices and explain how to prepare for welding.

a. Describe basic FCAW processes.
b. Identify FCAW-related safety practices.
c. Explain how to safely set up the equipment and work area for welding.

Performance Tasks

1. Make multiple-pass FCAW-G/GM (gas-shielded) and/or FCAW-S (self-shielded) fillet welds on carbon steel plate coupons in all four 1F through 4F positions.
2. Make multiple-pass FCAW-G/GM (gas-shielded) and/or FCAW-S (self-shielded) V-groove welds on carbon steel plate coupons in all four 1G through 4G positions, with or without backing.

Trade Terms

Degassing: The escape of gases formed during the welding process that can cause imperfections in the weld.

FCAW is based on the use of a flux-cored metal electrode and, in some cases, a shielding gas. This section describes FCAW equipment and how it works, identifies safety practices related to FCAW, and examines how to prepare for FCAW.

1.1.0 Welding Processes

FCAW processes use the intense heat produced by an electrical arc between a filler metal electrode and a workpiece to melt the metals and cause them to merge, or coalesce. Flux-cored arc welding, or FCAW, is one common form of arc welding used extensively in industry. Performing FCAW safely and effectively requires a good understanding of how the process and the equipment work. It is also critical to understand the safety practices that are necessary during an FCAW operation.

1.1.1 The FCAW Process

In the FCAW process, a continuous, consumable, flux-cored electrode is used as a filler metal to form a weld (*Figure 1*). Keep in mind that the proper term for the consumable material used for this filler metal is "electrode," although it is commonly referred to as wire. A shielding gas that protects the molten weld metal from contamination by the atmosphere can be generated entirely from the flux in the electrode core (self-shielded FCAW, or FCAW-S) or from both the flux in the electrode core and an external shielding gas (gas-shielded FCAW, or FCAW-G).

In all types of flux-cored electrodes, the flux contains agents for **degassing** the weld, stabilizing the arc, and forming a protective slag over the weld bead. Other elements are often added to the flux-cored electrode to alloy with the base metal and improve its chemical and mechanical characteristics. The FCAW process is basically limited to welding low-carbon steels, some low-alloy steels, and some stainless steels. FCAW is an excellent process for making fillet and groove welds on carbon steel plate.

1.1.2 FCAW Equipment

The equipment used for gas-shielded FCAW (FCAW-G) or self-shielded FCAW (FCAW-S) is basically the same except for the components needed for the gas used in the gas-shielding process. This additional equipment includes guns equipped with a gas nozzle, gas hoses, a gas control solenoid valve, a gas regulator/flowmeter, and a shielding gas cylinder or other source. *Figure 2* is an example of typical FCAW equipment.

FCAW-S is particularly useful in situations with high winds, because the shielding gas is created inside the welding arc as the flux vaporizes. The molten puddle has a constant supply of shield gas displacing the air from the arc outward, and the puddle is immediately covered by the residual slag, similar to SMAW. An external shield gas, in contrast, attempts to force gas down from the nozzle to displace the air, and the gas can be blown away from the weld puddle by strong winds.

1.2.0 Safety Summary

The following is a summary of safety procedures and practices that must be observed while cutting or welding. Keep in mind that this is only a summary. Complete safety coverage is provided in the Welding Level One module, *Welding Safety*. That module should be completed before continuing with this one. Above all, be sure to wear appropriate protective clothing and equipment when welding or cutting.

1.2.1 Protective Clothing and Equipment

Welding activities can cause injuries unless you wear all of the protective clothing and equipment that is designed specifically for the welding industry. The following information includes important safety guidelines about protective clothing and equipment:

- Wear a face shield over snug-fitting cutting goggles or safety glasses for gas welding or cutting. Either the face shield or the lenses of the welding goggles must be an approved shade for the application. A welding hood equipped with a properly tinted lens is best for all forms of welding.
- Wear proper protective leather and flame-retardant clothing along with welding gloves that will protect you from flying sparks, molten metal, and heat.
- Wear high-top safety shoes or boots. Make sure that the tongue and lace area of the footwear will be covered by a pant leg. If the tongue and lace area is exposed or if the footwear must be protected from burn marks, wear leather spats under the pants or chaps over the front and top of the footwear.
- Wear a 100-percent cotton cap with no mesh material included in its construction. The bill of the cap points to the rear. If a hard hat is required for the environment, use one that allows the attachment of rear deflector material and a face shield. A hard hat with a rear deflector is generally preferred when working overhead, and may be required by some employers and job sites.
- Wear earmuffs or at least earplugs to protect your ear canals from sparks.

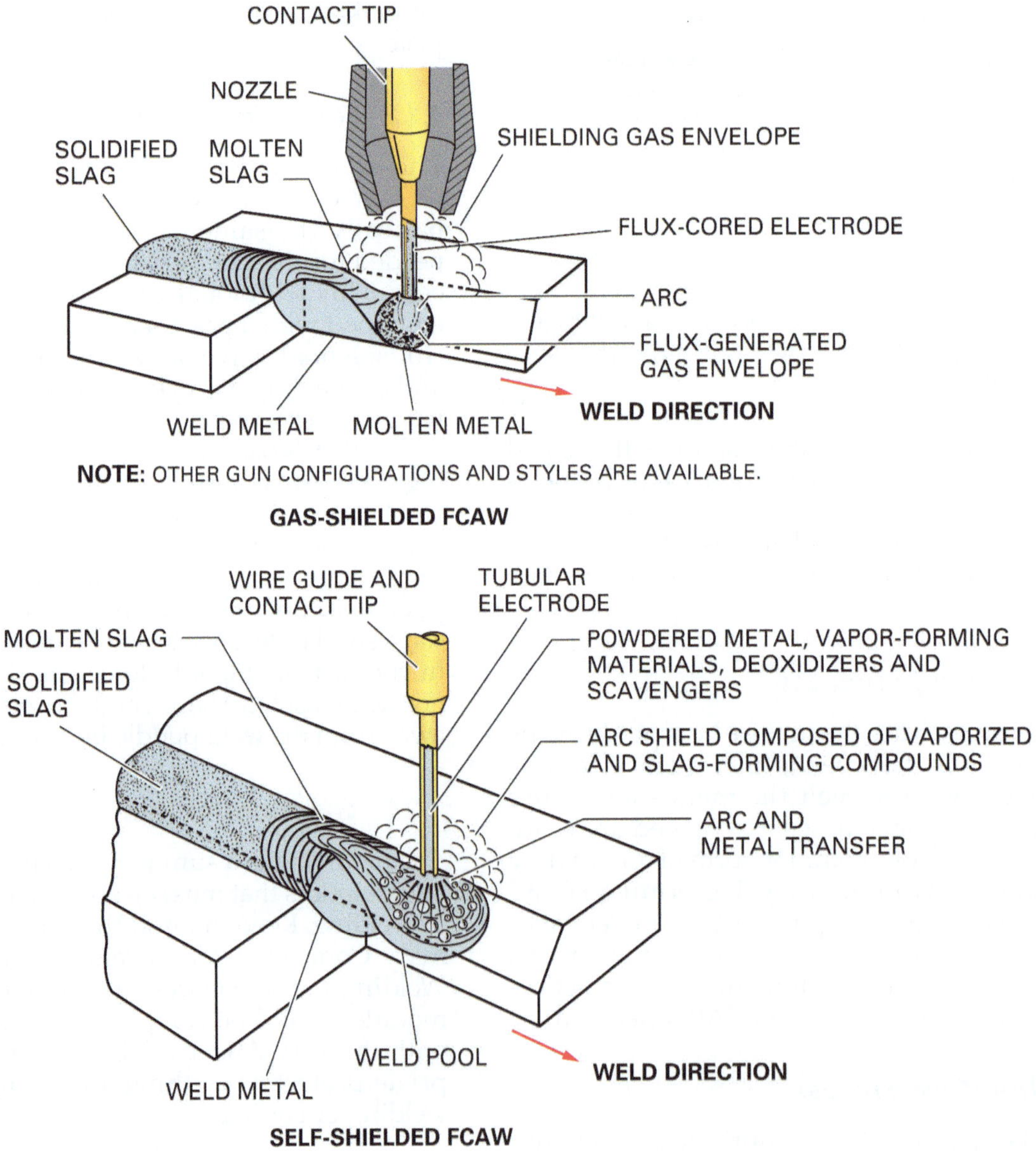

Figure 1 FCAW process.

Figure 2 Self-shielded FCAW (FCAW-S) equipment.

1.2.2 Fire/Explosion Prevention

Welding activities usually involve the use of fire or extreme heat to melt metal. Whenever fire is used, it must be controlled and contained. Welding or cutting activities are often performed on vessels that may once have contained flammable or explosive materials. Residues from those materials can catch fire or explode when a welder begins work on such a vessel. The following are fire and explosion prevention guidelines associated with welding:

- Never carry matches or gas-filled lighters in your pockets. Sparks can cause the matches to ignite or the lighter to explode, causing serious injury.
- Never use oxygen to blow dust or dirt from clothing. The oxygen can remain trapped in the fabric for a time. If a spark hits the clothing during this time, the clothing can burn rapidly and violently out of control.
- Make sure that any flammable material in the work area is moved or shielded by a fire-resistant covering. Approved fire extinguishers must be available before attempting any heating, welding, or cutting operations. If a hot-work permit and a fire watch are required, be sure those items are in place before beginning and ensure that all site requirements have been observed.
- Never release a large amount of oxygen or use oxygen as compressed air. Its presence around flammable materials or sparks can cause rapid and uncontrolled combustion. Keep oxygen away from oil, grease, and other petroleum products.
- Never release a large amount of fuel gas, especially acetylene. Propane tends to concentrate in and along low areas and can ignite at a considerable distance from the release point. Acetylene is lighter than air, but is even more dangerous than propane. When mixed with air or oxygen, acetylene will explode at much lower concentrations than any other fuel.
- To prevent fires, maintain a neat and clean work area, and make sure that any metal scrap or slag is cold before disposal.
- Before cutting containers such as tanks or barrels, check to see if they have contained any explosive, hazardous, or flammable materials, including petroleum products, citrus products, or chemicals that decompose into toxic fumes when heated. As a standard practice, always clean and then fill any tanks or barrels with water, or purge them with a flow of inert gas such as nitrogen to displace any oxygen. Containers must be cleaned by steam cleaning, flushing with water, or washing with detergent until all traces of the material have been removed.

WARNING!

Welding or cutting must never be performed on drums, barrels, tanks, vessels, or other containers until they have been emptied and cleaned thoroughly, eliminating all flammable materials and all substances (such as detergents, solvents, greases, tars, or acids) that might produce flammable, toxic, or explosive vapors when heated. Clean containers only in well-ventilated areas. Vapors can accumulate during cleaning, causing explosions or injury.

Proper procedures for cutting or welding hazardous containers are described in the *American Welding Society (AWS) F4.1, Safe Practices for the Preparation of Containers and Piping for Welding and Cutting.*

1.2.3 Work Area Ventilation

Vapors and fumes tend to rise in the air from their sources. Welders often have to work above the welding area where the fumes are being created. Welding fumes can cause personal injuries. Good work area ventilation helps to remove the vapors and to protect the welder. The following is a list of work area ventilation guidelines to consider before and during welding activities:

- Make sure to follow confined space procedures before conducting any welding or cutting in the confined space.
- Never use oxygen to provide ventilation in confined spaces.
- Always perform cutting or welding operations in a well-ventilated area. Cutting or welding operations involving zinc or cadmium materials or coatings result in toxic fumes. For long-term cutting or welding of such materials, always wear an approved full-face, supplied-air respirator that uses breathing air supplied from outside the work area. For occasional, very short-term exposure, a HEPA-rated or metal-fume filter may be used on a standard respirator.

> **Machine Bevels and Root Faces for FCAW**
>
> For V-groove joints made in thick materials, machined bevels and root faces provide the cleanest and most uniform way to prepare materials for satisfactory FCAW welds. SMAW, FCAW, and GTAW can tolerate some poor fitup for root passes.

- Make sure confined spaces are properly ventilated for cutting or welding purposes. Use powered extraction systems when available. *Figure 3* shows mobile and portable extraction equipment used to remove cutting or welding fumes and vapors from work areas.

1.3.0 Welding Equipment Setup

Before welding can take place, the area must be made ready, the welding equipment must be set up, and the metal to be welded must be prepared. The following sections will explain how to set up equipment for welding.

MOBILE UNIT

PORTABLE UNIT

Figure 3 Mobile and portable extraction equipment.

1.3.1 Preparing the Welding Area

To practice welding, a welding table, bench, or stand is needed. The welding surface must be steel, and provisions must be made for placing weld coupons out of position (*Figure 4*).

Follow these steps to set up the area for welding:

Step 1 Make sure that the area is properly ventilated. Make use of doors, windows, and fans.

Step 2 Check the area for fire hazards. Remove any flammable materials before proceeding.

Step 3 Know the location of the nearest fire extinguisher. Know how to use the fire extinguisher and do not proceed unless the extinguisher is charged.

Step 4 Set up flash shields around the welding area.

1.3.2 Preparing the Practice Weld Coupons

If possible, the practice weld coupons should be ⅜" (10 mm plate) thick carbon steel. If this size is not readily available, ¼" to ¾" (6 mm to 20 mm plate) thick steel can be used to practice welding. Use a wire brush or grinder to remove heavy mill scale or corrosion. Prepare weld coupons for practicing the following:

- *Running beads* – The coupons can be any size or shape that can be easily handled.
- *Overlapping beads* – The coupons can be any size or shape that can be easily handled.
- *Fillet welds* – Cut the metal into 2" × 6" (51 mm × 152 mm) rectangles for the base and 1" × 6" (25 mm × 152 mm) rectangles for the web. *Figure 5* shows the weld coupons for fillet welding.
- *V-groove welds* – As shown in *Figure 6*, cut the metal into 3" × 7" (76 mm × 178 mm) rectangles with one (or both) of the 7" (178 mm) lengths beveled at 30 to 37½ degrees for carbon steel or low-alloy plate. Note that the bevel angle for stainless steel coupons is 22½ degrees. Grind a 0" to ⅛" (0 to 3.2 mm) root face on the bevel as chosen and directed by your instructor.

Figure 4 Welding station.

NOTE

Open V-groove welds are rarely done using the FCAW process. It is assumed that the root pass will be done using another process better suited for open-root passes, such as GMAW, GTAW, or SMAW. If the root pass is to be done using the FCAW process, a backing bar should be added.

Follow these steps to prepare each V-groove weld coupon:

Step 1 Check the bevel face of each coupon. There should be no dross and a 0" to ⅛" (0 mm to 3.2 mm) root face. The bevel angle should be 30 to 37½ degrees. An example of this is shown in *Figure 7*.

Step 2 Align the beveled coupons on a flat surface with a root opening determined by the instructor. Place tack welds on the back side of the joint. Use from three to four ¼" (6.4 mm) tack welds.

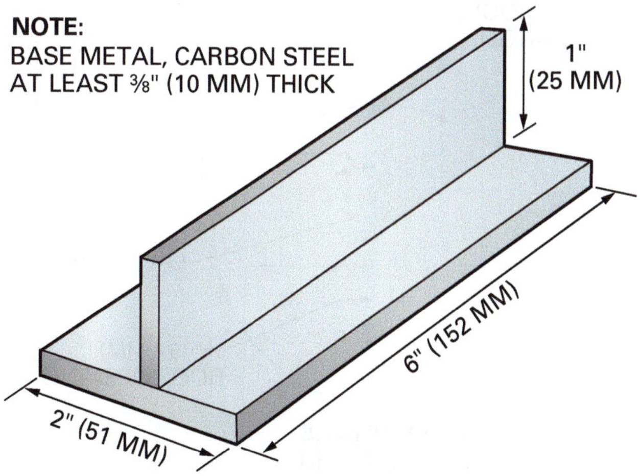

Figure 5 Fillet weld coupons.

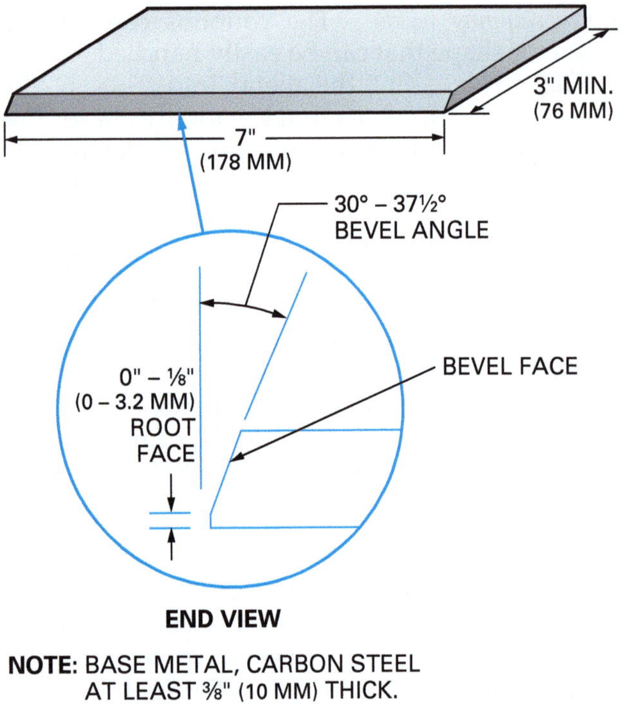

Figure 6 Metal cut for V-groove weld coupons.

1.3.3 The Welding Machine

Identify the proper welding machine for FCAW use, and follow these steps to set it up for welding:

Step 1 Verify that the welding machine can be used for FCAW, with or without gas shielding as required.

Step 2 Verify the location of the primary disconnect.

Step 3 Configure the welding machine for gas-shielded FCAW or self-shielded FCAW as directed by the instructor (*Figure 8*). Configure the gun polarity, and equip the gun with the correct liner material and contact tip for the diameter of the electrode being used and the correct nozzle for the application.

Step 4 In accordance with the manufacturer's instructions, load the wire feeder and gun with a flux-cored electrode of the proper diameter as directed by the instructor.

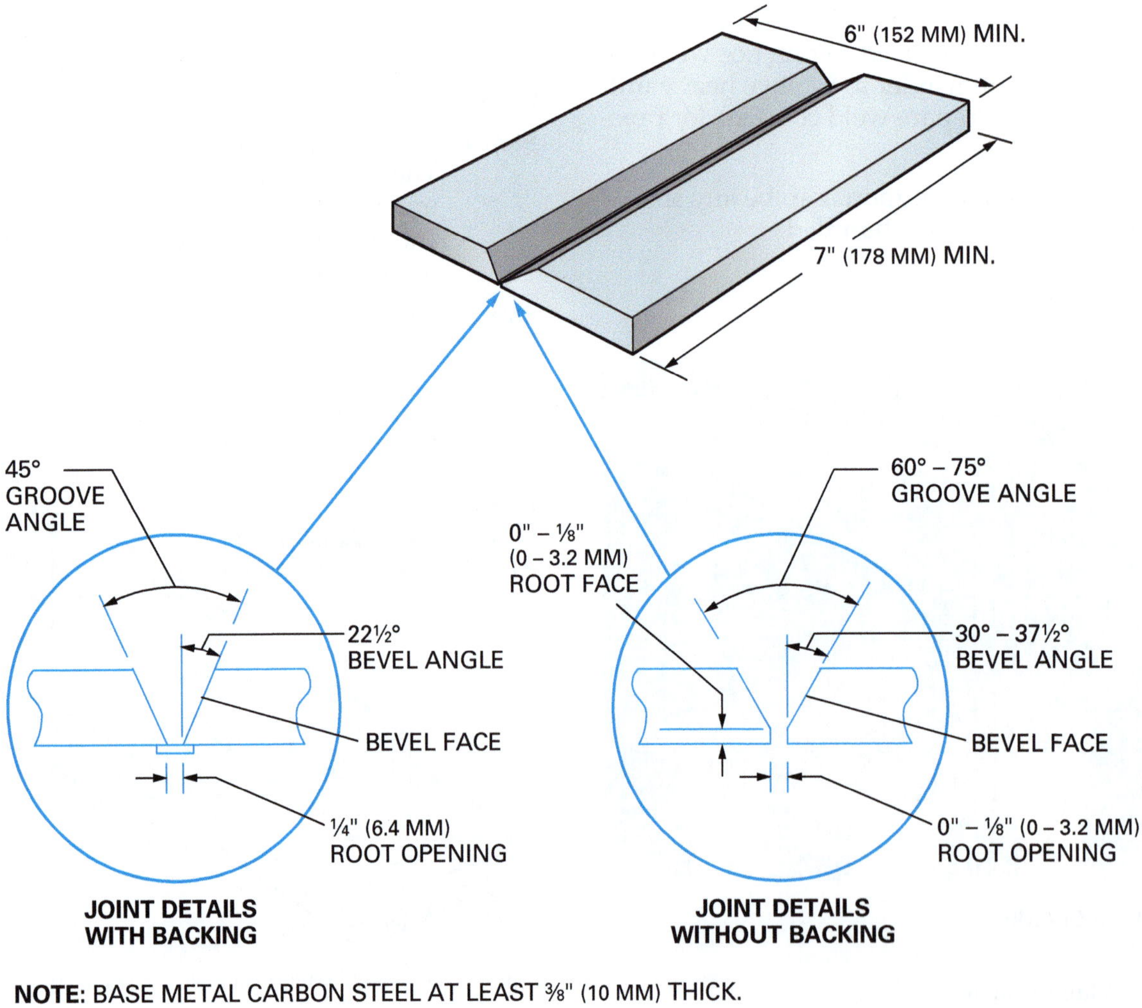

Figure 7 V-groove weld coupon.

Step 5 If required, connect the proper shielding gas for the application as specified by the electrode manufacturer, the WPS, the site quality standards, or the instructor.

Step 6 Connect the clamp of the workpiece lead to the table or workpiece.

Step 7 Turn on the welding machine and, if necessary, purge the gun as directed by the gun manufacturer's instructions.

Step 8 Set the initial welding voltage and wire feed speed for the type and size of electrode as recommended by the manufacturer, directed by the instructor, or provided in the WPS.

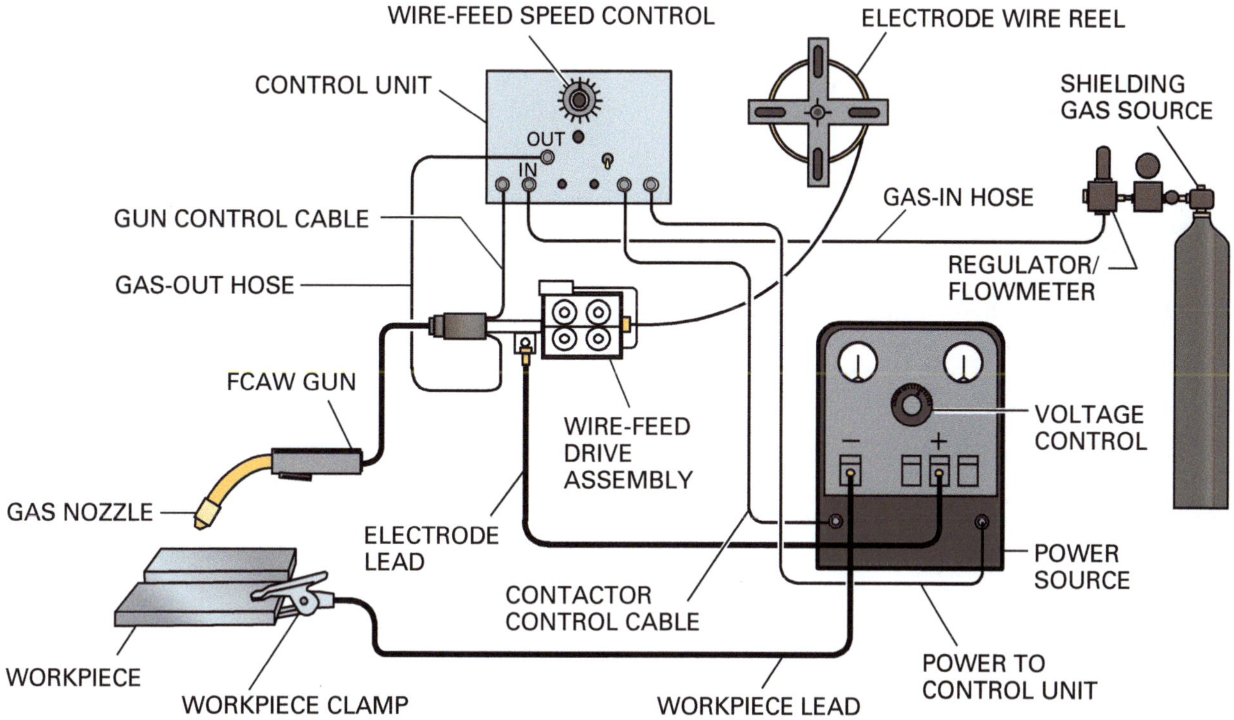

NOTE: The polarity of the gun and workpiece lead is determined by the type of filler metal and application.

Figure 8 Configuration diagram of a typical FCAW welding machine.

Submerged Arc Welding

No, this is not underwater welding. Submerged arc welding (SAW) was first patented in 1935. The process involves covering the electric arc with a bed of flux granules. Like FCAW, the electrode is in wire form and may be tubular or solid. The arc zone and weld are protected from atmospheric contamination by being covered by a blanket of flux that becomes molten due to the heat of the arc. Once liquefied, the flux becomes conductive and provides a current path between the electrode and the work. The flux also significantly reduces the UV radiation and spatter emitted from the weld site. SAW is normally done using automated equipment, but it can be done by hand. The process is usually limited to the flat or horizontal welding positions. As shown here on these quick demonstration beads, the cooled flux material snaps off easily in large pieces.

Additional Resources

AWS B1.10M/B1.10 Guide for the Nondestructive Examination of Welds. Miami, FL: American Welding Society.

AWS B1.11 Guide for the Visual Examination of Welds. Miami, FL: American Welding Society.

AWS D3.5-93R Guide for Steel Hull Welding. Miami, FL: American Welding Society.

Lincoln Electric website: **http://www.lincolnelectric.com** offers sources for products and training.

OSHA 1910.269, Appendix C, Protection from Step and Touch Potentials. Current Edition. Washington, DC: Occupational Safety & Health Administration (OSHA).

OSHA 1926.351, Arc Welding and Cutting. Current Edition. Washington, DC: Occupational Safety and Health Administration (OSHA).

The Procedure Handbook of Arc Welding. 14th Edition. Cleveland, OH: The James F. Lincoln Arc Welding Foundation, 2000.

Welding Handbook. Volume 1. Welding Science & Technology. Miami, FL: American Welding Society, 2001.

Welding Handbook. Volume 2. Part 1: Welding Processes. Miami, FL: American Welding Society, 2004.

1.0.0 Section Review

1. In the FCAW process, the shielding gas that is used to protect the weld zone typically comes from a(n) _____.

   a. flux-cored electrode
   b. electrode lead
   c. solid-cored wire
   d. gaseous filler metal

2. After a vessel has been thoroughly cleaned to prepare it for welding, it should be filled with water or _____.

   a. oxygen
   b. compressed air
   c. sand or soil
   d. a purging gas

3. When FCAW equipment is being used for practice welding, the welding area should be enclosed with _____.

   a. shielding gas
   b. flash shields
   c. gas extinguishers
   d. gun liner material

Section Two

2.0.0 Welding Procedures and Basic Beads

Objective

Describe equipment control and welding procedures for FCAW and explain how to produce basic weld beads.

a. Describe equipment control and welding techniques related to FCAW.
b. Explain how to produce basic FCAW weld beads.

Performance Tasks

1. Make multiple-pass FCAW-G/GM (gas-shielded) and/or FCAW-S (self-shielded) fillet welds on carbon steel plate coupons in all four 1F through 4F positions.
2. Make multiple-pass FCAW-G/GM (gas-shielded) and/or FCAW-S (self-shielded) V-groove welds on carbon steel plate coupons in all four 1G through 4G positions, with or without backing.

Trade Terms

Synergic: Welding technology that controls voltage, current, and wire feed using a program optimized for each welding situation.

Trim: A control found on synergic welding machines that is similar to the voltage control on regular machines and acts as a kind of override to the program.

In order to weld effectively, welders must know how to properly control the equipment being used, as well as how to use various welding techniques to produce strong and acceptable weld beads. This section examines factors that affect FCAW and how these factors are controlled. It also describes how to produce basic FCAW weld beads.

2.1.0 Welding Procedures

Welding can be affected by factors such as the welding voltage, the welding amperage, the position and travel speed of the gun, and other aspects of the welding equipment and process. This section describes aspects of equipment control and welding techniques that pertain to FCAW.

2.1.1 Welding Voltage

Arc length is determined by voltage, which is set at the power source. Arc length is the distance from the electrode tip to the base metal or molten pool at the base metal. If voltage is set too high, the arc will be too long, which could cause the electrode to fuse to the tip. Too high a voltage also causes porosity and excessive spatter. Voltage must be increased or decreased as wire feed speed is increased or decreased. Set the voltage so that the arc length is just above the surface of the base metal (*Figure 9*).

Synergic welding machines control voltage, current, and wire feed moment by moment using a program that's customized for each welding situation. These machines will have a **trim** control that works similarly to the voltage control on other machines. Essentially, it overrides the program.

2.1.2 Welding Amperage

With a standard power source, the wire feed speed control on the power source controls the welding amperage after the welder has set the initial recommended setting. The welding power source provides the amperage necessary to melt the electrode while maintaining the selected welding voltage. Within limits, when the wire feed speed is increased, the welding amperage and deposition rate also increase. This results in higher welding heat, deeper penetration, and higher beads. When the wire feed speed is decreased, the welding amperage automatically decreases. With lower welding amperage and less heat, the deposition rate drops and the weld beads are smaller and have less penetration.

Note that some constant-voltage (CV) welding machines used for FCAW provide varying degrees of current modification, such as slope or induction adjustments. Welding machines with a slope adjustment allow the amount of amperage change to vary in relation to voltage range of the

Electrode Manufacturer's Recommendations

Always obtain and follow the manufacturer's recommendations or WPS for shielding gas (if required) and for initially setting the welding voltage, current (if variable), and wire feed speed parameters. These balanced parameters are critical and are based on the welding position, size, and composition of the flux-cored electrode being used.

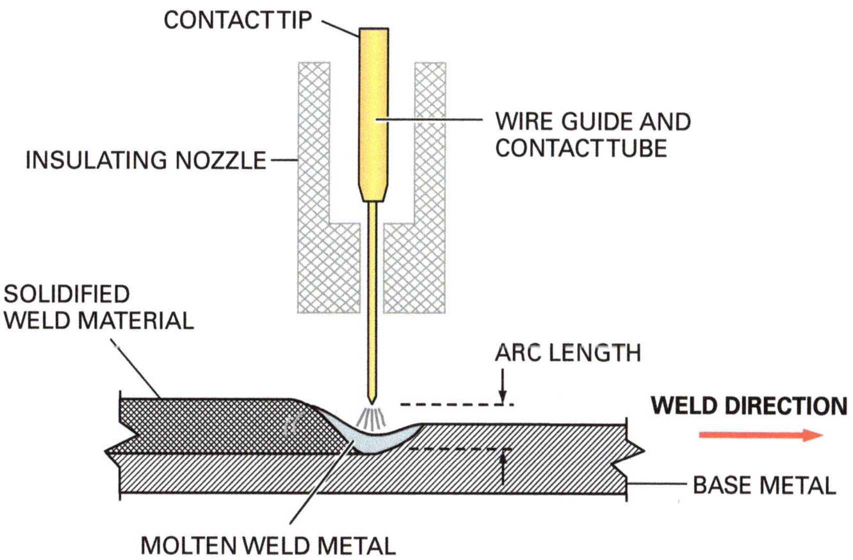

Figure 9 Arc length.

unit. This allows the correct short-circuiting current to be established for certain welding applications and electrode requirements.

Standard CV FCAW units have current slope that is fixed by the manufacturer for general welding applications and conditions. Other units allow adjustment of the response time for short-circuit current by using an induction control. In an instantaneous short-circuit current draw mode of operation, the machine draws a high level of current as soon as the electrode touches the workpiece. In the adjustable induction mode of operation, the machine gradually increases the welding current after the initial contact of the electrode to the workpiece. This evens out the arc characteristics during short-circuiting transfer.

Pulse-transfer welding machines allow the adjustment of the peak current for the pulse and the background current between pulses to match specific welding applications and electrode requirements. In other cases, where a voltage-sensing wire feeder is used with a constant-current (CC) welding machine, the current is adjustable and the wire feeder varies the wire feed speed to maintain the proper arc voltage at the electrode tip. Note that some codes prohibit the use of CC machines for FCAW.

Arc Voltage

A minimum arc voltage is needed to maintain spray transfer. However, penetration is not directly related to voltage. Penetration will increase with voltage for a time, but beyond an optimum voltage, it will actually decrease as the voltage increases further.

2.1.3 Welding Travel Speed

Weld travel speed is the speed of the electrode tip across the base metal in the direction of the weld, measured in inches per minute (ipm) or a metric equivalent. Travel speed has a great effect on penetration and bead size. Slower travel speeds build higher beads with deeper penetration. Faster travel speeds build smaller beads with less penetration. Ideally, the welding parameters should be adjusted so that the electrode tip is positioned at the leading edge of the weld puddle during travel.

2.1.4 Gun Position

Gun position for FCAW has an effect on weld penetration along with voltage and travel speed. The gun work angle (*Figure 10*) is an angle less than 90 degrees between a line perpendicular to the major workpiece surface at the point of electrode contact, and a plane determined by the electrode axis and the weld axis. For a T-joint or corner joint, the line is perpendicular to the non-butting member.

Travel Speed and Wire Feed Speed

Beginners are tempted to turn down the wire feed speed if they experience difficulty controlling the weld puddle. Electrodes must be run at certain balanced parameters that cannot be changed individually. Voltage, wire feed speed, and travel speed are adjusted and balanced together to control the weld puddle.

For pipe, the plane is determined by the electrode axis and a line tangent to the pipe surface at the same point.

The gun travel angle (also shown in *Figure 10*) is defined as an angle less than 90 degrees between the electrode axis and a line perpendicular to the weld axis at the point of electrode contact in a plane determined by the electrode axis and the weld axis. For pipe, the plane is determined by the electrode axis and a line tangent to the pipe surface at the same point.

- *Neutral angle* – The term *neutral angle* describes the travel angle when the center line of the gun electrode is exactly perpendicular (90 degrees) to the welding surface. The term is nonstandard for a 0-degree travel angle. Weld penetration is moderate when the gun electrode is used in this position.
- *Push angle* – A push angle is generally used for welding thin materials or when shallow penetration is required. A push angle is the travel angle created when the welding gun is tilted back so that the electrode tip precedes the gun in the direction of the weld. In this position, the electrode tip and shielding gas are being directed ahead of the weld bead. Welding with this angle produces a smaller weld bead with less penetration. Push angles greater than 15 degrees are not recommended for manual welding.
- *Drag angle* – A drag angle is the travel angle created when the gun is tilted forward so that the electrode tip trails the gun in the direction of the weld. In this position, the electrode tip and shielding gas are being directed back toward the weld bead. Welding with this angle produces a higher and wider weld bead with more penetration. Maximum penetration is achieved with a drag angle of approximately 25 degrees.

2.1.5 Electrode Extension, Stickout, and Standoff Distance

Electrode extension is the length of the electrode that extends beyond the contact tip of the welding gun. Changing the extension influences the welding current for low-conductivity metal electrodes

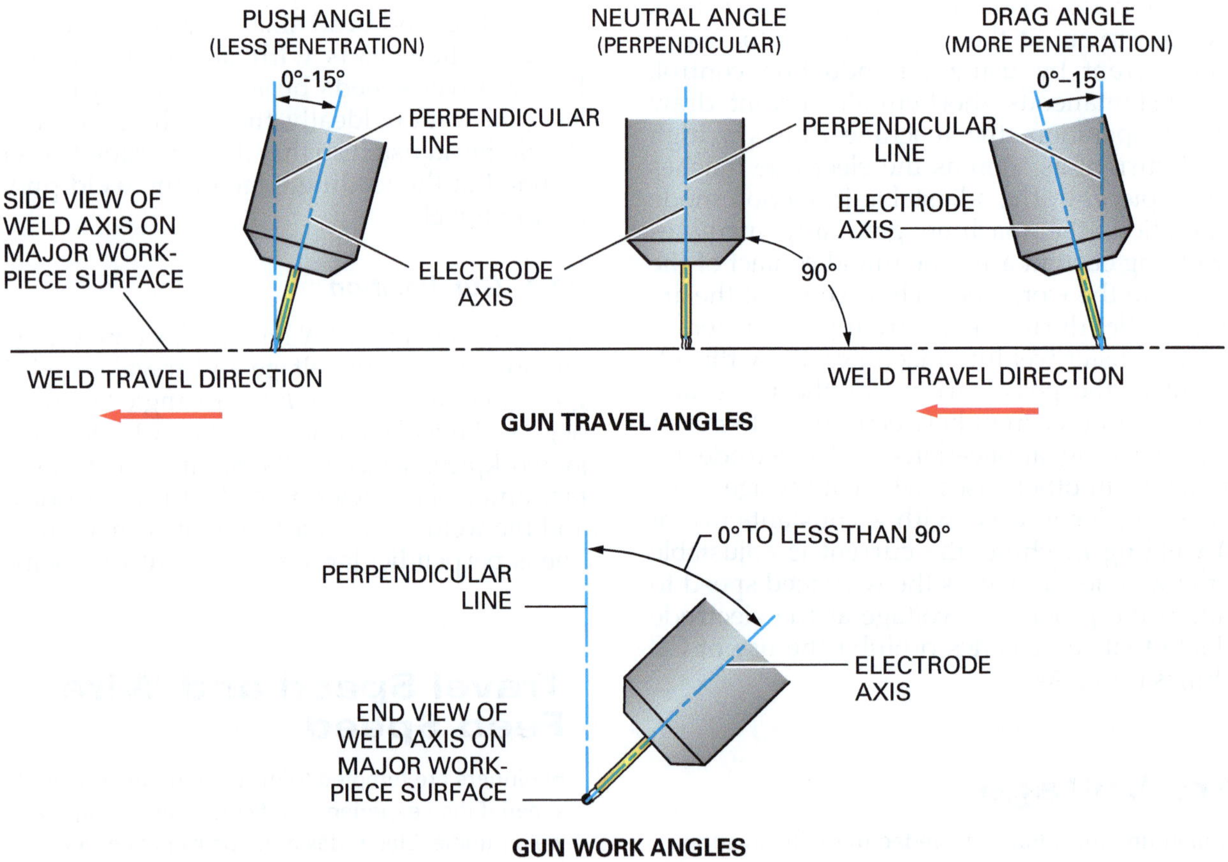

Figure 10 Gun travel and work angles.

because it changes the preheating of the electrode caused by the resistance of the electrode to current flow. As the extension increases, preheating of the electrode increases.

Since the welding power source is self-regulating, it does not have to furnish as much welding current to melt the electrode, so the current output automatically decreases. This results in less penetration. Increasing extension is useful for bridging gaps and compensating for mismatch, but it can cause overlap or lack of fusion and a ropy bead appearance. When the extension is decreased, the power source is forced to increase its current output to burn off the electrode. Too little extension can cause the electrode to weld to the contact tip or develop porosity in the weld. For metal electrodes which are highly conductive, the preheating effect of electrode resistance is minimal, and electrode speed and voltage settings have a more direct effect.

FCAW-G extensions vary widely. Always check the manufacturer's recommendations. The extension for FCAW-S and an insulating nozzle is ⅝" to 1" (16 mm to 25 mm), depending on the welding position and electrode size. *Figure 11* shows a typical electrode extension for an FCAW gun configuration.

Additional gun nomenclature is defined in *Figure 11*. Stickout is the distance from the insulating nozzle to the end of the electrode. Standoff distance is the distance from the insulating nozzle to the workpiece. Contact tip extension or setback is usually dependent on the transfer mode for the application.

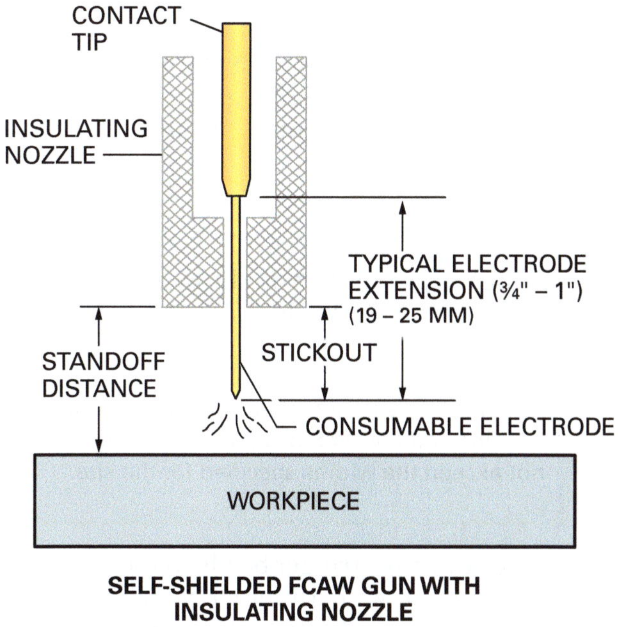

Figure 11 FCAW gun configuration.

2.1.6 Gas Nozzle Cleaning

As an FCAW-G welding machine is being used, weld spatter accumulates on the gas nozzle and the contact tip. If the gas nozzle is not properly cleaned, it will restrict the shielding gas flow and cause porosity in the weld.

Clean the gas nozzle with a reamer or round file or with the tang of a standard file. After cleaning, the nozzle can be sprayed or dipped in a special anti-spatter compound. The anti-spatter compound helps prevent the spatter from sticking to the nozzle.

Gas Marks and Electrode Extension

The FCAW process sometimes results in marks called gas marks on the weld surface. Also known as worm tracks, these marks are the result of excessive amounts of gases being trapped under the flux as it solidifies. If the flux solidifies before the gas escapes, pockets are left behind in the molten weld metal, causing the marks.

This surface discontinuity can be caused by an electrode extension that is too short. An extension that is too short allows dissolved hydrogen to remain trapped beneath the slag as it forms.

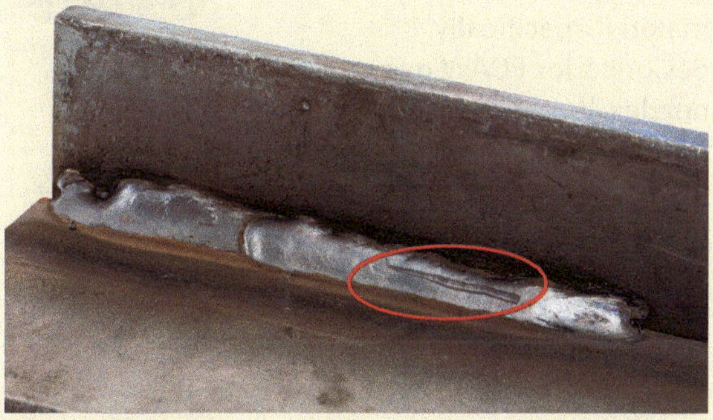

There are other potential causes as well. Gas marks are more likely with small-diameter electrodes operating at lower current levels than when using larger electrodes in the flat and horizontal position operating at higher currents. An excessive arc voltage, which increases the arc length, is also a common cause. Another possible cause is moisture trapped in the weld area, possibly entering from the surrounding air or from condensation in the shielding gas.

2.2.0 Bead Types

The following sections will explain how to create weld beads using FCAW equipment. The two basic bead types are stringer beads and weave beads.

2.2.1 Stringer Beads

Stringer beads are made with little or no side-to-side motion of the electrode.

> **CAUTION**
>
> The width of stringer beads is specified in the welding code or WPS being used at a site. Do not exceed the widths specified for the site.

Practice running stringer beads in the flat position. Experiment with different voltages, travel angles, and stickouts. Change the wire feed speed and observe the bead size. Try different travel angles and observe the bead shape and size.

Follow these steps to run stringer beads:

Step 1 Hold the gun at the desired travel angle (drag angle), with the electrode tip directly over the point where the weld is to begin, and pull the gun trigger.

Step 2 Hold the arc in place until the weld puddle begins to form.

Step 3 Slowly advance the arc while maintaining the gun angle. Stay in the leading edge of the puddle to prevent incomplete fusion.

Step 4 Continue to weld until a bead about 2" to 6" (50 mm to 152 mm) long is formed.

Step 5 Stop, raise the gun to a neutral (0-degree) travel angle, pause until the crater is filled, and then release the trigger.

Step 6 Inspect the bead for the following:

- Straightness of the bead
- Uniform appearance of the bead face
- Smooth, flat transition with complete fusion at the toes of the weld
- No porosity
- No undercut
- No overlap
- No lack of fusion
- Complete penetration
- No pinholes
- No inclusions
- Crater filled
- No cracks

Step 7 Continue practicing stringer beads until acceptable welds are made every time.

> **Anti-Spatter Materials**
>
> Use only anti-spatter material specifically designed for FCAW gas nozzles. Water-based products may cause hydrogen embrittlement in hydrogen-sensitive steel. Always refer to the SDS/MSDS for specific safety considerations.

2.2.2 Weave Beads

Weave beads are made with wide, side-to-side motions of the electrode. The width of a weave bead is determined by the amount of side-to-side motion. *Figure 12* shows an example of a weave bead.

When making a weave bead, use care at the toes to be sure there is proper tie-in to the base metal. To ensure proper tie-in at the toes, slow down or pause slightly at the edges. The pause at the edges will also flatten out the weld, giving it the proper profile.

Practice running weave beads about ⅝" (16 mm) wide in the flat position. Experiment with different weave motions, travel angles, and stickouts. Change the wire feed speed and observe the bead size. Try different gun angles and observe the bead shape and size.

Figure 12 Close-up view of FCAW weave beads.

Follow these steps to run weave beads:

Step 1 Hold the gun at the desired travel angle with the electrode tip directly over the point where the weld is to begin.

Step 2 Pull the gun trigger.

Step 3 Hold the arc in place until the weld puddle begins to form.

Step 4 Slowly advance the arc in a weaving motion while maintaining the gun angle (*Figure 13*).

Step 5 Continue to weld until a bead about 2" to 6" (50 mm to 152 mm) long is formed.

Step 6 Stop, raise the gun to a neutral (0-degree) travel angle, pause until the crater is filled, and then release the trigger.

Step 7 Inspect the bead for the following:
- Straightness of the bead
- Uniform appearance of the bead face
- Smooth, flat transition with complete fusion at the toes of the weld
- No porosity
- No undercut
- No overlap
- No lack of fusion
- Complete penetration
- No pinholes
- No inclusions
- Crater filled
- No cracks

Step 8 Continue practicing weave beads until acceptable welds are made every time.

2.2.3 Weld Restarts

A restart is the junction where a new weld connects to and continues the bead of a previous weld. Restarts are important because an improperly made restart creates a weld discontinuity.

Weave Beads and Stainless Steel

Weave beads are not normally used with stainless steel because of gas shielding problems and excessive heat input to the base metal. However, some fully automatic processes can make reasonable weave beads using pulse transfer techniques.

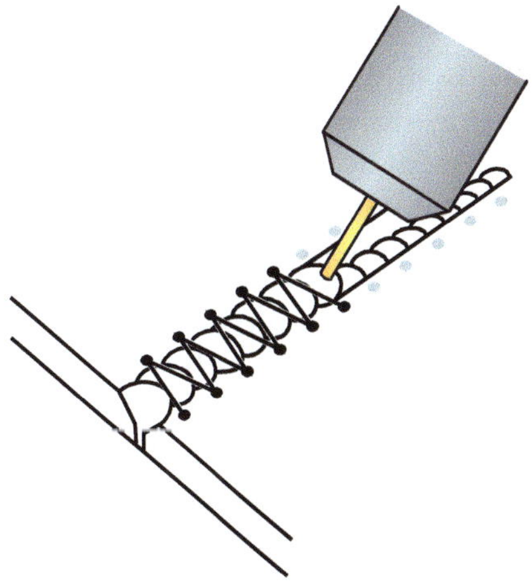

Figure 13 Weave motion.

One advantage of the semi-automatic FCAW process is that restarts can be minimized. Plan the weld to eliminate restarts, if possible. If a restart has to be made, it must be made so that it blends smoothly with the previous weld and does not stand out. The technique for making a restart is the same for both stringer and weave beads. Practice weld restarts using FCAW as directed by the instructor. Remember that the welding codes do not allow arc strikes outside the area of permanent welds.

Follow these steps to make a restart:

Step 1 With the weld clean, hold the gun at the proper travel angle while restarting the arc directly over the center of the crater.

Step 2 Pull the gun trigger.

Step 3 Move the electrode tip over the crater in a small circular motion to fill the crater with a molten puddle.

Step 4 Move to the leading edge of the puddle; as soon as the puddle fills the crater, continue the stringer or weave bead pattern being used.

Step 5 Inspect the restart. A properly made restart will blend into the bead, making it hard to detect. If the restart has undercut, not enough time was spent in the crater to fill it. If the undercut is on one side or the other, use more of a side-to-side motion while moving back into the crater. If the restart has a lump, it was overfilled; too much time was spent in the crater before resuming the forward motion.

Step 6 Continue to practice restarts until they are correct.

2.2.4 Weld Terminations

A weld termination is made at the end of a weld. A termination normally leaves a crater. When making a termination, the welding codes require that the crater must be filled to the full cross section of the weld. This can be difficult, since most terminations are at the edge of a plate, where welding heat tends to build up, making filling the crater more difficult. *Figure 14* shows the proper technique for terminating a weld.

Follow these steps to make a weld termination:

Step 1 Start to bring the gun up to a 0-degree travel angle, and slow the forward travel as the end of the weld is approached.

Step 2 Stop the forward movement about ⅛" (3.2 mm) from the end of the plate.

Step 3 With the gun at a 0-degree travel angle, stop and begin moving backwards until the crater is filled. An ⅛ to ¼ inch of backward movement is typical, depending upon the size of the crater.

Step 4 Release the trigger when the crater is completely filled.

Step 5 Inspect the termination. The crater should be filled to the full cross section of the weld.

2.2.5 Overlapping Beads

Overlapping beads are made by depositing connective weld beads parallel to one another. The parallel beads overlap, forming a flat surface. This is also called padding. Overlapping beads are used to build up a surface and to make multiple-pass welds. Both stringer and weave beads can be overlapped. When viewed from the end, properly overlapped beads will have a relatively flat surface. *Figure 15* shows proper and improper overlapping string and weave beads.

Bead Cleaning

There is no such thing as a self-cleaning electrode or shield gas. Welders must make sure that surfaces to be welded are clean and free of contaminants. It is important that multiple-pass welds are cleaned before each successive bead is laid down. Attempting to lay multiple-pass beads without properly cleaning the previous bead will result in inclusions in the welded joint. Shown here is a radiograph of a butt weld that has two slag lines easily visible in the weld root.

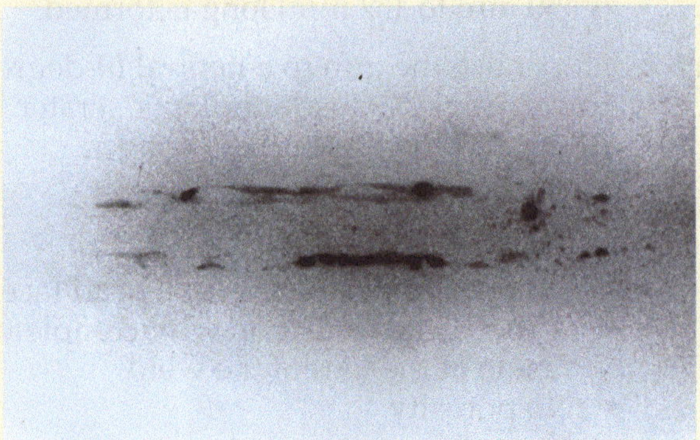

REPRODUCED BY PERMISSION TWI LTD.

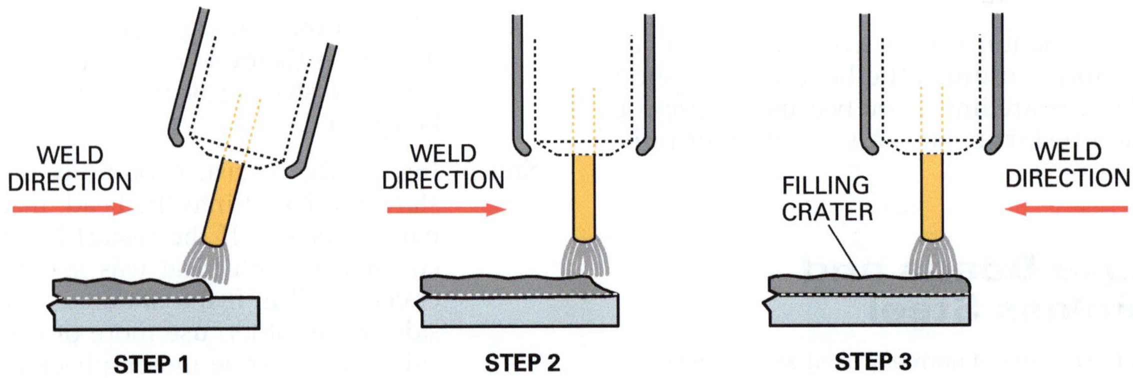

Figure 14 Terminating a weld.

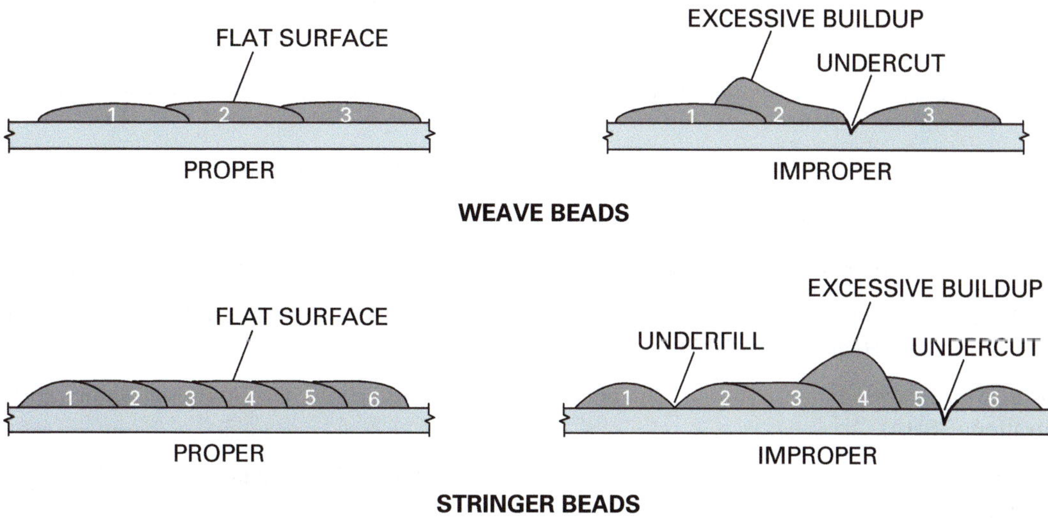

Figure 15 Proper and improper overlapping beads.

Follow these steps to weld overlapping stringer or weave beads, using a specific diameter flux-cored electrode and, if necessary, the proper shielding gas as directed by the instructor. Remember to chip and brush each bead before running the next bead.

Step 1 Mark out a square on a piece of steel.

Step 2 Weld a stringer or weave bead along one edge at a 0-degree work angle.

Step 3 Clean the weld.

Step 4 Position the gun at a work angle of 10 to 15 degrees toward the side of the previous bead to get proper tie-in (*Figure 16*), and pull the trigger.

Step 5 Continue running overlapping stringer or weave beads until the square is covered.

Step 6 Continue building layers of stringer or weave beads, one on top of the other, until perfecting the technique.

Build a pad using stringer beads. Repeat building the pad using weave beads, as shown in *Figure 17*.

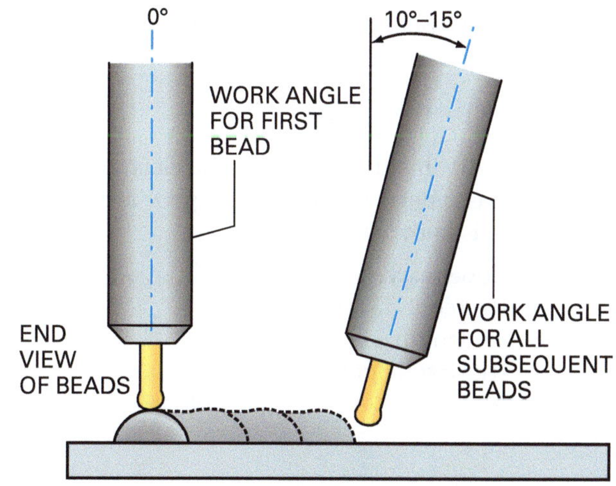

Figure 16 Work angles for overlapping beads.

Figure 17 Pad built with weave beads.

Additional Resources

AWS B1.10M/B1.10 Guide for the Nondestructive Examination of Welds. Miami, FL: American Welding Society.

AWS B1.11 Guide for the Visual Examination of Welds. Miami, FL: American Welding Society.

AWS D3.5-93R Guide for Steel Hull Welding. Miami, FL: American Welding Society.

Lincoln Electric website: **http://www.lincolnelectric.com** offers sources for products and training.

OSHA 1910.269, Appendix C, Protection from Step and Touch Potentials. Current Edition. Washington, DC: Occupational Safety & Health Administration (OSHA).

OSHA 1926.351, Arc Welding and Cutting. Current Edition. Washington, DC: Occupational Safety and Health Administration (OSHA).

The Procedure Handbook of Arc Welding. 14th Edition. Cleveland, OH: The James F. Lincoln Arc Welding Foundation, 2000.

Welding Handbook. Volume 1. Welding Science & Technology. Miami, FL: American Welding Society, 2001.

Welding Handbook. Volume 2. Part 1: Welding Processes. Miami, FL: American Welding Society, 2004.

2.0.0 Section Review

1. In FCAW, the distance from the electrode tip to the base metal or the molten pool at the base metal is known as _____.
   a. slope reach
   b. strikeout distance
   c. arc length
   d. stickout length

2. The width of a weave bead is determined by the amount of _____.
   a. forward and backward gun movement
   b. stringer bead padding overlap
   c. filler material in the flux core
   d. side-to-side motion of the electrode

SECTION THREE

3.0.0 FCAW FILLET AND V-GROOVE WELDS

Objective

Describe the welding procedures needed to produce proper fillet and V-groove welds using FCAW welding techniques.

a. Describe the welding procedures needed to produce proper fillet welds using FCAW welding techniques.
b. Describe the welding procedures needed to produce proper V-groove welds using FCAW welding techniques.

Performance Tasks

1. Make multiple-pass FCAW-G/GM (gas-shielded) and/or FCAW-S (self-shielded) fillet welds on carbon steel plate coupons in all four 1F through 4F positions.
2. Make multiple-pass FCAW-G/GM (gas-shielded) and/or FCAW-S (self-shielded) V-groove welds on carbon steel plate coupons in all four 1G through 4G positions, with or without backing.

Fillet welds and V-groove welds are two common types of welds produced with FCAW. This section examines the procedures used to make each of these welds in four different positions: flat, horizontal, vertical, and overhead.

3.1.0 Fillet Welds

Fillet welds require little base metal preparation, except for the cleaning of the weld area and the removal of any excess material from cut surfaces. Any dross from cutting will cause porosity in the weld. For this reason, the codes require that this material is entirely removed prior to welding.

The most common fillet welds are made in lap and T-joints. The weld position for plate is determined by the weld axis and the orientation of the workpiece. The positions for fillet welding on plate are flat, or 1F (the letter F stands for fillet); horizontal, or 2F; vertical, or 3F; and overhead, or 4F. In the 1F and 2F positions, the weld axis can be inclined up to 15 degrees. Any weld axis inclination for the other positions varies with the rotational position of the weld face as specified in AWS standards. *Figure 18* shows the four fillet welding positions for plate.

Fillet welds can vary from concave to convex, depending on the WPS or site quality standards. The welding codes require a fillet weld to have a uniform concave or convex face, although a slightly non-uniform face is acceptable. Convexity is the distance the weld extends above a line drawn between the toes of the weld. The convexity of a fillet weld or individual surface bead must not exceed that permitted by the applicable code or standard. In single-pass, weave-bead fillet welds where two workpieces are being joined at an angle (not lap joints), flat or slightly convex faces are usually preferred

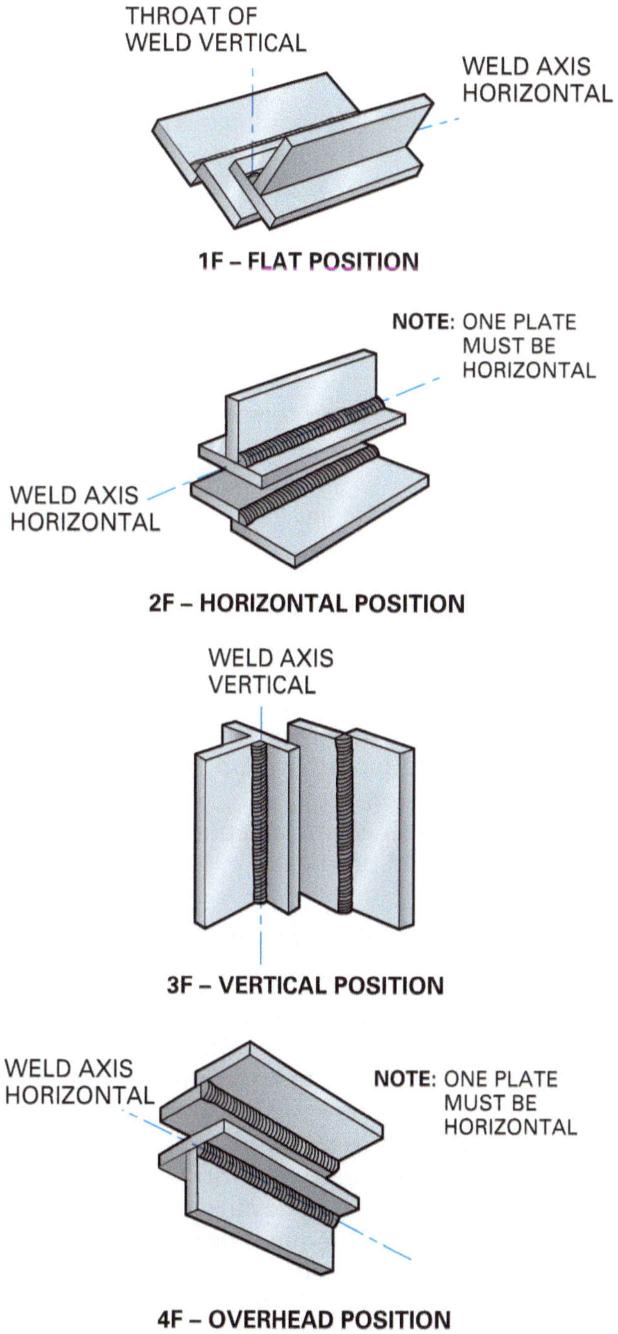

Figure 18 Fillet weld positions.

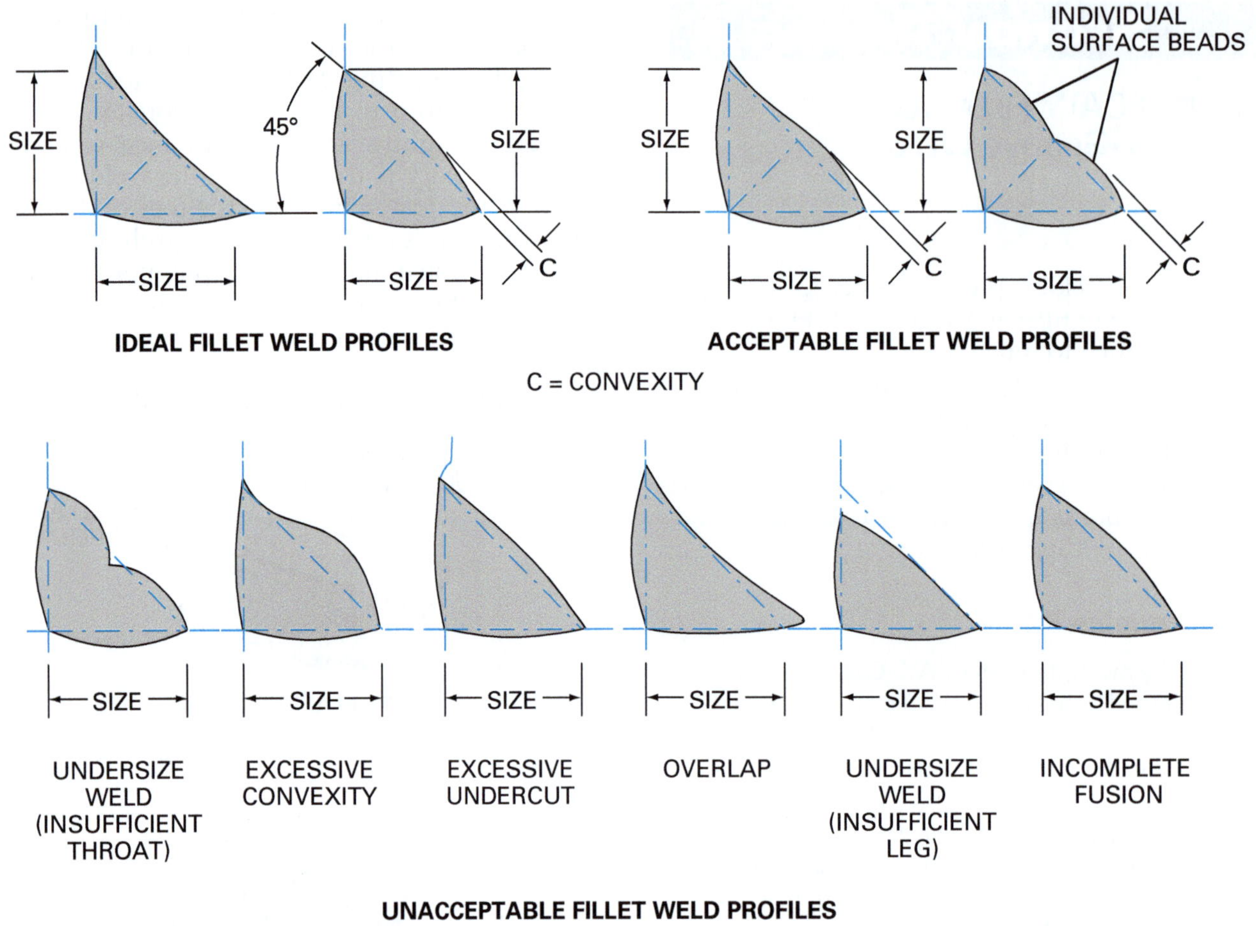

Figure 19 Ideal, acceptable, and unacceptable fillet weld profiles.

because weld stresses are more uniformly distributed through the fillet weld and workpieces. A fillet weld must be repaired if the profile has defects.

A fillet weld is unacceptable and must be repaired if the profile has insufficient throat, excessive convexity, undercut, overlap, insufficient leg, or incomplete fusion.

Figure 19 shows ideal, acceptable, and unacceptable fillet weld profiles.

3.1.1 Practicing Flat (1F) Position Fillet Welds

Practice flat (1F) position fillet welds by welding multiple-pass (six-pass) convex fillet welds in a T-joint. Use an instructor-specified size of flux-cored electrode wire with appropriate shielding gas as required. When making flat fillet welds, pay close attention to the work angle and travel speed. For the first bead, the work angle is vertical (45 degrees to both plate surfaces). The angle is adjusted for all subsequent beads. Increase or decrease the travel speed to control the amount of weld metal buildup. *Figure 20* shows the work angles for 1F position fillet welds.

Follow these steps to make a flat fillet weld. Remember to clean all weld beads before beginning the next bead.

Step 1 Tack two plates together to form a T-joint for the fillet weld coupon (*Figure 21*). Clean the tack welds.

Step 2 Clamp or tack weld the coupon in the flat position.

Step 3 Run the first bead along the root of the joint, using a work angle of 45 degrees and a 5- to 10-degree drag angle. Use a slight side-to-side motion (oscillation).

Step 4 Run the second bead along a toe of the first weld, overlapping about 75 percent of the first bead. Alter the work angle as shown in *Figure 20*, and use a 5- to 10-degree drag angle with a slight oscillation.

Step 5 Run the third bead along the other toe of the first weld, filling the groove created when the second bead was run. Use the work angle shown in *Figure 20* and a 5- to 10-degree drag angle with a slight oscillation.

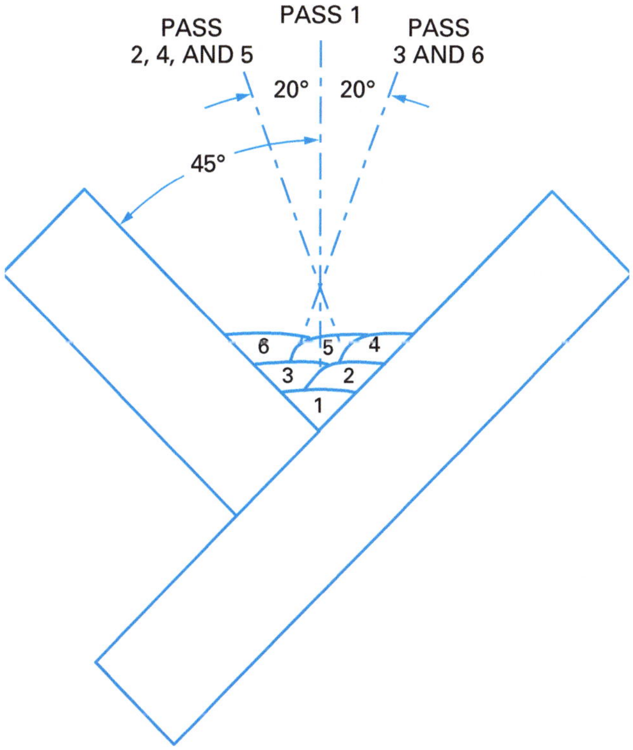

Figure 20 Multiple-pass 1F bead sequence and work angles.

Step 6 Run the fourth bead along the outside toe of the second weld, overlapping about half the second bead. Use the work angle shown in *Figure 20* and a 5- to 10-degree drag angle with a slight oscillation.

Step 7 Run the fifth bead along the inside toe of the fourth weld, overlapping about half the fourth bead that was run. Use the work angle shown in *Figure 20* and a 5- to 10-degree drag angle with a slight oscillation.

Step 8 Run the sixth bead along the toe of the fifth weld, filling the groove created when the fifth bead was run. Use the work angle shown in *Figure 20* and a 5- to 10-degree drag angle with a slight oscillation.

Step 9 Inspect the weld. The weld is acceptable if it has the following features:

- Uniform appearance on the bead face
- Craters and restarts filled to the full cross section of the weld
- Uniform weld size ±1⁄16" (1.6 mm)
- Acceptable weld profile in accordance with the applicable code or standard
- Smooth transition with complete fusion at the toes of the weld
- No porosity
- No undercut
- No overlap
- No inclusions
- No cracks

3.1.2 Practicing Horizontal (2F) Position Fillet Welds

Practice horizontal (2F) fillet welding by placing multiple-pass fillet welds in a T-joint using flux-cored electrodes and, if required, appropriate shielding gas as directed by the instructor. When making horizontal fillet welds, pay close attention to the gun angles and travel speed. For the first bead, the electrode work angle is 45 degrees. The work angle is adjusted for all other welds. Increase or decrease the travel speed to control the amount of weld metal buildup.

Follow these steps to make a horizontal fillet weld:

Step 1 Tack two plates together to form a T-joint for the fillet weld coupon. Clean the tack welds.

Step 2 Clamp or tack weld the coupon in the horizontal position.

Step 3 Run the first bead along the root of the joint using a work angle of approximately 45 degrees with a 5- to 10-degree drag angle and a slight side-to-side oscillation (*Figure 22*).

Step 4 Clean the weld.

Figure 21 Fillet weld coupon.

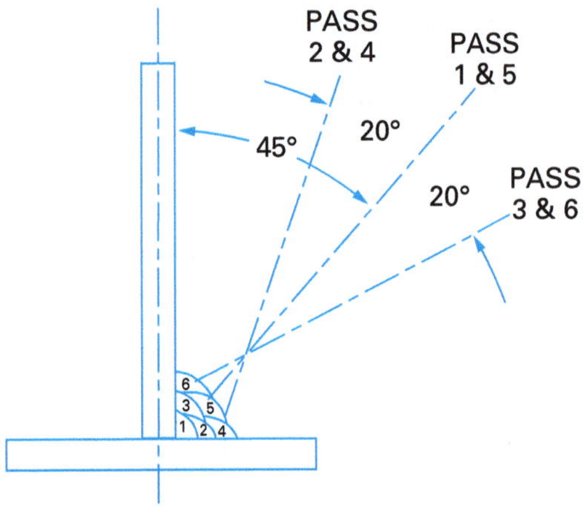

Figure 22 Multiple-pass 2F bead sequence and work angles.

Step 5 Run the remaining passes at the appropriate work angles using a 5- to 10-degree drag angle and a slight oscillation at the appropriate work angles. Overlap each previous pass and clean the weld after each pass.

Step 6 Have the instructor inspect the weld. The weld is acceptable if it has the following features:

- Uniform appearance on the bead face
- Craters and restarts filled to the full cross section of the weld
- Uniform weld size $\pm 1/16$" (1.6 mm)
- Acceptable weld profile in accordance with the applicable code or standard
- Smooth transition with complete fusion at the toes of the weld
- No porosity
- No undercut
- No overlap
- No inclusions
- No cracks

3.1.3 Practicing Vertical (3F) Position Fillet Welds

Practice vertical (3F) fillet welding by placing multiple-pass fillet welds in a T-joint using flux-cored electrodes and, if required, appropriate shielding gas as directed by the instructor. Normally, vertical welds are made welding uphill from the bottom to the top, using a gun push angle (angled up). However, downhill welding can also be accomplished using a slight drag angle. When vertical welding, either stringer or weave beads can usually be used. However, weave beads are avoided on stainless steel or in downhill welding. On the job, the site WPS or quality standard will specify which technique to use.

Follow these steps to make an uphill fillet weld:

Step 1 Tack two plates together to form a T-joint for the fillet weld coupon.

Step 2 Clamp or tack weld the coupon in the vertical position.

Step 3 Starting at the bottom, run the first bead along the root of the joint using a work angle of approximately 45 degrees and a 10- to 15-degree push angle. Pause in the weld puddle to fill the crater.

Step 4 Clean the weld.

Step 5 Run the remaining passes using a 5- to 10-degree push angle and a side-to-side weave technique with a 45-degree work angle (*Figure 23*). Use a slow motion across the face of the weld, pausing at each toe for penetration and to fill the crater. Adjust the travel speed across the

T-Joint Heat Dissipation

In T-joints, the welding heat dissipates more rapidly in the thicker or non-butting member. On various bead passes, the arc may have to be concentrated slightly more on the thicker or the non-butting member to compensate for the heat loss.

Shielding Gas Blender

This wall-mounted Harris US-S10 gas blender is designed to blend together two gases from different cylinders. The mixture ratio is easily adjusted on the panel. Gases can be mixed and routed to a manifold providing the same mixture to a number of welding stations. Another model is available that allows the blending of three different gases.

face of the weld to control the buildup. Overlap each previous pass as required, and clean the weld after each pass.

Step 6 Have the instructor inspect the weld. The weld is acceptable if it has the following features:

- Uniform appearance on the bead face
- Craters and restarts filled to the full cross section of the weld
- Uniform weld size ±1⁄16" (1.6 mm)
- Acceptable weld profile in accordance with the applicable code or standard
- Smooth transition with complete fusion at the toes of the weld
- No porosity
- No undercut
- No overlap
- No inclusions
- No cracks

Repeat vertical fillet (3F) welding using stringer beads. Use a 0- to 10-degree push angle and the work angles shown in *Figure 23* as required. Depending on the electrode type, note that a drag angle may be required. Follow the manufacturer's guidance for the electrode type.

3.1.4 Practicing Overhead (4F) Position Fillet Welds

Practice overhead (4F) fillet welding by welding multiple-pass fillet welds in a T-joint using flux-cored electrodes and, if required, appropriate shielding gas as directed by the instructor. When making overhead fillet welds, pay close attention to the gun angles and travel speed. For the first bead, the work angle is approximately 45 degrees. The work angle is adjusted for all other welds. Increase or decrease the travel speed to control the amount of weld metal buildup.

Follow these steps to make an overhead fillet weld:

Step 1 Tack two plates together to form a T-joint for the fillet weld coupon.

Step 2 Clamp or tack weld the coupon so it is in the overhead position.

Step 3 Run the first bead along the root of the joint, using a work angle of approximately 45 degrees with a 5- to 10-degree drag angle. Use a slight oscillation to tie-in the weld at the toes.

Vertical Fillet Welds

When making vertical fillet welds, pay close attention to the gun angles and travel speed. For the first bead, the work angle is approximately 45 degrees. The work angle is adjusted for all other welds. Increase or decrease the travel speed to control the amount of weld metal buildup.

Step 4 Clean the weld.

Step 5 Using a slight oscillation, run the remaining passes using a 5- to 10-degree drag angle and the work angles shown in *Figure 24*. Overlap each previous pass and clean the weld after each pass.

Step 6 Have the instructor inspect the weld. The weld is acceptable if it has the following features:

- Uniform appearance on the bead face
- Craters and restarts filled to the full cross section of the weld
- Uniform weld size ±1⁄16" (1.6 mm)
- Acceptable weld profile in accordance with the applicable code or standard
- Smooth transition with complete fusion at the toes of the weld
- No porosity
- No undercut
- No overlap
- No inclusions
- No cracks

3.2.0 Open V-Groove Welds

The open V-groove weld is a common groove weld normally made on plate and pipe. Practicing the open V-groove weld on plate will help with making the more difficult pipe welds covered in the next level of the Welding curriculum.

3.2.1 Root Pass

The most difficult part of making an open V-groove weld is the root pass. The root pass is made from the V-groove side of the joint and must have complete penetration, but not an excessive amount of root reinforcement. The penetration is controlled with a side-to-side motion or oscillation.

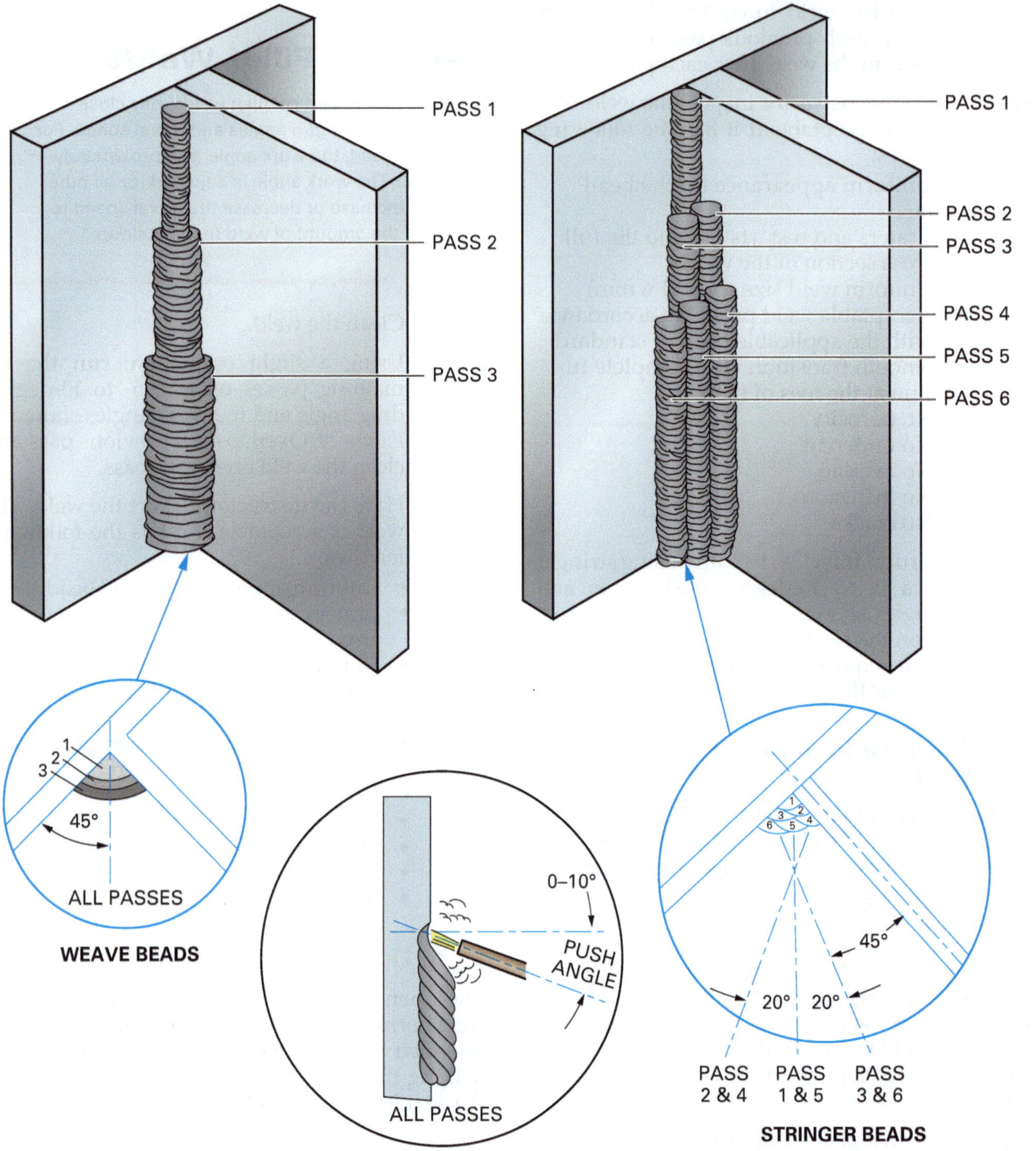

Figure 23 Multiple-pass 3F bead sequences and work angles for weave and stringer beads.

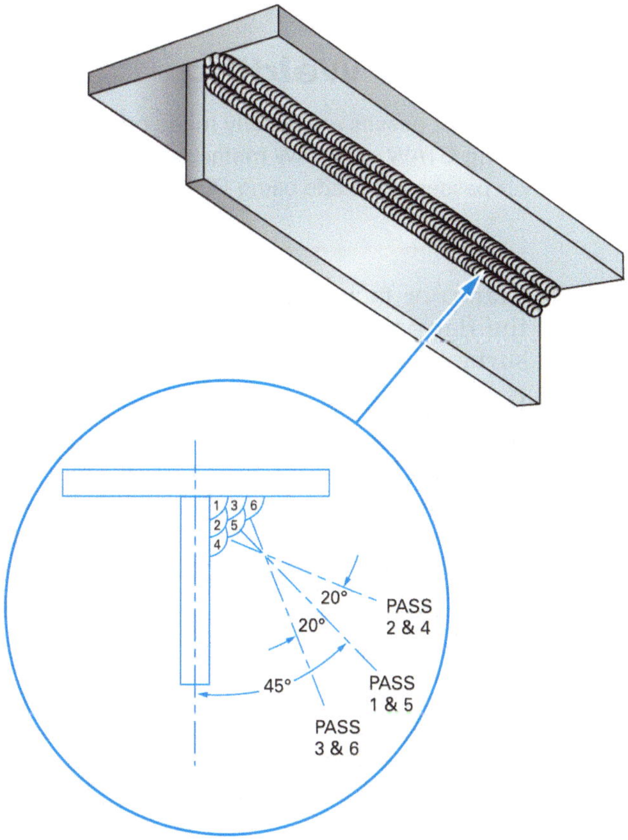

Figure 24 Multiple-pass 4F bead sequence and work angles.

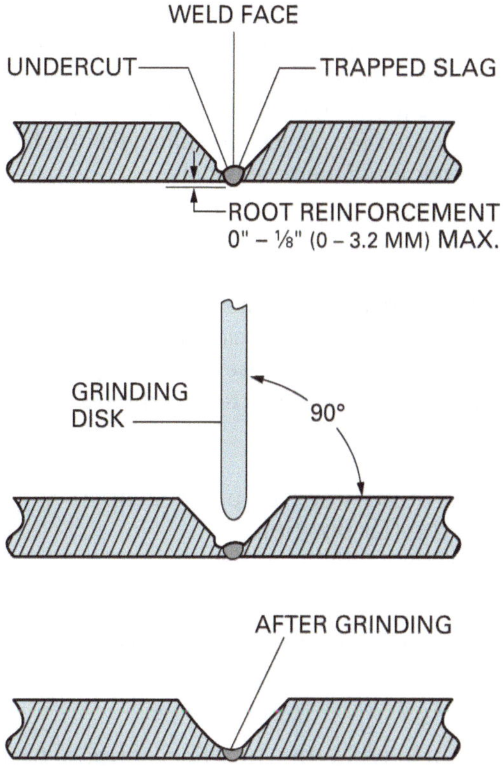

Figure 25 Grinding a root pass.

Note that open-root passes are not typically done with FCAW. Short-circuit GMAW, GTAW, or SMAW processes are typically used for root passes on open V-groove welds. For this module, you have the option of using other processes for the root pass of an open V-groove weld, or use the FCAW process with a backing bar.

Clean and inspect the root pass after it has been placed. Inspect the face of the weld for excess buildup or undercut (*Figure 25*), which could trap slag if an FCAW root pass is run. Remove excess buildup or undercut with a hand grinder by grinding the face of the root pass with the edge of a grinding disk. Use a slight oscillation to remove any undercut. Take care not to grind through the root pass.

3.2.2 Groove Weld Positions

As shown in *Figure 26*, groove welds can be made in all positions. The weld position for plate is determined by the axis of the weld and the orientation of the workpiece. Groove weld positions for plate are flat, or 1G (groove); horizontal, or 2G; vertical, or 3G; and overhead, or 4G.

In the 1G and 2G positions, the weld axis can be inclined up to 15 degrees. Any weld axis inclination for the other positions varies with the rotational position of the weld face as specified in the WPS.

3.2.3 Acceptable and Unacceptable Groove Weld Profiles

Groove welds should be made with slight reinforcement and a gradual transition to the base metal at each toe. As shown in *Figure 27*, groove welds must not have excessive reinforcement or undercut, underfill, or overlap. If a groove weld has any of these defects, it must be repaired.

3.2.4 Practicing Flat (1G) Position V-Groove Welds

Practice flat (1G) V-groove welds, as shown in *Figure 28*. Use flux-cored electrodes and, if required, appropriate shielding gas as directed by the instructor. Except for a root pass, use weave beads and keep the gun angle at 90 degrees to the plate surface (0-degree work angle) with a 5- to 10-degree drag angle. When using stringer beads, use a 5- to 10-degree drag angle (except for the root pass), but adjust the work angle to tie-in and fill the bead. Pay special attention at the termination of the weld to fill the crater. If the practice coupon gets too hot between passes, it may have to be quenched with water.

> **CAUTION**
>
> Cooling with water (quenching) is done only on practice coupons. Never cool test coupons or on-the-job welds with water. Cooling with water can cause weld cracks and affect the mechanical properties of the base metal.

> **WARNING!**
>
> Use pliers to handle the hot practice coupons. Wear gloves when placing the practice coupon in water. Steam will rise off the coupon and can burn or scald unprotected hands and arms.

Follow these steps to practice open V-groove welds in the flat position:

> **NOTE**
>
> Avoid using weave beads for stainless steel.

Step 1 Tack weld the practice coupon together as explained earlier.

> ## Root Pass Welding
>
> In practice, root passes are usually made on various metals using GTAW or SMAW methods; then the fill and cover passes are made using FCAW or GMAW.

Step 2 Clamp or tack weld the weld coupon in the flat position above the welding table surface.

Step 3 If using FCAW and a backing bar, run the root pass using a 5- to 10-degree drag angle and a slight side-to-side motion to control penetration. For an open-root V-groove weld, run the root pass using GTAW or SMAW before proceeding.

Step 4 Clean the root pass and grind if necessary (and if grinding is allowed).

Step 5 Run the remaining passes using a 5- to 10-degree drag angle and applicable work angles. Overlap each previous pass and clean the weld after each pass.

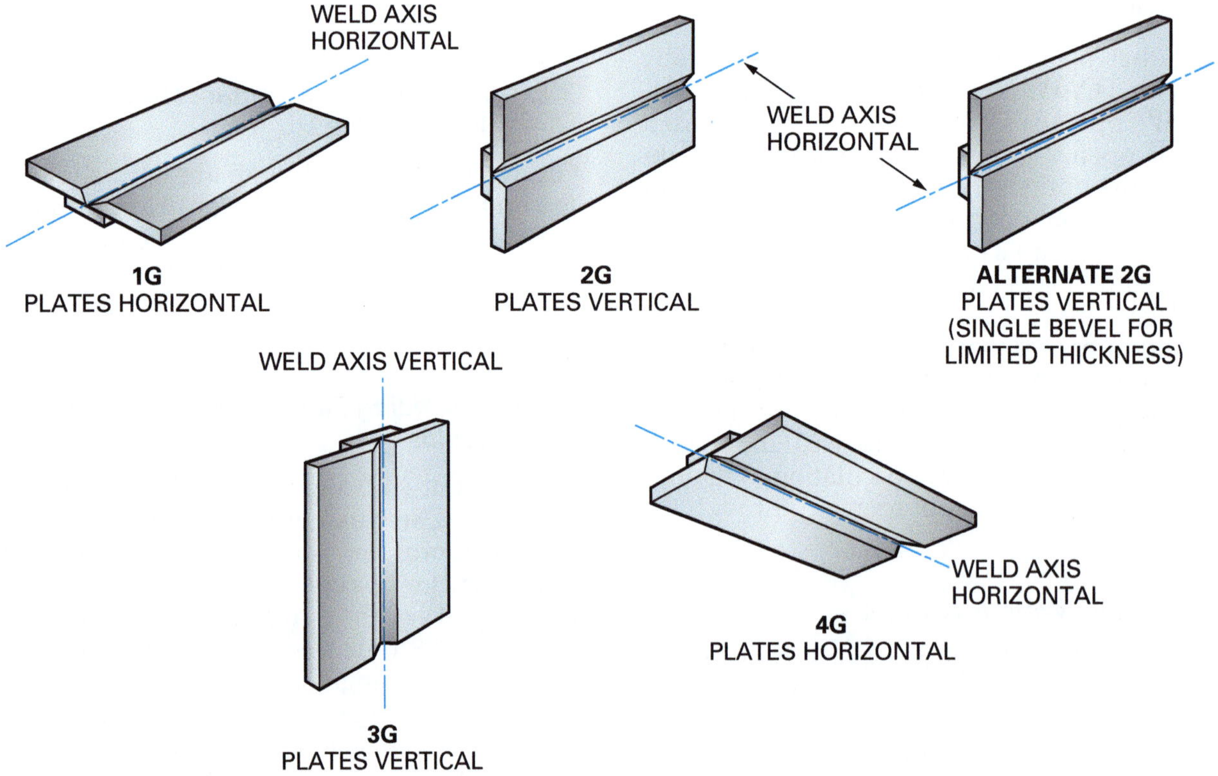

Figure 26 Groove weld positions for plate.

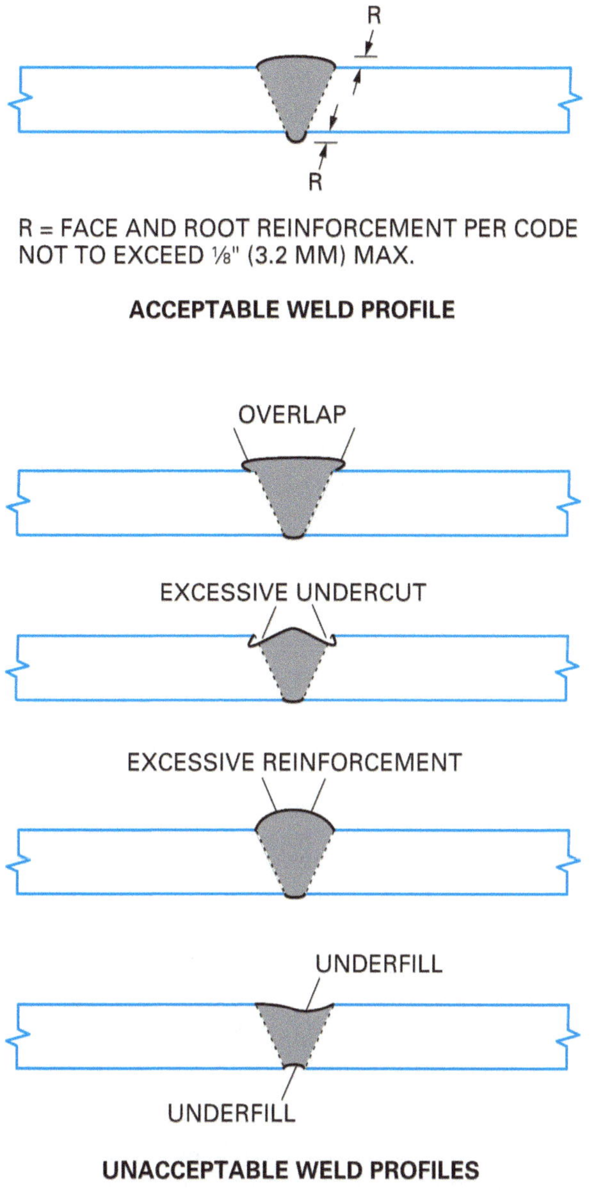

Figure 27 Acceptable and unacceptable groove weld profiles.

Step 6 Inspect the weld. The weld is acceptable if it has the following:

- Uniform appearance on the bead face
- Craters and restarts filled to the full cross section of the weld
- Uniform weld width ±1⁄16" (1.6 mm)
- Acceptable weld profile in accordance with the applicable code or standard
- Complete uniform root and face reinforcement at least flush with the base metal to a maximum buildup of ⅛" (3.2 mm)
- Smooth, flat transition with complete fusion at the toes of the weld
- No porosity
- No undercut
- No inclusions
- No cracks

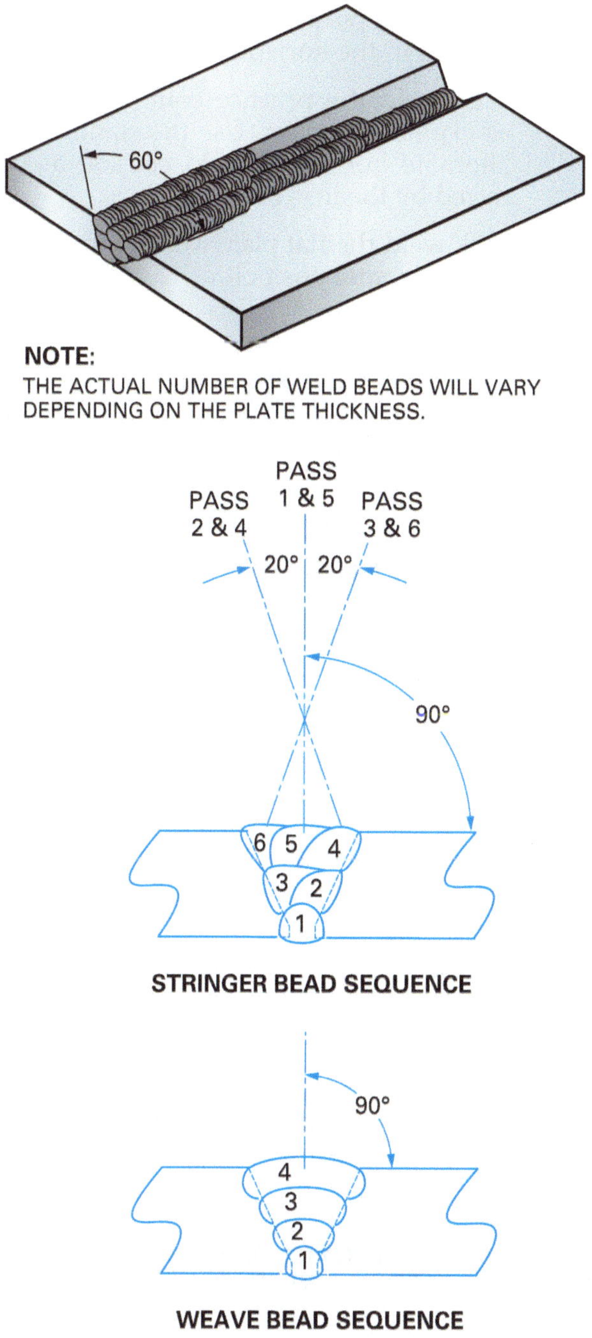

Figure 28 Multiple-pass 1G bead sequences and work angles.

3.2.5 Practicing Horizontal (2G) Position V-Groove Welds

A horizontal (2G) V-groove weld is shown in *Figure 29*. Practice horizontal open V-groove welds using flux-cored electrodes and, if required, appropriate shielding gas as directed by the instructor. For the gun angles, use a 5- to 10-degree drag angle (except for the root pass) and adjust the work angle as required. Pay particular attention to fill the crater at the termination of the weld.

Follow these steps to practice welding open V-groove welds in the horizontal position:

Step 1 Tack weld the practice coupon together as explained earlier. Use the standard or alternate horizontal weld coupon as directed by the instructor.

Step 2 Tack weld the flat plate weld coupon into the horizontal position.

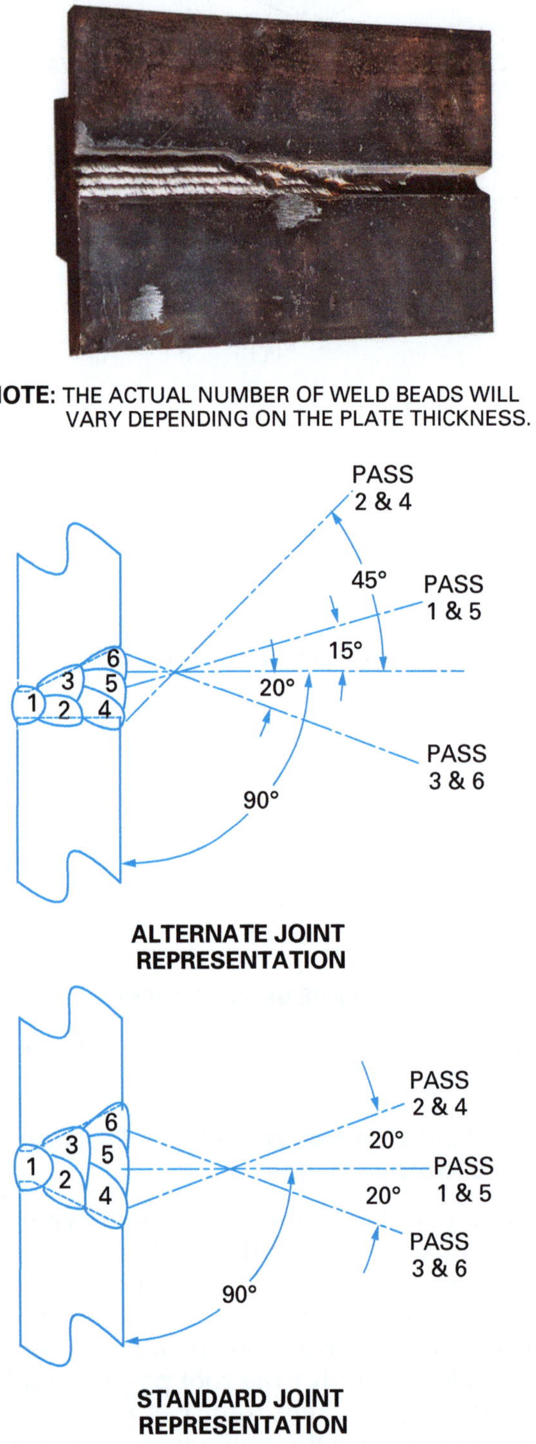

Figure 29 Multiple-pass 2G bead sequences and work angles.

Step 3 If using FCAW and a backing bar, run the root pass using a 5- to 10-degree drag angle and a slight side-to-side motion to control penetration. For an open-root V-groove weld, run the root pass using GTAW or SMAW before proceeding.

Step 4 Clean the weld.

Step 5 Run the remaining passes using a 5- to 10-degree drag angle and appropriate work angles. Overlap each previous pass and clean the weld after each pass.

Step 6 Inspect the weld. The weld is acceptable if it has the following:

- Uniform appearance on the bead face
- Craters and restarts filled to the full cross section of the weld
- Uniform weld width $\pm\frac{1}{16}$" (1.6 mm)
- Acceptable weld profile in accordance with the applicable code or standard
- Complete uniform root and face reinforcement at least flush with the base metal to a maximum buildup of $\frac{1}{8}$" (3.2 mm)
- Smooth, flat transition with complete fusion at the toes of the weld
- No porosity
- No undercut
- No inclusions
- No cracks

3.2.6 Practicing Vertical (3G) Position V-Groove Welds

Practice vertical (3G) multiple-pass groove welds using stringer beads or weave beads. Set the welding parameters at the lower ends of the suggested ranges. Out-of-position welds are made at lower welding voltages and currents than flat or horizontal welds to prevent the molten weld metal from sagging or running.

When welding vertically, either stringer or weave beads can be used. However, weave beads are not normally used on stainless steel. On the job, the WPS or site quality standards will specify which technique to use. Check with the instructor to see if stringer or weave beads or both techniques should be practiced. The root bead can be run either uphill or downhill. The fill and cover beads can be run uphill or downhill as well. However, downhill weave beads are not normally used. Use the direction preferred at the site or as specified in the WPS or site quality standards.

Run the root bead, using a work angle of 0 degrees (nozzle center line bisects the V-groove) and a 5- to 15-degree push angle. The work angle is adjusted for all other welds.

Practice 3G open V-groove welds using FCAW as directed by the instructor. Also use flux-cored electrodes and appropriate shielding gas as directed by the instructor.

Follow these steps to make a vertical groove weld with stringer beads:

Step 1 Tack weld the practice coupon together as explained earlier.

Step 2 Clamp or tack weld the coupon in the vertical position.

Step 3 If using FCAW and a backing bar, run the root pass uphill with a work angle of 0 degrees (nozzle center line bisects the V-groove) and a 0- to 10-degree push angle. For an open-root V-groove weld, run the root pass using GTAW or SMAW before proceeding.

Step 4 Clean the weld.

Step 5 Run the remaining passes at the work angles shown in *Figure 30*, using a 5- to 15-degree push angle. Overlap each previous pass and use a slight side-to-side oscillation to tie-in the welds at the toes. Clean the weld after each pass.

Step 6 Inspect the weld. The weld is acceptable if it has the following features:

- Uniform appearance on the bead face
- Craters and restarts filled to the full cross section of the weld
- Uniform weld width $\pm\frac{1}{16}$"(1.6 mm)
- Acceptable weld profile in accordance with the applicable code or standard
- Complete uniform root and face reinforcement at least flush with the base metal to a maximum buildup of $\frac{1}{8}$" (3.2 mm)
- Smooth, flat transition with complete fusion at the toes of the weld
- No porosity
- No undercut
- No inclusions
- No cracks

Follow these steps to make a vertical groove weld with weave beads:

Step 1 Tack weld the practice coupon together as explained earlier.

Step 2 Clamp or tack weld the coupon in the vertical position.

NOTE: THE ACTUAL NUMBER OF WELD BEADS WILL VARY WITH THE METAL THICKNESS.

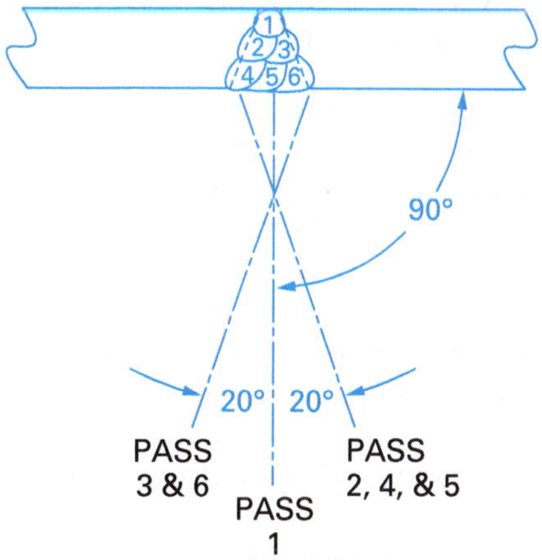

Figure 30 Multiple-pass 3G bead sequence and work angles for stringer beads.

Step 3 Run the root pass uphill with a work angle of 0 degrees (90 degrees from the plate) and a 0- to 10-degree push angle. Use a slight side-to-side oscillation to control the penetration.

Step 4 Clean the weld.

Step 5 Run the remaining passes as uphill weave beads across the root bead at a 0-degree work angle. Clean the weld after each pass. Use a zigzag weave pattern across the root bead and pause at each toe for good fusion. Use a push angle of 5 to 15 degrees. Increase or decrease the travel speed to control the amount of metal buildup. *Figure 31* shows the zigzag pattern and bead sequence for weave beads.

Step 6 Inspect the weld. The weld is acceptable if it has the following features:

- Uniform appearance on the bead face
- Craters and restarts filled to the full cross section of the weld
- Uniform weld width ±⅛" (3.2 mm)
- Acceptable weld profile in accordance with the applicable code or standard
- Complete uniform root and face reinforcement at least flush with the base metal to a maximum buildup of ⅛" (3.2 mm)
- Smooth, flat transition with complete fusion at the toes of the weld
- No porosity
- No undercut
- No inclusions
- No cracks

3.2.7 Practicing Overhead (4G) Position V-Groove Welds

Practice overhead (4G) V-groove welds using FCAW as directed by the instructor. Use flux-cored electrodes and, if required, appropriate shielding gas as directed by the instructor. To reduce heat and prevent the molten weld metal from sagging or running, FCAW overhead welds are made with lower welding voltages and currents than flat or horizontal welds.

Use a work angle of 0 degrees (90 degrees to plate) and a 5- to 15-degree drag angle for the root bead.

Follow these steps to make an overhead groove weld with stringer beads:

Step 1 Tack weld the practice coupon together as explained earlier.

Step 2 Clamp or tack weld the coupon in the overhead position.

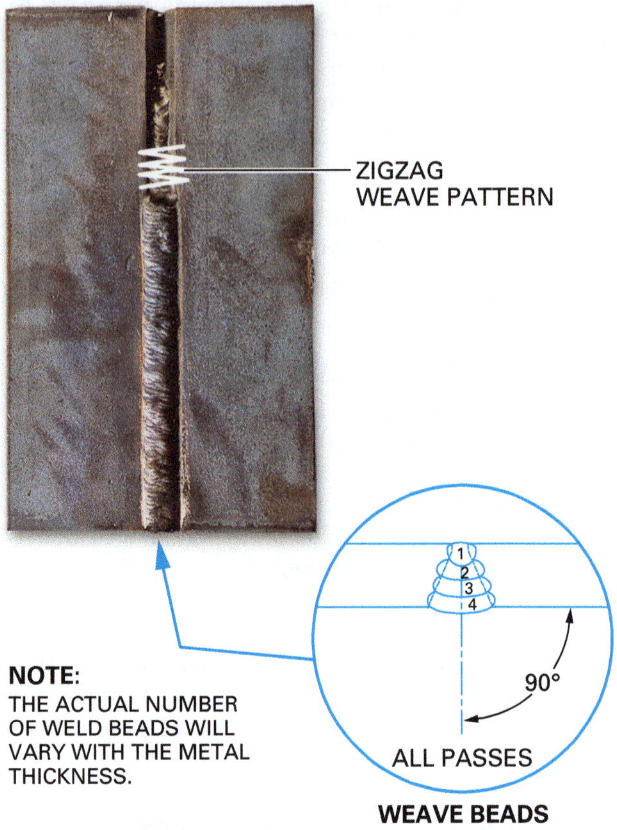

Figure 31 Multiple-pass 3G bead sequence and work angle for weave beads.

Step 3 Run the root pass, using a work angle of 0 degrees and a 5- to 15-degree drag angle. Use a slight side-to-side oscillation to control the penetration.

Step 4 Clean the weld.

Step 5 Run the remaining passes at the work angles shown in *Figure 32* using a 5- to 10-degree drag angle. Overlap each previous pass, and use a slight side-to-side motion to tie-in the welds at the toes. Each bead should be cleaned before running the next bead.

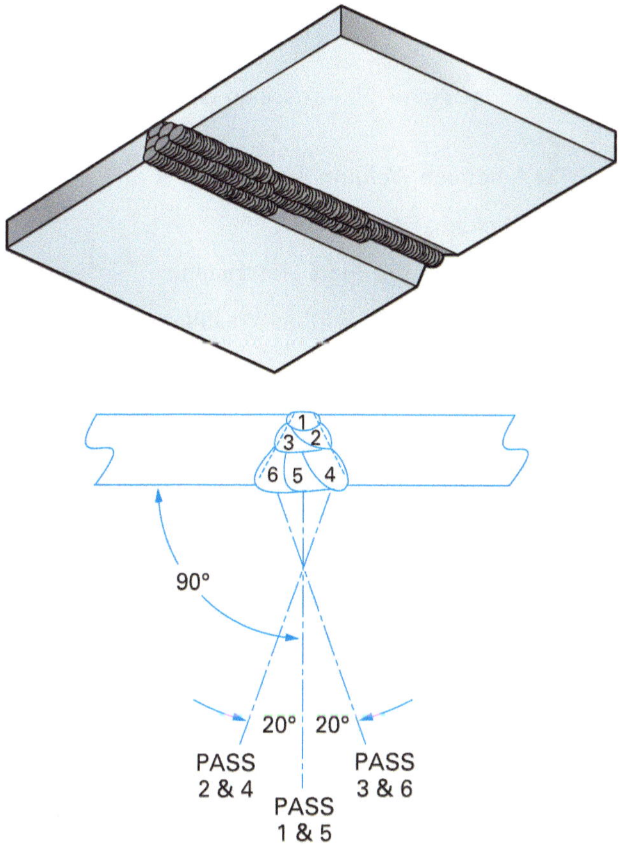

Figure 32 Multiple-pass 4G bead sequence and work angles.

Step 6 Inspect the weld. The weld is acceptable if it has the following features:

- Uniform appearance on the bead face
- Craters and restarts filled to the full cross section of the weld
- Uniform weld width ±1⁄16" (1.6 mm)
- Acceptable weld profile in accordance with the applicable code or standard
- Complete uniform root and face reinforcement at least flush with the base metal to a maximum buildup of 1⁄8" (3.2 mm)
- Smooth, flat transition with complete fusion at the toes of the weld
- No porosity
- No undercut
- No inclusions
- No cracks

Additional Resources

AWS B1.10M/B1.10 Guide for the Nondestructive Examination of Welds. Miami, FL: American Welding Society.

AWS B1.11 Guide for the Visual Examination of Welds. Miami, FL: American Welding Society.

AWS D3.5-93R Guide for Steel Hull Welding. Miami, FL: American Welding Society.

Lincoln Electric website: **http://www.lincolnelectric.com** offers sources for products and training.

OSHA 1910.269, Appendix C, Protection from Step and Touch Potentials. Current Edition. Washington, DC: Occupational Safety & Health Administration (OSHA).

OSHA 1926.351, Arc Welding and Cutting. Current Edition. Washington, DC: Occupational Safety and Health Administration (OSHA).

The Procedure Handbook of Arc Welding. 14th Edition. Cleveland, OH: The James F. Lincoln Arc Welding Foundation, 2000.

Welding Handbook. Volume 1. Welding Science & Technology. Miami, FL: American Welding Society, 2001.

Welding Handbook. Volume 2. Part 1: Welding Processes. Miami, FL: American Welding Society, 2004.

3.0.0 Section Review

1. When flat (1F) fillet welds are being made, the travel speed can be increased or decreased to control the amount of _____.
   a. base metal fusion
   b. weld metal buildup
   c. root penetration
   d. flux-core shielding

2. A 4G weld is a(n) _____.
   a. vertical plate weld
   b. overhead groove weld
   c. flat groove weld
   d. horizontal plate weld

Summary

A welder must possess a number of essential skills in order to perform basic welding jobs and to progress to more difficult welding procedures. These skills include setting up FCAW equipment, preparing the welding work area, running stringer and weave beads, and making acceptable fillet welds and V-groove welds on plate in all positions. The ability to consistently produce acceptable welds requires an understanding of the necessary welding techniques as well as dedicated practice.

Review Questions

1. The filler metal in an FCAW process is provided by _____.
   a. individual rods, or fillets, of steel
   b. a continuous flux-cored electrode
   c. the expansion of molten base metal
   d. the material in the flux-cored wiring

2. The lenses used in safety glasses, face shields, or helmets worn during FCAW procedures must be _____.
   a. made of Plexiglas
   b. properly tinted
   c. spatter resistant
   d. opaque and blued

3. A gas that is commonly used to purge areas or containers prior to FCAW operations is _____.
   a. nitrogen
   b. propane
   c. oxygen
   d. acetylene

4. If possible, practice weld coupons for FCAW should be _____.
   a. ¼" (6.4 mm) thick stainless steel
   b. ⅜" (10 mm) thick carbon steel
   c. ¾" (19.1 mm) thick low-alloy plate
   d. ⅞" (22.2 mm) thick composite steel

5. Setting up a gas-shielded FCAW machine requires connecting the proper shielding gas for the application as well as _____.
   a. locking out and tagging out the primary disconnect
   b. removing any liner material from the feeder and gun
   c. attaching the ground lead to the gun handle
   d. loading the wire feeder and gun with the proper electrode

6. After a welder has made the initial recommended setting, the welding amperage of an FCAW system is typically controlled by the _____.
   a. wire feed speed control
   b. arc length and porosity
   c. shielding gas flow regulator
   d. gun position and travel

7. When weave beads are made, the motion of the electrode is _____.
   a. quickly forward and backward
   b. slightly side-to-side
   c. widely side-to-side
   d. slowly circular and forward

8. An improperly made FCAW restart will create a _____.
   a. bead overlap
   b. weave bead
   c. stringer bead
   d. weld discontinuity

9. Unless the proper technique is used, a weld termination normally leaves _____.
   a. a crack
   b. porosity
   c. spatter
   d. a crater

10. Weld beads that are used to build up a surface and to make multiple-pass welds are known as _____.
    a. fillet beads
    b. layer beads
    c. crater beads
    d. overlapping beads

11. The most common fillet welds are made in _____.
    a. horizontal piping
    b. root passes
    c. lap and T-joints
    d. vertical piping

12. What work angle should be used for the first bead in a horizontal (2F) fillet weld?
    a. 25 degrees
    b. 45 degrees
    c. 60 degrees
    d. 90 degrees

13. With FCAW, groove welds can be made in _____.
    a. only the 1G and 2G positions
    b. only the 2G and 3G positions
    c. only the 1G, 2G, and 3G positions
    d. all positions

14. A flat position open V-groove weld is acceptable if it has complete fusion at the toes of the weld and a smooth, flat _____.
    a. cross section
    b. transition
    c. overlap
    d. crater

15. When making a 3G open V-groove weld with weave beads, use a zigzag pattern across the root bead and pause at each toe _____.
    a. for complete penetration
    b. to prevent excessive spatter
    c. for good fusion
    d. to fill in the crater

PERFORMANCE ACCREDITATION TASKS

The Performance Accreditation Tasks (PATs) correspond to and support the learning objectives in *AWS EG2.0, Curriculum Guide for the Training of Welding Personnel: Level 1 – Entry Welder.*

PATs provide specific acceptable criteria for performance and help to ensure a true competency-based welding program for students.

The following tasks are designed to evaluate your ability to run fillet and groove welds with FCAW equipment. Perform each task when you are instructed to do so by your instructor. As you complete each task, show it to your instructor for evaluation. Do not proceed to the next task until instructed to do so by your instructor.

Have trainees complete PAT 1 through PAT 8, according to the acceptance criteria.

PAT 1, Make a Fillet Weld in the (1F) Flat Position. This task corresponds to *AWS EG2.0, Module 6, Flux-Cored Arc Welding (FCAW-G, FCAW)*, Key Indicators 5, 7, 10, and 12.

PAT 2, Make a Fillet Weld in the (2F) Horizontal Position. This task corresponds to *AWS EG2.0, Module 6, Flux-Cored Arc Welding (FCAW-G, FCAW)*, Key Indicators 5, 7, 10, and 12.

PAT 3, Make a Fillet Weld in the (3F) Vertical Position. This task corresponds to *AWS EG2.0, Module 6, Flux-Cored Arc Welding (FCAW-G, FCAW)*, Key Indicators 5, 7, 10, and 12.

PAT 4, Make a Fillet Weld in the (4F) Overhead Position. This task corresponds to *AWS EG2.0, Module 6, Flux-Cored Arc Welding (FCAW-G, FCAW)*, Key Indicators 5, 7, 10, and 12.

PAT 5, Make a Groove Weld, with or without Backing, in the (1G) Flat Position. This task corresponds to *AWS EG2.0, Module 6, Flux-Cored Arc Welding (FCAW-G, FCAW)*, Key Indicators 6 and 11.

PAT 6, Make a Groove Weld, with or without Backing, in the (2G) Horizontal Position. This task corresponds to *AWS EG2.0, Module 6, Flux-Cored Arc Welding (FCAW-G, FCAW)*, Key Indicators 6 and 11.

PAT 7, Make a Groove Weld, with or without Backing, in the (3G) Vertical Position. This task corresponds to *AWS EG2.0, Module 6, Flux-Cored Arc Welding (FCAW-G, FCAW)*, Key Indicators 6 and 11.

PAT 8, Make a Groove Weld, with or without Backing, in the (4G) Overhead Position. This task corresponds to *AWS EG2.0, Module 6, Flux-Cored Arc Welding (FCAW-G, FCAW)*, Key Indicators 6 and 11.

MAKE A FILLET WELD IN THE (1F) FLAT POSITION

As directed by the instructor, use FCAW with flux-cored electrodes and, if required, the appropriate shielding gas to make a six-pass fillet weld using stringer beads on carbon steel plate, as shown.

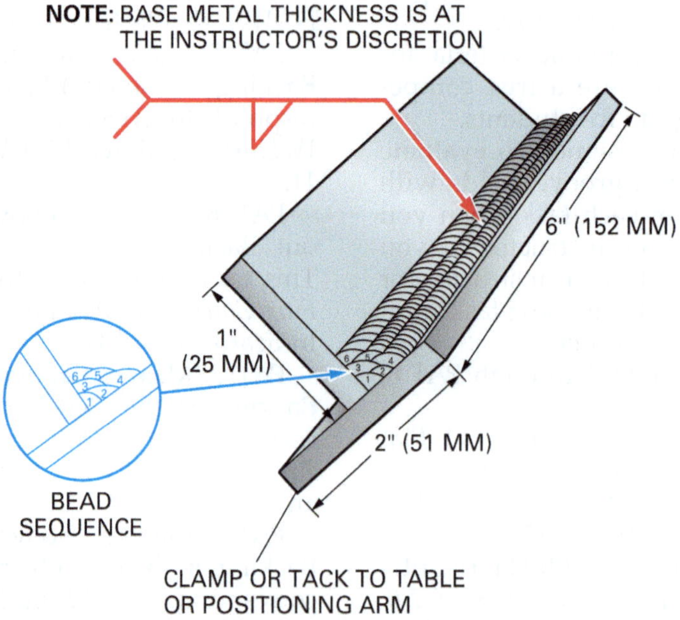

Criteria for Acceptance:

- Uniform rippled appearance on the bead face ____________
- Craters and restarts filled to the full cross section of the weld ____________
- Uniform weld width ±1⁄16" (1.6 mm) ____________
- Smooth flat transition with complete fusion at the toes of the welds ____________
- Acceptable weld profile in accordance with the applicable code or standard ____________
- No porosity ____________
- No undercut ____________
- No overlap ____________
- No inclusions ____________

MAKE A FILLET WELD IN THE (2F) HORIZONTAL POSITION

As directed by the instructor, use FCAW with flux-cored electrodes and, if required, the appropriate shielding gas to make a six-pass fillet weld using stringer beads on carbon steel plate, as shown.

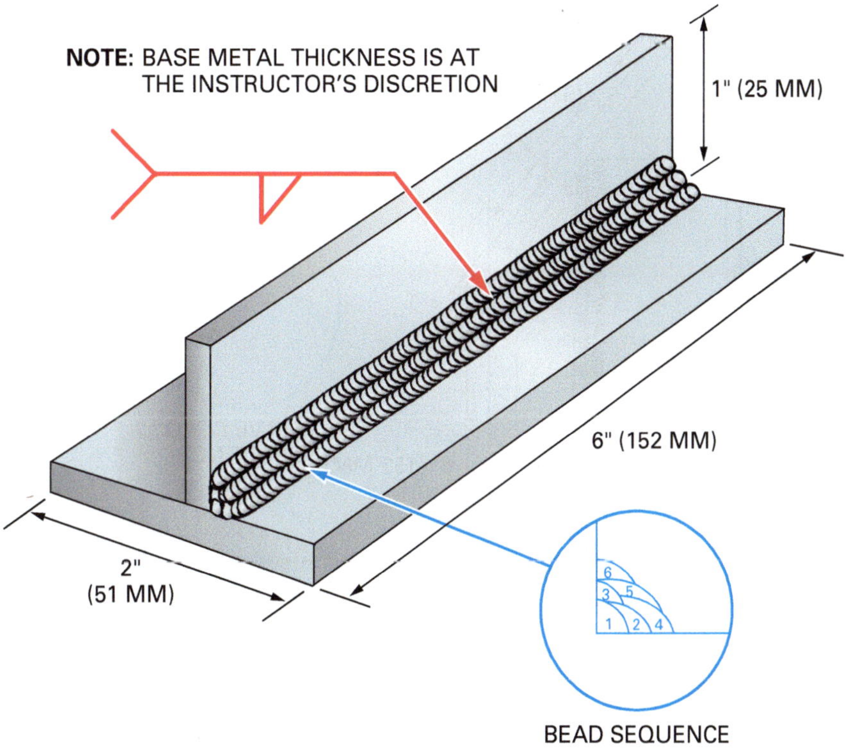

Criteria for Acceptance:

- Uniform rippled appearance on the bead face ____________
- Craters and restarts filled to the full cross section of the weld ____________
- Uniform weld width ±1⁄16" (1.6 mm) ____________
- Smooth flat transition with complete fusion at the toes of the welds ____________
- Acceptable weld profile in accordance with the applicable code or standard ____________
- No porosity ____________
- No undercut ____________
- No overlap ____________
- No inclusions ____________

MAKE A FILLET WELD IN THE (3F) VERTICAL POSITION

As directed by the instructor, use FCAW with flux-cored electrodes and, if required, the appropriate shielding gas to make a vertical fillet weld on carbon steel plate, as shown.

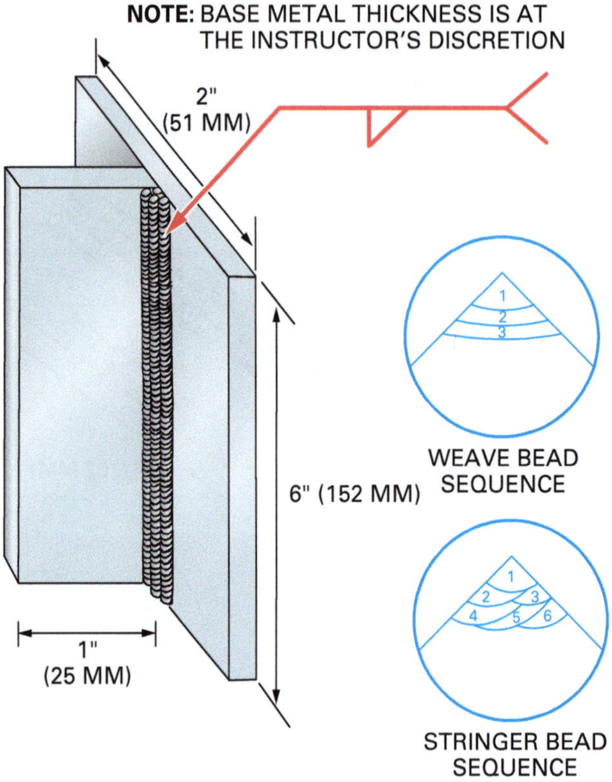

Criteria for Acceptance:

- Uniform rippled appearance on the bead face ________
- Craters and restarts filled to the full cross section of the weld ________
- Uniform weld width ±1⁄16" (1.6 mm) ________
- Acceptable weld profile in accordance with the applicable code or standard ________
- Smooth flat transition with complete fusion at the toes of the weld ________
- No porosity ________
- No undercut ________
- No inclusions ________
- No cracks ________

MAKE A FILLET WELD IN THE (4F) OVERHEAD POSITION

As directed by the instructor, use FCAW with flux-cored electrodes and, if required, appropriate shielding gas to make a six-pass fillet weld using stringer beads on carbon steel plate, as shown.

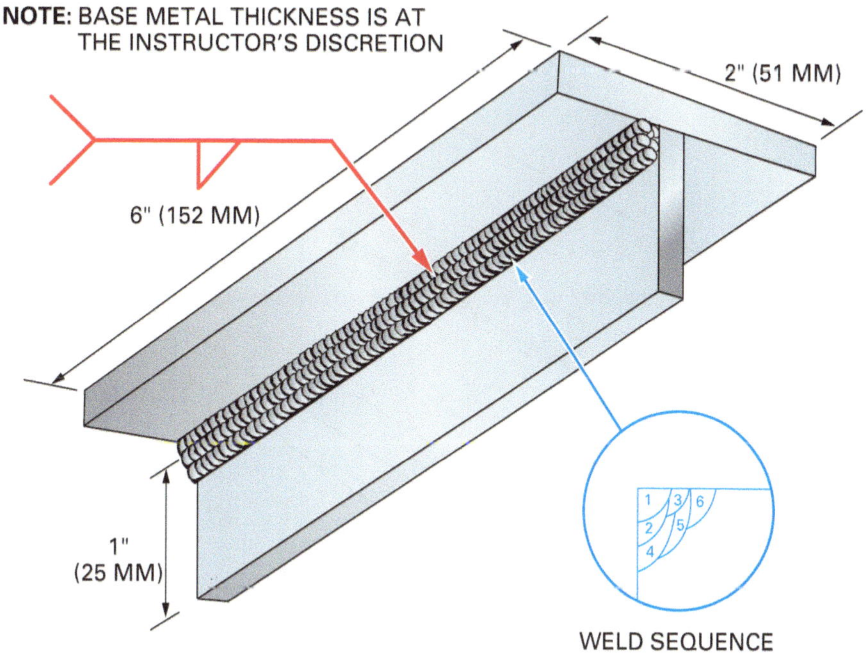

Criteria for Acceptance:

- Uniform rippled appearance on the bead face ____________
- Craters and restarts filled to the full cross section of the weld ____________
- Uniform weld width ±1⁄16" (1.6 mm) ____________
- Acceptable weld profile in accordance with the applicable code or standard ____________
- Smooth flat transition with complete fusion at the toes of the weld ____________
- No porosity ____________
- No undercut ____________
- No inclusions ____________
- No cracks ____________

MAKE A GROOVE WELD, WITH OR WITHOUT BACKING, IN THE (1G) FLAT POSITION

As directed by the instructor, use FCAW with flux-cored electrodes and, if required, the appropriate shielding gas to make a multiple-pass groove weld using stringer beads on carbon steel plate, with or without backing, as shown.

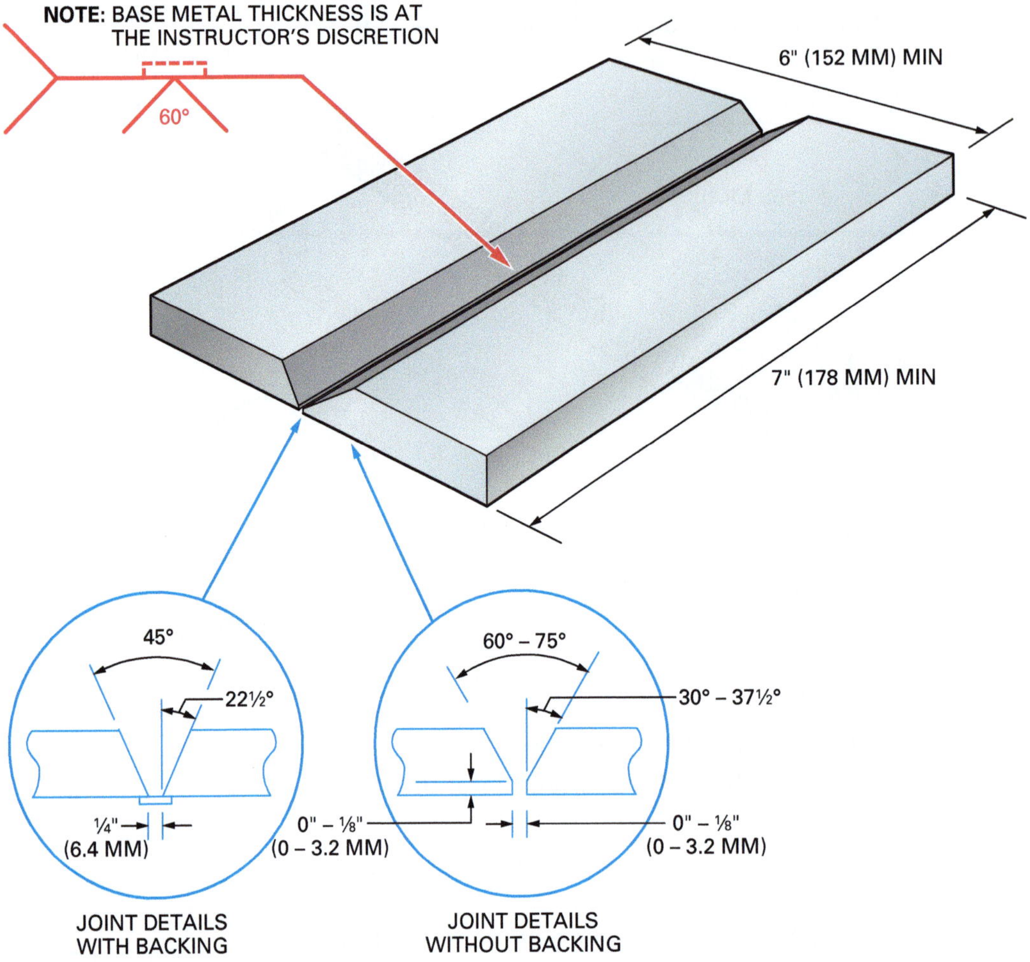

Criteria for Acceptance:

- Uniform rippled appearance on the bead face __________
- Craters and restarts filled to the full cross section of the weld __________
- Uniform weld width ±1⁄16" (1.6 mm) __________
- Acceptable weld profile in accordance with the applicable code or standard __________
- Smooth transition with complete fusion at the toes of the weld __________
- Complete joint penetration with uniform root and face reinforcement at least flush with the base metal to a maximum buildup of ⅛" (3.2 mm), if applicable __________
- No porosity __________
- No undercut __________
- No inclusions __________
- No cracks __________

MAKE A GROOVE WELD, WITH OR WITHOUT BACKING, IN THE (2G) HORIZONTAL POSITION

As directed by the instructor, use FCAW with flux-cored electrodes and, if required, the appropriate shielding gas to make a multiple-pass groove weld on carbon steel plate, with or without backing, as shown.

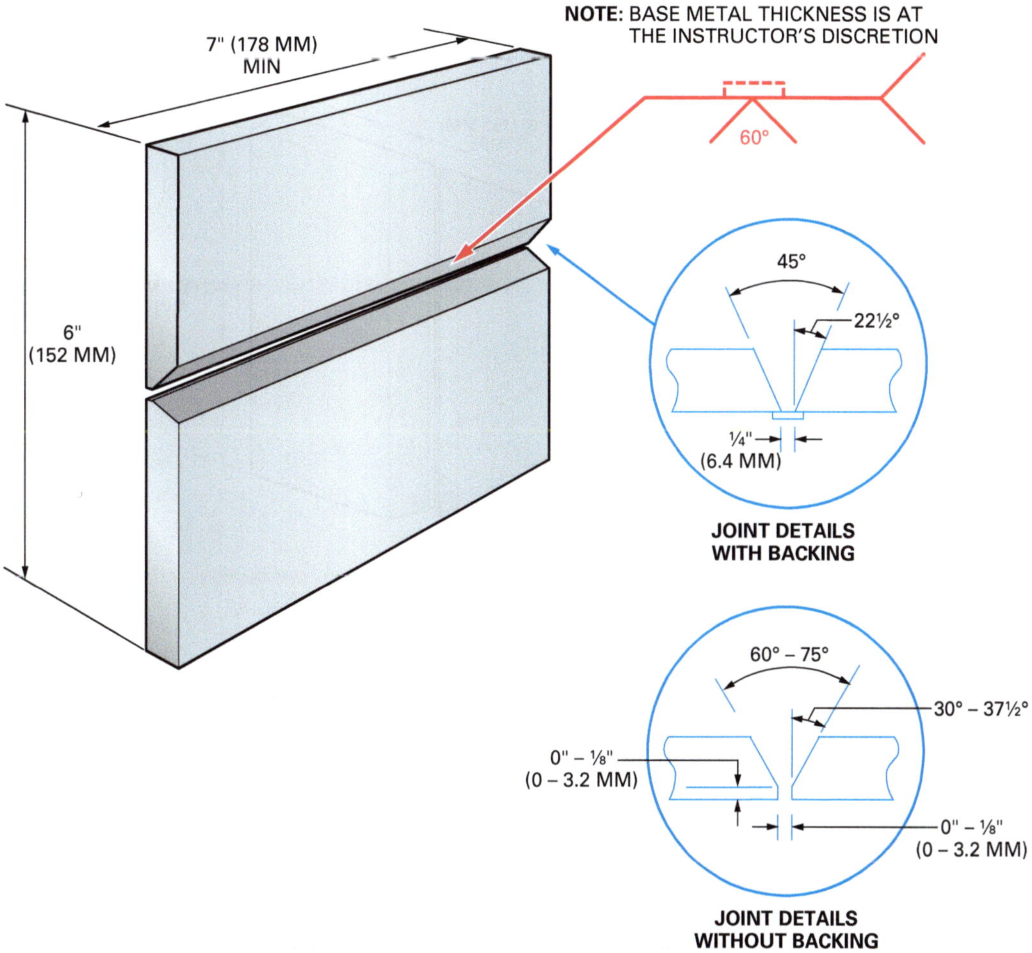

Criteria for Acceptance:

- Uniform rippled appearance on the bead face ________
- Craters and restarts filled to the full cross section of the weld ________
- Uniform weld width ±1⁄16" (1.6 mm) ________
- Acceptable weld profile in accordance with the applicable code or standard ________
- Complete joint penetration with uniform root and face reinforcement at least flush with the base metal to a maximum buildup of ⅛" (3.2 mm), if applicable ________
- Smooth transition with complete fusion at the toes of the weld ________
- No porosity ________
- No undercut ________
- No inclusions ________
- No cracks ________

MAKE A GROOVE WELD, WITH OR WITHOUT BACKING, IN THE (3G) VERTICAL POSITION

As directed by the instructor, use FCAW with flux-cored electrodes and, if required, the appropriate shielding gas to make a multiple-pass groove weld on carbon steel plate, with or without backing, as shown.

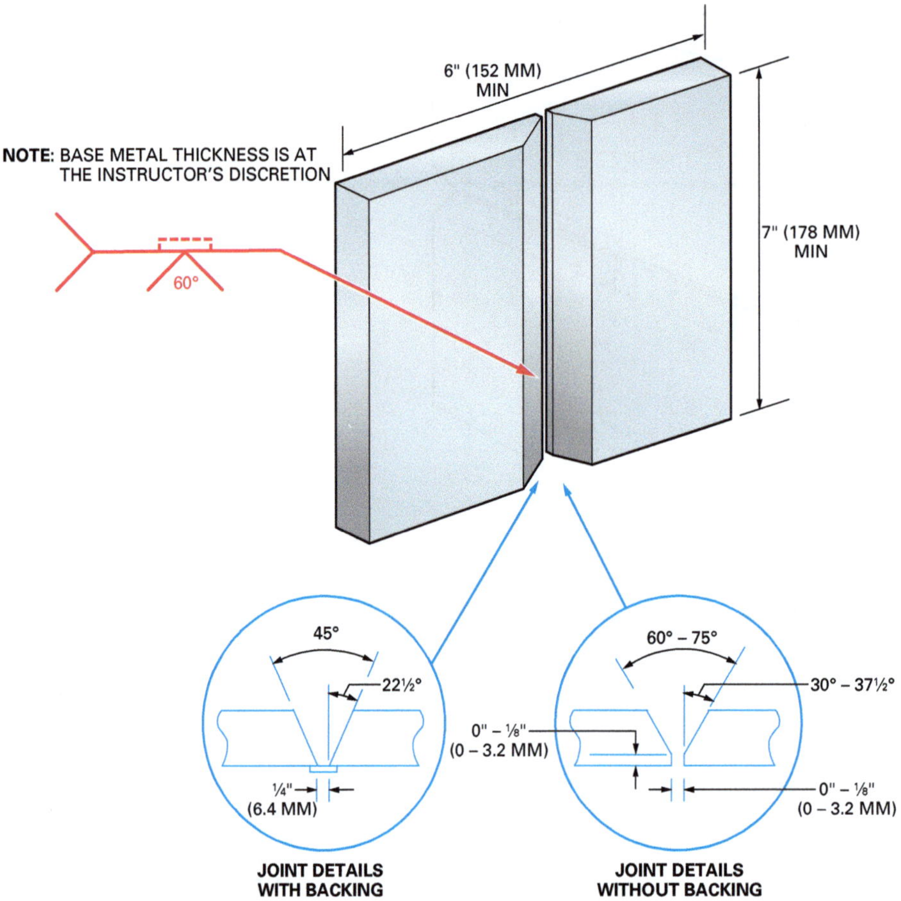

Criteria for Acceptance:

- Uniform rippled appearance on the bead face ____________
- Craters and restarts filled to the full cross section of the weld ____________
- Uniform weld width ±1⁄16" (1.6 mm) ____________
- Acceptable weld profile in accordance with the applicable code or standard ____________
- Complete joint penetration with uniform root and face reinforcement at least flush with the base metal to a maximum buildup of ⅛" (3.2 mm), if applicable ____________
- Smooth transition with complete fusion at the toes of the weld ____________
- No porosity ____________
- No undercut ____________
- No inclusions ____________
- No cracks ____________

MAKE A GROOVE WELD, WITH OR WITHOUT BACKING, IN THE (4G) OVERHEAD POSITION

As directed by the instructor, use FCAW with flux-cored electrodes and, if required, the appropriate shielding gas to make a multiple-pass groove weld on carbon steel plate, with or without backing, as shown.

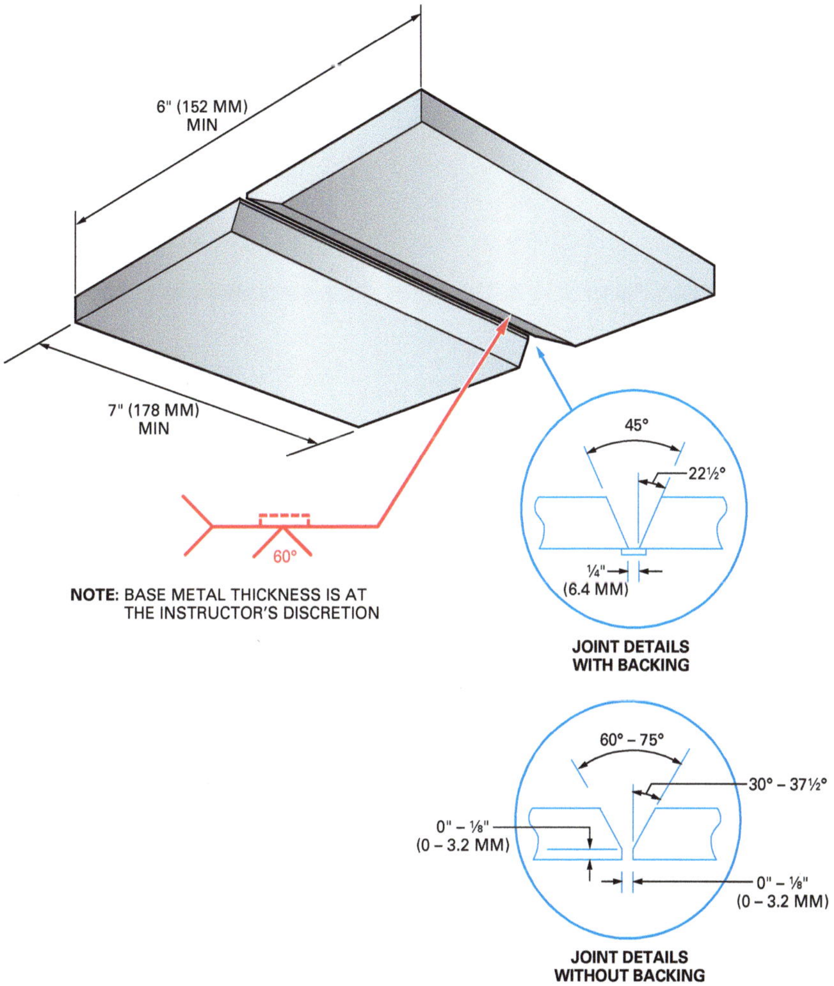

Criteria for Acceptance:

- Uniform rippled appearance on the bead face ________
- Craters and restarts filled to the full cross section of the weld ________
- Uniform weld width ±1⁄16" (1.6 mm) ________
- Acceptable weld profile in accordance with the applicable code or standard ________
- Complete joint penetration with uniform root and face reinforcement at least flush with the base metal to a maximum buildup of ⅛" (3.2 mm), if applicable ________
- Smooth transition with complete fusion at the toes of the weld ________
- No porosity ________
- No undercut ________
- No inclusions ________
- No cracks ________

Trade Terms Introduced in This Module

Degassing: The escape of gases formed during the welding process that can cause imperfections in the weld.

Synergic: Welding technology that controls voltage, current, and wire feed using a program optimized for each welding situation.

Trim: A control found on synergic welding machines that is similar to the voltage control on regular machines and acts as a kind of override to the program.

Additional Resources

This module is intended as a thorough resource for task training. The following reference works are suggested for further study.

AWS B1.10M/B1.10 Guide for the Nondestructive Examination of Welds. Miami, FL: American Welding Society.

AWS B1.11 Guide for the Visual Examination of Welds. Miami, FL: American Welding Society.

AWS D3.5-93R Guide for Steel Hull Welding. Miami, FL: American Welding Society.

Lincoln Electric website: **http://www.lincolnelectric.com** offers sources for products and training.

OSHA 1910.269, Appendix C, Protection from Step and Touch Potentials. Current Edition. Washington, DC: Occupational Safety & Health Administration (OSHA).

OSHA 1926.351, Arc Welding and Cutting. Current Edition. Washington, DC: Occupational Safety and Health Administration (OSHA).

The Procedure Handbook of Arc Welding. 14th Edition. Cleveland, OH: The James F. Lincoln Arc Welding Foundation, 2000.

Welding Handbook. Volume 1. Welding Science & Technology. Miami, FL: American Welding Society, 2001.

Welding Handbook. Volume 2. Part 1: Welding Processes. Miami, FL: American Welding Society, 2004.

Figure Credits

The Lincoln Electric Company, Cleveland, OH, USA, Module opener, Figures 2, 3, SA02, SA03, Figure 17

Terry Lowe, Figures 4, 29– 31 (photos)

Dan Sterry, TIC, Figure 12

Holley Thomas, SME, Figure 21

Topaz Publications, Inc., SA01, SA05

Reproduced by permission TWI Ltd., SA04

Section Review Answer Key

Answer	Section Reference	Objective
Section One		
1. a	1.1.1	1a
2. d	1.2.2	1b
3. b	1.3.1	1c
Section Two		
1. c	2.1.1	2a
2. d	2.2.2	2b
Section Three		
1. b	3.1.1	3a
2. b	3.2.2	3b

NCCER CURRICULA — USER UPDATE

NCCER makes every effort to keep its textbooks up-to-date and free of technical errors. We appreciate your help in this process. If you find an error, a typographical mistake, or an inaccuracy in NCCER's curricula, please fill out this form (or a photocopy), or complete the online form at **www.nccer.org/olf**. Be sure to include the exact module ID number, page number, a detailed description, and your recommended correction. Your input will be brought to the attention of the Authoring Team. Thank you for your assistance.

Instructors – If you have an idea for improving this textbook, or have found that additional materials were necessary to teach this module effectively, please let us know so that we may present your suggestions to the Authoring Team.

NCCER Product Development and Revision

13614 Progress Blvd., Alachua, FL 32615

Email: curriculum@nccer.org
Online: www.nccer.org/olf

❑ Trainee Guide ❑ Lesson Plans ❑ Exam ❑ PowerPoints Other ____________________

Craft / Level: Copyright Date:

Module ID Number / Title:

Section Number(s):

Description:

Recommended Correction:

Your Name:

Address:

Email: Phone:

GTAW – Equipment and Filler Metals

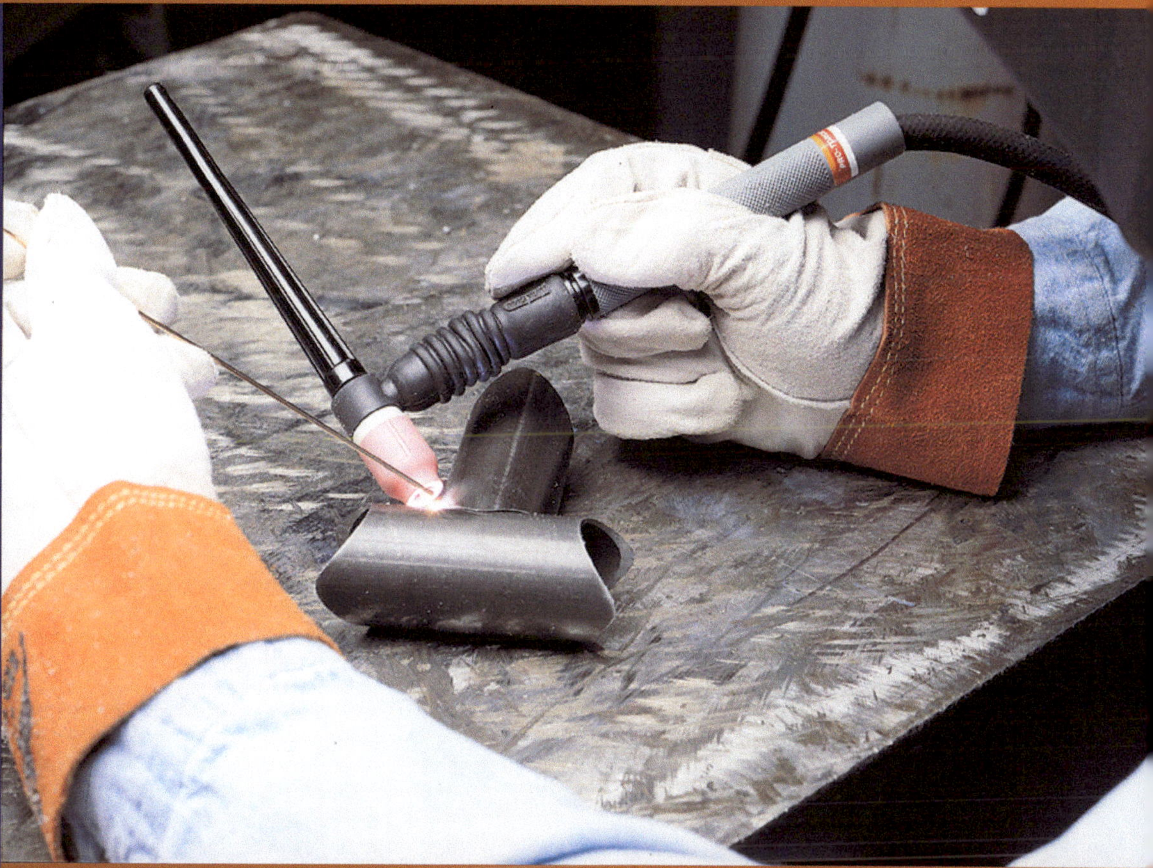

Overview

To produce consistent, high-quality welds, welders must select, install, and use the equipment and materials involved in the GTAW process properly. This module provides an overview of the equipment and safety concerns that are associated with GTAW. It describes GTAW equipment and consumables, including power sources, torches, nozzles, electrodes, shielding gases, and filler metals. It also explains how to select, prepare, and set up GTAW equipment safely and efficiently.

Module 29207

Trainees with successful module completions may be eligible for credentialing through the NCCER Registry. To learn more, go to **www.nccer.org** or contact us at **1.888.622.3720**. Our website has information on the latest product releases and training, as well as online versions of our *Cornerstone* magazine and Pearson's product catalog.

Your feedback is welcome. You may email your comments to **curriculum@nccer.org**, send general comments and inquiries to **info@nccer.org**, or fill in the User Update form at the back of this module.

This information is general in nature and intended for training purposes only. Actual performance of activities described in this manual requires compliance with all applicable operating, service, maintenance, and safety procedures under the direction of qualified personnel. References in this manual to patented or proprietary devices do not constitute a recommendation of their use.

29207 V5

29207
GTAW – Equipment and Filler Materials

Objectives

When you have completed this module, you will be able to do the following:

1. Identify GTAW-related safety practices and describe the electrical characteristics that affect GTAW.
   a. Identify GTAW-related safety practices.
   b. Describe the electrical characteristics that affect GTAW.
2. Identify and describe GTAW equipment and consumables.
   a. Identify and describe GTAW welding machines.
   b. Identify and describe GTAW torches.
   c. Identify and describe GTAW torch nozzles and electrodes.
   d. Identify and describe GTAW shielding gases.
   e. Identify and describe GTAW filler metals.
3. Explain how to set up for GTAW welding.
   a. Explain how to select and position the welding machine.
   b. Explain how to connect and set up the shielding gas flow rate.
   c. Explain how to select and prepare the tungsten electrode.
   d. Explain how to select and install the nozzle along with the tungsten electrode.

Performance Tasks

Under the supervision of your instructor, you should be able to do the following:

1. Select a GTAW shielding gas.
2. Select a GTAW filler metal.
3. Connect the shielding gas and set the flow rate.
4. Select and prepare the tungsten electrode.
5. Break down and reassemble a GTAW torch.

Trade Terms

Lift-arc technology
Potentiometer
Truncated
Tungsten

Industry Recognized Credentials

If you are training through an NCCER-accredited sponsor, you may be eligible for credentials from NCCER's Registry. The ID number for this module is 29207. Note that this module may have been used in other NCCER curricula and may apply to other level completions. Contact NCCER's Registry at 888.622.3720 or go to **www.nccer.org** for more information.

Contents

Topics to be presented in this module include:

Figures and Tables

Figures and Tables (Continued)

SECTION ONE

1.0.0 GTAW

Objective

Identify GTAW-related safety practices and describe the electrical characteristics that affect GTAW.

a. Identify GTAW-related safety practices.
b. Describe the electrical characteristics that affect GTAW.

Trade Terms

Lift-arc technology: A type of GTAW technology in which the tungsten electrode may be touched to the work, but the arc will not strike until the tip is lifted away to an appropriate distance.

Truncated: Having the pointed top (vertex) of a conical shape cut off, leaving only a plane section parallel to the base of the cone.

Tungsten: A nonstandard common name for the tungsten electrode used in GTAW; also, the principal metal from which the electrodes are constructed.

Gas **tungsten** arc welding (GTAW) is an arc welding process that uses a welding power source to produce an electric arc between a tungsten electrode and the base metal. The arc melts and fuses the base metal to form the weld. A filler metal is typically added. In processes where a filler metal is not added, it is referred to as autogenous welding. In manual welding (*Figure 1*), the filler metal is usually a handheld metal rod or wire that is fed into the leading edge of the weld metal pool. An inert gas or gas mixture is used to shield the electrode and molten weld metals to prevent oxidation and contamination from the atmosphere. The finished weld has no slag because no flux is used. Also, since no filler metal is transported within the arc, the process produces little or no spatter.

GTAW is used to weld a variety of metals. The process is slower than gas metal arc welding (GMAW); however, it produces high-quality welds. GTAW has been commonly used for root pass welds on pipe where the subsequent fill and cover passes are performed with SMAW, GMAW, or FCAW. In the past, GTAW was called tungsten inert gas (TIG) welding. Today, the nonstandard term TIG is still used, including in the control nomenclature marked on some welding machines.

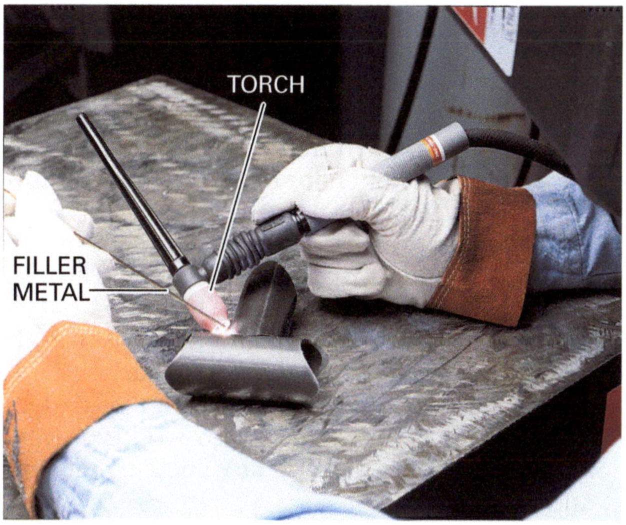

Figure 1 Performing manual GTAW.

1.1.0 Safety Summary

The following is a summary of safety procedures and practices that welders must observe when cutting or welding. Keep in mind that this is only a summary. Complete safety coverage is provided in the Welding Level One module, *Welding Safety*. That module should be completed before continuing with this one. Above all, be sure to wear appropriate protective clothing and equipment when welding or cutting.

1.1.1 Protective Clothing and Equipment

Welding activities can cause injuries unless you wear all of the protective clothing and equipment that is designed specifically for the welding industry. The following information includes important safety guidelines about protective clothing and equipment:

- Wear a face shield over snug-fitting cutting goggles or safety glasses for gas welding or cutting. Either the face shield or the lenses of the welding goggles must be an approved shade for the application. A welding hood equipped with a properly tinted lens is best for all forms of welding.
- Wear proper protective leather and/or flame-retardant clothing along with welding gloves that protect the welder from flying sparks, molten metal, and heat.

- Wear high-top safety shoes or boots. Make sure that the tongue and lace area of the footwear will be covered by a pant leg. If the tongue and lace area is exposed or the footwear must be protected from burn marks, wear leather spats under the pants or chaps over the front and top of the footwear.
- Wear a 100-percent cotton cap with no mesh material included in its construction. The bill of the cap points to the rear. If a hard hat is required for the environment, use one that allows the attachment of rear deflector material and a face shield. A hard hat with a rear deflector is generally preferred when working overhead, and may be required by some employers and job sites.
- Wear earmuffs or at least earplugs to protect your ear canals from sparks.

WARNING!

Using proper personal protective equipment (PPE) for the hands and eyes is particularly important. The most common injuries that welders experience during GTAW operations are injuries to the fingers and eyes. Welders who fail to use proper gloves, safety glasses, and face shields when they are holding and grinding the tungsten electrode by hand can cut or burn their fingers, have splintered tungsten electrode lodge in their hands or fingers, or get small slivers of the electrode stuck in their eyes.

1.1.2 Fire/Explosion Prevention

Welding activities usually involve the use of fire or extreme heat to melt metal. Whenever fire is used, it must be controlled and contained. Welding or cutting activities are often performed on vessels that may once have contained flammable or explosive materials. Residues from those materials can catch fire or explode when a welder begins work on such a vessel. The following are fire and explosion prevention guidelines associated with welding:

- Never carry matches or gas-filled lighters in your pockets. Sparks can cause the matches to ignite or the lighter to explode, causing serious injury.
- Never use oxygen to blow dust or dirt from clothing. The oxygen can remain trapped in the fabric for a time. If a spark hits clothing during this time, the clothing can burn rapidly and violently out of control.
- Make sure that any flammable material in the work area is moved or shielded by a fire-resistant covering. Approved fire extinguishers must be available before attempting any heating, welding, or cutting operations. If a hot-work permit and a fire watch are required, be sure those items are in place before beginning and ensure that all site requirements have been observed.
- Never release a large amount of oxygen or use oxygen as compressed air. The presence of oxygen around flammable materials or sparks can cause rapid and uncontrolled combustion. Keep oxygen away from oil, grease, and other petroleum products.
- Never release a large amount of fuel gas, especially acetylene. Propane tends to concentrate in and along low areas and can ignite at a considerable distance from the release point. Acetylene is lighter than air, but is even more dangerous than propane. When mixed with air or oxygen, acetylene will explode at much lower concentrations than any other fuel.
- To prevent fires, maintain a neat and clean work area, and make sure that any metal scrap or slag is cold before disposing of it.
- Before cutting containers such as tanks or barrels, check to see if they have contained any explosive, hazardous, or flammable materials, including petroleum products, citrus products, or chemicals that decompose into toxic fumes when heated. As a standard practice, always clean and then fill any tanks or barrels with water, or purge them with a flow of inert gas such as nitrogen to displace any oxygen. Containers must be cleaned by steam cleaning, flushing with water, or washing with detergent until all traces of the material have been removed.

WARNING!

Welding or cutting must never be performed on drums, barrels, tanks, vessels, or other containers until they have been emptied and cleaned thoroughly, eliminating all flammable materials and all substances (such as detergents, solvents, greases, tars, or acids) that might produce flammable, toxic, or explosive vapors when heated. Clean containers only in well-ventilated areas. Vapors can accumulate during cleaning, causing explosions or injury.

Proper procedures for cutting or welding hazardous containers are described in the *American Welding Society (AWS) F4.1, Safe Practices for the Preparation of Containers and Piping for Welding and Cutting.*

1.1.3 Work Area Ventilation

Vapors and fumes tend to rise in the air from their sources. Welders often have to work above the welding area where the fumes are being created. Welding fumes can cause personal injuries. Good work area ventilation helps to remove the vapors and to protect the welder. The following is a list of work area ventilation guidelines to consider before and during welding activities:

- Follow confined space procedures before conducting any welding or cutting in a confined space.
- Never use oxygen for ventilation in confined spaces.
- Always perform cutting or welding operations in a well-ventilated area. Cutting or welding operations involving zinc or cadmium materials or coatings result in toxic fumes. For long-term cutting or welding of such materials, always wear an approved full-face, supplied-air respirator that uses breathing air supplied from outside of the work area. For occasional, very short-term exposure, a HEPA-rated or metal-fume filter may be used on a standard respirator.
- Make sure confined spaces are properly ventilated for cutting or welding purposes. Use powered extraction systems when available.

1.1.4 GTAW-Specific Safety

GTAW, also referred to as heliarc or TIG welding, is a process that is different from the processes used in shielded metal arc welding (SMAW), GMAW, and flux-cored arc welding (FCAW). In other forms of welding, the electrode or wire is consumed in the welding process. In GTAW, the electrode itself is not consumed, but a separate rod of filler metal is consumed.

Vapors from the melted base metal, the filler metal, and from the shielding gases used in the GTAW process can create respiratory issues. As always, welders performing any kind of work must be aware of the vapors generated from the welding processes. Argon gas is often used for shielding in the GTAW process. A combination of the arc and the gas generates substantial levels of ultraviolet (UV) light, which reacts with the oxygen in the area of the weld. UV radiation can ionize oxygen in the atmosphere near the weld, which can lead to a rise in ozone and nitrogen dioxide levels. As long as the work is being performed in well-ventilated areas, the risks are minimal. If work must be performed in confined spaces, consider using additional ventilation or respirators. Follow your employer's policies in this matter.

Because GTAW work is often performed on delicate welds, the welder is frequently positioned very close to the work. The risk of UV radiation being emitted from the electrical arc is higher because the shielding gas used with GTAW is more transparent than gases used with other forms of welding. This transparency is due to the minimal smoke produced from the weld. Because of the brightness of the GTAW process, safety representatives recommend that the welder wear darker clothing to minimize UV reflections under the welding helmet. They also recommend that the walls of the work area be painted with a paint that reduces reflections from the UV light generated by the GTAW process. Because aluminum and stainless steel reflect more light than common carbon steels, welders need to take extra precautions to protect their vision when performing GTAW work on these metals.

The GTAW process requires the base metal and the filler metal to be extremely clean. Cleaning solvents are used to prepare the metals. Welders must read all Safety Data Sheet (SDS) information about any cleaning agent before using it. Some cleaning agents are flammable, while others are nonflammable. Make sure that you know which kind you are using. Pay special attention to any storage requirements for such cleaning agents.

WARNING!

Chlorinated cleaning solvents can produce toxic gases. For this reason, they should never be used for cleaning metals in a welding setting.

The tungsten electrodes used for GTAW become contaminated over time and with use. When electrodes lose their original shapes, welders must grind the tungsten tips to reshape them and make them suitable for continued use. The tungsten dust and flying particles from these grinding activities can damage the welder's vision and present a respiratory hazard. Read and understand the SDS that comes with each batch of electrodes.

1.2.0 Welding Current

The welding arc required for GTAW is produced by electrical current from a constant-current (CC) welding machine. Electrical current is the flow of electrons along a conductor. The current flows through two welding cables, one connected to the workpiece and the other connected to the torch, which holds a tungsten electrode.

An arc is established in the gap between the end of the tungsten electrode and the work. The gap has a high electrical resistance. When the electrical current meets resistance, heat is generated. When the arc is struck across the gap, the resistance generates intense heat of 6,000°F to 10,000°F (≈3,300°C to 5,500°C), melting the base metal and forming a weld. The area surrounding the arc is flooded with a shielding gas to prevent oxidation of the tungsten electrode, filler rod, and base metal surface during the welding process. *Figure 2* shows the GTAW welding process.

1.2.1 Characteristics of Welding Current

The current produced by a welding machine to perform welding has different characteristics than the current flowing through utility power lines. Welding current has low voltage and high amperage, while the power line current has high voltage and low amperage.

Voltage is the measure of the electromotive force or pressure that causes current to flow in a circuit. There are two voltage levels associated with welding current: open-circuit voltage and operating voltage. Open-circuit voltage is the voltage that is present when the machine is on but no welding is taking place. For GTAW, there are usually ranges of open-circuit voltages that can be selected, up to about 80V. Operating voltage, or arc voltage, is the voltage after the arc has been struck. With GTAW, arc voltage is generally much lower than the open-circuit voltage. The arc voltage drops rapidly with increasing current flow.

Amperage is the electrical current flow in a circuit. The unit of measurement for amperage is the ampere, or amp. The number of amps produced by the welding machine determines the intensity of the arc and the amount of heat generated. The higher the amperage, the hotter the arc. Usually, welders use a finger control or a foot pedal to control the GTAW welding machine current.

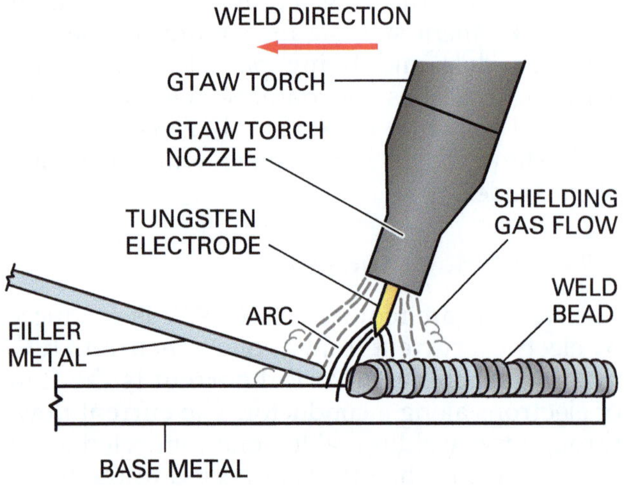

Figure 2 Diagram of the GTAW welding process.

1.2.2 DC Welding Current

Direct current (DC) is electrical current that flows in one direction only. The direction in which the current is flowing is called the polarity.

The direction in which the current flows in a circuit determines polarity. Direct current electrode negative (DCEN), also referred to by the nonstandard term *straight polarity*, is almost always used for performing GTAW on ferrous and nonferrous metals.

When the welding circuit is set up as DCEN, the tungsten electrode of the GTAW torch is connected to the negative terminal of the welding machine and the workpiece lead is connected to the positive terminal. DCEN produces the greatest amount of heat in the base metal, compared to other types of welding current. About two-thirds of the arc heat goes to the base metal and one-third goes to the tungsten electrode. Welders usually use a tungsten electrode with a pointed end that is slightly **truncated**, or blunted, for DCEN. The weld bead produced is narrow but with deep penetration. DCEN is used to weld steels, stainless steels, nickel, and titanium. Because DCEN provides limited cleaning action, the base metal surfaces must first be thoroughly cleaned of any contaminants such as rust, mill scale, or oxides.

Direct current electrode positive (DCEP), also referred to by the nonstandard term *reverse polarity*, produces the least amount of heat in the base metal. About one-third of the arc heat goes to the base metal and two-thirds goes to the tungsten electrode. Because of this, DCEP usually requires a large tungsten electrode with a rounded end. Unlike DCEN, the weld bead produced is wide but with shallow penetration. This current type also has a strong cleaning action on the surface of oxidized base metals. DCEP is only used for special applications in which shallow penetration is required. *Figure 3* shows typical DCEN and DCEP welding circuits.

Polarity Switches

Many welding machines have a polarity switch. Instead of physically having to disconnect the welding leads and reconnect them to change polarity, the welder changes polarity by turning a switch.

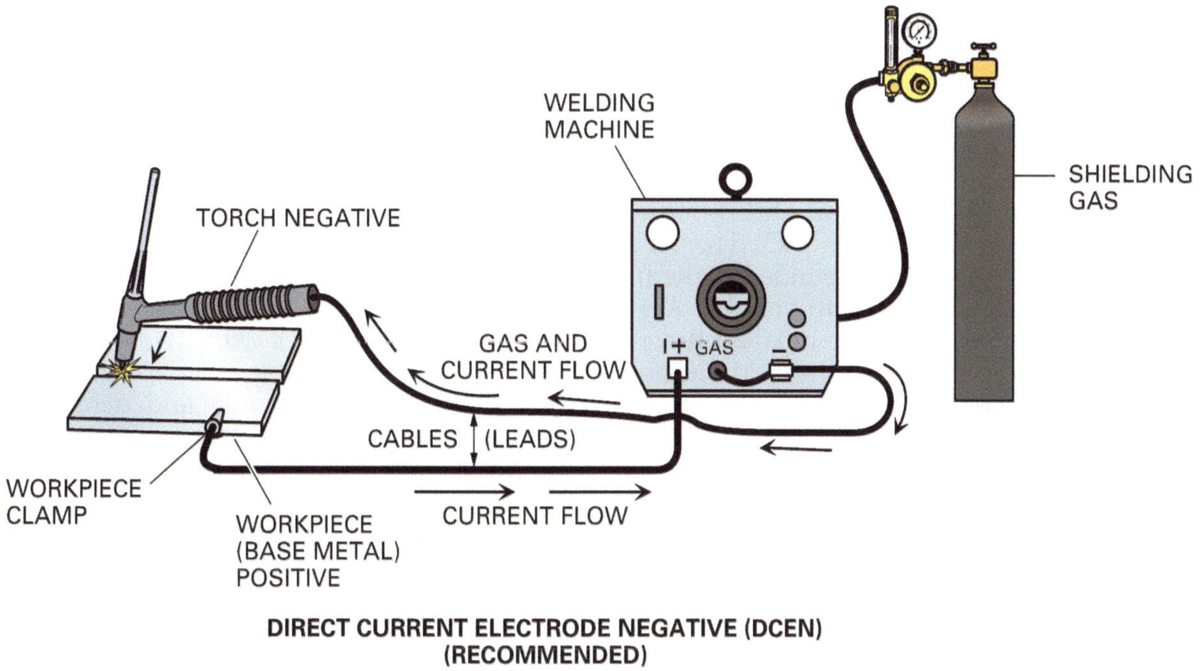

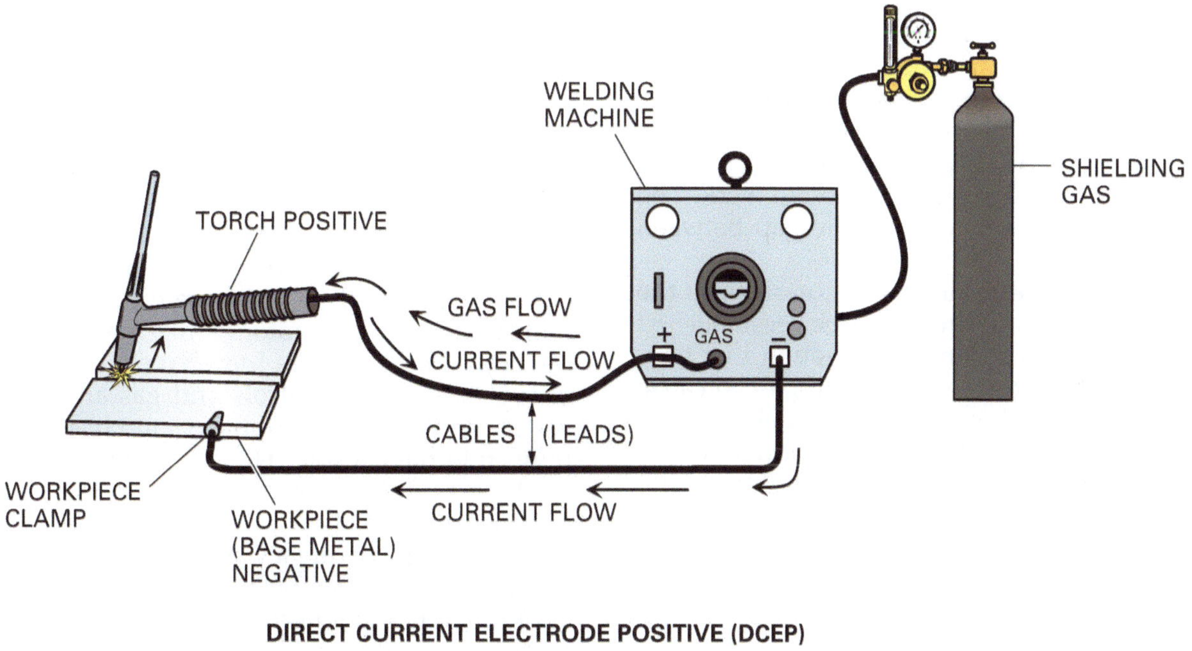

Figure 3 Representative DCEN and DCEP welding circuits.

1.2.3 AC Welding Current

Alternating current (AC) is electrical current that alternates between positive and negative values. In the positive half of the cycle, the current flows in one direction; during the negative half, the current reverses itself. The number of cycles completed in one second is called the frequency. In the United States, AC is almost always 60 cycles per second, or 60 Hertz (Hz). In other countries, the frequency is often 50 Hz. *Figure 4* shows one cycle of AC.

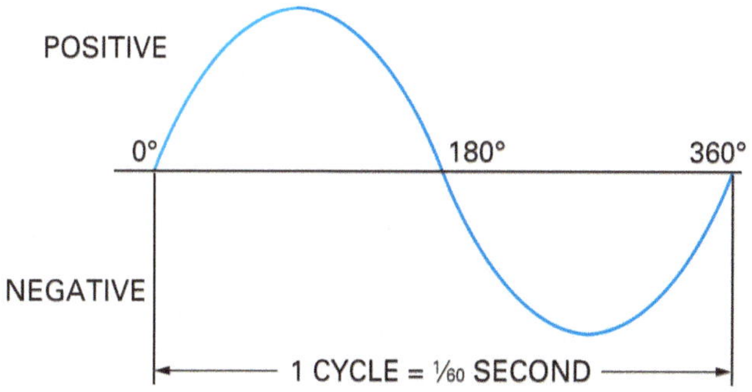

Figure 4 AC cycle.

AC is DCEN for half the time and DCEP for the remainder of the time. For this reason, AC characteristics fall in between those of DCEN and DCEP. About half of the heat goes to the base metal and the other half to the tungsten electrode. The weld bead size and penetration are midway between the beads produced by DCEN and DCEP. Some cleaning action does occur with AC. The AC tungsten electrode has a small, rounded (balled) tip. Newer welding machines with variable AC-frequency controls do not require a balled tungsten tip. AC is used with GTAW to weld aluminum, magnesium, and their alloys.

1.2.4 High Frequency

High frequency is an electrical current with a very high voltage (3,000V to 5,000V) and very low amperage. It is called high frequency because it cycles at millions of times per second compared to standard current, which has a frequency of 60 Hz. This high-voltage, high-frequency current creates a very stable arc that is able to jump a gap of about ½ inch (≈13 mm). Because high-frequency current has such low amperage, it creates very little heat.

High frequency has two purposes: to stabilize AC and to allow a DC arc to be started without touching the tungsten electrode to the work. The high voltage allows it to jump the gap between the two.

When AC is used with GTAW, each time the arc cycles, it goes out and must be reestablished in the opposite direction. Even though this happens very fast (60 times per second), it can cause the arc to extinguish on the reverse polarity part of the AC cycle when the electron flow is weakest. If this happens, oxides are absorbed into the weld, causing porosity, which weakens the weld. To prevent the AC arc from extinguishing, a high-frequency current is added to the alternating welding current. The high-frequency current, with its high voltage and low amperage, establishes a path on which the alternating welding current can travel, preventing the arc from extinguishing as it cycles. When non-squarewave AC is being used, the high frequency must be on continuously.

When using DC for GTAW without high frequency, welders must physically touch the tungsten to the work to establish the arc. Except for newer GTAW machines with **lift-arc technology**, this may not be desirable since it can cause contamination of the electrode and defects in the weld. To prevent this, high-frequency current can be added to the direct welding current. When the tungsten electrode is brought close to the work, the high-voltage, high-frequency arc jumps the gap, establishing a path for the direct welding current to travel without touching the tungsten to the work. Typically, a timer will automatically shut off the high frequency about two to three seconds after the welding arc has been established. Since DC flows in only one direction, it is very stable. The high voltage and frequency is only needed for starting the arc.

Welding power sources specifically designed for GTAW have a high-frequency generator built into them. If conventional power sources are used for GTAW, a separate auxiliary high-frequency generator can be attached if required. The generators require 120VAC to operate. Some auxiliary high-frequency generators only have high-frequency controls; others may have a combination of high-frequency controls and gas and cooling-water controls. Most will have a torch switch that plugs into the console. The switch attaches to the torch and can be used to turn the power, shielding gas, and cooling water (if used) on or off. The auxiliary high-frequency generator is connected to the power source by the electrode cable. The GTAW torch is connected to the auxiliary high-frequency generator (*Figure 5*).

Starting DC GTAW without High Frequency

DC GTAW is often performed without a high-frequency start for noncritical welds. The arc is touch-started in the weld zone or on a piece of copper. Copper is less likely to contaminate the tungsten. Some machines have a built-in selectable no high-frequency start mode.

High-Frequency Operation Precaution

When operating most transformer and transformer-rectifier machines in SMAW mode, welders must turn off the high-frequency generators.

Figure 5 Auxiliary high-frequency generator with a GTAW torch.

Additional Resources

AWS B1.10, Guide for the Nondestructive Examination of Welds. Miami, FL: American Welding Society.

AWS B1.11, Guide for the Visual Examination of Welds. Miami, FL: American Welding Society.

AWS D3.5, Guide for Steel Hull Welding. Miami, FL: American Welding Society.

AWS D3.7, Guide for Aluminum Hull Welding. Miami, FL: American Welding Society.

Lincoln Electric website: **http://www.lincolnelectric.com**

Modern Welding Technology. 6th Edition. Howard B. Cary. Englewood Cliffs, NJ: Prentice Hall, Inc.

OSHA 1910.269, Appendix *C, Protection from Step and Touch Potentials*. Current Edition. Washington, DC: Occupational Safety and Health Administration (OSHA).

OSHA 19260.351, Arc Welding and Cutting. Current Edition. Washington, DC: Occupational Safety and Health Administration (OSHA).

Welding Handbook. Vol. 1. Welding Science & Technology. 9th Edition. Miami, FL: American Welding Society, 2001.

Welding Handbook. Vol. 2, Part 1: Welding Processes. Miami, FL: American Welding Society, 2004.

1.0.0 Section Review

1. When performing long-term cutting or welding of zinc or cadmium materials or coatings, always _____.
   a. ventilate the area with a continuous oxygen supply
   b. connect the welding nozzle to a nitrogen purge
   c. wear an approved full-face, supplied-air respirator
   d. use a water-cooled, multi-flame (rosebud) torch

2. Direct current electrode negative (DCEN) produces the greatest amount of heat in the _____.
   a. base metal
   b. filler metal
   c. electrode
   d. torch

Section Two

2.0.0 GTAW Equipment

Objective

Identify and describe GTAW equipment and consumables.

a. Identify and describe GTAW welding machines.
b. Identify and describe GTAW torches.
c. Identify and describe GTAW torch nozzles and electrodes.
d. Identify and describe GTAW shielding gases.
e. Identify and describe GTAW filler metals.

Performance Tasks

1. Select a GTAW shielding gas.
2. Select a GTAW filler metal.
5. Break down and reassemble a GTAW torch.

Trade Terms

Potentiometer: A three-terminal resistor with an adjustable center connection; used for variable voltage control.

GTAW operations involve the use of a number of different types of equipment and consumables. Chief among these are GTAW welding machines, torches, nozzles, electrodes, shielding gases, and filler metals.

2.1.0 Welding Power Source Types

Welders use CC welding machines to perform GTAW. The type of welding current (DCEN, DCEP) depends on the type and thickness of the base metal.

Many types of welding machines may supply welding current for GTAW. Usually, any welding machine that produces constant current, such as those used for SMAW, can be used. In addition, there are welding machines specially designed for GTAW. These special machines have built-in high frequency and controls for shielding gas and cooling water. Types of welding machines that can be used for GTAW include the following:

- Transformer welding machines
- Transformer-rectifier welding machines
- Engine-driven generator and alternator welding machines
- Inverter power sources

DC welding machines are usually designed to produce either constant-current or constant-voltage direct welding current. Constant current means that welding current is produced over a wide voltage range regardless of arc length. This is typical of an SMAW or GTAW welding machine. A constant-voltage machine maintains a constant voltage as the output current varies. This is typical of a GMAW or FCAW welding machine. The output voltages and output currents of a welding machine can be plotted on a graph to form a curve. These curves show how the output voltage relates to the output current as either of these factors changes (*Figure 6*).

2.1.1 Transformer Welding Machines

Transformer welding machines produce alternating welding current only. They use a voltage step-down transformer, which converts high-voltage, low-amperage current from commercial power lines to low-voltage, high-amperage welding current. The primary electrical power input required for a transformer welder can be 120VAC, 208-230VAC single- or three-phase, or 460V three-phase. Special light-duty transformer welding machines used for sheet metal work are designed to be plugged into a 120VAC outlet. However, most light-duty transformer welding machines require 230VAC primary current. Heavy-duty industrial transformer welders require 460V three-phase primary current.

> **WARNING!**
>
> Coming in contact with the primary current of a welding machine can cause electrocution. Make sure that welding machines are properly grounded to prevent injury.

Transformer welders are not as common as other types of welding machines on the job site, but they are used for special jobs. A transformer welder has an On/Off switch, amperage control, and terminals for connecting the electrode lead and the workpiece lead. Transformer welding machines used for GTAW must have high-frequency capabilities built in or they must use an auxiliary high-frequency generator.

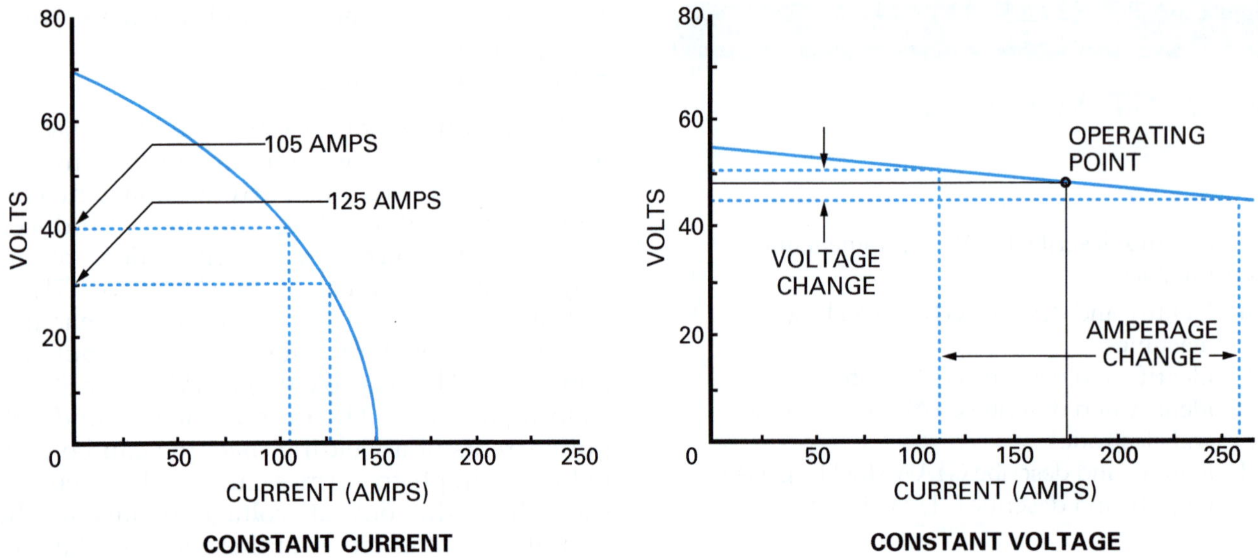

Figure 6 Constant-current and constant-voltage output curves.

2.1.2 Transformer Rectifier Welding Machines

The transformer-rectifier welding machine uses a transformer to transform the high-voltage, low-amperage primary current to low-voltage, high-amperage welding current and a rectifier to convert the AC to DC. Transformer-rectifier welding machines are designed to produce either direct welding current only or both alternating and direct welding currents. Small transformer-rectifier machines that produce both alternating and direct welding currents are usually lighter-duty than those that produce DC only. Transformer-rectifier welding machines that produce direct welding current only are sometimes simply referred to as rectifiers. Depending on their size, transformer-rectifier welding machines may require 120VAC, 208-230VAC single- or three-phase, or 460V three-phase power to operate.

Transformer-rectifiers used for GTAW have an On/Off switch and an amperage control. The welding cables (electrode lead and workpiece lead) are connected to terminals. They often have selector switches to select DCEN or DCEP. If there is no selector switch, welders must change the cables on the machine terminals manually to select the type of current desired. Most high-amperage machines designed specifically for GTAW are the transformer-rectifier type. Transformer rectifiers designed specifically for GTAW generally have the following:

- High-frequency generator
- High-frequency selector switch for DC Start, Continuous (for AC), or Off (no high frequency)
- Shielding gas preflow timer
- Control to automatically start cooling water flow when an arc is struck
- Shielding gas postflow timer control to prevent weld contamination (keeps shielding gas flowing for a set time after the arc has been terminated)
- Hand- or foot-operated remote current control
- Remote/local selector switch for remote control
- Start amperage control

Welders often use GTAW for critical welds on difficult-to-weld materials. For this reason, many special advanced features and controls have been developed and are available on machines designed for GTAW. These advanced features and controls include the following:

- High-frequency intensity control for better arc starting
- High-frequency stabilizer control to control tungsten spitting
- Crater fill timer to automatically taper the welding current from the selected setting to a minimum current at the end of a weld
- Balanced wave adjustment for penetration/cleaning control (sets percentage of time AC remains on the straight or reverse side of the cycle)

Regardless of the machine used, welders should always consult the manufacturer's documentation for specific operating information. *Figure 7* shows a typical industrial, high-amperage, heavy-duty transformer-rectifier welding machine.

KEY CONTROLS

A. MINIMUM OUTPUT PRESET CONTROL & DISPLAY SWITCH
B. SET-UP MENU
C. DIGITAL AMPERAGE METER
D. LOCAL/REMOTE CURRENT SWITCH
E. MAXIMUM OUTPUT PRESET CONTROL
F. POST FLOWTIME
G. THERMAL SHUTDOWN LIGHT
H. DOWNSLOPE TIME
I. PULSE BACKGROUND CURRENT CONTROL
J. PULSE % ON TIME CONTROL
K. SPOT TIME
L. PULSE FREQUENCY CONTROL
M. POLARITY SWITCH
N. POWER SWITCH
O. PULSE/SPOT TIME MODE SWITCH
P. TRIGGER SWITCH
Q. AC BALANCE CONTROL
R. MODE SWITCH
S. REMOTE RECEPTACLE (NOT SHOWN)

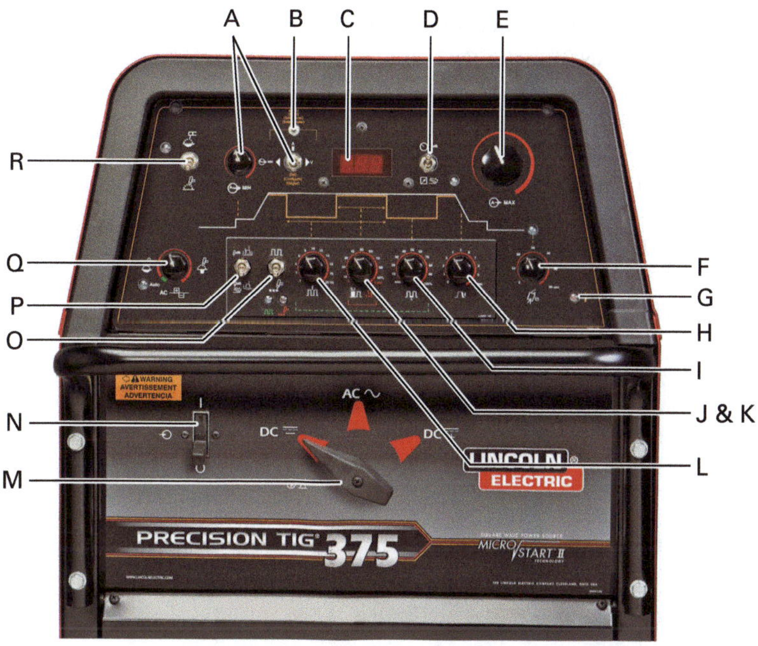

SETUP MENU

- DC TIG START MODES: HIGH FREQUENCY, SCRATCH START, AND TOUCH START TIG™
- ADJUSTABLE PREFLOW TIME
- ADJUSTABLE START PULSE FOR SOFT OR FORCEFUL STARTS
- ADJUSTABLE TIG HOT START
- ADJUSTABLE UPSLOPE TIME
- ADJUSTABLE STICK HOT START
- ADJUSTABLE STICK ARC FORCE

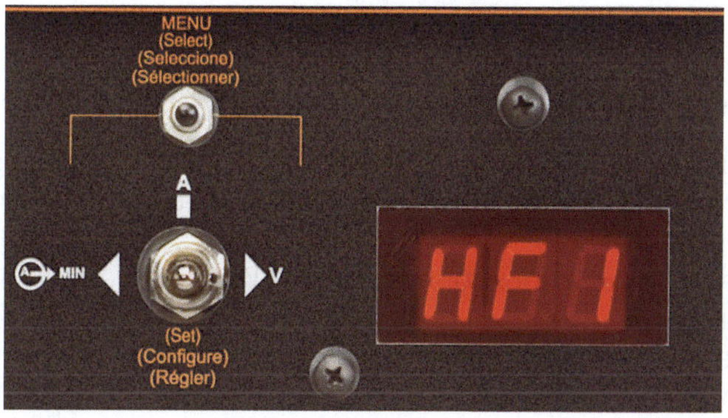

Figure 7 Transformer-rectifier welding machine designed for SMAW, GTAW, and pulsing.

2.1.3 Inverter Power Sources

Inverter power sources (*Figure 8*) increase the frequency of the incoming primary power. This provides a smaller, lighter power source with a faster response time and much more waveform control for pulse welding. Welders generally use inverter power sources in applications where space is limited and portability is important. The controls on these machines vary according to size and application.

The power source shown in *Figure 8* weighs less than 30 pounds (14 kg), and is a full-function inverter intended for alloy fabrication or critical maintenance work. It produces up to 90 amps with 100 percent duty cycle for GTAW, and 60 amps with 100 percent duty cycle for SMAW on an input power supply of 120VAC. However, it can produce up to 130 amps for both SMAW and GTAW when the input power is 208–230VAC.

Many production inverter welding machines used today are referred to as advanced inverter welding machines. They usually include all the controls listed for transformer-rectifier machines and basic inverter machines plus the following:

- Pulse arc control that allows the welding current to be pulsed between a high-current setting and a low-current setting for puddle control
- Hot-start current control to enable a controlled surge of welding current to establish a puddle quickly

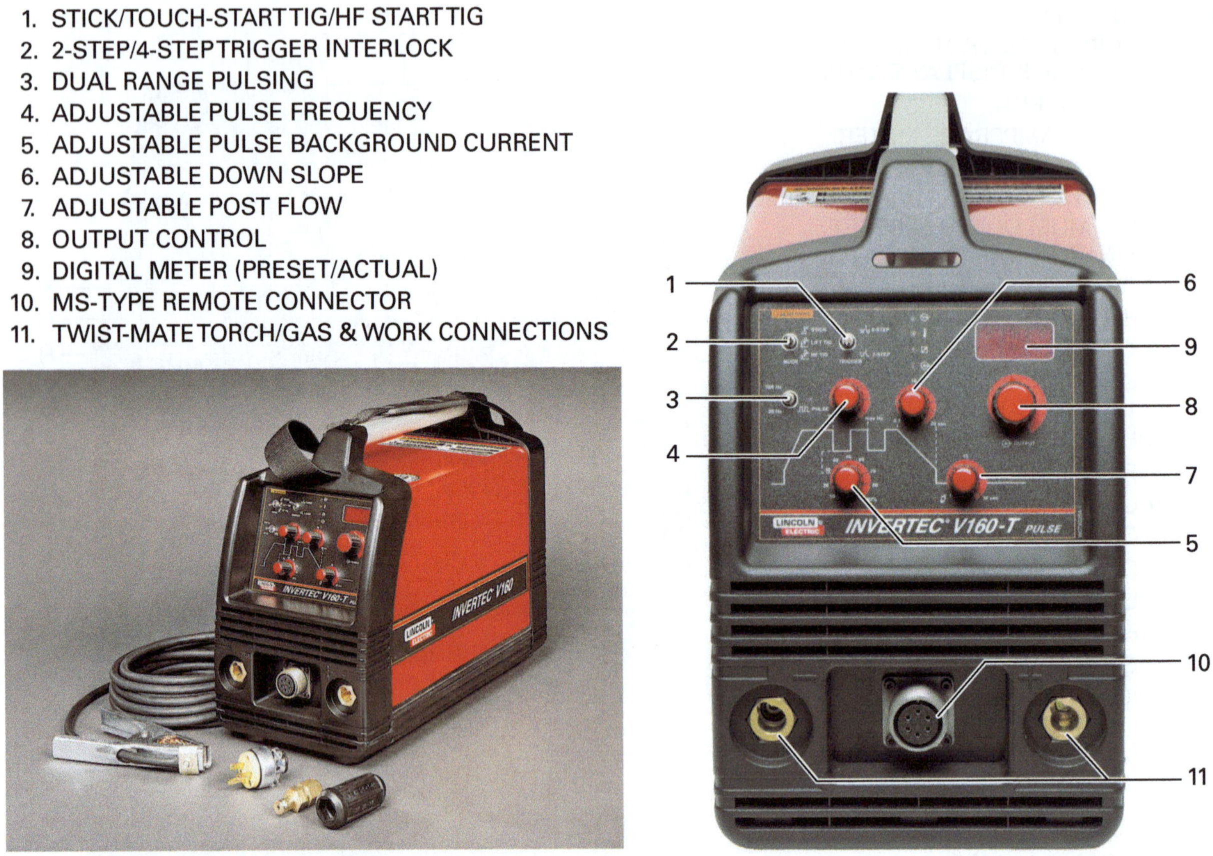

Figure 8 Basic GTAW inverter welding machine.

- Slope-up control for direct hot-start current, which controls how quickly the current rises to the hot-start setting
- Optional spot arc timer control for GTAW spot welding
- Variable AC-frequency control
- AC balance-wave control

Advanced inverter power sources are available in capacities of up to 500 amps that weigh approximately 100 pounds (45 kg).

2.1.4 Engine-Driven Generator Welding Machines

Welding machines can also be powered by gasoline or diesel engines. The engine can be connected to a generator or to an alternator. Engine-driven generators (*Figure 9*) produce direct welding current. Engine-driven alternators produce AC that is fed through a rectifier to produce direct welding current. The size and type of engine used depend on the size of the welding machine.

To produce welding current, the generator must turn at a required number of revolutions per minute. The engines powering the generators have governors that control the engine speed. Most governors have a welding speed switch that can be set to idle the engine when no welding is taking place. When the electrode is touched to the base metal, the governor automatically increases the speed of the engine to the required rpm for welding. If no welding takes place for about 15 seconds, the engine automatically returns to idle. The switch can also be set for the engine to run continuously at the welding speed.

Figure 9 Engine-driven generator welding machine.

Engine-driven generators often have an auxiliary power unit that produces 120VAC current for lighting, power tools, and other common equipment. When 120VAC current is required, the engine-driven generator must run continuously at the welding speed.

Engine-driven generators have engine controls and welding current controls. The engine controls vary with the type and size of the unit, but normally include the following:

- Starter
- Voltage gauge
- Temperature gauge
- Fuel gauge
- Hour meter
- Oil pressure indicator

The following are the common welding current controls:

- Amperage control
- Current range switch
- Amperage and voltage gauge
- Polarity switch

The advantage of engine-driven generators is that they are portable and can be used in the field where electricity is not available to power other types of welding machines. The disadvantage is that engine-driven generators are more costly to purchase, operate, and maintain.

2.1.5 Power Source Ratings

The rating of a welding machine is determined by the amperage output of the machine at a given duty cycle. The duty cycle of a welding machine is based on a 10-minute period. It is the percentage of ten minutes during which the machine can continuously produce its rated amperage without overheating. For example, a machine with a rated output of 300 amps with a 60-percent duty cycle can deliver 300 amps of welding current for 6 minutes out of every 10 without overheating.

The duty cycle of a welding machine will generally be 20, 30, 40, 60, or 100 percent. A welding machine having a duty cycle of 20 to 40 percent is considered a light- to medium-duty machine; welding can only be maintained for 2 to 4 minutes out of each 10-minute period. Most industrial, heavy-duty machines for manual welding have a 60- or 100-percent duty cycle rating. Machines designed for automatic welding operations have a 100-percent duty cycle rating.

With the exception of the 100-percent duty cycle machines, the maximum amperage that a welding machine can produce is always higher than its rated capacity. A welding machine rated for 300 amps with a 60-percent duty cycle generally puts out a maximum of 375 to 400 amps. However, since the duty cycle is a function of its rated capacity, the duty cycle will decrease as the amperage is raised over 300 amps. Welding at 375 amps with a welding machine rated for 300 amps with a 60-percent duty cycle will lower the duty cycle to about 30 percent. If welding continues for more than 3 out of 10 minutes under these conditions, the machine will overheat. Note that most welding machines have a heat-activated circuit breaker that will shut off the machine automatically when it overheats. The machine cannot be turned back on until it has cooled below a preset temperature.

Conversely, if the amperage is set below the rated amperage, the duty cycle increases. Setting the amperage at 200 amps for a welding machine rated at 300 amps with a 60-percent duty cycle will increase the duty cycle to 100 percent.

Figure 10 is a graph that shows the relationship between amperage and duty cycle.

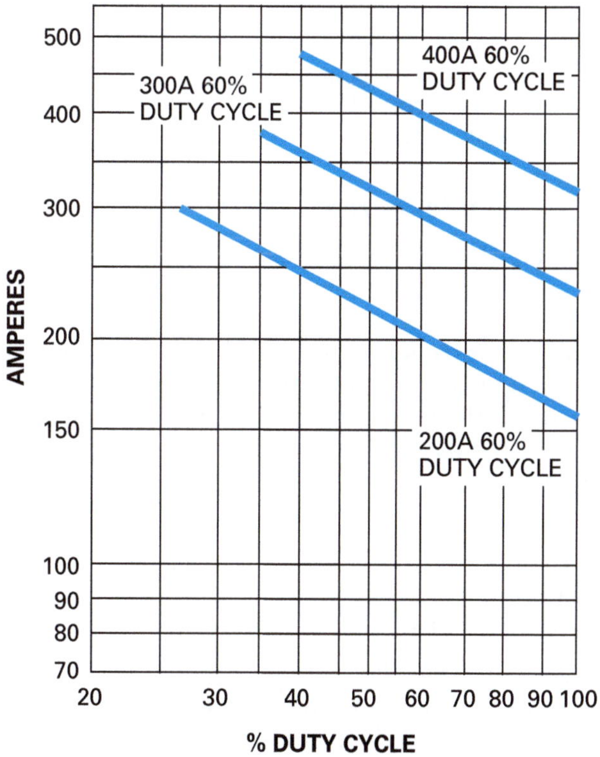

Figure 10 Relationship between amperage and duty cycle.

2.1.6 Welding Cable

Cables used to carry welding current are designed for maximum strength and flexibility. The conductors inside the cable are made of fine strands of copper wire. The copper strands are covered with layers of rubber reinforced with nylon or Dacron® cord. *Figure 11* shows the construction of a welding cable.

The size of a welding cable is based on the number of copper strands it contains. Large-diameter cable has more copper strands and can carry more welding current. Typically, the smallest cable size is number 4 and the largest is number 3/0 (spoken as "3 aught").

When selecting a welding cable size, consider the amperage load as well as the distance the current will travel. The longer the distance the current has to travel, the larger the cable must be to reduce voltage drop and heating caused by the electrical resistance in the welding cable. When selecting welding cable, use the rated capacity of the welding machine for the cable amperage requirement. For the distance, measure both the electrode and workpiece leads, and add the two lengths together. To identify the welding cable size required, refer to a table of recommended welding cable sizes. These tables are furnished by most welding cable manufacturers. Check all cable connections for tightness and repair loose or overheated lugs. If quick-disconnects are used, they must be properly sized to the cable or overheating may occur.

2.1.7 Remote Current Control

With GTAW, control of the welding current is often performed with remote controls such as foot- or hand-operated switches and potentiometers. Remote controls are generally designed for use with a specific welding machine, so they cannot be used universally on all machines. Remote controls with only a switch can control one or more of the following:

- High frequency
- Shielding gas
- Cooling water
- Welding current (on and off only)

Remote controls equipped with a potentiometer can be used to continuously vary the welding current as needed, from zero to the maximum current set on the power source.

Figure 11 Construction of a welding cable.

The most useful remote control has both switches and a potentiometer, and it controls the following:

- High frequency
- Shielding gas
- Cooling water
- Welding current (on and off)
- Welding current up or down as needed (zero to maximum setting on the power source)

Figure 12 shows various remote control units.

2.2.0 GTAW Torches

In addition to a CC welding machine (and a high-frequency generator for AC welding), GTAW requires the following:

- GTAW torch
- Shielding gas
- Optional torch cooling system

In addition, a remote amperage control is often used with GTAW. *Figure 13* shows a typical GTAW welding system.

Aluminum Welding with a Balled-Tip versus a Pointed-Tip Electrode

Traditional AC welding of aluminum using a balled-tip electrode is required for most applications using conventional power sources. However, for inverter power sources that use squarewave technology with AC balance and variable AC frequency control, welding thinner materials and creating smaller beads is accomplished much more easily using the same pointed and blunted 2 percent thoriated or ceriated electrodes as used for DCEN welding along with increased electrode-negative time.

(A) FOOT CONTROL

(B) WIRELESS FOOT CONTROL

(C) FINGER CONTROL

(D) HAND CONTROL

Figure 12 Typical foot-, hand-, and finger-operated remote controls.

The GTAW torch provides an electrical path for the welding current between the welding current lead and the tungsten electrode. There are many different manufacturers of torches, but all torches are basically the same even though their parts are not often interchangeable. All torch bodies have a built-in or replaceable collet body and electrode collet. The electrode collet and collet body come in a variety of sizes to match the size of the electrode being used. The collet holds the tungsten electrode in place and is secured by a back cap. The back cap, which is threaded onto the torch, is loosened to change the electrode or to adjust its stickout. The reserve length of electrode that extends from the back of the collet is insulated and protected by the back cap. Back caps are made in several lengths to cover tungstens of various lengths. The shorter back caps are used with short tungstens in confined spaces where torch clearance is inadequate.

The torch also contains passages for the shielding gas or cooling water, if needed. The shielding gas exits the torch through an insulated gas nozzle, sometimes called a cup, which surrounds the tungsten. *Figure 14* provides some GTAW torch terminology.

Heat is removed from GTAW torches either by air cooling with a shielding gas or by water cooling. *Figure 15* shows examples of both air-cooled and water-cooled GTAW torches. Two styles of air-cooled torches are common. One style uses a single plastic hose that has a power cable inside it, as shown in *Figure 15*. The shielding gas flows around the power cable, cooling it and the torch before being discharged to shield the electrode and weld. A special power cable adapter attaches to the power terminal on the power source.

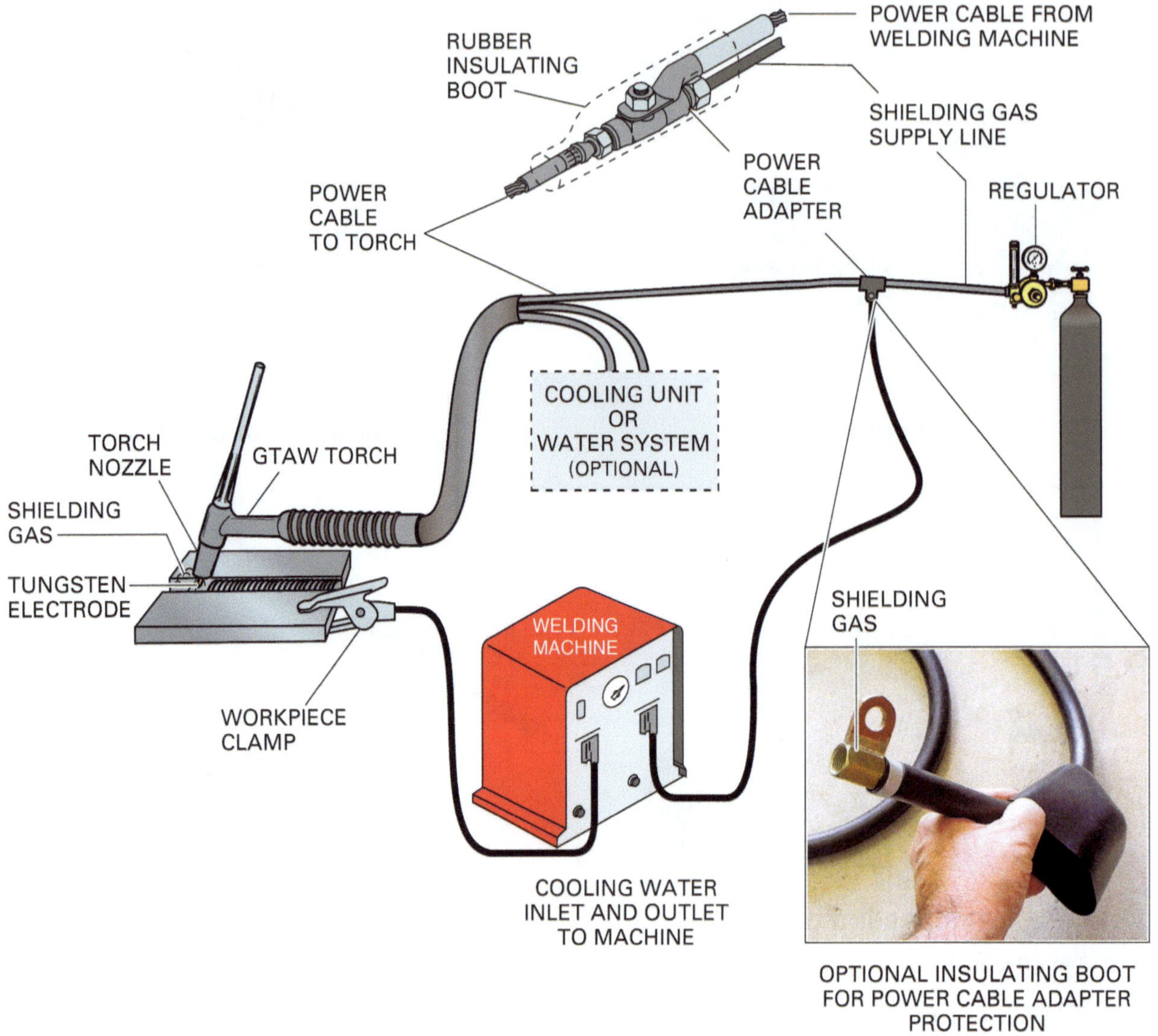

Figure 13 Basic GTAW welding system.

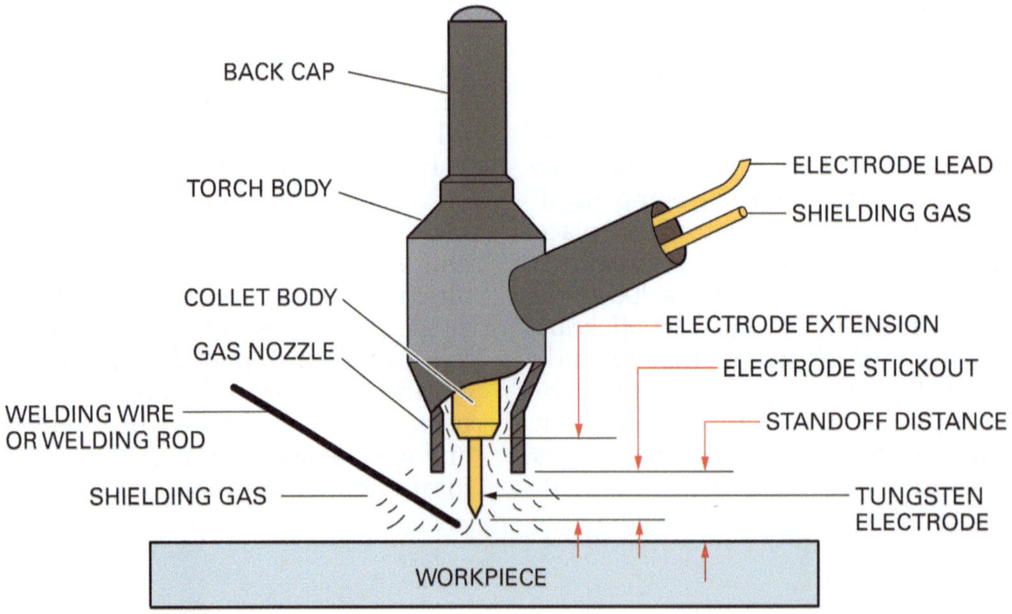

Figure 14 GTAW torch terminology.

This connector needs to be covered by a rubber insulating boot to prevent shorts. The single hose connects to one side of the adapter. This connection also makes the electrical connection. The shielding gas supply hose connects to the other side of the adapter.

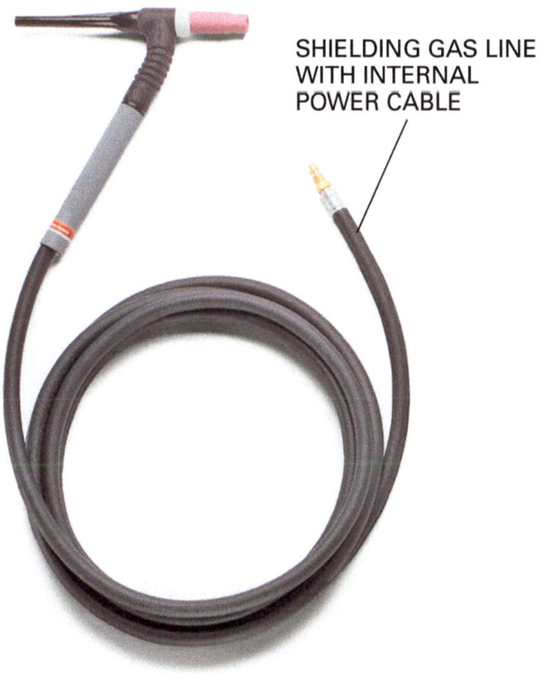

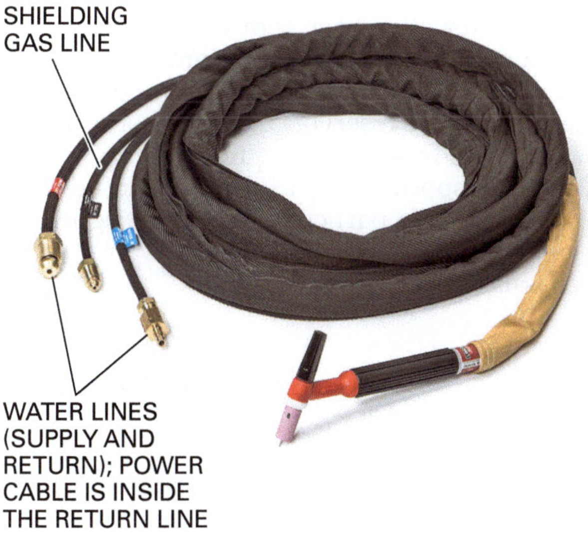

Figure 15 GTAW torches.

The second type of air-cooled torch has a separate power cable and shielding gas hose. Either the single- or double-hose torch can be purchased with a gas control valve in the torch to manually control the shielding gas flow. This is necessary when the power source being used does not have a solenoid valve to automatically start and stop the shielding gas flow. *Figure 16* shows the various parts of an air-cooled GTAW torch.

Water-cooled torches have three hoses. One hose supplies shielding gas; one delivers cooling water to the torch; the third hose contains the power cable and the cooling water return or discharge. The cooling water may be supplied from the domestic water system or from a closed-loop cooling system. Regardless of the system used, it should include a fusible link in the torch cooling line. A fusible link is a temperature-based safety device that will open and stop the welding current flow if the coolant flow to the torch is interrupted. *Figure 17* shows the parts of a water-cooled GTAW torch.

Cable Covers

To prevent damage from hot or molten metal to GTAW torch cables containing hoses for gas or water, leather or fire-resistant cable covers can be added as protection. These covers zip or snap around the torch cable for a distance of 10" to 20" (25 cm to 50 cm) behind the torch handle.

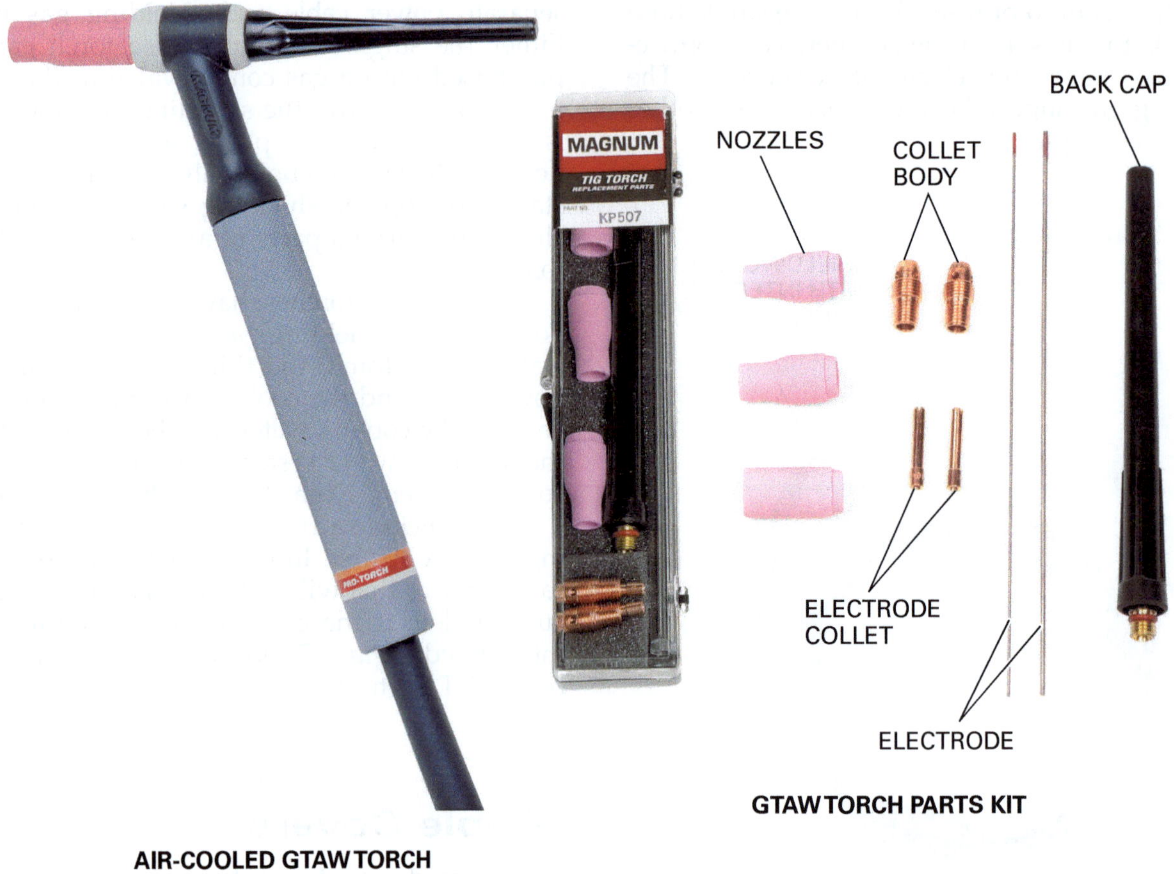

Figure 16 Air-cooled GTAW torch.

Closed-loop cooling systems recirculate demineralized water or special cooling fluids that will not corrode internal torch surfaces or plug the cooling passages with mineral scale. After cooling the torch, the water is typically returned to the system through a hose that also contains the torch power cable. A typical closed-loop cooling system usually consists of a reservoir, circulating pump, heat-transfer equipment to reject any collected heat to the atmosphere, and the connecting lines. The reservoir may also serve as the portable unit's base and be equipped with wheels for easy transport. *Figure 18* shows a closed-loop cooling unit. This particular unit is designed to stack with welding machines that have a matching chassis.

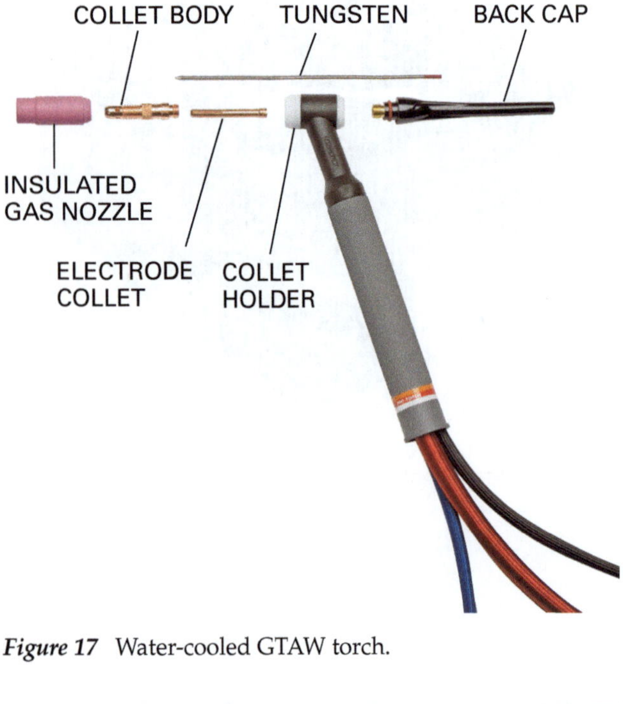

Figure 17 Water-cooled GTAW torch.

Figure 18 Typical closed-loop cooling unit.

2.3.0 Nozzles, Collets, and Electrodes

GTAW collets and electrodes are designed to work together, with the tungsten electrode fitting precisely into the collet. For this reason, welders must take both of the devices into consideration when selecting either the proper collet or electrode for a particular job.

2.3.1 Gas Nozzles

The GTAW gas nozzle shapes and directs the shielding gas flow as it exits the torch. Nozzles, sometimes called cups, are usually threaded onto the torch to form a gas-tight joint. They are made of ceramic material, chrome-plated steel, Pyrex®, or glass. Ceramic nozzles can be used up to about 300 amperes. Above that amperage, water-cooled, metal-coated ceramic nozzles or water-cooled ceramic nozzles must be used.

Gas nozzles are available in different lengths and diameters (refer to *Figures 16* and *17*). Nozzle length is determined by the job requirements, such as limited clearance or deep grooves. Nozzle size is the exit orifice inside diameter; it depends on the torch type and size and the electrode diameter. Nozzle size may be specified in fractions of an inch (¼", ⅜", 7⁄16", ½", ¾") or in millimeters (6 mm, 10 mm, 11 mm, 13 mm, 19 mm). Metric size nozzles are not, however, designed to be equivalent in size to imperial versions, and metric size and imperial size nozzles cannot be substituted for each other. Sometimes nozzle diameter is given as a size number, such as 4, 6, 7, 8, or 12. These numbers are the nozzle opening diameters in multiples of 1⁄16". For example, a Number 6 nozzle is 6 × 1⁄16", or ⅜" in diameter.

A GTAW torch may also be equipped with a gas lens collet body. A gas lens collet body contains an assembly of fine screens that straightens the gas flow from the nozzle and eliminates turbulence, which causes the gas to flow smoothly past the tungsten electrode. This reduces the chance that the turbulence will pull atmosphere into the weld zone and cause contamination and weld defects. The use of a gas lens allows longer tungsten stickout. This is useful for welding in tight quarters. Increasing tungsten stickout decreases the current-carrying capacity of the tungsten. *Figure 19* shows a typical gas lens collet body and torch with a nozzle, collets, and electrodes.

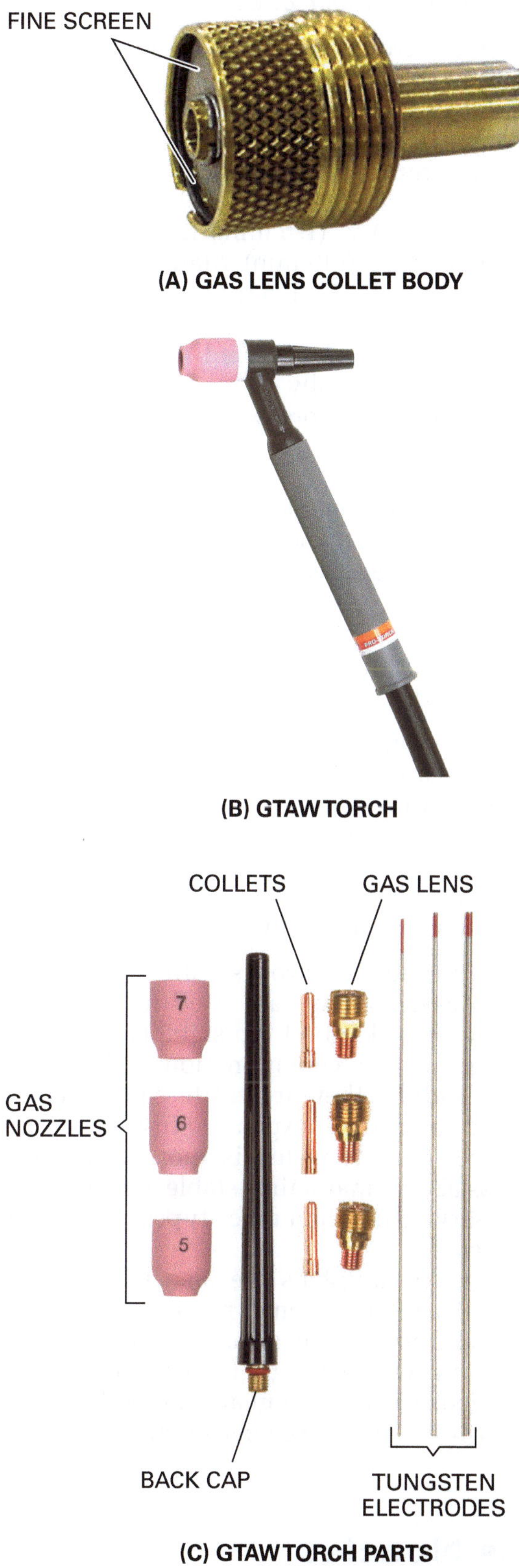

Figure 19 Typical gas lens collet body and torch with nozzle.

2.3.2 Tungsten Electrodes

Tungsten electrodes used for GTAW are manufactured in different formulations and sizes to meet the requirements of *AWS Specification A5.12*. Electrode diameters range from 0.01" (0.3 mm) up to ¼" (6.4 mm) for current ratings of 5 to 1,000 amperes. Electrodes are made in lengths of 3" (76 mm), 6" (152 mm), 7" (175 mm), 12" (305 mm), 18" (457 mm), and 24" (610 mm). Manual welding is usually done with 7" (175 mm) or shorter electrodes.

Electrodes are manufactured with two different finishes—chemically cleaned or ground. The ground finish is the more expensive option.

CAUTION

Always handle tungsten electrodes with clean gloves to prevent contamination from body oils and other materials.

A number of different tungsten electrodes are manufactured. They include the following:

- Pure tungsten
- Zirconiated tungsten
- Thoriated tungsten
- Ceriated tungsten
- Lanthanated tungsten
- Rare earth tungsten
- Cryogenically treated thoriated tungsten

Pure tungsten electrodes, often simply referred to as tungstens, are used for AC welding. They provide good arc stability and good resistance against contamination. When used for DC welding, they are easily contaminated. Although relatively inexpensive in terms of per unit price, pure tungsten is more difficult to start, less able to maintain a stable arc, and has a shorter service life than other types of tungsten electrodes.

Zirconiated tungstens are also used for AC welding. They have a small percentage of zirconium added to the tungsten. They are used for AC welding where tungsten inclusions cannot be tolerated. The zirconium also gives these electrodes easy arc-starting characteristics.

Thoriated tungsten electrodes are used for DC welding. They have a small percentage of thorium added to do the following:

- Make the arc easier to start
- Increase the current range
- Help prevent tip melt
- Reduce the tendency to stick or freeze to the work
- Increase resistance to contamination, when properly used

Thoriated tungsten electrodes are most commonly manufactured in these two concentrations of thorium dioxide (thoria): EWTh-1 (approximately 1 percent thoria) and EWTh-2 (approximately 2 percent thoria).

WARNING!

Thoriated tungsten electrodes contain low-level radioactive thorium. The grinding dust from these electrodes is considered a hazard. *AWS Safety and Health Fact Sheet 27* provides further information about this subject. If possible, always use tungsten electrodes containing cerium or lanthanum instead of thorium.

Ceriated and lanthanated tungsten electrodes have the advantage of not being radioactive. Ceriated tungsten electrodes can be balled and used for AC welding.

EWG electrodes that combine three non-radioactive rare earth materials into one electrode are increasingly used as replacements for 2-percent thoriated electrodes. Besides their advantage of not being radioactive, these mixed tungsten electrodes start and re-ignite well and have very good service life for applications in which welding cycles of at least 15 minutes are used.

Thoriated electrodes that have been treated with a multi-step cryogenic process are another good replacement for 2 percent thoriated electrodes. The cryogenic treatment maximizes grain structure and improves electron flow. The cryogenically treated electrodes are easier to ignite and have longer tip life, longer overall electrode life, and higher amperage tolerance than non-treated 2-percent thoriated tungsten.

Gas Nozzles

Some manufacturers recommend that the inside diameter of the nozzle should be a minimum of three times the electrode diameter.

Think About It

Nozzle Size

What size nozzle would a welder select if a ½" diameter opening was needed?

Other tungsten electrodes continue to be developed. For example, tungsten electrodes with less common oxides, such as yttrium and magnesium, are also becoming available.

Tungsten electrodes are identified by a color band at one end. *Table 1* shows GTAW electrode color codes and AWS classifications.

2.4.0 Shielding Gas

The GTAW process always uses inert shielding gases to displace the atmosphere from the weld zone and prevent oxidation and contamination of the tungsten electrode, weld puddle, and filler metal. The two principal shielding gases used for GTAW are argon (Ar) and helium (He). The use of helium was what prompted the naming of GTAW initially as heliarc welding. These gases may be used alone or as mixtures. Occasionally, small percentages of other gases are added for special conditions.

Each shielding gas and shielding gas mixture has distinct performance characteristics. Each affects the arc differently and produces different weld characteristics. Mixtures of gases often have the best features of the individual gases within them. The following sections explain shielding gas characteristics and their principal uses.

Table 1 GTAW Electrode Color Codes and AWS Classifications

Electrode	Color Band	Electrode AWS Classification
Pure Tungsten	Green	EWP (Minimum 99.5% Tungsten)
1% Thoriated	Yellow	EWTh-1 (1% Thoria)
2% Thoriated	Red	EWTh-2 (2% Thoria)
Zirconiated	Brown	EWZr (0.15%–0.40% Zirconium Oxide)
Zirconiated	White	EWZr (8% Zirconium Oxide)
Ceriated	Gray	EWCe-2 (2% Ceria)
Lanthanated	Black	EWLa-1.0 (1% Lanthana)
Lanthanated	Gold	EWLa-1.5 (1.5% Lanthana)
Lanthanated	Blue	EWLa-2.0 (2% Lanthana)
EWG	*Not Standard	Undefined (*e.g.*, various combinations of rare earth alloys, yttrium oxide, magnesium oxide, etc.)

*Manufacturers may select any color not already in use.

2.4.1 Argon

Argon is the most common shielding gas used with GTAW. Argon provides a smooth, quiet arc that requires a lower arc voltage than other shielding gases for a given arc length. This means that it gives the welder the greatest tolerance for arc gap variation. Argon also works well with AC, and it provides better base metal cleaning than helium. Argon is ten times heavier than helium, so it forms a better gas shield than helium, which tends to rise at the same flow rates. Argon is used both alone and in combination with other shielding gases.

When compared with helium, argon has the following advantages:

- A smoother, quieter arc
- Lower arc voltage for a given arc length
- Easier arc starting
- Better cathode cleaning on aluminum and magnesium with AC
- Lower shielding flow rate
- Better shielding in cross drafts
- Least expensive shielding gas

Tungsten Electrode Packaging

Most tungsten electrodes are packaged as shown to prevent contamination or damage to the electrodes. An inner divider containing the electrodes slides out of the package. Note the orange color band (may appear reddish due to printing color variations) on the end of each electrode, identifying the type in agreement with *Table 1*. All color bands typically discolor after use due to the heat produced. The color band end is not used when sharpening an electrode.

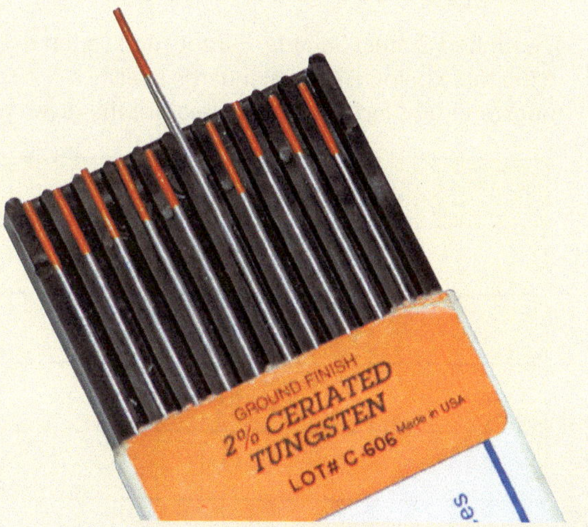

Small amounts of hydrogen are sometimes mixed with argon for deeper penetration in welding stainless steel. Hydrogen cannot be used with aluminum or carbon steels because it produces porosity and underbead cracking. The most common argon-hydrogen mixture is 85 percent argon and 15 percent hydrogen.

Nitrogen is sometimes added to argon to stabilize austenitic stainless steel and to increase penetration in welding copper.

2.4.2 Helium

Helium is used when deeper penetration and higher travel speed are required. Arc stability is not as good as it is with argon. However, helium is capable of delivering more heat on the base metal than is possible with argon, although it does require a higher voltage. This makes helium better for welding thick sections of highly heat-conductive metals such as aluminum and copper.

When compared with argon, helium has the following advantages:

- Deeper penetration into the weld joint
- Increased welding speed
- Welding of highly heat conductive metals

Helium is often mixed with argon. A common mixture is 75 percent helium and 25 percent argon. This mixture gives good cathodic cleaning action and deep weld penetration, which are good characteristics of both gases.

2.4.3 Cylinder Safety

Shielding gases may be supplied in bulk liquid tanks, in liquid cylinders, or in high-pressure cylinders of various sizes. The most common container is the high-pressure cylinder, which is portable and can be easily moved as needed.

When transporting and handling cylinders, observe the following rules:

- Always install the safety cap over the valve, except when the cylinder has been connected for use.
- Secure the cylinder to prevent it from falling when it is in use. Chain or clamp it to the welding machine or to a post, beam, or pipe.
- Always use a cylinder cart to transport cylinders.
- Never hoist cylinders with a sling or magnets. Cylinders can slip out of the sling or fall from the magnets. Always use a hoisting basket or similar device.
- Always open the cylinder valve slowly, and open it completely to prevent leakage around the valve stem.

WARNING!

Do not remove the valve's protective cap until the cylinder has been secured. If the cylinder falls over and the valve breaks off, the cylinder will shoot like a rocket, causing severe injury or death to anyone in its way.

Around the World

Tungsten Standards

The United States, Canada, Europe, and Japan each have their own published standards for tungsten, as shown in the table below. These standards specify the dimensions, packaging, and manufacturing requirements that tungsten manufacturers must meet for the given market area.

MARKET AREA	STANDARD NAME
UNITED STATES	ANSI/AWS A5.12
CANADA	ASME/SFA
EUROPE	ISO 6848
JAPAN	JIS

Always Use the Correct Gas

Make sure to always use the correct gas or gas mixture. Using an incorrect gas mixture, like a common one used for GMAW, which is 75 percent argon/25 percent CO^2, will immediately cause the tungsten electrode to be consumed and deposited in the weld as a contaminant. This is definitely not the kind of result you are looking for.

2.4.4 Gas Regulators/Flowmeters

A gas regulator and a flowmeter are required to supply the shielding gas to the torch at the proper pressure and flow rate (*Figure 20*). A typical regulator/flowmeter has a preset pressure regulator with a cylinder valve spud, a flow-metering needle valve, and a flow-rate gauge. The pressure regulator is usually equipped with a cylinder pressure gauge to indicate the cylinder gas pressure. The metering valve is used to adjust the gas flow to the torch nozzle. The flow gauge indicates the gas flow rate in cubic feet per hour (cfh) or liters per minute (L/min) as shown by a ball-type flow indicator. Although the regulator and flowmeter are shown separately in *Figure 20*, their functions may also be combined into a single device.

Gas flow to the torch is started and stopped either by a manually operated valve or by an electric solenoid valve. The manual valve may be on the torch, or it may be in the gas line to the torch. An electric solenoid valve is controlled automatically by the welding current flow or by a manual switch.

2.5.0 GTAW Filler Metals

Filler metal for manual GTAW is generally supplied in 36" (914.4 mm) lengths in diameters from 1⁄16" (1.6 mm) to 1⁄4" (6.4 mm). There are some automatic and manually operated wire feeders used with manual GTAW, but they are usually found only in high-production facilities. Filler metals for GTAW are drawn from high-grade pure alloys compounded for specific applications. Generally, the rods are not coated (bare, without a flux component added) except for a corrosion-resistant copper electroplating on some carbon steel rods. However, some special-purpose rods may have flux cores or coatings.

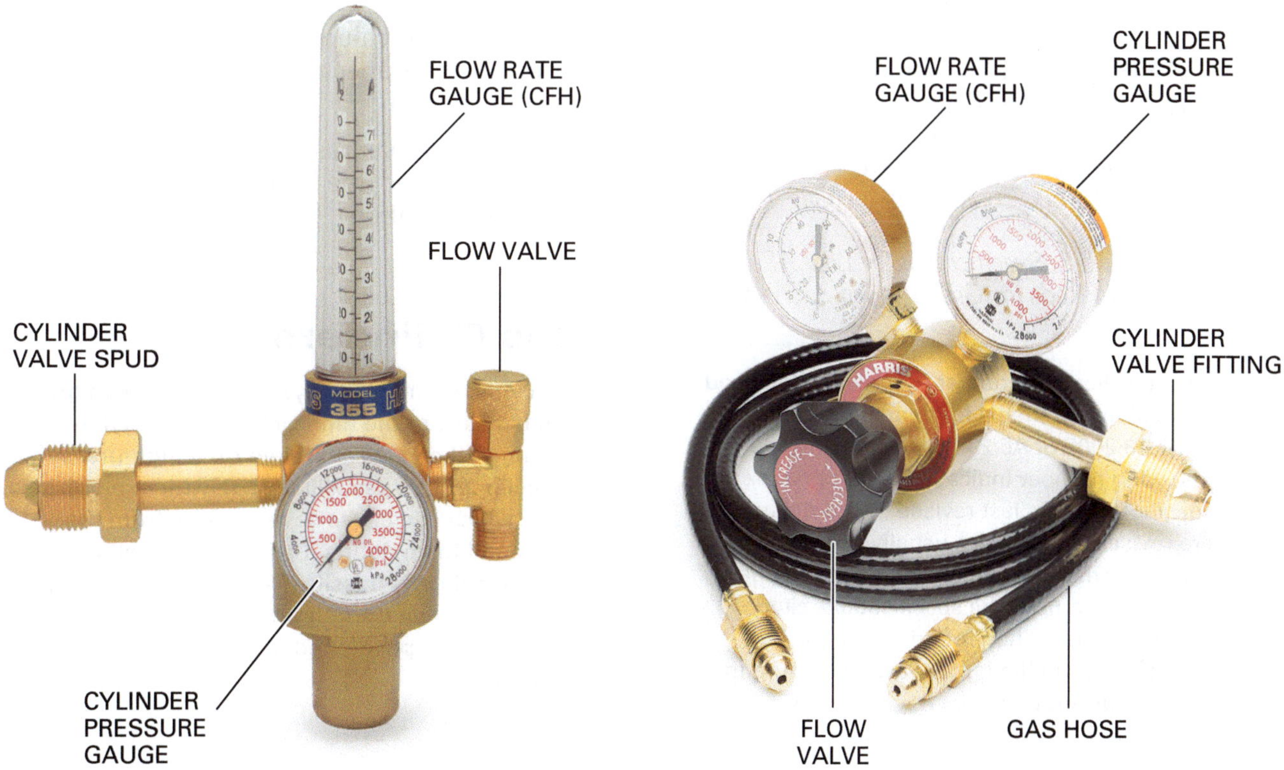

Figure 20 Shielding-gas flowmeter and gas regulator.

Several industry organizations and the United States government publish specification standards for filler metals. The most common standards are those published by the American Welding Society. The purpose of the AWS specifications is to set standards that all manufacturers must follow when manufacturing welding consumables. This ensures consistency for the user regardless of who manufactured the product. The specifications set standards for the following:

- Classification system, identification, and marking
- Chemical composition of the deposited weld metal
- Mechanical properties of the deposited weld metal

Examples of AWS specifications that apply to GTAW filler metals are the following:

- *AWS A5.7, Specification for Copper and Copper Alloy Bare Welding Rods and Electrodes*
- *AWS A5.9/A5.9M, Specification for Bare Stainless Steel Welding Electrodes and Welding Rods*
- *AWS A5.10/A5.10, Specification for Bare Aluminum and Aluminum Alloy Welding Electrodes and Rods*
- *AWS A5.14/A5.14M, Specification for Nickel and Nickel Alloy Base Welding Electrodes and Rods*
- *AWS A5.16/A5.16M, Specification for Titanium and Titanium Alloy Electrodes and Rods*
- *AWS A5.18/A5.18M, Specification for Carbon Steel Electrodes and Rods for Gas Shielded Arc Welding*
- *AWS A5.19, Specification for Magnesium Alloy Welding Electrodes and Rods*
- *AWS A5.22, Specification for Stainless Steel Electrodes for Flux Cored Arc Welding and Stainless Steel Flux Cored Rods for Gas Tungsten Arc Welding*
- *AWS A5.28/A5.2, Specification for Low Alloy Steel Electrodes and Rods for Gas Shielded Arc Welding*

NOTE

The specification number is generally followed by a colon and then a four-digit number, such as A5.18:2005 or A5.28:2015. The colon and following number indicate the year that the specification was last revised. These documents are revised periodically, usually every five years. For that reason, the year numbers have been omitted from the specifications cited in this module. When referring to an AWS specification on the job, always make sure to use the current version.

Filler metals and wire are graded for three major areas of use:

- *General use* – The wire or rod meets specifications. No record of chemical composition, strength, etc., is supplied to the user with the wire purchase.
- *Rigid control fabrication* – A Certificate of Conformance is supplied with the wire or rod at purchase. The stock is identified by heat numbers or code numbers located on the roll package.
- *Critical use* – A Certified Chemical Analysis report is supplied, and records are kept by the fabricator of welds and processes for later reference on aircraft, nuclear reactors, and pressure vessels.

Factors that affect the selection of a GTAW filler metal include the following:

- Base metal chemical composition
- Base metal mechanical properties
- Weld joint design
- Service or specification requirements
- Shielding gas used

The following are commonly used filler metal types:

- Carbon steel
- Stainless steel
- Aluminum and aluminum alloy
- Copper and copper alloy
- Low-alloy steel

Table 2 summarizes the GTAW filler metal specifications and the major AWS class covered by each specification.

Gas Cylinders

Gas cylinders must always be handled with great care because of the high-pressure gases they contain. If the valve is broken off a high-pressure gas cylinder, the gas will escape with explosive force, which can cause severe injury from the blast and blown debris. The cylinder itself may become a deadly missile. The tanks must always be chained in an upright position when being transported or positioned in their storage or use locations.

Table 2 GTAW Filler Metal Specifications and Classification System

Material	Filler Metal		X Designator Description	Example
	AWS Spec.	AWS Class*		
Carbon steel	A5.18	ERXXS-Y EXXC-Y	Tensile strength × 1,000 (psi)	ER70S-3 E70C-3
Low-alloy steel	A5.28	ERXXS-Y EXXC-Y	Tensile strength × 1,000 (psi)	ER80S-B2 E80C-B2
Stainless steel	A5.9	ERXXXY	Stainless alloy (308, 410, etc.)	ER308L EC308L
Stainless steel (flux-cored)	A5.22	RXXXTi-5	Stainless alloy (308, 309, etc.)	R309LT1-5
Aluminum	A5.10	ERXXXX-Y	Aluminum alloy (4043, 5083, etc.)	ER4043
Nickel	A5.14	ERNiXX-Y	Major alloying elements (Cr, Fe, Mo, etc.)	ERNiCr-3
Copper	A5.7	ERCuXX-Y	Major alloying elements (Al, Ni, Si, etc.)	ERCuAl-A2
Magnesium	A5.19	ERXXYYY	Major alloying elements (Al, Zn, etc.)	ERAZ92A
Titanium	A5.16	ERTi-Y		ERTi-5

*Legend

E – Filler metal may be used as an electrode.
R – Filler metal may be used as a rod.
S – Solid filler metal.
C – Composite or stranded filler metal.
Y – Designator (or combination of designators) that describes specific alloy, shielding gas to be used, diffusible hydrogen limit, etc.
Refer to the appropriate AWS Filler Metal Specification shown in table for explanation.

The AWS classification of metal filler rod is identified on its container. In addition, each rod is usually marked either by stamping the AWS classification into the rod or by a paper or plastic tab wrapped around the rod (*Figure 21*). Consume filler metal rods so that the ID marker is not destroyed. Markers are located at each end. Cutting a rod in half allows you to retain the marker on the used stub. Retaining the stubs with the markings until the weld is complete provides evidence that the correct filler metal was used.

2.5.1 Carbon Steel and Low-Alloy Steel Filler Metals

Carbon steel filler metals are identified by *AWS Specification A5.18*. Low-alloy steel filler metals are identified by *AWS Specification A5.28*. The rod classification number is found on a label on the packaging. *Figures* 22 and 23 show the AWS classifications for carbon steel and low-alloy steel filler metals.

All steel filler metals contain alloys such as silicon, manganese, aluminum, and carbon. Other alloys such as nickel, chromium, and molybdenum are also often added. The purpose of the alloys is as follows:

- *Silicon (Si)* – Concentrations of 0.40 to 1.00 percent are used to deoxidize the puddle and to strengthen the weld. Silicon above 1 percent may make the welds crack-sensitive.
- *Manganese (Mn)* – Concentrations of 1 to 2 percent are also used as a deoxidizer and to strengthen the weld. Manganese also decreases hot-crack sensitivity.
- *Aluminum (Al), Titanium (Ti), and Zirconium (Zr)* – One or more of these elements may be added in very small amounts for deoxidizing. These elements may also increase strength.
- *Carbon (C)* – Concentrations of 0.05 to 0.12 percent are used to add strength without adversely affecting ductility, porosity, or toughness.
- *Nickel (Ni), Chromium (Cr), and Molybdenum (Mo)* – These elements may be added in small amounts to improve corrosion resistance, strength, and toughness.

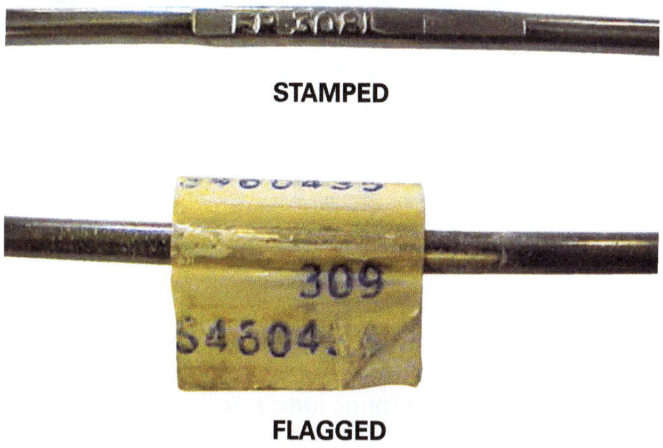

Figure 21 Filler rod markings.

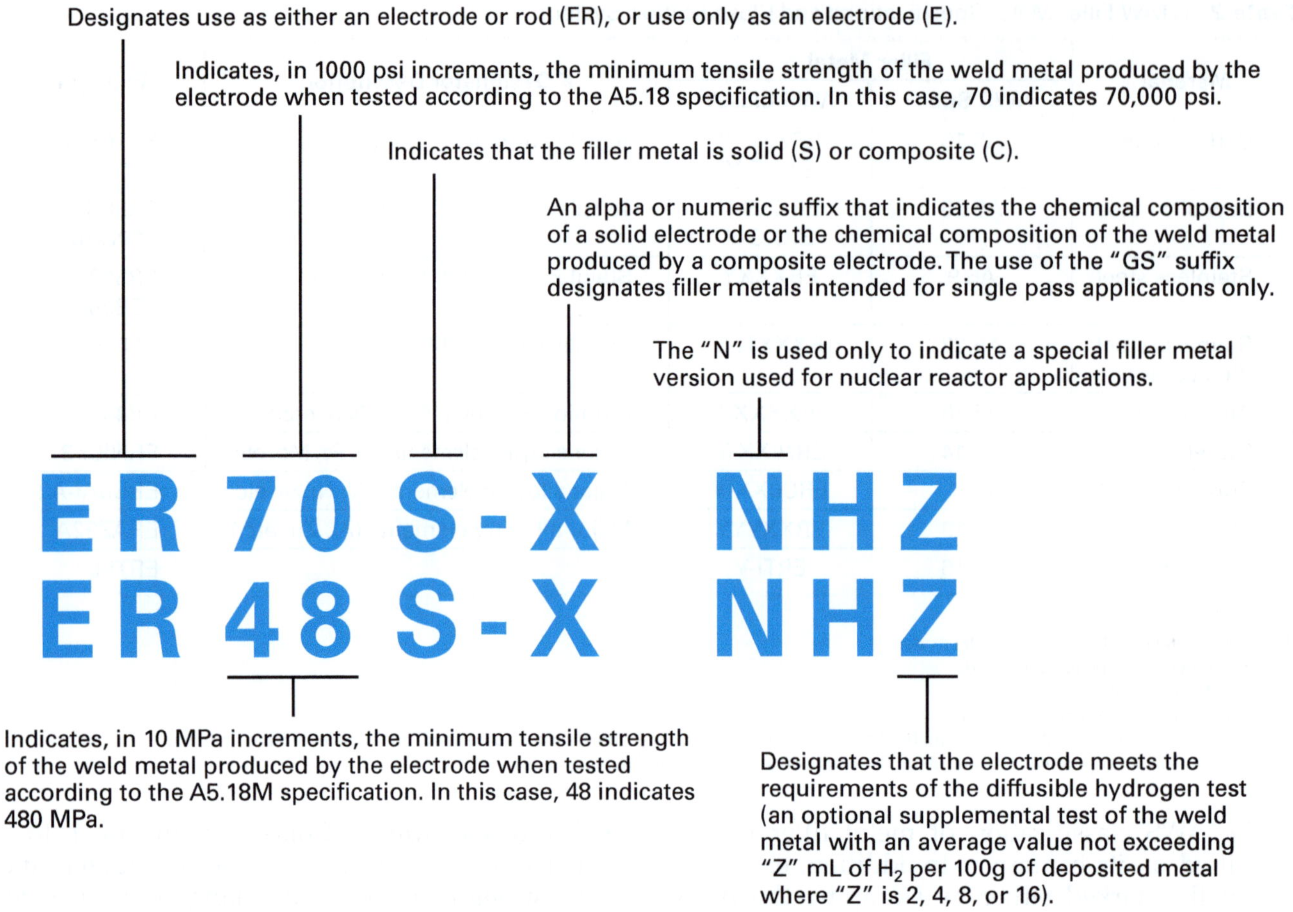

Figure 22 AWS classification for carbon steel filler metals.

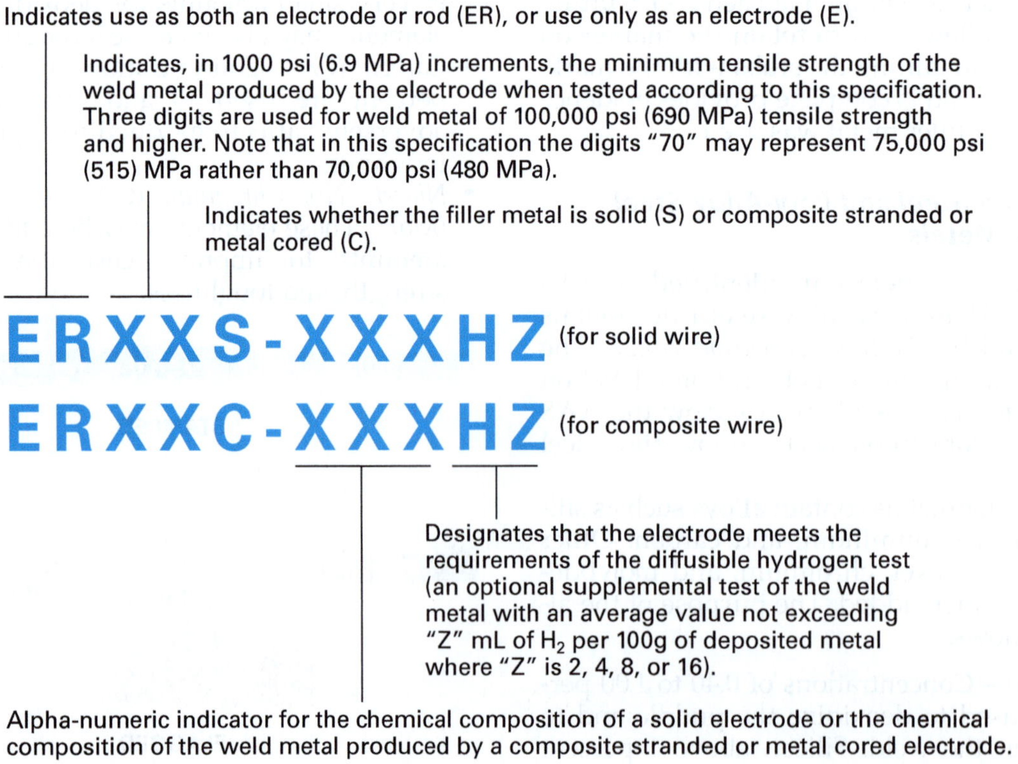

Figure 23 AWS classification for low-alloy steel filler metals.

2.5.2 Stainless Steel Filler Metals

Stainless steel filler metals are identified by *AWS Specification A5.9.* Designations for a typical AWS stainless steel electrode rod classification are shown in *Figure 24.*

Select stainless steel filler metal to closely match the alloy composition of the base metal.

2.5.3 Aluminum and Aluminum-Alloy Filler Metals

Aluminum electrode rods are covered by *AWS Specification A5.10* as shown in *Figure 25.* Aluminum filler metals usually contain magnesium, manganese, zinc, silicon, or copper for increased strength. Corrosion resistance and ease of being welded are also considerations. Aluminum filler metals are designed to weld various types of aluminum and should be selected for compatibility. The most widely used aluminum rods are ER4043 (contains silicon) and ER5356 (contains magnesium).

2.5.4 Copper and Copper Alloy Filler Metals

Copper electrode rods are covered by *AWS Specification A5.7.* Most copper filler metals contain other elements to increase strength, deoxidize the weld metal, and match the base metal composition (*Figure 26*).

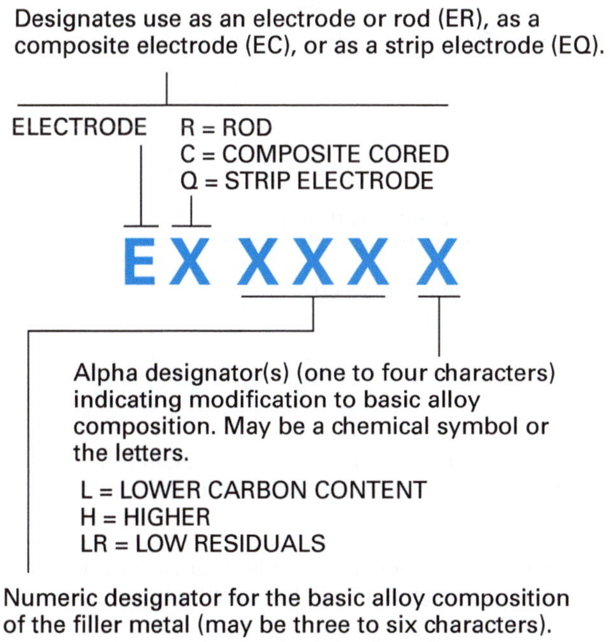

Figure 24 Typical AWS stainless steel rod classification.

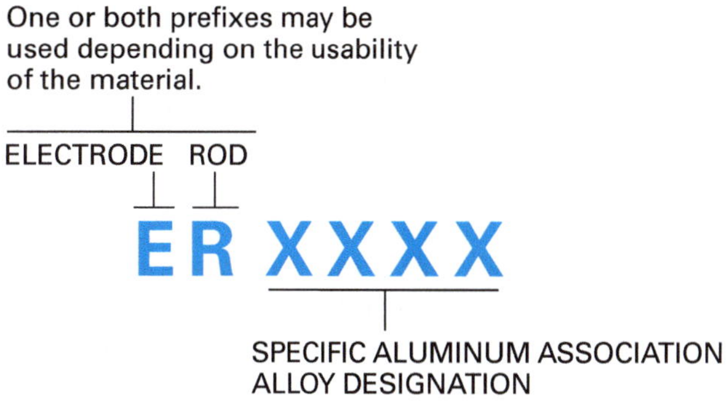

Figure 25 Typical AWS aluminum filler metal classification.

2.5.5 Nickel and Nickel Alloy Filler Metals

Nickel-based filler metals are covered by *AWS Specification A5.14.* These filler metals contain other elements to match base metal applications and to increase the strength and quality of the weld metal. For GTAW, DCEN is used with high-purity argon, helium, or both argon and helium used as a shielding gas. *Figure 27* shows the AWS classification for nickel filler metals.

2.5.6 Magnesium Alloy Filler Metals

Magnesium alloy filler metals (*Figure 28*) are covered by *AWS Specification A5.19.* These filler metals are usually used with GTAW, GMAW, and plasma arc welding (PAW) processes. For GTAW welding, the techniques and equipment are similar to those for aluminum. Argon, helium, or both are used for shielding. Alternating current is preferred for arc cleaning and penetration. Direct current is also employed with DCEP, used for thin materials; and DCEN, used for mechanized welding with helium, for deep penetration. GTAW is recommended for the defect repair of clean magnesium castings.

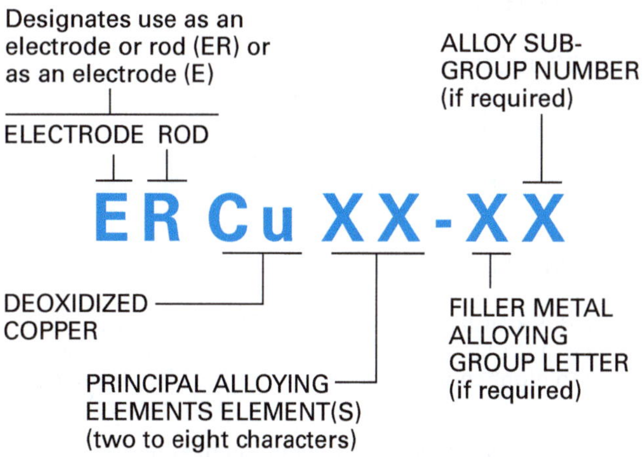

Figure 26 Typical AWS copper filler metal classification.

2.5.7 Titanium and Titanium Alloy Filler Metals

Titanium and titanium alloy filler metals (*Figure 29*) are covered by *AWS Specification A5.16*. These filler metals are generally used with GTAW, GMAW, SAW, and PAW processes. Titanium is sensitive to embrittlement by oxygen, nitrogen, and hydrogen at temperatures above 500°F (260°C). Like aluminum, titanium requires weld cleaning and high-purity gas shielding, and an especially adequate postflow shielding, to avoid embrittlement. Titanium can be successfully fusion-welded to zirconium, tantalum, niobium, and vanadium. Titanium should not be fusion-welded to copper, iron, nickel, or aluminum. Like magnesium, titanium fines are also flammable, so don't allow them to accumulate.

Rod and Electrode Designations

The letter R that follows the E in an AWS classification number indicates that the classification is used as a metal filler rod as well as an electrode. Electrode wire is used in GMAW. Since a tungsten-based electrode is used in GTAW, filler metal classifications with an ER must be used.

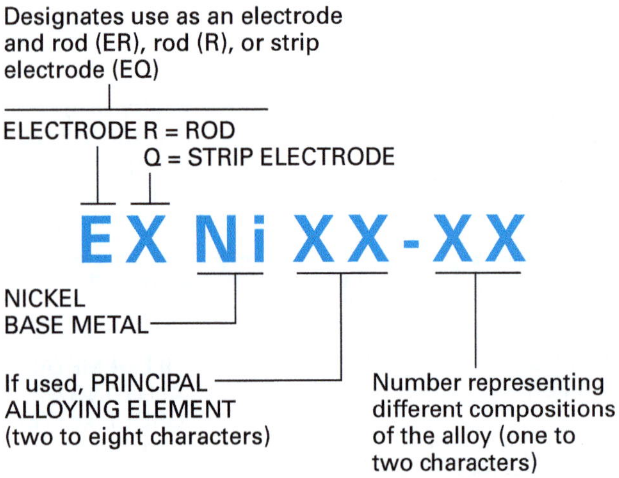

Figure 27 AWS classification for nickel filler metals.

2.5.8 Stainless Steel Flux-Cored Electrodes and Rods

Figure 30 shows the *AWS A5.22* specification format for stainless steel flux-cored electrodes and rods.

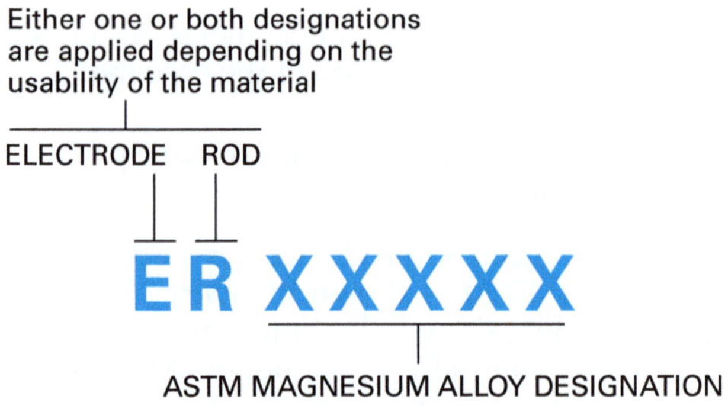

Figure 28 AWS classification for magnesium filler metals.

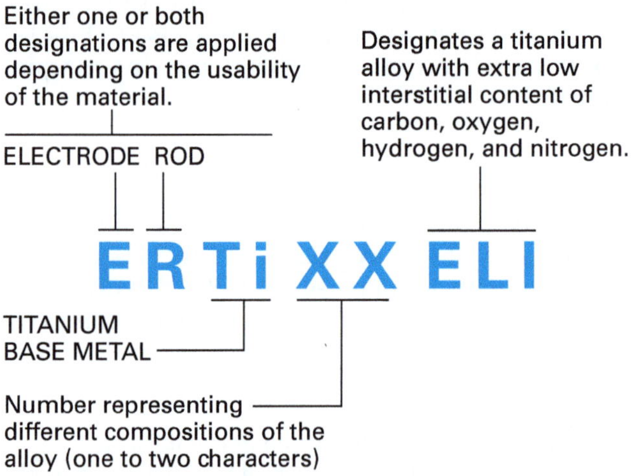

Figure 29 AWS classification for titanium filler metals.

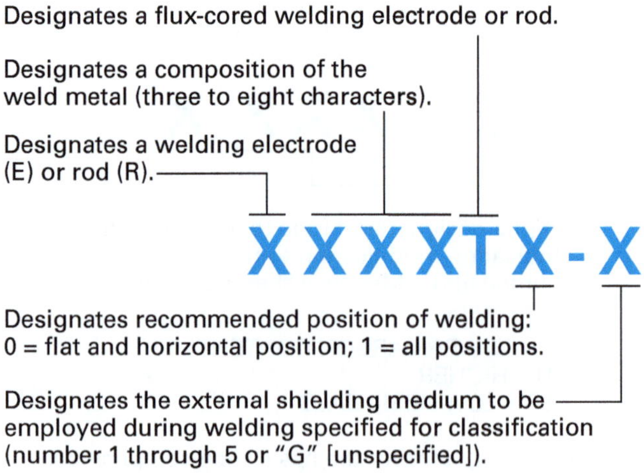

Figure 30 AWS classification for stainless steel flux-cored electrodes and rods.

Magnesium Fires

While ignition of magnesium is a very remote possibility when welding, the fire will cease when the heat source is removed. Ignition of the weld pool is prevented by the gas shielding used in GTAW, GMAW, and PAW processes. Most magnesium fires occur when fines from grinding or filing, or chips from machining, are allowed to accumulate. Take care to prevent the accumulation of fines on clothing. Graphite-based or salt-based powders recommended for extinguishing magnesium fires should be stored in the work area. If large amounts of fines are produced, collect them in a waterwash-type of dust collector designed for use with magnesium. Follow special precautions for the handling of wet magnesium fines.

Additional Resources

AWS B1.10, Guide for the Nondestructive Examination of Welds. Miami, FL: American Welding Society.

AWS B1.11, Guide for the Visual Examination of Welds. Miami, FL: American Welding Society.

AWS D3.5, Guide for Steel Hull Welding. Miami, FL: American Welding Society.

AWS D3.7, Guide for Aluminum Hull Welding. Miami, FL: American Welding Society.

AWS Safety and Health Fact Sheet 27. Miami, FL: American Welding Society.

Lincoln Electric website: **http://www.lincolnelectric.com**

Modern Welding Technology. 6th Edition. Howard B. Cary. Englewood Cliffs, NJ: Prentice Hall, Inc.

OSHA 1910.269, Appendix *C, Protection from Step and Touch Potentials*. Current Edition. Washington, DC: Occupational Safety and Health Administration (OSHA).

OSHA 19260.351, Arc Welding and Cutting. Current Edition. Washington, DC: Occupational Safety and Health Administration (OSHA).

Tungsten Electrodes: An **Arc-Zone.com**® Technical Focus Paper (pdf). Carlsbad, CA: **Arc-Zone.com**, Inc. **https://www.arc-zone.com/azc-scip/TungstenElectrodes.pdf**

Tungsten Guidebook (pdf). Newbury Park, CA: Diamond Ground Products, Inc. **http://www.diamond-ground.com/TungstenGuidebook2013.pdf**

Welding Handbook. Vol. 1. Welding Science & Technology. 9th Edition. Miami, FL: American Welding Society, 2001.

Welding Handbook. Vol. 2, Part 1: Welding Processes. Miami, FL: American Welding Society, 2004.

2.0.0 Section Review

1. A type of welding machine that is designed to produce either direct welding current only or both alternating and direct welding currents is a(n) _____.
   a. transformer welding machine
   b. transformer-rectifier welding machine
   c. alternator welding machine
   d. inverter power source

2. How many hoses does a water-cooled GTAW torch have?
   a. One
   b. Two
   c. Three
   d. Four

3. The device that contains an assembly of fine screens that eliminate turbulence by straightening the gas flow from a GTAW gas nozzle is called a _____.
   a. laminar flow gas lens
   b. shielding gas supply hose
   c. closed-loop gas rectifier
   d. gas lens collet body

4. In GTAW, small amounts of hydrogen are sometimes mixed with argon for deeper penetration when welding _____.
   a. titanium
   b. aluminum
   c. carbon steel
   a. stainless steel

5. To increase strength, deoxidize the weld metal, and match the base metal composition, most copper electrodes contain _____.
   a. other elements
   b. gas lens filters
   c. magnesium castings
   d. anodized quenchers

SECTION THREE

3.0.0 WELDING EQUIPMENT SETUP

Objective

Explain how to set up for GTAW welding.

a. Explain how to select and position the welding machine.
b. Explain how to connect and set up the shielding gas flow rate.
c. Explain how to select and prepare the tungsten electrode.
d. Explain how to select and install the nozzle along with the tungsten electrode.

Performance Tasks

3. Connect the shielding gas and set the flow rate.
4. Select and prepare the tungsten electrode.
5. Break down and reassemble a GTAW torch.

To weld safely and efficiently, all the necessary welding equipment must be properly set up. The following sections explain how to properly set up GTAW equipment.

3.1.0 The GTAW Welding Machine

Select the welding machine and current type for the base metal to be welded, considering the following factors:

- GTAW requires a constant-current power source. SMAW power sources and welding machines are commonly used for GTAW.
- Consider the welding current type required, AC or DC; generally, carbon steels, stainless steels, and alloy steels are welded with DC, while aluminum and magnesium are welded with AC at high frequency.
- Determine the maximum amperage required.
- Determine if there is an electrical power source and appropriate receptacles in which to plug a welding machine. Otherwise, an engine-driven welding machine is needed.

Because of the limited length of GTAW torch cables, the shielding gas supply must be located reasonably close to the welding site. The welding machine can be some distance away, but unless remote controls are installed, it will not be convenient to operate.

Select a site where the GTAW equipment will not be in the way but will be protected from welding, grinding, or cutting sparks. There should be good air circulation to keep the welding machine cool. The environment should be free from explosive or corrosive fumes and as free as possible from dust and dirt. Welding machines have internal cooling fans that will pull these particles into the machines if they are present. The site should also be free of standing water or water leaks. If an engine-driven generator is used, locate it so that it can be easily refueled and serviced. Also, make sure that the area around the generator is well ventilated.

There should be easy access to the site so that the equipment can be started, stopped, or adjusted as needed. If the machine will be plugged into an outlet, be sure that the outlet has been properly installed and properly grounded by a licensed electrician. Also, be sure to identify the location of the electrical power disconnect switch or circuit breaker before plugging the welding machine into the outlet.

3.2.0 The Shielding Gas

The hose from the shielding gas regulator and flowmeter connects to the welding machine gas solenoid if the welding machine is designed for GTAW. If a standard welding machine is used, the shielding gas hose connects to the torch cable. To connect the shielding gas, perform the following steps:

Step 1 Identify the shielding gas required by referring to the WPS or site quality standard.

Step 2 Locate a cylinder of the correct gas or mixture, and secure it nearby.

WARNING!

Be sure to secure the gas cylinder first so it cannot fall over. Do not remove the cylinder cap until the cylinder is secured.

Step 3 Remove the cylinder's protective cap, momentarily crack open the cylinder valve to blow out any dirt, and then close it again.

Step 4 Using a regulator or combination regulator/flowmeter with the correct CGA connection for the cylinder, mount the regulator on the cylinder.

Step 5 Connect the gas hose to the flowmeter and to the gas solenoid on the welding machine or to the end of the torch cable.

NOTE

If the gas hose is connected to the torch cable and the torch does not contain a gas shutoff valve, install a valve in the line between the torch cable and the regulator/flowmeter.

Step 6 Check to be sure the flowmeter adjusting valve is closed.

Step 7 Very slowly open the cylinder valve slightly. After the pressure gauge indicates the cylinder pressure and is stable, open the valve completely.

Figure 31 shows how to connect the shielding gas.

3.2.1 Setting the Shielding Gas Flow Rate

Gas flow rate is measured in cubic feet per hour (ft^3/hr or cfh) or liters per minute (L/min), which represents the volume of gas flowing from the torch nozzle. Flow rate is important because it affects the quality and the cost of a weld. A flow rate that is too low will not shield the weld zone adequately and will result in a poor-quality weld. An excessively high gas flow rate wastes expensive gas and can generate turbulence in the gas column above the weld. This turbulence can pull air into the weld zone and cause oxidation and weld contamination. The nozzle must be large enough to gently flood the weld pool with inert gas. As a general rule, larger nozzles require higher flow rates. Welding specifications contain nozzle sizes and shielding gas flow rates.

Factors that may affect the shielding gas flow rate include the following:

- *Drafts* – The flow rate must be increased in a drafty location to maintain the gas shield around the weld zone.
- *Specific gas used* – For example, helium usually requires a higher flow rate than argon because helium rises much faster because of its low density.
- *Welding current* – High welding currents require higher flow rates.
- *Nozzle size (exit opening)* – Larger nozzles require higher flow rates.
- *Weld joint type* – Welds on flat surfaces require higher flow rates than welds in deep grooves or fillets.
- *Welding speed* – Fast advance speeds require higher flow rates than slower advance speeds.
- *Weld position* – Vertical- and horizontal-position welds require higher flow rates than flat or overhead welds.

Adjust the flow for the type and thickness of the base metal being welded, or per applicable WPS. To set the shielding gas flow rate, perform the following steps.

Step 1 Start the gas flow by depressing the remote control foot pedal or hand-operated switch to open the gas solenoid on the welder, or manually open the hand valve on the torch.

CAUTION

Some flowmeters are preset for a specific gas. These flowmeters do not require any adjustment, but they can only be used with the specific shield gases indicated on the flowmeter. Due to density differences in the various shielding gases, using this type of flowmeter with the wrong gas will result in incorrect flow rate readings.

Step 2 Set the gas flow by turning the flowmetering adjustment knob out until the ball inside the glass tube rises to the desired flow rate (*Figure 32*). The flow rate is usually read from the top of the floating ball. However, always check the manufacturer's instructions to determine the correct adjustment method.

Step 3 When the proper gas flow has been achieved, release the remote foot pedal or hand switch, or manually close the valve on the torch to stop the gas flow.

NOTE

Some flowmeters are equipped with several scales of different calibrations around the same sight tube for monitoring the flows of different types (densities) of gases. Be sure to rotate the scales or read the correct side for the gas type being used.

Step 4 Check for leaks before starting to weld.

The following sections provide general recommendations for flow rates for various types of base metals. Note that these flow rates can vary; always check the WPS for the appropriate flow rate.

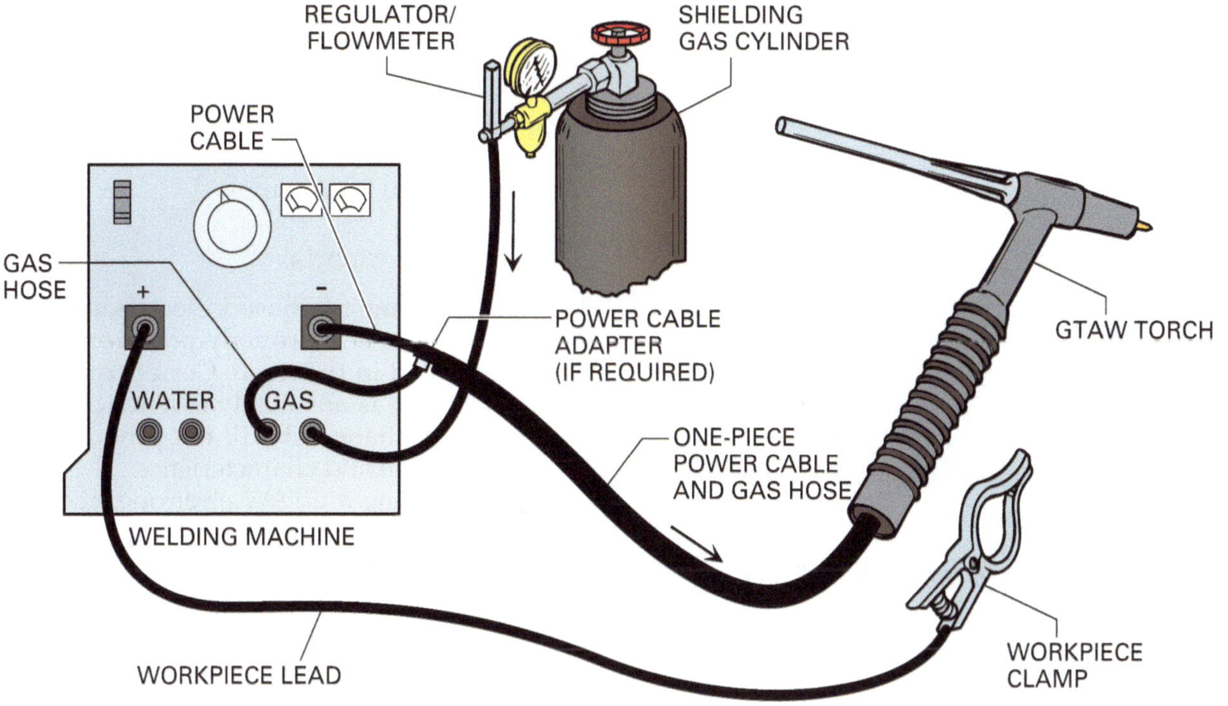

WARNING: IF A POWER CABLE ADAPTER IS USED IT MUST BE COVERED BY AN INSULATING RUBBER BOOT.

SINGLE-HOSE, GAS-COOLED TORCH SETUP

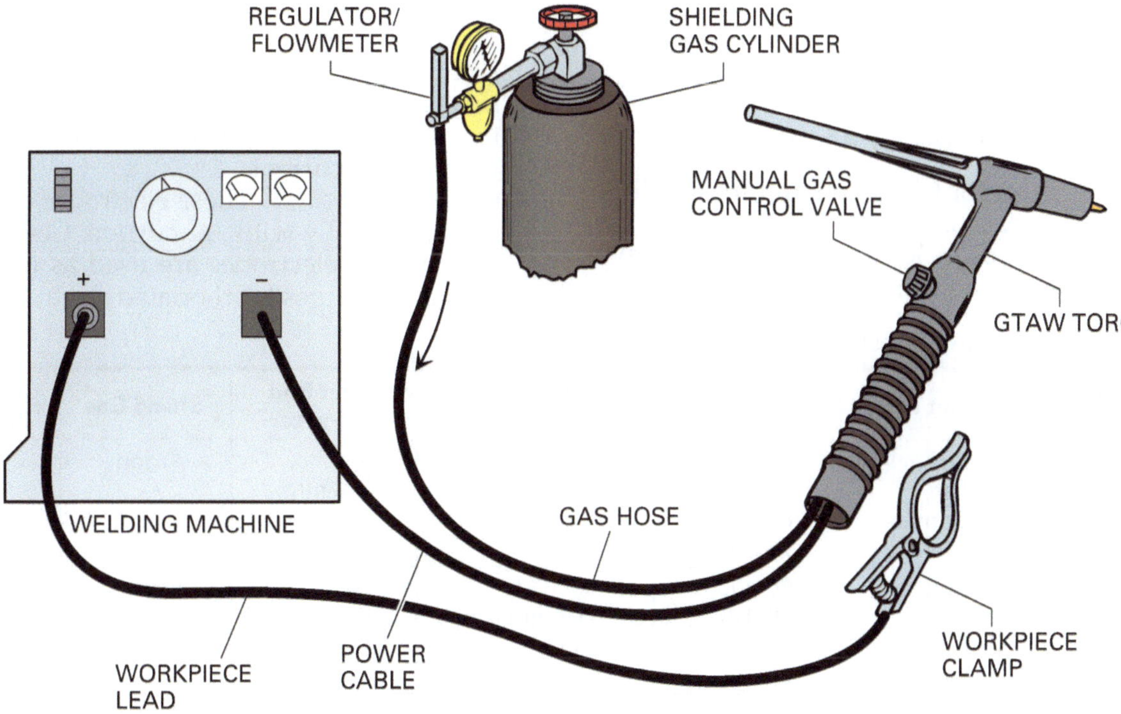

DOUBLE-HOSE, GAS-COOLED TORCH SETUP

Figure 31 Connecting the shielding gas.

Figure 32 Adjusting the flowmeter.

3.2.2 Gas Flow Rates for Carbon Steel

Argon flow rates for GTAW on low- and medium-carbon steels and low-alloy steels (*Table 3*) typically vary from 15 cfh (7 L/min) for light welding at 60 amps, to 20 cfh (9 L/min) for heavier welding at 200 or more amps.

3.2.3 Gas Flow Rates for Stainless Steel

Argon flow rates for GTAW on stainless steel (*Table 4*) typically vary from 10 cfh (5 L/min) for light welding at 80 amps to 15 cfh (7 L/min) for heavy welding at 350 or more amps.

3.2.4 Gas Flow Rates for Aluminum

Argon flow rates for GTAW on aluminum (*Table 5*) typically vary from 15 cfh (7 L/min) for light welding at 60 amps to 30 cfh (14 L/min) for heavy welding at 350 or more amps.

3.2.5 Gas Flow Rates for Copper

A typical argon flow rate for GTAW on deoxidized copper up to ⅛" (3.2 mm) thick is 15 cfh (7 L/min) from 110 to 250 amps. With copper over ⅛" (3.2 mm) thick, use helium at 30 to 40 cfh (14 to 19 L/min) from 190 to 525 amps.

3.3.0 The Electrode

The GTAW tungsten-based electrode must be selected and the end properly prepared before it can be installed in the torch. Correct preparation of the electrode is absolutely essential. If the tip is improperly shaped, it will not produce the required arc shape and characteristics.

When selecting a GTAW electrode, choose the type recommended for the welding current required by the base metal. The type of electrode will be specified in the WPS or site quality standards. General recommendations are as follows:

- *Pure tungsten (EWP)* – These electrodes are used for welding aluminum and magnesium with AC and high frequency.
- *Zirconiated tungsten (EWZr)* – Used for welding aluminum and magnesium with AC and high frequency for welds where tungsten inclusions are not tolerated and higher current capacity is desired.
- *Thoriated tungsten (EWTh-1 or EWTh-2), ceriated tungsten (EWCe-2), or lanthanated tungsten (EWLa-1, -1.5, or -2)* – These electrodes are used for mild steels, alloy steels, and stainless steels with DCEN current.
- *Ceriated electrodes* – These electrodes also operate successfully with AC current. Ceriated and lanthanated electrodes are used as non-radioactive substitutes for thoriated electrodes.

Table 3 Carbon Steel—Direct Current Electrode Negative (DCEN)

Metal Thickness	Joint Type	Amperage	Electrode Diameter	Filler Rod Diameter	Shield Gas	Cubic Feet per Hour
1⁄16" (1.6 mm)	Butt Lap Corner Fillet	60–70 70–90 60–70 70–90	1⁄16" (1.6 mm)	1⁄16" (1.6 mm)	Argon	15
⅛" (3.2 mm)	Butt Lap Corner Fillet	80–100 90–115 80–100 90–115	1⁄16"–3⁄32" (1.6 mm–2.4 mm)	3⁄32" (2.4 mm)	Argon	15
3⁄16" (5 mm)	Butt Lap Corner Fillet	115–135 140–165 115–135 140–170	3⁄32" (2.4 mm)	⅛" (3.2 mm)	Argon	20
¼" (6 mm)	Butt Lap Corner Fillet	160–175 170–200 160–175 175–210	⅛" (3.2 mm)	5⁄32" (4 mm)	Argon	20

Table 4 Stainless Steel—Direct Current Electrode Negative (DCEN)

Metal Thickness	Joint Type	Amperage (Flat Position)	Electrode Diameter	Filler Rod Diameter	Shield Gas	Cubic Feet per Hour
1⁄16" (1.6 mm)	Butt Lap Corner Fillet	80–100 100–120 80–100 90–110	1⁄16" (1.6 mm)	1⁄16" (1.6 mm)	Argon	10
1⁄8" (3.2 mm)	Butt Lap Corner Fillet	120–140 130–150 120–140 130–150	1⁄16" (1.6 mm)	3⁄32" (2.4 mm)	Argon	10
3⁄16" (5 mm)	Butt Lap Corner Fillet	200–250 225–275 200–250 225–275	3⁄32" or 1⁄8" (2.4 mm or 3.2 mm)	1⁄8" (3.2 mm)	Argon	15
1⁄4" (6 mm)	Butt Lap Corner Fillet	275–350 300–375 275–350 300–375	1⁄8" (3.2 mm)	5⁄32" (4 mm)	Argon	15
1⁄2" (12.5 mm)	Butt Lap Fillet	350–450 375–475 375–475	1⁄8" or 3⁄16" (3.2 mm or 4.8 mm)	3⁄16" (4.8 mm)	Argon	15

Table 5 Aluminum—Alternating Current with High Frequency Stabilization

Metal Thickness	Joint Type	Amperage (Flat Position)	Electrode Diameter	Filler Rod Diameter	Shield Gas	Cubic Feet per Hour
1⁄16" (1.6 mm)	Butt Lap Corner Fillet	60–85 70–90 60–85 75–100	1⁄16" (1.6 mm)	1⁄16" (1.6 mm)	Argon	15
1⁄8" (3.2 mm)	Butt Lap Corner Fillet	125–150 130–160 120–140 130–160	3⁄32" (2.4 mm)	3⁄32" or 1⁄8" (2.4 mm or 3.2 mm)	Argon	20
3⁄16" (5 mm)	Butt Lap Corner Fillet	180–225 190–240 180–225 190–240	1⁄8" (3.2 mm)	1⁄8" (3.2 mm)	Argon	20
1⁄4" (6 mm)	Butt Lap Corner Fillet	240–280 250–320 240–280 250–320	3⁄16" (4.8 mm)	1⁄8" or 3⁄16" (3.2 mm or 4 mm)	Argon	25
1⁄2" (12.5 mm)	Butt Lap Corner Fillet	400–450 400–450 400–450 420–470	3⁄16" or 1⁄4" (4.8 mm or 6.4 mm)	3⁄16" or 1⁄4" (4.8 mm or 6.4 mm)	Argon	30

- *Rare earth electrodes (EW)* – These electrodes are used as non-radioactive substitutes for 2-percent thoriated electrodes. They start and re-ignite well and have very good service life for applications in which welding cycles of at least 15 minutes are used.
- *Cryogenically treated thoriated tungsten electrodes* – These electrodes are used as non-radioactive substitutes for 2-percent thoriated electrodes. The cryogenically treated electrodes are easier to ignite and have longer tip life, longer overall electrode life, and higher amperage tolerance than non-treated 2-percent thoriated tungsten.

Select the size of electrode rated for the amperage to be used. *Table 6* lists examples of electrode current ranges by electrode size and type when used with argon shielding gas under various GTAW modes.

3.3.1 Preparing the Electrode

The different types of GTAW welding current require different-shaped electrode tips. For manual DCEN welding, a taper-ground and truncated point is the best shape. For manual DCEP welding, a rounded end is the best shape. For manual AC welding, a tapered and balled end is the best shape. *Figure 33* shows GTAW electrode end shapes.

Electrode tips must be formed on all new electrodes and on used electrodes that have damaged or contaminated tips. The contaminated section must be reshaped. Do not prepare the end of the electrode that contains the color-code band; otherwise you will not be able to identify the electrode's type in the future.

The degree of taper on an electrode tip affects weld penetration and bead width. A long, narrow

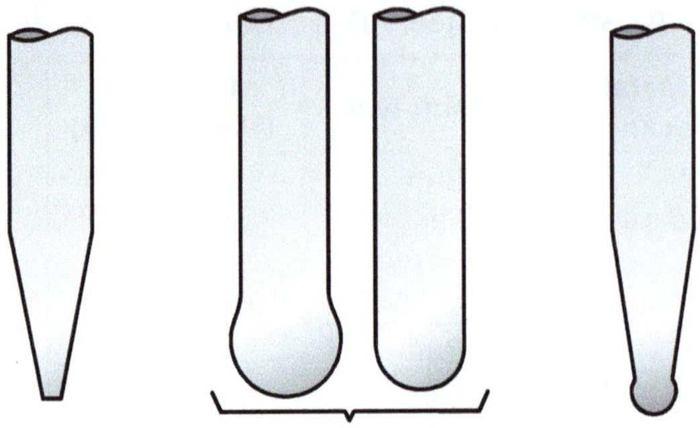

Figure 33 GTAW electrode end shapes for different current types.

taper (small angle, more pointed) produces a wide bead with shallow penetration. A short, wide taper (large angle, less pointed) produces a narrow bead and deep penetration. During any particular welding operation, it is important to maintain the same taper on the electrode during sharpening or replacement so that the bead characteristics remain the same.

3.3.2 Pointing the Electrode Tip

For DCEN welding, electrodes can be made pointed by dipping the red-hot end into a special chemical powder, or they can be tapered by grinding or sanding.

Perform the following steps to use a chemical powder to sharpen a tungsten electrode:

Step 1 Place the tungsten electrode in the torch so that it extends about 1 inch (≈25 mm) from the gas nozzle.

Table 6 Electrode Current Ranges

Electrode Diameter	GTAW Modes			
	AC*		Pure Tungsten, Rare Earth, or Thoriated Tungsten Electrodes	
	Pure Tungsten	Thoriated or Rare Earth Electrodes**	DCEN	DCEP
0.020" (0.5 mm)	5–15	8–20	8–20	–
0.040" (1.0 mm)	10–60	15–80	15–80	–
1⁄16" (1.6 mm)	50–100	70–150	70–150	10–20
3⁄32" (2.4 mm)	100–160	140–235	150–250	15–30
1⁄8" (3.2 mm)	150–210	225–325	250–400	25–40
5⁄32" (4.0 mm)	200–275	300–425	400–500	40–55
3⁄16" (4.8 mm)	250–350	400–525	500–800	55–80
1⁄4" (6.4 mm)	325–475	500–700	800–1,000	80–125

* Maximum values shown have been determined using an unbalanced AC power source. If a balanced AC power source is used, either reduce these values by about 30 percent or use the next size electrode. This is necessary because of the higher heat input to the electrode in a balanced AC setup.

** Balled electrode tip ends can best be sustained at these current levels.

Step 2 Heat the tungsten electrode by shorting it on a copper plate or by striking an arc. Be sure shielding gas is flowing to protect the tungsten electrode.

Step 3 When the tip of the tungsten electrode is cherry-red for a distance of about four times the electrode diameter, place the end of the electrode in the chemical powder. For example, if the electrode diameter is ⅛" (3.2 mm), about ½" (13 mm) of the length should be cherry-red.

Step 4 Hold the end of the tungsten electrode in the powder for the time recommended on the container for the diameter of the electrode being pointed. The chemical powder will dissolve the tungsten electrode and form a sharp point at the proper angle. For an un-tapered (new or broken off) tungsten, the procedure may have to be repeated a number of times to obtain the desired shape.

Reading Torch Flow Rate

Welders must maintain accurate and consistent torch flow rates to create consistent welds and adhere to a WPS. The device shown here can be used to quickly measure shielding gas flow directly at the torch tip. This simple peashooter style is durable and convenient. Digital models offer a bit more sophistication. They can also be mounted on the gas line, providing a constant digital display of flow rate.

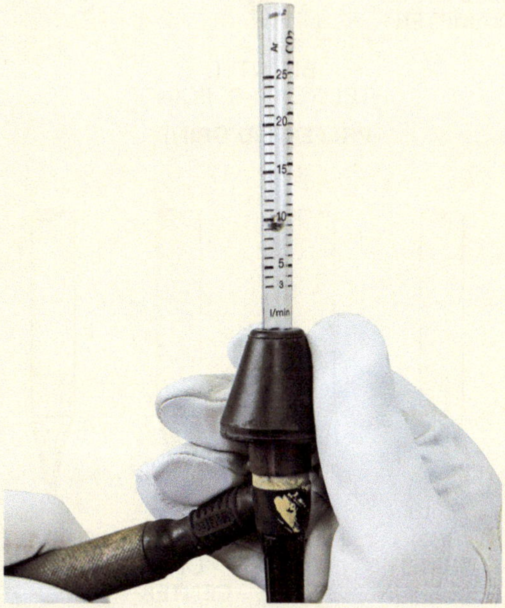

There Is No Universal Electrode Size

Contrary to some popular beliefs, the 3⁄32" (2.4 mm) electrode is not a universal size that can be used for everything. Always use the electrode size recommended for the amperage required. Using an electrode that's too big will result in arc starting and stability problems. Using an electrode that is too small can result in tungsten spitting (tungsten melting and migration into the weld).

Figure 34 shows the chemical powder used for sharpening.

CAUTION

Always wear clean gloves when handling tungsten electrodes to avoid contaminating the electrode.

More often, tungsten electrodes are sharpened by grinding or sanding. Since tungsten electrodes are very hard, they are best ground with diamond-impregnated silicon carbide or alumina oxide grinding wheels or sanding belts. Use 80 grit for fast shaping and 120 grit for finishing.

WARNING!

Observe proper respiratory, hazard, and environmental contamination procedures when using a grinder. Dust from tungsten electrode grinding can be hazardous and may contain radioactive elements. Refer to the manufacturer's SDS/MSDS to determine any specific hazards. Safety glasses must be worn when using a grinder.

Figure 34 Tungsten sharpening powder.

Follow these steps to point the end of a tungsten electrode using a grinder:

> **CAUTION**
>
> Use a grinding wheel designated only for tungsten. Be careful to never use the grinding wheel for any other material, or contaminants will be deposited on the tungsten electrode and the grinding wheel will have to be discarded. Always wear safety glasses when using a grinder.

Step 1 With the color band end held farthest from the grinder, hold the tungsten electrode against the grinder so that the tungsten electrode is in the plane of the wheel and makes an angle of about 30 degrees with the wheel or belt surface (*Figure 35*).

Step 2 Slowly rotate the tungsten electrode as it is ground to keep the point on the center line of the axis. Orient the grind marks toward the point along the long dimension of the tungsten electrode, not around the point. For DCEN electrodes, the length of the taper should be between two and three times the electrode diameter. The point should be truncated to the approximate tip diameter shown in *Table 7*.

Step 3 Check the ground end to make sure the tapered length is correct and that the point is in the center of the tungsten electrode and not toward one side. Correct the shape if necessary.

Figure 36 shows preferred and incorrect grinds for prepared pointed electrode ends.

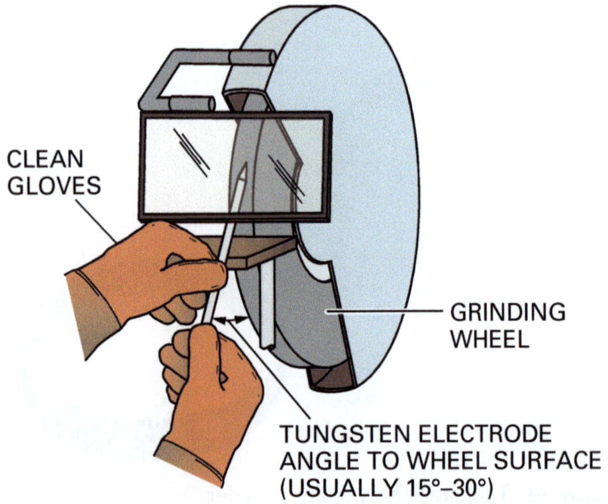

Figure 35 Proper grinding angle and tungsten electrode orientation.

Table 7 Tip Diameters

Electrode Diameter	Diameter at Tip
0.040" (1.0 mm)	0.005" (0.13 mm)
0.040" (1.0 mm)	0.010" (0.25 mm)
0.062" (1.6 mm)	0.020" (0.51 mm)
0.062" (1.6 mm)	0.030" (0.76 mm)
0.093" (2.4 mm)	0.030" (0.76 mm)
0.093" (2.4 mm)	0.045" (1.14 mm)
0.125" (3.2 mm)	0.045" (1.14 mm)
0.125" (3.2 mm)	0.060" (1.52 mm)

3.3.3 Balling the Electrode Tip

Only pure tungsten (EWP) or zirconiated tungsten (EWZr) can be balled, or spherical. The thorium in thoriated tungstens or other rare earth (except cerium) tungstens inhibits the formation of a ball. If welders attempt to ball most rare earth tungstens, the tungsten forms a number of small protrusions at the tip instead of a ball. These protrusions cause the arc to be unstable.

If the electrode is to be used for lower-amperage DCEP welding, the electrode end should be tapered about two diameters in length and be quite blunt. A ball will form on the blunt end when welding.

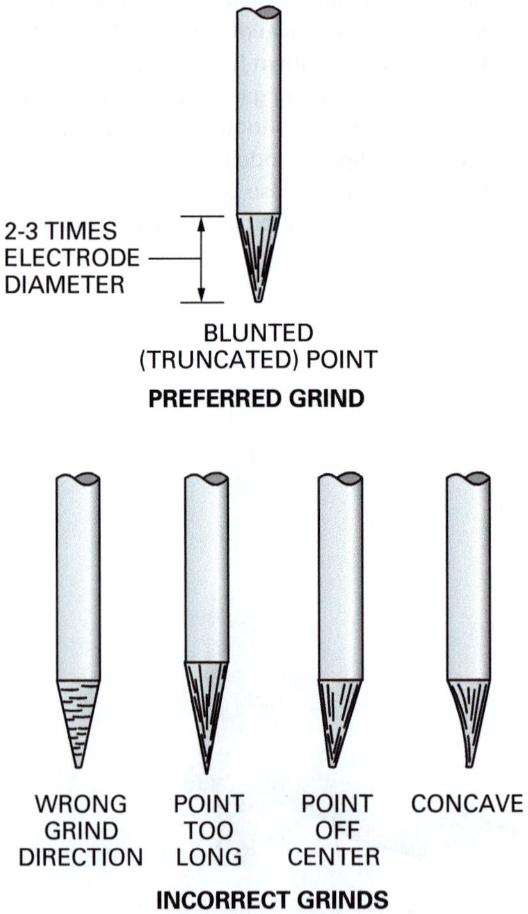

Figure 36 Properly and improperly prepared pointed electrode ends.

For AC welding, the end is balled with or without tapering. Balling is done by arcing the electrode over clean copper or another clean metal. Do not arc over carbon, because it will contaminate the electrode.

AC or DCEP can be used to ball an electrode. When balling an electrode, start off with a low current and gradually increase it until the end of the electrode starts to melt and form a hemispherical (half-round) end or slight ball. The ball should be no more than 1½ times the electrode diameter. Using DCEP will cause a ball to form at a much lower amperage setting than it will be formed using AC current. Using DCEP, only about 35 to 45 amps of welding current will be required to ball ³⁄₃₂" (2.4 mm) to ⅛" (3.2 mm) diameter electrodes. *Figure 37* shows ideal, acceptable, and unacceptable prepared balled electrode ends.

> **CAUTION**
>
> Using excess amperage will cause the ball to melt and drop off the electrode. Be especially careful when using DCEP as it will generate much more heat on the electrode than the same AC setting.

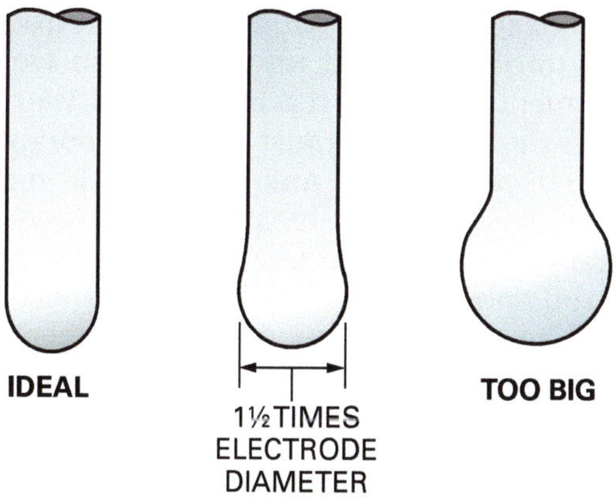

Figure 37 Properly and improperly prepared balled, non-tapered electrode ends for AC welding.

3.4.0 Selecting and Installing the Nozzle

Typically, a range of nozzle sizes and styles is available for a given torch in order to accommodate different electrode sizes and welding applications. Nozzles are made for close clearances, for long reaches, for reaching deep into narrow places, and for covering wide beads. Discard nozzles with chips, cracks, or metal buildup on the end because they can affect the gas flow pattern and produce weld defects.

Diamond-Wheel Grinders

Diamond wheel grinders are used to prepare tungsten electrodes. Both stationary and handheld, portable grinders are available. A stationary water-bath diamond-wheel grinder (*A*) can be set to different electrode grind angles as necessary. A water bath captures the tungsten dust that is generated during grinding so the dust can be disposed of later. A handheld, portable diamond-wheel grinder (*B*), can also be set to different electrode grind angles as necessary. The dust generated during grinding is captured inside the grinder's fully enclosed grind head. The grinder also comes with an attachment that allows the grinder to be connected to a shop vacuum so accumulated dust can be removed from the grinding chamber.

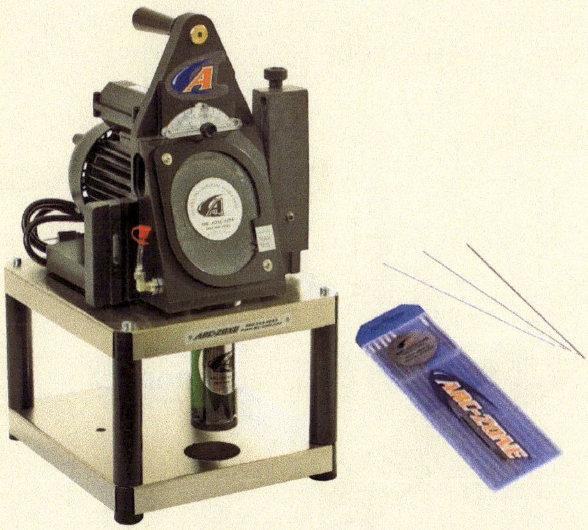

(A) STATIONARY GRINDER

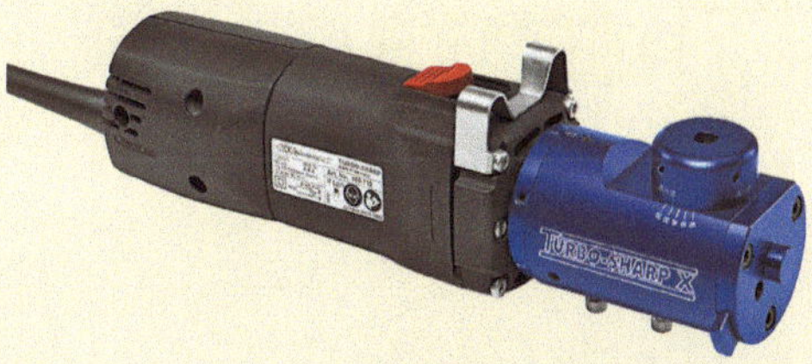

(B) HANDHELD GRINDER

The nozzle must be sized for the electrode diameter to be used and for any unique welding requirements, such as reaching into a V-groove. The nozzle diameter must be large enough to cover the entire weld area with shielding gas. A wide, easy flow of shielding gas is generally preferable to a narrow jet, which is apt to create turbulence and draw air into the weld zone. Small-diameter nozzles tend to overheat and break easily. However, a small-diameter nozzle is useful for steadying the torch and arc for the root pass in a V-groove weld.

To select the proper size and shape nozzle, always refer to the manufacturer's nozzle recommendations for the torch being used. In general, the nozzle's inside diameter should be about three times the diameter of the electrode being used.

3.4.1 Installing the Electrode

The electrode is usually installed after the nozzle because the electrode stickout (ESO) is measured from the end of the nozzle. The electrode is clamped in place in the torch body by the collet assembly. The collet is a cylindrical clamp that tightens as it is pulled or pushed (depending on the torch design) into a tapered holder as the threaded electrode cap is screwed on and tightened. To adjust the electrode stickout or to remove the electrode, loosen the cap assembly. Collet designs vary with torch manufacturers (*Figure 38*).

CAUTION

Handle tungsten electrodes with clean gloves to avoid contamination of the electrode.

Follow these steps to install the electrode:

Step 1 Remove the electrode cap from the rear of the torch.

Step 2 Check the collet size and collet body. If it is not the correct size for the electrode to be installed, remove and replace it with the correct size.

Step 3 Replace the electrode cap, but do not tighten it. Screw the cap only about one turn to leave the collet clamp loose.

Step 4 Insert the electrode into the collet from the nozzle end of the torch, and adjust it so that it extends from the nozzle about 1 to 1½ times the electrode diameter. Insert the color-coded end first to prevent destroying the electrode markings.

CAUTION

If the electrode cap is overtightened, the threads will be damaged as the torch heats with use. If the cap is too loose, the collet and electrode will overheat and damage the torch.

Step 5 Handtighten the electrode cap to lock the electrode in place.

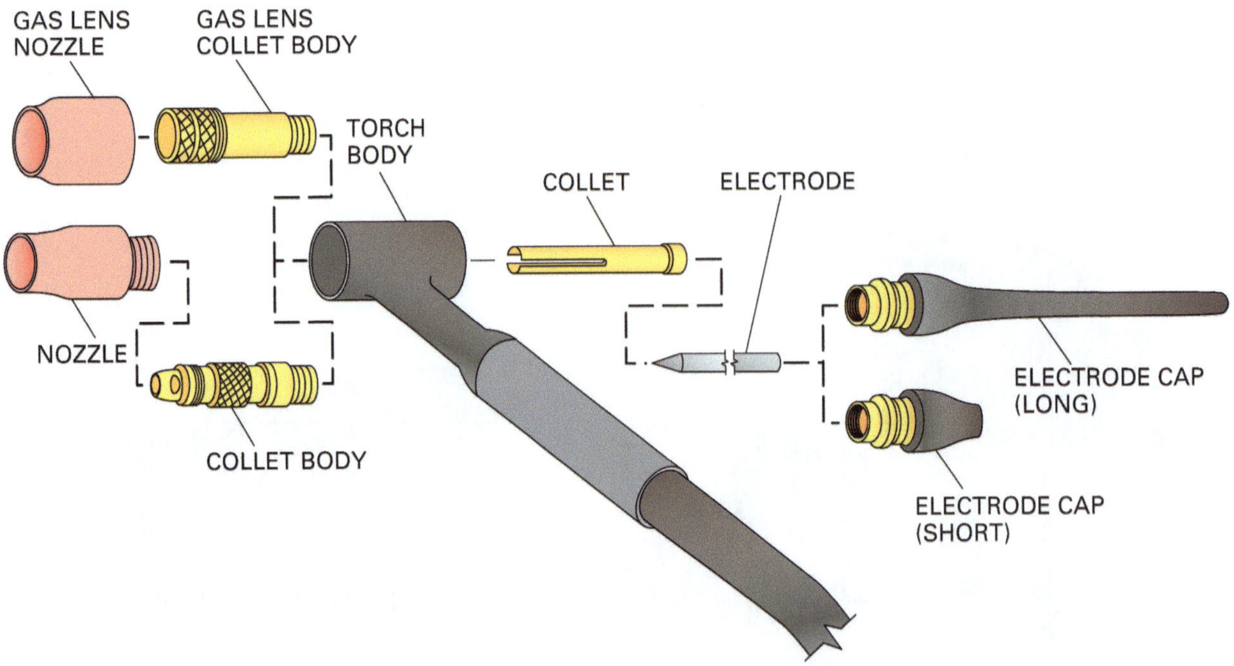

Figure 38 Collet assembly of a typical GTAW torch.

Electrode Stickout

Various manufacturers recommend stickout distances that differ from flush at the end of the cup to distances equal to the inside diameter of the cup, depending on the application and task. For instance, a root pass in a deep, narrow groove can be accomplished more easily with a narrow cup and a stickout that is about equal to the inside diameter of the cup. However, the use of a gas lens allows a longer tungsten stickout for welding under the same circumstances.

Additional Resources

AWS B1.10, Guide for the Nondestructive Examination of Welds. Miami, FL: American Welding Society.

AWS B1.11, Guide for the Visual Examination of Welds. Miami, FL: American Welding Society.

AWS D3.5, Guide for Steel Hull Welding. Miami, FL: American Welding Society.

AWS D3.7, Guide for Aluminum Hull Welding. Miami, FL: American Welding Society.

Lincoln Electric website: **http://www.lincolnelectric.com**

Modern Welding Technology. 6th Edition. Howard B. Cary. Englewood Cliffs, NJ: Prentice Hall, Inc.

OSHA 1910.269, Appendix *C, Protection from Step and Touch Potentials.* Current Edition. Washington, DC: Occupational Safety and Health Administration (OSHA).

OSHA 19260.351, Arc Welding and Cutting. Current Edition. Washington, DC: Occupational Safety and Health Administration (OSHA).

Welding Handbook. Vol. 1. Welding Science & Technology. 9th Edition. Miami, FL: American Welding Society, 2001.

Welding Handbook. Vol. 2, Part 1: Welding Processes. Miami, FL: American Welding Society, 2004.

3.0.0 Section Review

1. Before plugging a welding machine into an electrical outlet, be sure to identify the location of the _____.
   a. supplemental oxygen source
   b. backup generator
   c. alternator back cap
   d. electrical disconnect

2. When connecting the shielding gas for GTAW, if the gas hose is connected to the torch cable and the torch does not contain a gas shutoff valve, install a valve in the line between the torch cable and the _____.
   a. welding machine's gas solenoid
   b. gas cylinder's regulator/flowmeter
   c. gas cylinder's protective cap
   d. remote control foot pedal

3. Welders should select the size of GTAW electrode rated for the _____.
   a. amperage to be used
   b. voltage of the welding machine
   c. temperature at the work site
   d. required type of shielding gas

4. In general, the inside diameter of the nozzle should be about how many times the diameter of the GTAW electrode that is being used?
   a. One
   b. Two
   c. Three
   d. Four

SUMMARY

This module provided an overview of GTAW safety practices, equipment, shielding gas characteristics, electrode types and preparation, filler metal types and classifications, and equipment setup used for GTAW. A good understanding of these topics is essential for selecting and setting up GTAW equipment safely and efficiently, and it provides the foundation for making high-quality GTAW welds.

Review Questions

1. When making GTAW welds, always use safety glasses with a(n) _____.
   a. flame-retardant coating
   b. air-supplied respirator
   c. full face shield or a helmet
   d. OSHA-approved dust mask

2. When welding, the gas that should never be released in a large amount as compressed air around flammable materials is _____.
   a. oxygen
   b. argon
   c. carbon dioxide
   d. nitrogen

3. When a GTAW arc is struck across a gap, the resistance generates heat in the range of _____.
   a. 1,000°F to 2,000°F (550°C to 1,100°C)
   b. 2,000°F to 5,000°F (1,100°C to 2,750°C)
   c. 6,000°F to 10,000°F (3,300°C to 5,500°C)
   d. 15,000°F to 18,000°F (8,300°C to 10,000°C)

4. The duty cycle of a welding machine is based on a period of how many minutes?
   a. 5
   b. 10
   c. 15
   d. 20

5. What percent duty cycle do most heavy-duty industrial machines used for manual welding have?
   a. 50
   b. 100
   c. 50 to 60
   d. 60 to 100

6. An air-cooled GTAW torch that contains a manual gas control valve is typically used if the power source for the torch does not have a(n) _____.
   a. backup generator for the shielding gas supply
   b. electric solenoid valve to automatically control the shielding gas flow
   c. backflow prevention valve in the line leading to the electrode
   d. overload connection to a heat exchanger or cooling fan

7. Which of the following types of electrodes is radioactive?
   a. Lanthanated
   b. Thoriated
   c. Ceriated
   d. Zirconiated

8. How many times heavier is argon shielding gas than helium?
   a. Five
   b. Eight
   c. Ten
   d. Twenty

9. The filler metal used for manual GTAW is generally supplied in diameters of ¹⁄₁₆" (1.6 mm) to ¼" (6.4 mm) and lengths of _____.
   a. 48" (1,219 mm)
   b. 36" (914 mm)
   c. 24" (610 mm)
   d. 12" (305 mm)

10. The aluminum alloy filler metal rod ER4043, widely used in GTAW, also contains _____.
    a. iron
    b. flux
    c. silicon
    d. tungsten

11. When setting up GTAW equipment, the shielding gas supply must be located reasonably close to the welding site because of the _____.
    a. limited length of GTAW torch cables
    b. risk of acetylene dispersion at the work site
    c. weight and immobility of the gas cylinders
    d. need to refuel the torch frequently

12. When a standard welding machine is used for a GTAW operation, the shielding gas hose connects to the _____.
    a. torch cable
    b. electrode back cap
    c. retaining cup
    d. nozzle gasket

13. The type of GTAW electrode to be used for a welding operation is specified in _____.
    a. OSHA regulations
    b. ASTM and ANSI appendices
    c. SDS/MSDS and NFPA requirements
    d. WPS or site quality standards

14. When using sanding to taper the point on a tungsten electrode, what size grit should a welder use for finishing?
    a. 60
    b. 80
    c. 120
    d. 200

15. A GTAW electrode is usually installed after the nozzle because the _____.
    a. nozzle grounds the entire torch assembly
    b. electrode stickout (ESO) is measured from the end of the nozzle
    c. electrode back cap threads onto the nozzle to form a seal
    d. nozzle determines the maximum outer diameter of the electrode

Trade Terms Introduced in This Module

Lift-arc technology: A type of GTAW technology in which the tungsten electrode may be touched to the work, but the arc will not strike until the tip is lifted away to an appropriate distance.

Potentiometer: A three-terminal resistor with an adjustable center connection; used for variable voltage control.

Truncated: Having the pointed top (vertex) of a conical shape cut off, leaving only a plane section parallel to the base of the cone.

Tungsten: A nonstandard common name for the tungsten electrode used in GTAW; also, the principal metal from which the electrodes are constructed.

Additional Resources

This module is intended as a thorough resource for task training. The following reference works are suggested for further study.

AWS B1.10, Guide for the Nondestructive Examination of Welds. Miami, FL: American Welding Society.

AWS B1.11, Guide for the Visual Examination of Welds. Miami, FL: American Welding Society.

AWS D3.5, Guide for Steel Hull Welding. Miami, FL: American Welding Society.

AWS D3.7, Guide for Aluminum Hull Welding. Miami, FL: American Welding Society.

AWS Safety and Health Fact Sheet 27. Miami, FL: American Welding Society.

Modern Welding Technology. 6th Edition. Howard B. Cary. Englewood Cliffs, NJ: Prentice Hall, Inc., 2004.

OSHA 1910.269, Appendix *C, Protection from Step and Touch Potentials*. Current Edition. Washington, DC: Occupational Safety and Health Administration (OSHA).

OSHA 1926.351, Arc Welding and Cutting. Current Edition. Washington, DC: Occupational Safety and Health Administration (OSHA).

Tungsten Electrodes: an **Arc-Zone.com** ® Technical Focus Paper (pdf). Carlsbad, CA: **Arc-Zone.com**, Inc. **https://www.arc-zone.com/azc-scip/TungstenElectrodes.pdf**

Tungsten Guidebook (pdf). Newbury Park, CA: Diamond Ground Products, Inc. **http://www.diamond-ground.com/TungstenGuidebook2013.pdf**

Welding Handbook. Vol. 1. Welding Science & Technology. 9th Edition. Miami, FL: American Welding Society, 2001.

Welding Handbook. Vol. 2, Part 1: Welding Processes. Miami, FL: American Welding Society, 2004.

Lincoln Electric website: **http://www.lincolnelectric.com** offers sources for products and training.

Figure Credits

The Lincoln Electric Company, Cleveland, OH, USA, Module opener, Figures 1, 5, 7–9, 12, 15–18, 19, 20, SA01

Topaz Publications, Inc., Figures 13 (photo), 34, SA02

AWS A3.0:2001, Figure 36, adapted with permission from the American Welding Society (AWS), Miami, FL, USA, Figure 14

Zachry Industrial, Inc., Figures 19A, 21

AWS A.18/A5.18M:2005, Figure A.1, Reproduced with permission from the American Welding Society (AWS), Miami, FL, USA, Figure 22

Terry Lowe, Figure 32, SA04

Arc-Zone.com®, Inc., SA05A

CK Worldwide, Inc., SA05B

Section Review Answer Key

Answer	Section Reference	Objective
Section One		
1. c	1.1.3	1a
2. a	1.2.2	1b
Section Two		
1. b	2.1.2	2a
2. c	2.2.0	2b
3. d	2.3.1	2c
4. d	2.4.1	2d
5. a	2.5.4	2e
Section Three		
1. d	3.1.0	3a
2. b	3.2.0	3b
3. a	3.3.0	3c
4. c	3.4.0	3d

NCCER CURRICULA — USER UPDATE

NCCER makes every effort to keep its textbooks up-to-date and free of technical errors. We appreciate your help in this process. If you find an error, a typographical mistake, or an inaccuracy in NCCER's curricula, please fill out this form (or a photocopy), or complete the online form at **www.nccer.org/olf**. Be sure to include the exact module ID number, page number, a detailed description, and your recommended correction. Your input will be brought to the attention of the Authoring Team. Thank you for your assistance.

Instructors – If you have an idea for improving this textbook, or have found that additional materials were necessary to teach this module effectively, please let us know so that we may present your suggestions to the Authoring Team.

NCCER Product Development and Revision

13614 Progress Blvd., Alachua, FL 32615

Email: curriculum@nccer.org
Online: www.nccer.org/olf

❑ Trainee Guide ❑ Lesson Plans ❑ Exam ❑ PowerPoints Other ______________________

Craft / Level: Copyright Date:

Module ID Number / Title:

Section Number(s):

Description:

____

____

____

____

Recommended Correction:

____

____

____

____

Your Name:

Address:

____

Email: Phone:

GTAW – Plate

Overview

Gas tungsten arc welding (GTAW) is one of several types of arc welding processes that uses the heat from an electric arc to join a filler metal and a base metal. The arc is generated between a tungsten electrode and the base metal, and a handheld rod provides the filler metal. This module identifies GTAW-related safety practices, describes common GTAW equipment, and examines different techniques used to produce various types of weld beads.

Module 29208

Trainees with successful module completions may be eligible for credentialing through the NCCER Registry. To learn more, go to **www.nccer.org** or contact us at **1.888.622.3720.** Our website has information on the latest product releases and training, as well as online versions of our *Cornerstone* magazine and Pearson's product catalog.

Your feedback is welcome. You may email your comments to **curriculum@nccer.org,** send general comments and inquiries to **info@nccer.org**, or fill in the User Update form at the back of this module.

This information is general in nature and intended for training purposes only. Actual performance of activities described in this manual requires compliance with all applicable operating, service, maintenance, and safety procedures under the direction of qualified personnel. References in this manual to patented or proprietary devices do not constitute a recommendation of their use.

29208 V5

29208
GTAW – PLATE

Objectives

When you have completed this module, you will be able to do the following:

1. Identify GTAW-related safety practices and explain how to set up for welding.
   a. Identify GTAW-related safety practices.
   b. Explain how to safely set up the equipment and work area for welding.
2. Describe welding techniques for GTAW and explain how to produce basic weld beads.
   a. Describe welding techniques related to GTAW.
   b. Explain how to produce basic GTAW weld beads.
3. Describe the welding techniques needed to produce proper fillet and open V-groove welds using GTAW welding techniques.
   a. Describe the welding techniques needed to produce proper fillet welds using GTAW.
   b. Describe the welding techniques needed to produce proper open V-groove welds using GTAW.

Performance Tasks

Under the supervision of your instructor, you should be able to do the following:

1. Build a pad with stringer beads on carbon steel plate coupons in the flat position, using GTAW equipment and carbon steel filler metal.
2. Perform multiple-pass fillet welds on carbon steel plate coupons in all four 1F through 4F positions, using GTAW equipment and carbon steel filler metal.
3. Perform multiple-pass open V-groove welds on carbon steel plate coupons in all four 1G through 4G positions, using GTAW equipment and carbon steel filler metal.

Trade Terms

On-the-wire
Walking-the-cup

Industry Recognized Credentials

If you are training through an NCCER-accredited sponsor, you may be eligible for credentials from NCCER's Registry. The ID number for this module is 29208. Note that this module may have been used in other NCCER curricula and may apply to other level completions. Contact NCCER's Registry at 888.622.3720 or go to **www.nccer.org** for more information.

Contents

Topics to be presented in this module include:

Figures

SECTION ONE

1.0.0 GTAW PLATE WELDING

Objective

Identify GTAW-related safety practices and explain how to set up for welding.

a. Identify GTAW-related safety practices.
b. Explain how to safely set up the equipment and work area for welding.

In GTAW, an arc is generated between a non-consumable tungsten electrode and the base metal to melt together the base metal and a filler metal (*Figure 1*). The filler metal is a rod of similar composition to the base metal that is usually hand-held and manually fed into the leading edge of the weld puddle. The GTAW process is often called tungsten inert gas (TIG) welding because the weld area is protected from atmospheric contamination by a flow of inert gas from the torch nozzle, and a tungsten-based electrode. In GTAW, the electrode and the filler metal are not one and the same.

GTAW produces high-quality welds without slag or oxidation. Since there is no flux, there can be no corrosion due to flux entrapment, so no postweld cleaning is necessary. An exception is the slag left by some flux-cored rods, which must be removed.

1.1.0 Safety Summary

The following is a summary of safety procedures and practices that welders must observe when cutting or welding. Keep in mind that this is only a summary. Complete safety coverage is provided in the Welding Level One module, *Welding Safety*.

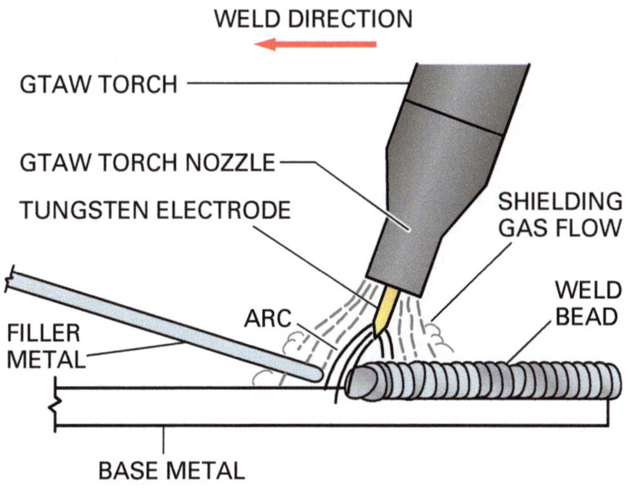

Figure 1 GTAW process.

That module should be completed before continuing with this one. Above all, be sure to wear appropriate protective clothing and equipment when welding or cutting.

1.1.1 Protective Clothing and Equipment

Welding activities can cause injuries unless you wear all of the protective clothing and equipment that is designed specifically for the welding industry. The following information includes important safety guidelines about protective clothing and equipment:

- Wear a face shield over snug-fitting cutting goggles or safety glasses for gas welding or cutting. Either the face shield or the lenses of the welding goggles must be an approved shade for the application. A welding hood equipped with a properly tinted lens is best for all forms of welding.
- Wear proper protective leather and flame-retardant clothing along with welding gloves, which will provide protection from flying sparks, molten metal, and heat.
- Wear high-top safety shoes or boots. Make sure that the tongue and lace area of the footwear will be covered by a pant leg. If the tongue and lace area is exposed or if the footwear must be protected from burn marks, wear leather spats under the pants or chaps over the front and top of the footwear.
- Wear a 100-percent cotton cap with no mesh material included in its construction. The bill of the cap points to the rear. If a hard hat is required for the environment, use one that allows the attachment of rear deflector material and a face shield. A hard hat with a rear deflector is generally preferred when working overhead, and may be required by some employers and job sites.
- Wear earmuffs or at least earplugs to protect your ear canals from sparks.

1.1.2 Fire/Explosion Prevention

Welding activities usually involve the use of fire or extreme heat to melt metal. Whenever fire is used, it must be controlled and contained. Welding or cutting activities are often performed on vessels that may once have contained flammable or explosive materials. Residues from those materials can catch fire or explode when a welder begins work on such a vessel. The following are fire and explosion prevention guidelines associated with welding:

- Never carry matches or gas-filled lighters in pockets. Sparks can cause the matches to ignite or the lighter to explode, causing serious injury.

- Never use oxygen to blow dust or dirt from clothing. The oxygen can remain trapped in the fabric for a time. If a spark hits the clothing during this time, the clothing can burn rapidly and violently out of control.
- Make sure that any flammable material in the work area is moved or shielded by a fire-resistant covering. Approved fire extinguishers must be available before attempting any heating, welding, or cutting operations. If a hot-work permit and a fire watch are required, be sure those items are in place before beginning and ensure that all site requirements have been observed.
- Never release a large amount of oxygen or use oxygen as compressed air. The presence of oxygen around flammable materials or sparks can cause rapid and uncontrolled combustion. Keep oxygen away from oil, grease, and other petroleum products.
- Never release a large amount of fuel gas, especially acetylene. Propane tends to concentrate in and along low areas and can ignite at a considerable distance from the release point. Acetylene is lighter than air, but is even more dangerous than propane. When mixed with air or oxygen, acetylene will explode at much lower concentrations than any other fuel.
- To prevent fires, maintain a neat and clean work area, and make sure that any metal scrap or slag is cold before disposal.
- Before cutting containers such as tanks or barrels, check to see if they have contained any explosive, hazardous, or flammable materials, including petroleum products, citrus products, or chemicals that decompose into toxic fumes when heated. As a standard practice, always clean and then fill any tanks or barrels with water, or purge them with a flow of inert gas such as nitrogen to displace any oxygen. Containers must be cleaned by steam cleaning, flushing with water, or washing with detergent until all traces of the material have been removed.

WARNING!

Welding or cutting must never be performed on drums, barrels, tanks, vessels, or other containers until they have been emptied and cleaned thoroughly, eliminating all flammable materials and all substances (such as detergents, solvents, greases, tars, or acids) that might produce flammable, toxic, or explosive vapors when heated. Clean containers only in well-ventilated areas. Vapors can accumulate during cleaning, causing explosions or injury.

Proper procedures for cutting or welding hazardous containers are described in the *American Welding Society (AWS) F4.1, Safe Practices for the Preparation of Containers and Piping for Welding and Cutting.*

1.1.3 Work Area Ventilation

Vapors and fumes tend to rise in the air from their sources. Welders often have to work above the welding area where the fumes are being created. Welding fumes can cause personal injuries. Good work area ventilation helps to remove the vapors and to protect the welder. The following is a list of work area ventilation guidelines to consider before and during welding activities:

- Make sure to follow confined space procedures before conducting any welding or cutting in the confined space.
- Never use oxygen for ventilation purposes in confined spaces.
- Always perform cutting or welding operations in a well-ventilated area. Cutting or welding operations involving some materials or coatings result in toxic fumes. For long-term cutting or welding of such materials, always wear an approved respirator that uses breathing air supplied from outside the work area. For occasional, very short-term exposure, a high-efficiency, HEPA-rated or metal-fume filter may be used on a standard respirator.
- Make sure confined spaces are properly ventilated for cutting or welding purposes. Use powered extraction systems when available.

1.1.4 GTAW-Specific Safety

GTAW, also referred to as heliarc or TIG welding, is a process that is different from the processes used in shielded metal arc welding (SMAW), GMAW, and flux-cored arc welding (FCAW). In other forms of welding, the electrode or wire is consumed in the welding process. In GTAW, the electrode itself is not consumed, but a separate rod of filler metal is consumed.

Vapors from the melted base metal, the filler metal, and from the shielding gases used in the GTAW process can create respiratory issues. As always, welders performing any kind of work must be aware of the vapors generated from the welding processes. Argon gas is often used for shielding in the GTAW process.

A combination of the arc and the gas generates substantial levels of ultraviolet (UV) light, which reacts with the oxygen in the area of the weld. UV radiation can ionize oxygen in the atmosphere near the weld, which can lead to a rise in ozone and nitrogen dioxide levels. As long as the work is being performed in well-ventilated areas, the risks are minimal. If work must be performed in confined spaces, consider using additional ventilation or respirators. Follow your employer's policies in this matter.

Because GTAW work is often performed on delicate welds, the welder is frequently positioned very close to the work. The risk of UV radiation being emitted from the electrical arc is higher because the shielding gas used with GTAW is more transparent than gases used with other forms of welding. This transparency is due to the minimal smoke produced from the weld.

Because of the brightness of the GTAW process, safety representatives recommend that the welder wear darker clothing to minimize UV reflections under the welding helmet. They also recommend that the walls of the work area be painted with a paint that reduces reflections from the UV light generated by the GTAW process. Because aluminum and stainless steel reflect more light than common carbon steels, welders need to take extra precautions to protect their vision when performing GTAW work on these metals.

The GTAW process requires the base metal and the filler metal to be extremely clean. Cleaning solvents are used to prepare the metals. Welders must read all Safety Data Sheet (SDS) information about any cleaning agent before using it. Some cleaning agents are flammable, while others are nonflammable. Make sure that you know which kind you are using. Pay special attention to any storage requirements for such cleaning agents.

WARNING!

Chlorinated cleaning solvents can produce toxic gases. For this reason, they should never be used for cleaning metals in a welding setting.

The tungsten electrodes used for GTAW become contaminated over time and with use. When electrodes lose their original shapes, welders must grind the tungsten tips to reshape them and make them suitable for continued use. The tungsten dust and flying particles from these grinding activities can damage the welder's vision and present a respiratory hazard. Read and understand the SDS that comes with each batch of electrodes.

1.2.0 GTAW Welding Equipment Setup

Before welding can take place, the work area must be made ready, the welding equipment set up, and the metal to be welded prepared. The following sections explain how to prepare the area and set up the equipment that will be used to perform GTAW welding of carbon steel plate.

1.2.1 Setting Up the Welding Area

Practice welding on a welding table, bench, or stand. The welding surface must be non-flammable and electrically conductive, and provisions must be made for mounting practice welding coupons out of position. Follow these steps to set up the area for welding:

Step 1 Make sure the area is properly ventilated. Make use of doors, windows, and fans.

Step 2 Check the area for fire hazards. Remove any flammable materials before proceeding.

Step 3 Know the location of the nearest fire extinguisher. Know how to use the fire extinguisher and do not proceed unless the extinguisher is charged.

Step 4 Set up welding curtains around the welding area.

1.2.2 Preparing Practice Welding Coupons

Use the material thickness as directed by the instructor. If possible, the practice weld coupons should be ¼" (6 mm) thick minimum. Open V-groove welds generally have a bevel angle of 30 to 37½ degrees.

Clean the steel plate before welding by using a wire brush or grinder to remove mill scale in the weld zone and a minimum of ½" (13 mm) beyond the weld zone. This is done because the direct current electrode negative (DCEN) used in GTAW has little or no cleaning action. Prepare weld coupons for practicing the following welds:

- *Stringer and weave bead welds* – The coupons can be any size or shape that can be easily handled.
- *Open V-groove welds* – As shown in *Figure 2*, cut the metal into 3" × 7" (about 76 mm × 178 mm) rectangles with one (or both) of the lengths beveled at 30 to 37½ degrees. Grind a 0" to ⅛" (0 to 3.2 mm) root face on one bevel as directed by your instructor.

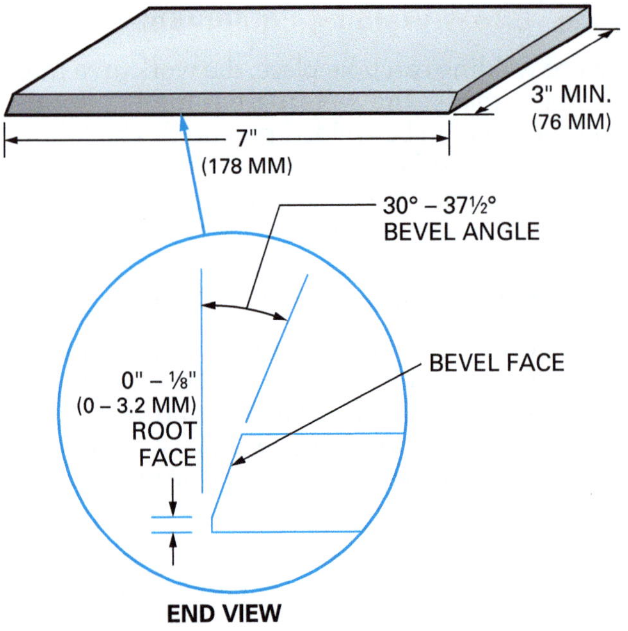

Figure 2 Metal cut for open V-groove weld coupons.

Follow these steps to prepare each open V-groove weld coupon:

Step 1 Check the bevel face. There should be no dross and a 0" to ⅛" (0 to 3.2 mm) root face. The bevel angle should be 30 to 37½ degrees (*Figure 3*).

Step 2 Center the beveled strips using a root opening determined by the instructor and tack weld them in place. Place the tack welds on the ends of the joint.

> GOING GREEN
>
> **Conserve Weld Coupons**
>
> Steel for practice welding is expensive and difficult to obtain. Every effort should be made to conserve the material that is available. Reuse welding coupons until all surfaces have been used for welding, then cut the weld coupon apart and reuse the pieces. Use material that cannot be cut into weld coupons to practice running beads.

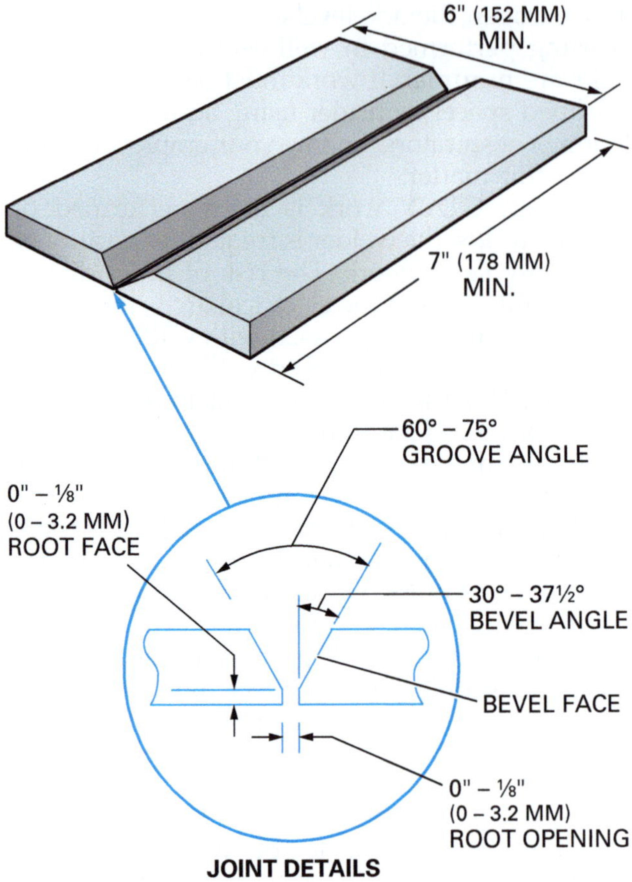

Figure 3 Open V-groove weld coupon.

1.2.3 Welding Equipment

Identify a proper welding machine for GTAW use. Follow these steps to set the machine up for welding:

Step 1 Verify that the welding machine can be used for GTAW with or without internal gas shielding control. If desired, identify an optional cooling unit.

Step 2 Identify an air-cooled or water-cooled GTAW torch. Make sure that it is compatible with the welding machine and any cooling unit.

Step 3 Verify the location of the primary disconnect.

Step 4 Configure the welding machine and torch for GTAW welding (*Figure 4*) as directed by the instructor. Configure the torch polarity and equip the torch with a tungsten electrode that is the correct diameter and the properly prepared type for the filler metal used for the application.

Step 5 Connect the proper shielding gas for the application as directed by the instructor, filler metal manufacturer, WPS, or site quality standards.

Step 6 Connect the clamp of the workpiece lead to the table or workpiece.

Step 7 Turn on the welding machine, and purge the torch as directed by the manufacturer's instructions.

Step 8 Set the initial welding current for the desired GTAW application.

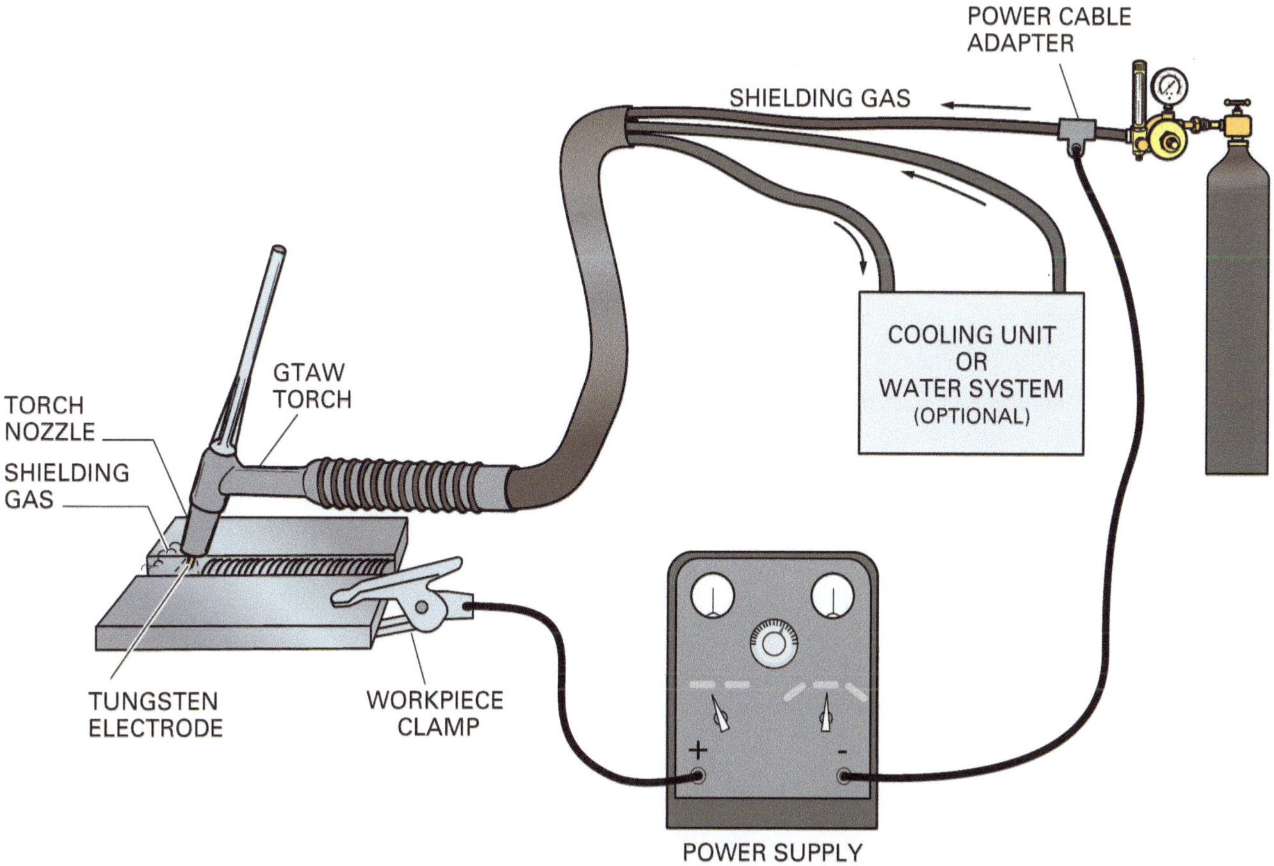

Figure 4 Configuration diagram of typical GTAW welding equipment.

Additional Resources

Lincoln Electric website: **www.lincolnelectric.com** offers sources for products and training.

OSHA Standard 1926.351, Arc Welding and Cutting. Current Edition. Washington, DC: Occupational Safety and Health Administration.

The Procedure Handbook of Arc Welding. Cleveland, OH: The Lincoln Electric Company, 2000.

Welding Handbook, Volume 1 (2001) and Volume 2 (2004). Miami, FL: The American Welding Society.

1.0.0 Section Review

1. For long-term cutting or welding of materials that may produce toxic fumes, always wear an approved respirator that uses _____.
   a. a HEPA-rated charcoal vacuum system
   b. breathing air from outside the work area
   c. compressed oxygen from an approved tank
   d. an inert gas that will displace the fumes

2. When configuring a welding machine for GTAW welding, the torch must be equipped with a _____.
   a. reel of tungsten wire
   b. polarity limiter
   c. reel of filler wire
   d. tungsten electrode

SECTION TWO

2.0.0 GTAW TECHNIQUES AND BASIC BEADS

Objective

Describe welding techniques for GTAW and explain how to produce basic weld beads.

a. Describe welding techniques related to GTAW.
b. Explain how to produce basic GTAW weld beads.

Performance Tasks

1. Build a pad with stringer beads on carbon steel plate coupons in the flat position, using GTAW equipment and carbon steel filler metal.
2. Perform multiple-pass fillet welds on carbon steel plate coupons in all four 1F through 4F positions, using GTAW equipment and carbon steel filler metal.
3. Perform multiple-pass open V-groove welds on carbon steel plate coupons in all four 1G through 4G positions, using GTAW equipment and carbon steel filler metal.

Trade Term

Walking-the-cup: A welding method in which the welder rests the edge of the torch nozzle (cup) against the base metal or groove edges to steady the torch and maintain a constant arc length.

Producing strong and acceptable weld beads requires welders to properly control the equipment being used, as well as understand how to use various welding techniques. This section identifies factors that affect GTAW, describes basic GTAW techniques, and explains how to produce basic GTAW weld beads.

2.1.0 GTAW Techniques

GTAW weld bead characteristics and quality are affected by several factors that result from the way the welder handles the torch. These factors include the following:

- Torch travel speed and arc length
- Torch angles
- Torch and filler metal handling techniques

2.1.1 Torch Travel Speed and Arc Length

Torch travel speed and arc length affect the GTAW weld puddle and the penetration of the weld. A slow travel speed allows more heat to concentrate and forms a larger, more deeply penetrating puddle. Faster travel speeds prevent heat buildup and form smaller, shallower puddles. As the torch is raised, arc length increases, voltage increases, and the bead width increases.

2.1.2 Torch Angles

The two basic torch angles that must be controlled when performing GTAW are the work angle and the travel angle. The definition of these angles is the same as for all other methods of arc welding.

The torch work angle (*Figure 5*) is an angle less than 90 degrees between a line perpendicular to the major workpiece surface at the point of electrode contact and a plane determined by the electrode axis and the weld axis. In other words, the torch work angle is the side-to-side angle in relation to the weld joint. This is more commonly referred to as the work angle. For a T-joint or corner joint, the line is 45 degrees to the non-butting member.

The torch travel angle (*Figure 6*) is an angle less than 90 degrees between the electrode axis and a line perpendicular to the weld axis at the point of electrode contact in a plane determined by the electrode axis and the weld axis. A push angle is used for GTAW. A push angle is the travel angle created when the torch is tilted back so that the electrode is pointing in the direction of weld progression. In other words, the electrode tip and the shielding gas are being directed ahead of the weld bead. Push angles of 15 to 20 degrees are normally used for GTAW. Too great a push angle will tend to draw air from under the back edge of the torch nozzle where it will mix with the shielding gas stream and contaminate the weld.

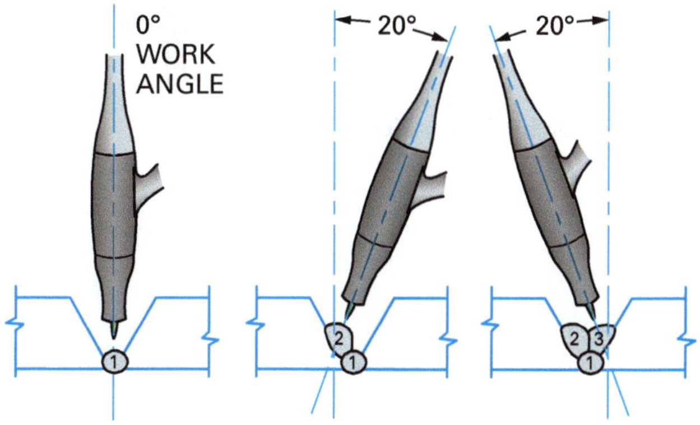

Figure 5 Typical torch work angles.

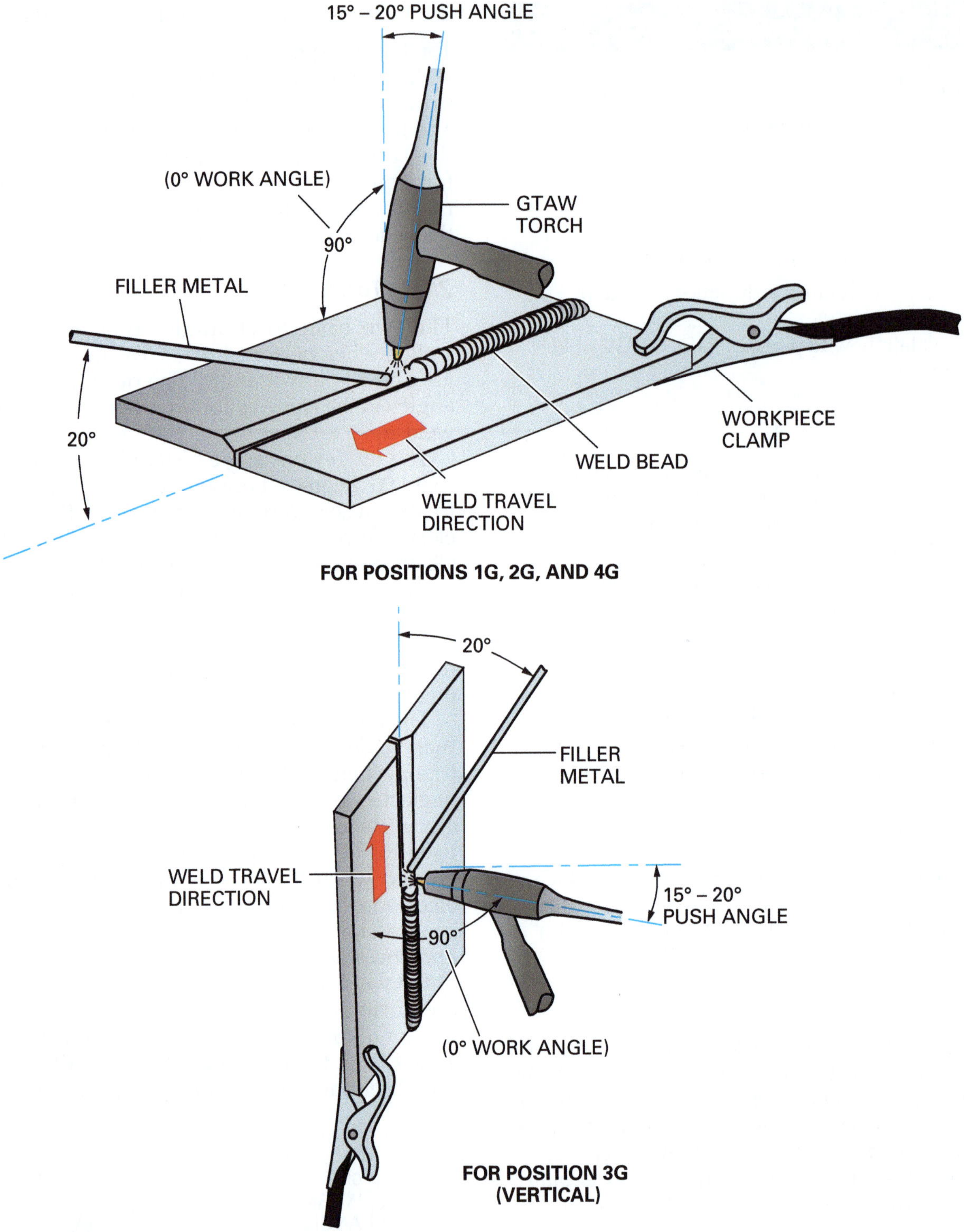

Figure 6 Torch travel angles.

2.1.3 Torch and Filler Metal Handling Techniques

The two basic handling techniques used to perform GTAW are the freehand method and walking-the-cup. Try both techniques, and use the one that gives the best results.

In the freehand technique, the torch electrode tip is held just above the weld puddle or base metal. The torch is supported by the welder's hand, which is usually steadied by resting some part of it on or against the base metal to maintain the proper arc length. If required, the welder can move the torch tip in a small circular motion within the molten puddle to maintain its size and advance the puddle. Filler metal is added as needed.

The GTAW filler metal is held in the hand that is not holding the torch. For all positions, the filler rod is held at an angle of about 20 degrees above the base metal surface and in line with the weld. The tip of the filler rod is always kept within the shielding gas envelope to protect the filler rod from atmospheric contamination and to keep it preheated.

Insert the filler metal into the leading edge of the weld puddle, using extreme care not to touch the tungsten electrode with the end of the filler metal. If the tungsten electrode touches the filler metal or weld puddle, it will become contaminated with filler metal. The electrode must then be removed and cleaned by grinding or chemical cleaning before proceeding. Also, do not insert the end of the filler metal rod into the molten puddle under the electrode and then attempt to melt it off. Even if it does not touch the tungsten, this technique can cause hard spots and weld defects. A technique for preventing contamination of the electrode is to coordinate the movement of the electrode toward the back edge of the weld puddle as the filler metal is dabbed into the leading edge of the weld puddle.

In the walking-the-cup method, the welder rests the edge of the torch nozzle (cup) against the base metal or groove edges to steady the torch and maintain a constant arc length. The torch is rocked from side to side on the edge of the cup as it is advanced to maintain the puddle size and heat both sides of the groove.

The filler metal is added in the same manner as in the freehand technique, using care not to contaminate the electrode.

2.1.4 Filler Metal Handling Techniques for Continuous Feed

Feeding the filler wire consistently as the weld progresses is important to ensure the weld puddle develops evenly and produces an even bead. Like all welding techniques, the technique described here requires practice to master but is well worth the effort.

Cradle the filler wire between the pointer finger and the middle finger, similar to the way you might hold a pencil. Then use the thumb to consistently push the wire forward in small motions. Once the thumb is extended pushing the wire forward, quickly bend and reposition the thumb to begin a new forward push. Consistent motion is important, so the thumb must be repositioned quickly to maintain the pace. The thumb can also be used to slowly roll the electrode as well during the pushing motion. This technique is easily practiced without actively welding; simply practice with a wire at home or during free time to become comfortable with the technique before trying it on a weld.

Another technique to consider is referred to as the finger-pull technique. The wire is secured between the pointer and middle fingers, which are used to pull the electrode forward. At the end of the forward stroke, the fingers are slipped backwards along the wire to reposition, without pulling the wire backwards and out of the weld puddle in the process. Once the two fingers are repositioned, the pulling motion begins again.

A Rare Feat

In 2012, a pipe welding team employed by Team Fabricators LLC, a Texas-based subsidiary of Team Industries, accomplished something quite rare. Working on a piping project to do full-penetration, open V-groove pipe welds, the team completed literally thousands of welds. After 847 radiographic examinations of randomly selected welds, done in accordance with the ASME code, not a single pipe weld had to be rejected.

2.2.0 GTAW Bead Types

The basic bead types that can be made with GTAW are stringer beads and weave beads.

2.2.1 Stringer Beads

Stringer beads (*Figure 7*) are made with little or no side-to-side movement (oscillation) of the torch. The bead is generally no more than three times the diameter of the filler rod being used.

Practice running stringer beads in the flat position using both the freehand and walking-the-cup techniques. Flat-position stringer beads should be run with a 15- to 20-degree push angle. Practice some stringer beads without filler metal until comfortable with the torch and you get rippled beads that are consecutively uniform.

To run stringer beads with filler metal, hold the filler metal in front of the weld bead, with the end of the filler metal within the torch gas envelope. The filler rod should be held at a 20-degree angle with the plate surface (*Figure 8*).

Follow these steps to run flat-position stringer beads:

> **WARNING!**
>
> Use pliers to handle the hot practice coupons. Wear gloves when quenching the practice coupons in water. Steam will rise off the coupons and can burn or scald unprotected hands and arms.

Step 1 Position the torch at a 15- to 20-degree push angle, with the electrode tip directly over the point where the weld is to begin.

Step 2 Press the remote control to energize the electrode and start the shielding gas flow.

> **NOTE**
>
> If the equipment being used does not have remote control, the electrode will be energized when the power source is turned on. Start the shielding gas flow by opening the manual shielding gas valve on the torch or in the line.

Step 3 Initiate the arc by bringing the electrode close to the base metal (with high frequency) or by lightly touching the electrode to the base metal (no high frequency).

Figure 7 Stringer bead (on stainless steel).

Step 4 Move the electrode tip in a small circular motion until the weld puddle begins to form.

Step 5 While using either the freehand or the walking-the-cup technique, slowly advance the torch and add filler metal as needed.

Step 6 Continue to weld until a bead about 2" to 6" (50 to 152 mm) long is formed, and then stop the arc without withdrawing the torch.

Step 7 Continue to hold the torch in place so that the postflow gas will protect the hot weld and filler metal until they are cool enough to prevent oxidation damage.

> **CAUTION**
>
> Do not remove the torch or filler metal from the torch-shielding gas flow until the puddle has solidified and cooled. This continued flow of gas will protect the molten metal and the tungsten electrode. Removing the torch and shielding gas too soon can cause crater porosity or cracks.
>
> If the postflow time period is too short, the tungsten electrode may oxidize (turn a bluish-black color), which will require repointing the electrode. If the filler metal becomes contaminated, the contaminated end will have to be discarded.

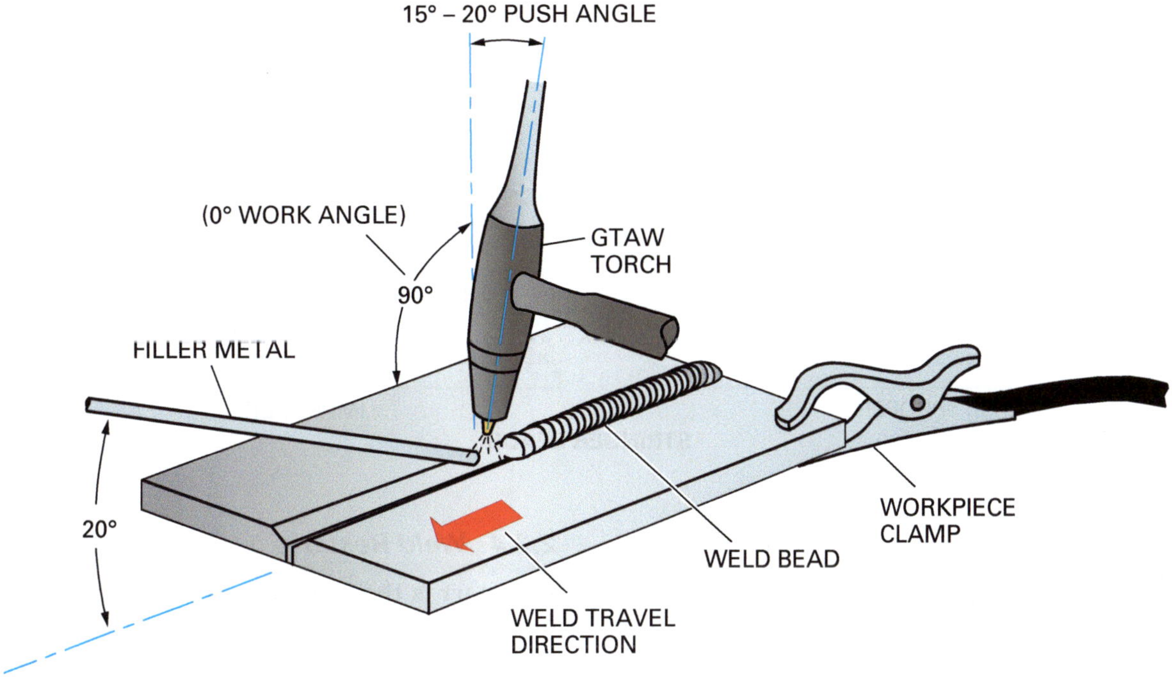

Figure 8 Flat-position torch and filler metal angles.

Step 8 Remove the torch, and close the shielding gas valve if the equipment does not have remote control.

Step 9 Inspect the bead for the following:

- Straightness
- Uniform rippled appearance of the bead face
- Smooth, flat transition with complete fusion at the toes of the weld
- No porosity
- No undercut

Step 10 Continue practicing stringer beads until acceptable welds are made every time.

2.2.2 Weave Beads

Weave beads are made by working the weld puddle back and forth across the axis of the weld. This produces a weld bead much wider than the puddle width. The width of a weave bead is determined by the amount of oscillation or cross motion.

Stopping the Arc Without Remote Control

If the equipment being used does not have remote control, slowly swing the torch up the groove face to break the arc, and then immediately return it to shield the weld and the end of the filler metal.

Weave beads put more heat into the base metal. For this reason, they are not recommended for welding metals sensitive to heat input, such as most types of stainless steels and aluminum. Always check the WPS or site quality standards to determine if stringer or weave beads should be used.

2.2.3 Overlapping Beads

Overlapping beads are made by depositing overlapping stringer (or weave) beads parallel to one another. When each successive bead overlaps the toe of the previous bead to form an approximately flat surface, the process is called padding. Overlapping beads are used to pad surfaces and to make multiple-pass welds.

Properly overlapped beads, when viewed from their ends, form a nearly flat surface. *Figure 9* shows both proper and improper overlapping stringer and weave beads.

A Few Suggestions

To improve your technique, practice with both hands in opposing directions when welding in the 1G, 2G, and 4G positions. Also, the first weld bead can be done up the center of the pad, with the remaining beads being worked from the center out. This tactic gives you practice overlapping beads on the pad as you might do in a groove weld.

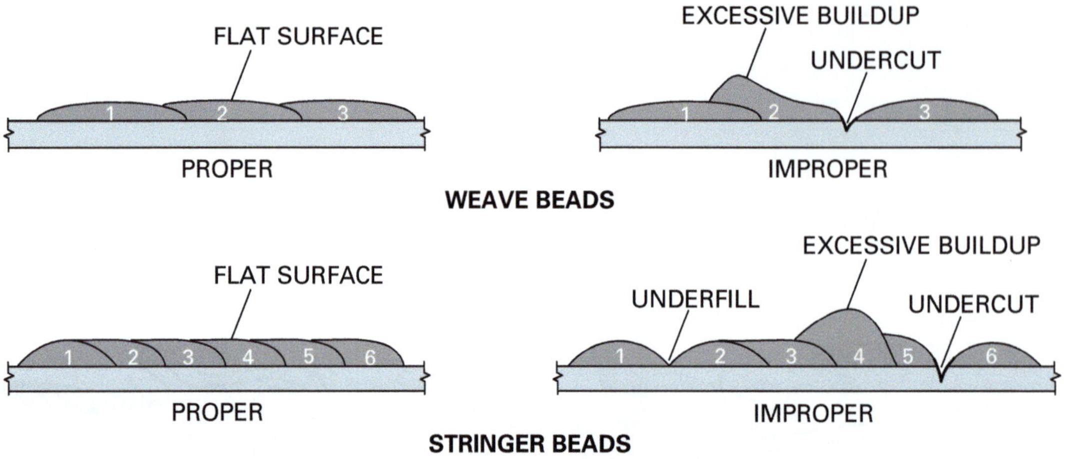

Figure 9 Proper and improper overlapping beads.

Follow these steps to make GTAW overlapping stringer beads using 3⁄32" or 1⁄8" (2.4 mm or 3.2 mm) carbon steel filler metal:

Step 1 Use a 2" × 6" × 1⁄4" (50 mm × 152 mm × 6 mm) piece of steel.

Step 2 Weld a stringer bead along one edge.

Step 3 Clean the weld.

Step 4 Run a second bead along the previous bead. Be sure to overlap the previous bead to obtain a good tie-in and produce a flat profile.

Step 5 Continue running overlapping stringer beads until the face of the metal square is covered.

Step 6 Continue building layers of stringer beads (pads), one on top of the other, until the technique is perfected.

Tacking and Aligning Longer Workpieces

When tacking longer workpieces together (longer workpieces than typical practice ones), both sides of the workpieces are usually tacked with welds that are about 1⁄2" (13 mm) long to position the workpieces and minimize distortion when the final welds are made. After the first tack weld, use a hammer or other tool to align the workpieces side-to-side and end-to-end, and then tack the opposite side. Tack the far ends of the workpieces in the same manner. Intermediate tack welds can be made every 5" to 6" (≈125 mm to 150 mm) as necessary to minimize lengthwise distortions.

2.2.4 Weld Restarts

A restart is the junction where a new weld connects to and continues the bead of a previous weld. Restarts are important because an improperly made restart may create a discontinuity. A restart must be made so that it blends smoothly with the previous weld and does not stand out. If possible, avoid restarts by running a bead the full length of the weld joint that is to be made.

Follow these steps to make a restart:

Step 1 Use a wire brush to clean the restart area.

CAUTION

Do not use the same brush on different metals.

Step 2 Hold the torch at the proper angle and arc distance, and then restart the arc directly over the center of the crater. Remember that the welding codes do not allow arc strikes outside the area to be welded.

Step 3 Move the electrode tip in a small circular motion over the crater, and add filler metal when the molten puddle is the same size as the crater.

Step 4 As soon as the puddle fills the crater, advance the puddle slightly, and continue to add filler metal as needed.

Step 5 Inspect the restart. A properly made restart will blend into the bead, making it hard or impossible to detect.

If the restart has underfill or undercut, not enough filler metal was added. If the restart is higher than the rest of the bead, too much filler metal was added.

Thin Materials

On thin materials, stop about 1" to 2" (25 mm to 50 mm) short of the edge without terminating the weld. Then back-weld from the edge to the crater, filling the crater.

Continue to practice restarts until they are correct. Use the same techniques for making restarts whenever performing GTAW.

2.2.5 Weld Terminations

A weld termination is made at the end of a weld. For terminating a weld, most welding codes require that the crater must be filled to the full cross section of the weld to prevent crater cracking. This can be difficult because most terminations are at the edge of a plate where welding heat tends to build up. This makes filling the crater more difficult. Filling the crater is much easier when remote control equipment with a potentiometer is used. The potentiometer allows the welding current to be reduced as the end of the weld is approached.

Follow these steps to make a termination using remote control equipment equipped with a potentiometer:

Step 1 As the end of the weld is approached, slowly reduce the welding current. The amount of current reduction is determined by the weld puddle width. If the puddle starts to become wider, there is too much heat; if it starts to become narrower, there is not enough heat.

Step 2 When adding filler metal, continue to reduce the welding current until the crater is filled.

Step 3 Stop the arc and hold the torch in place until the gas postflow cools the weld metal and filler metal.

Follow these steps to make a termination without a potentiometer:

Step 1 As the end of the weld is approached, add filler metal at a faster pace. The filler metal will help absorb excess heat as it melts.

Step 2 Continue to add filler metal until the crater is filled, and then stop the arc.

Step 3 Hold the torch in place until the gas postflow cools the weld metal and filler metal enough to prevent oxidation damage.

CAUTION

Do not remove the torch or filler metal from the torch-shielding gas flow until the puddle has solidified and cooled. This continued flow of gas will protect the molten metal and the tungsten electrode. Removing the torch and shielding gas too soon can cause crater porosity or cracks.

If the postflow time period is too short, the tungsten electrode may oxidize (turn a bluish-black color), which will require repointing the electrode. If the filler metal becomes contaminated, the contaminated end will have to be discarded.

Inspect the termination for the following:

- Crater filled to the full cross section of the weld (no underfill)
- No crater cracks
- No crater porosity
- No undercut

Additional Resources

Lincoln Electric website: **www.lincolnelectric.com** offers sources for products and training.

OSHA Standard 1926.351, Arc Welding and Cutting. Current Edition. Washington, DC: Occupational Safety and Health Administration.

The Procedure Handbook of Arc Welding. Cleveland, OH: The Lincoln Electric Company.

Welding Handbook, Volume 1 (2001) and Volume 2 (2004). Miami, FL: The American Welding Society.

2.0.0 Section Review

1. The two basic torch and filler metal handling techniques used to perform GTAW are ____.
   a. the travel angle method and walking-the-bead
   b. the push angle method and the freehand method
   c. the freehand method and walking-the-cup
   d. walking-the-cup and the push bead method

2. When running stringer beads using GTAW, the filler metal rod should be held ____.
   a. at a 20-degree angle with the plate surface
   b. at the toe padding behind the weld bead
   c. just outside the torch gas envelope
   d. at a 45-degree angle with the electrode

SECTION THREE

3.0.0 FILLET AND GROOVE WELDS

Objective

Describe the welding techniques needed to produce proper fillet and open V-groove welds using GTAW welding techniques.

a. Describe the welding techniques needed to produce proper fillet welds using GTAW.
b. Describe the welding techniques needed to produce proper open V-groove welds using GTAW.

Performance Tasks

1. Build a pad with stringer beads on carbon steel plate coupons in the flat position, using GTAW equipment and carbon steel filler metal.
2. Perform multiple-pass fillet welds on carbon steel plate coupons in all four 1F through 4F positions, using GTAW equipment and carbon steel filler metal.
3. Perform multiple-pass open V-groove welds on carbon steel plate coupons in all four 1G through 4G positions, using GTAW equipment and carbon steel filler metal.

Trade Term

On-the-wire: A root pass technique in which the filler metal is held in the bottom of the groove parallel to the plates and then melted and fused in place by moving the torch along the groove.

Two types of welds that are commonly made using GTAW are fillet welds and open V-groove welds. This section examines the procedures used to make both of these welds in the flat, horizontal, vertical, and overhead positions.

3.1.0 Fillet Welds

The most common fillet welds made with GTAW are made in lap and T-joints. The weld position is determined by the axis of the weld. The positions for fillet welding are 1F (flat), 2F (horizontal), 3F (vertical), and 4F (overhead). In the 1F and 2F positions, the weld axis can be inclined up to 15 degrees. Any weld axis inclination for the other positions varies with the rotational position of the weld face as specified in AWS Standards. *Figure 10* shows the four fillet welding positions.

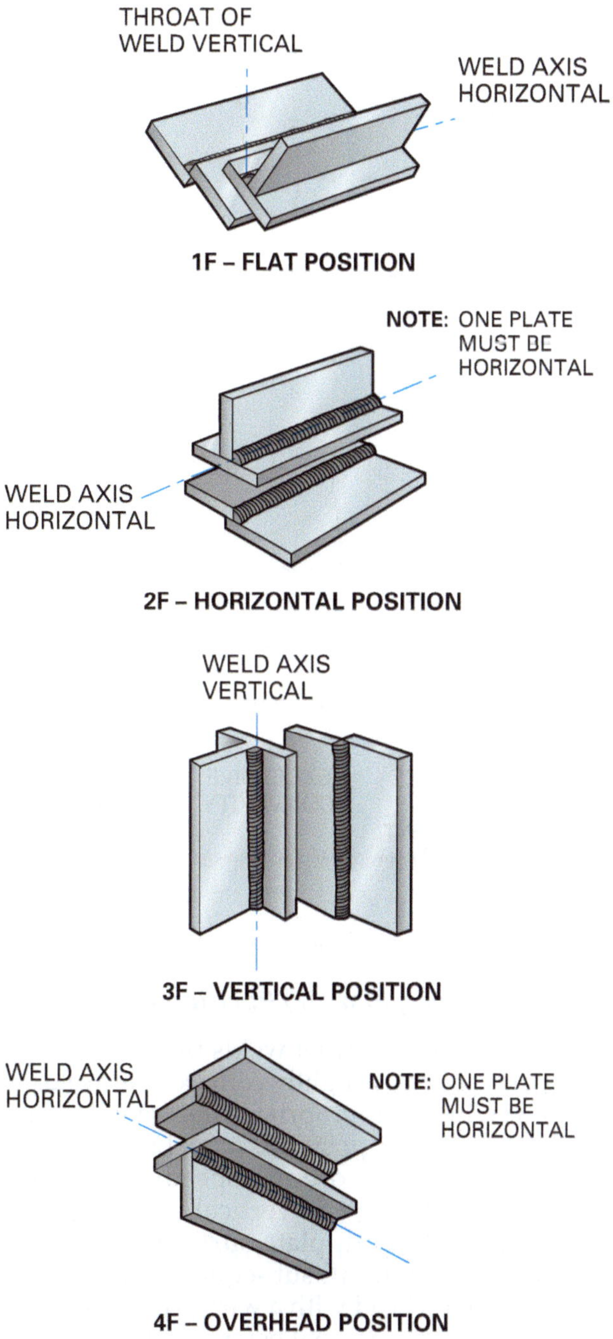

Figure 10 Fillet welding positions.

Fillet welds can vary between concave and convex, depending on the WPS or site quality standards. The welding codes require a fillet weld to have a uniform face, although a slightly non-uniform face is acceptable. The convexity of a fillet weld or individual surface bead must not exceed that permitted by the applicable code or standard. A fillet weld with profile defects is unacceptable and must be repaired. *Figure 11* shows a variety of ideal, acceptable, and unacceptable fillet weld profiles.

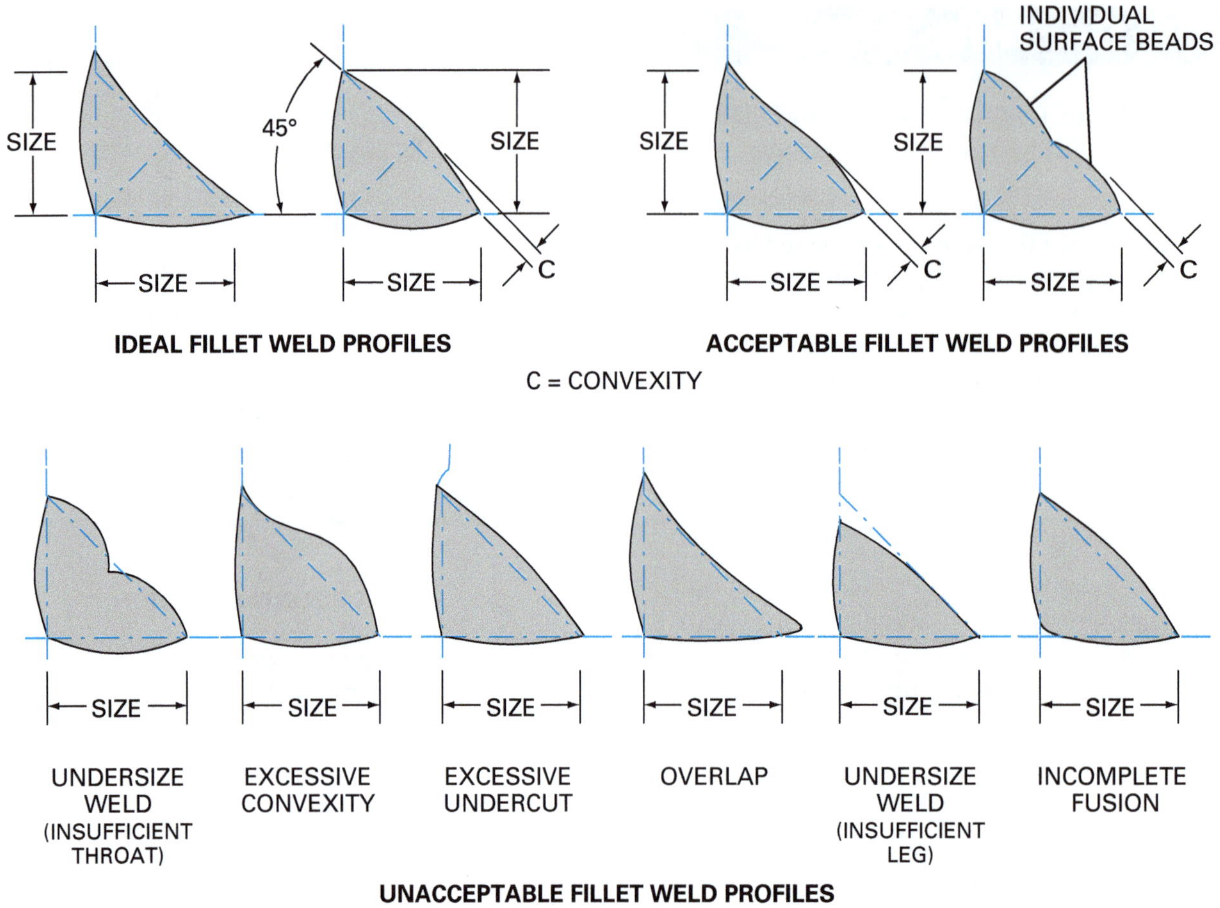

Figure 11 Ideal, acceptable, and unacceptable fillet weld profiles.

3.1.1 Practicing Flat (1F) Position Fillet Welds

Practice 1F-position fillet welds by making multiple-pass (six-pass) slightly convex fillet welds in a T-joint with an appropriate filler metal. When welding in the 1F position, pay close attention to the torch angle and travel speed. For the first bead, the torch angle is vertical, forming a 45-degree angle to both plate surfaces. The angle is then adjusted for all subsequent beads. Clean each completed bead with a wire brush used only on the same base material before starting the next bead.

Figure 12 shows the torch angles for 1F-position fillet welds.

> **T-Joint Heat Dissipation**
>
> In T-joints, the welding heat dissipates more rapidly in the thicker or non-butting member. On various bead passes, the arc may have to be concentrated slightly more on the thicker or the non-butting member to compensate for the heat loss.

Follow these steps to make a flat fillet weld. Be sure to clean each bead before beginning the next:

Step 1 Tack two plates together to form a T-joint for the fillet weld coupon (*Figure 13*). Clean the tack welds.

Step 2 Clamp or tack weld the coupon in the 1F position (flat).

Step 3 Run the first bead along the root of the joint, using a work angle of 45 degrees with a 10- to 15-degree push angle.

Step 4 Run the second bead along a toe of the first weld, overlapping about 75 percent of the first bead. Alter the work angle as shown in *Figure 12* and use a 10- to 15-degree push angle with a slight oscillation.

Step 5 Run the third bead along the other toe of the first weld, filling the groove created when the second bead was run. Use the work angle shown in *Figure 12* and a 10- to 15-degree push angle with a slight oscillation.

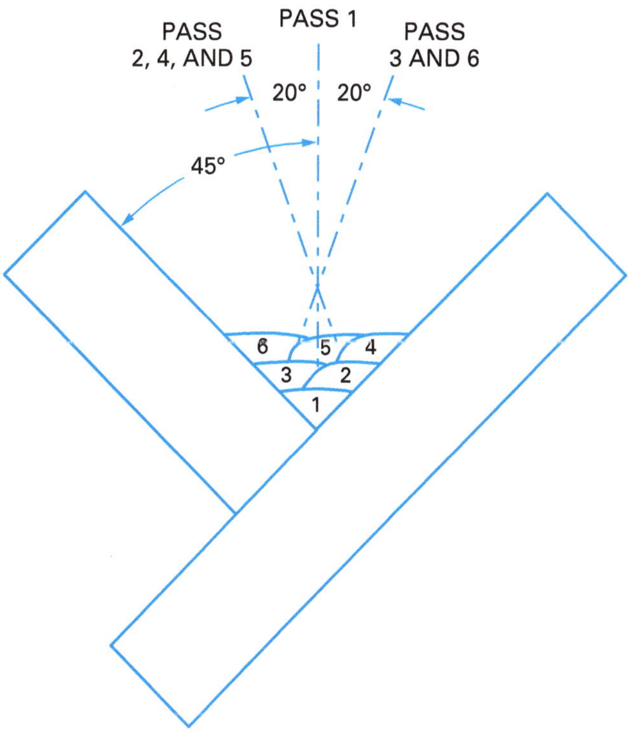

Figure 12 Multiple-pass 1F bead sequence and work angles.

Step 6 Run the fourth bead along the outside toe of the second weld, overlapping about half the second bead. Use the work angle shown in *Figure 12* and a 10- to 15-degree push angle with a slight oscillation.

Step 7 Run the fifth bead along the inside toe of the fourth weld, overlapping about half the fourth bead. Use the work angle shown in *Figure 12* and a 10- to 15-degree push angle with a slight oscillation.

Step 8 Run the sixth bead along the toe of the fifth weld, filling the groove created when the fifth bead was run. Use the work angle shown in *Figure 12* and a 10- to 15-degree push angle with a slight oscillation.

Raising the Working Height of a GTAW Torch

In some cases, raising the torch height (within limits, of course) can be used to widen second and subsequent weld beads without oscillating the torch.

Figure 13 Fillet weld coupon.

Step 9 Inspect the weld. The weld is acceptable if it has the following features:

- Uniform appearance on the bead face
- Craters and restarts filled to the full cross section of the weld
- Uniform weld size ±1⁄16" (±1.6 mm)
- Acceptable weld profile in accordance with the applicable code or standard
- Smooth transition with complete fusion at the toes of the weld
- No porosity
- No undercut
- No overlap
- No inclusions
- No cracks

3.1.2 Practicing Horizontal (2F) Position Fillet Welds

Practice horizontal (2F) fillet welding by placing multiple-pass fillet welds in a T-joint using an appropriate filler metal as directed by the instructor. When making horizontal fillet welds, pay close attention to the torch angles. For the first bead, the electrode work angle is 45 degrees. The work angle is adjusted for all other welds.

Follow these steps to make a horizontal fillet weld:

Step 1 Tack two plates together to form a T-joint for the fillet weld coupon. Clean the tack welds.

Step 2 Clamp or tack weld the coupon in the horizontal position.

Step 3 Run the first bead along the root of the joint using a work angle of approximately 45 degrees with a 10- to 15-degree push angle (*Figure 14*).

Step 4 Clean the weld.

Step 5 Run the remaining passes using a 10- to 15-degree push angle and a slight oscillation at the appropriate work angles (refer to *Figure 14*). Overlap each previous pass. Clean the weld after each pass.

Step 6 Have the instructor inspect the weld. The weld is acceptable if it has these features:

- Uniform appearance on the bead face
- Craters and restarts filled to the full cross section of the weld
- Uniform weld size ±1⁄16" (±1.6 mm)
- Acceptable weld profile in accordance with the applicable code or standard
- Smooth transition with complete fusion at the toes of the weld
- No porosity
- No undercut
- No overlap
- No inclusions
- No cracks

3.1.3 Practicing Vertical (3F) Position Fillet Welds

Practice vertical (3F) fillet welding by placing multiple-pass fillet welds in a T-joint using an appropriate filler metal as directed by the instructor. Normally, vertical welds are accomplished by welding uphill from the bottom to the top using a torch push angle. Because of the uphill welding and push angle, this type of weld is sometimes called vertical-up fillet welding. When vertical welding, either stringer or weave beads can be used. On the job, the site WPS or site quality standard will specify which technique to use. Check with the instructor to see if stringer beads or weave beads should be used here, or practice both techniques.

Follow these steps to make an uphill fillet weld:

Step 1 Tack two plates together to form a T-joint for the fillet weld coupon (*Figure 15*).

Step 2 Clamp or tack weld the coupon in the vertical position.

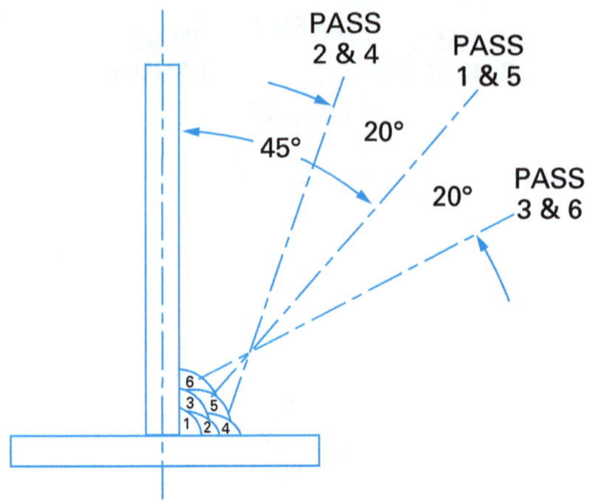

Figure 14 Multiple-pass 2F bead sequence and work angles.

Step 3 Run the first bead along the root of the joint, starting at the bottom and using a work angle of approximately 45 degrees with a 10- to 15-degree push angle. Pause in the weld puddle to fill the crater.

Step 4 Clean the weld.

Step 5 Run the remaining passes using a 10- to 15-degree push angle and a side-to-side weave technique with a 45-degree work angle as shown in *Figure 15*. Use a slow motion across the face of the weld, pausing at each toe for penetration and to fill the crater. Clean the weld after each pass.

Step 6 Have the instructor inspect the weld. The weld is acceptable if it has these features:

- Uniform appearance on the bead face
- Craters and restarts filled to the full cross section of the weld
- Uniform weld size ±1⁄16" (±1.6 mm)
- Acceptable weld profile in accordance with the applicable code or standard
- Smooth transition with complete fusion at the toes of the weld
- No porosity
- No undercut
- No overlap
- No inclusions
- No cracks

Repeat vertical fillet (3F) welding using stringer beads. Use a slight oscillation and side-to-side motion, pausing slightly at each toe to prevent undercut. For stringer beads, use a 10- to 15-degree push angle and the work angles, shown in *Figure 15*, as required.

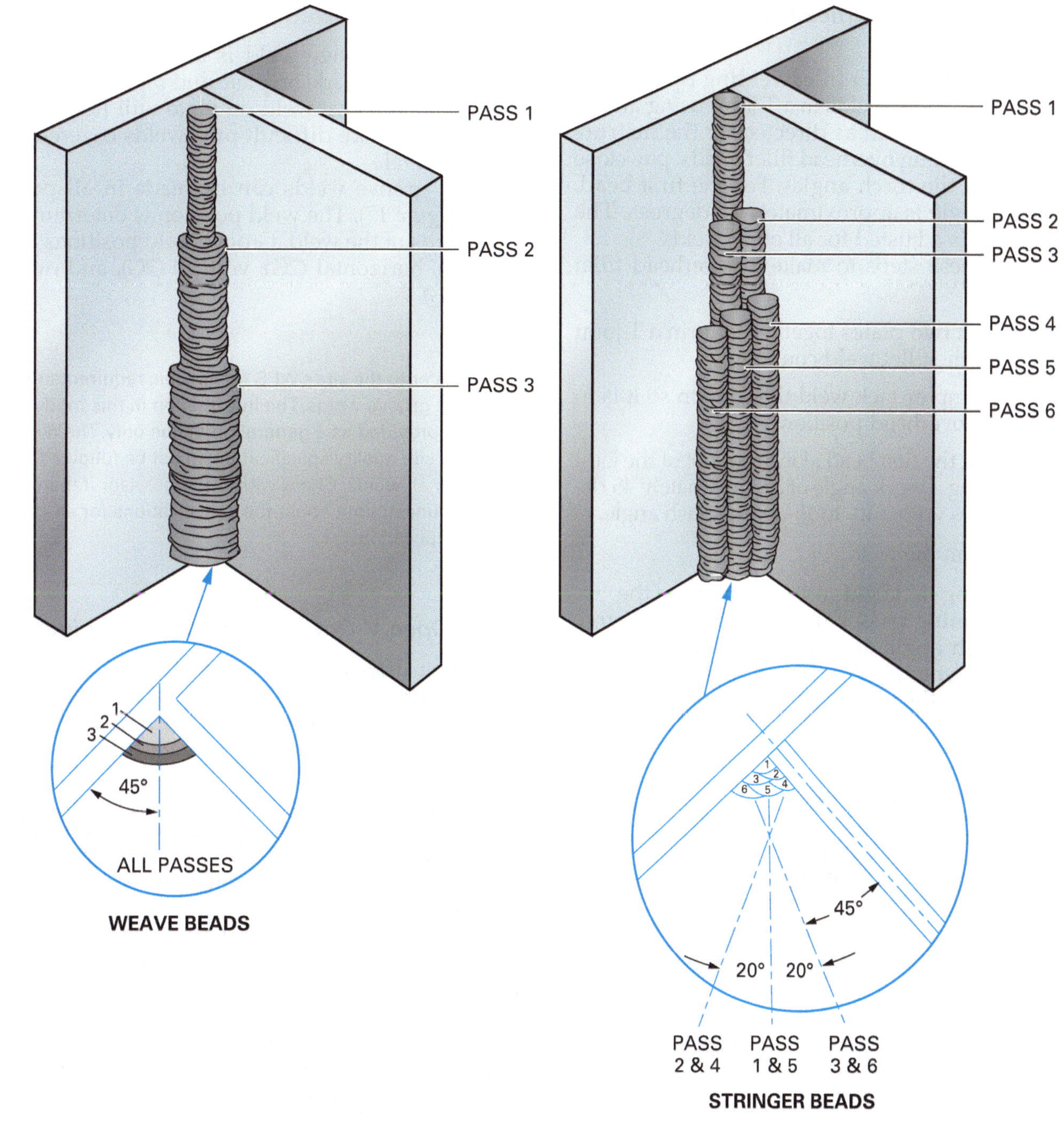

Figure 15 Multiple-pass 3F bead sequences and work angles for stringer and weave beads.

3.1.4 *Practicing Overhead (4F) Position Fillet Welds*

Practice overhead (4F) fillet welding by welding multiple-pass fillet welds in a T-joint using an appropriate filler metal as directed by the instructor. When making overhead fillet welds, pay close attention to the torch angles. For the first bead, the work angle is approximately 45 degrees. The work angle is adjusted for all other welds.

Follow these steps to make an overhead fillet weld:

Step 1 Tack two plates together to form a T-joint for the fillet weld coupon.

Step 2 Clamp or tack weld the coupon so it is in the overhead position.

Step 3 Run the first bead along the root of the joint using a work angle of approximately 45 degrees with a 10- to 15-degree push angle.

Step 4 Clean the weld.

Step 5 Using a slight oscillation, run the remaining passes using a 10- to 15-degree push angle and the work angles shown in *Figure 16*. Overlap each previous pass. Clean the weld after each pass.

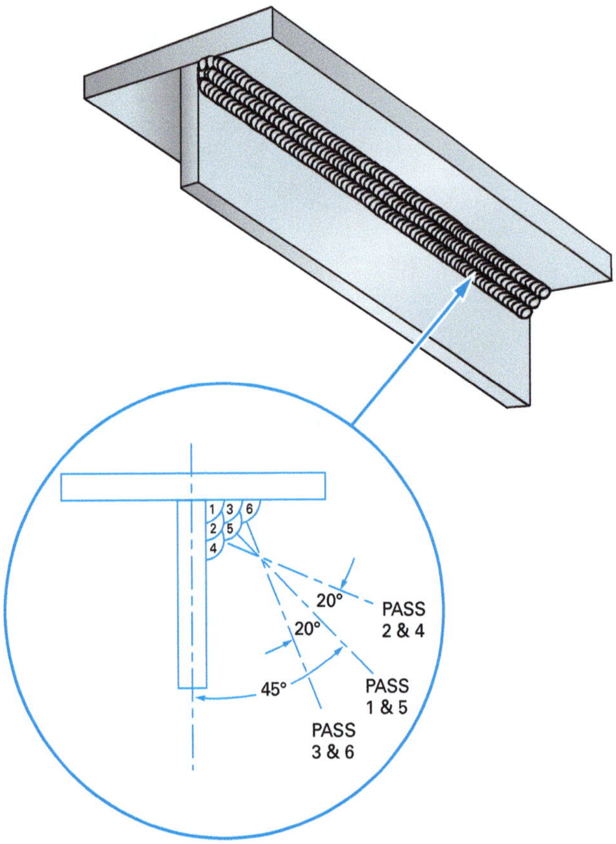

Figure 16 Multiple-pass 4F bead sequence and work.

3.2.0 Groove Welds

The open V-groove weld is a common groove weld normally made on plate and pipe. Practicing the open V-groove weld on plate will help with making the more difficult pipe welds covered in the next level.

Open groove welds can be made in all positions (*Figure 17*). The weld position is determined by the axis of the weld. Groove weld positions are flat (1G), horizontal (2G), vertical (3G), and overhead (4G).

CAUTION

Refer to the site's WPS for specific requirements on groove welds. The information in this module is provided as a general guideline only. The WPS or site quality specifications must be followed for all welds. Check with the supervisor if there is uncertainty about the specifications for an application.

3.2.1 *Open V-Groove Root Pass Techniques*

The most difficult part of welding an open V-groove butt joint is the root pass. The root must have complete penetration, but not have an excessive amount of penetration or melt-through root reinforcement. The penetration is controlled by using the correct amperage, filler metal application, and torch travel speed for the root opening and root face being used. Root reinforcement should be flush to ⅛" (3.2 mm), depending on the site specifications.

The two techniques for running the root pass are **on-the-wire** and keyhole. Regardless of the technique used, always start the root pass from a tack weld.

To perform the on-the-wire root pass technique, secure or hold the filler metal in the bottom of the root opening of the V-groove parallel to the plates. Fuse the filler metal in place by moving the torch along the groove using either the freehand or walking-the-cup torch technique. Advance the torch slowly enough to thoroughly fuse the filler metal with the plate edges, but not so slowly that it can burn through. This technique is used with a filler metal that is the same size as or slightly larger than the width of the root opening; the method is similar to pipe welding using a consumable insert. With this technique, to ensure a uniform bead thickness, the filler metal should not be moved into or out of the weld puddle. *Figure 18* shows the preparation for the on-the-wire root pass technique.

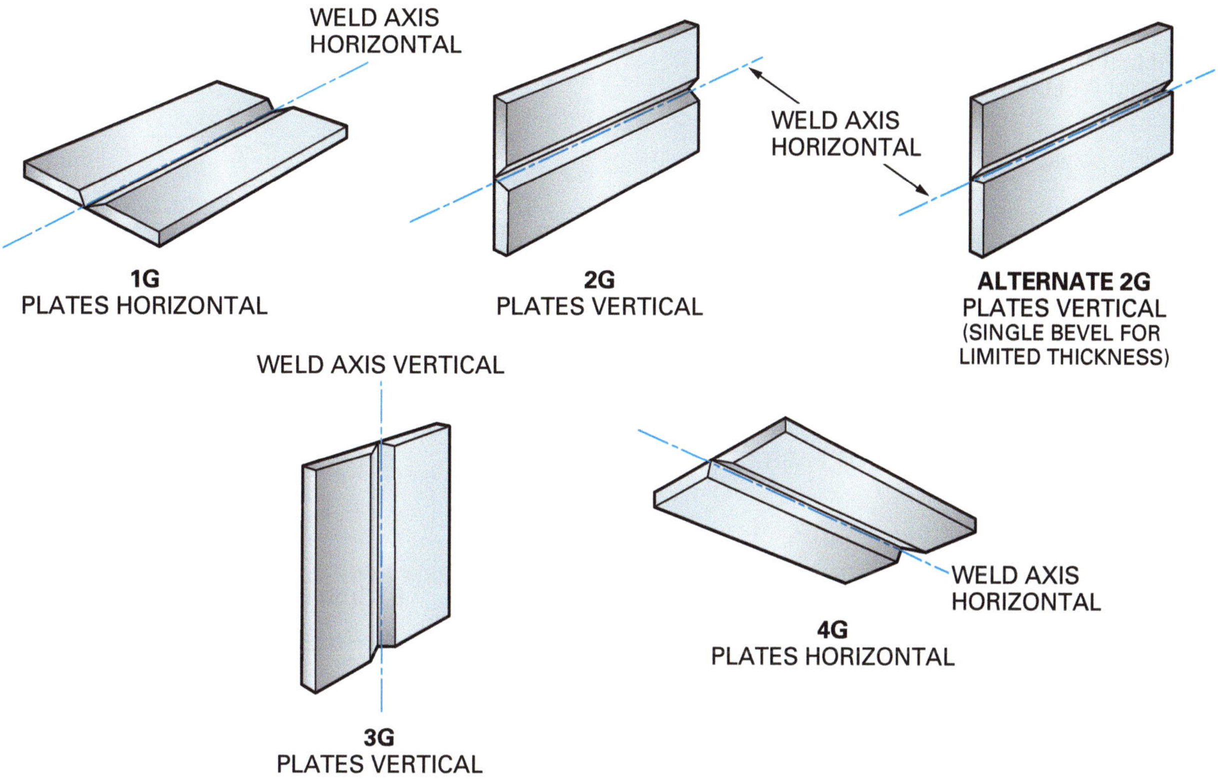

Figure 17 Groove weld positions.

To perform the keyhole root pass technique (*Figure 19*), hold the filler metal at an angle to the plate surfaces and feed it into the groove root from the weld side. This technique uses a filler metal rod diameter that is slightly larger than the root opening so that the rod can rest on the groove.

When making a root pass, start on a tack weld. When a keyhole weld puddle has been established by melting the root faces of both plates together, move the arc to the back of the puddle and insert the end of the filler metal into the leading edge of the weld puddle. Then move the filler metal away from the weld puddle. Advance the torch, using either the walking-the-cup or the freehand technique, maintaining the keyhole. Insert the filler metal back into the leading edge of the molten puddle and forward out of the molten puddle to add filler metal as needed. Be sure the filler metal is always kept in the shielding gas stream to keep the filler metal preheated and protected from atmospheric contamination.

After the root pass has been completed, clean and inspect it for the following:

- Uniform and smooth face and root
- Excessive root reinforcement
- Excessive face buildup
- Undercut
- Porosity
- Cracks

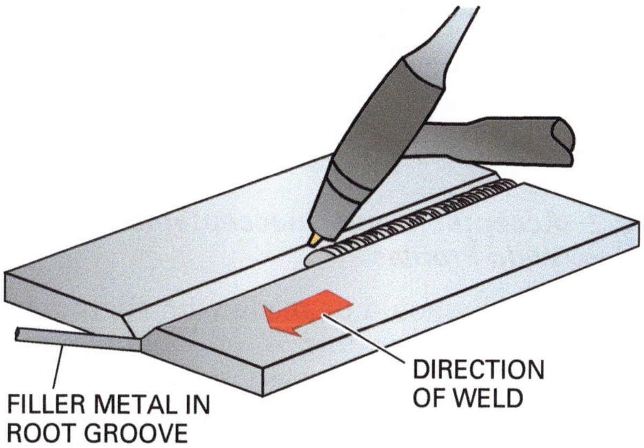

Figure 18 On-the-wire root pass technique.

Cutting Filler Rod

Snip the rod at a 45-degree angle. Then insert the rod into the puddle with the angled tip at the top of the feathered tack. This results in better penetration of the tie-in and a smoother transition from the previous weld into the new weld.

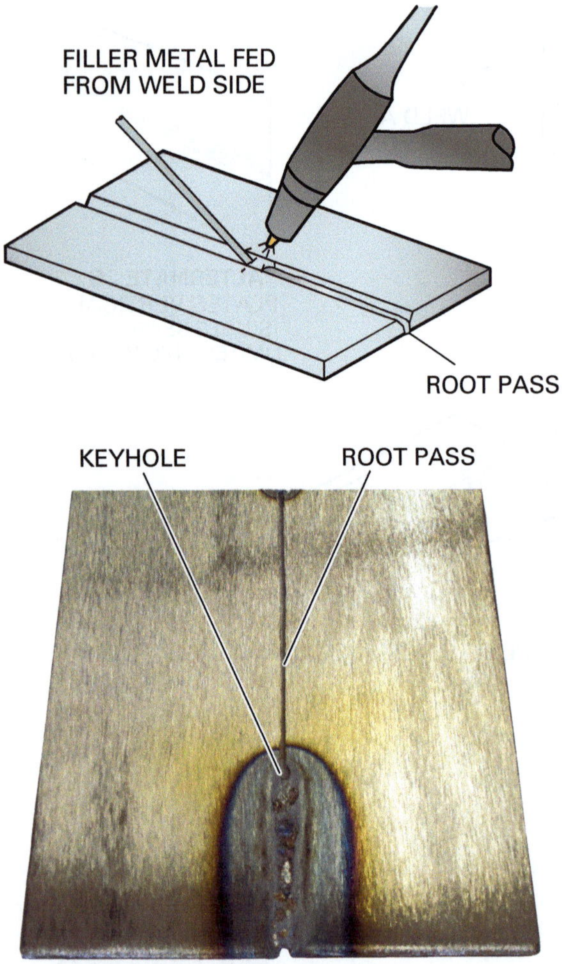

Figure 19 Root pass and keyhole.

3.2.2 Acceptable and Unacceptable Groove Weld Profiles

Groove welds should be made with slight reinforcement (not exceeding 1/8" or 3.2 mm) and a gradual transition to the base metal at each toe. Groove welds must not have excess reinforcement, underfill, undercut, or overlap (*Figure 20*). If a groove weld has any of these defects, it must be replaced.

3.2.3 Practicing Flat (1G) Position Open V-Groove Welds

Follow these steps to make open V-groove welds in the flat (1G) position (*Figure 21*):

Step 1 Tack weld the practice coupon together, following the example given in the section on preparing welding coupons.

Step 2 Clamp or tack weld the welding coupon in a flat position above the welding table surface.

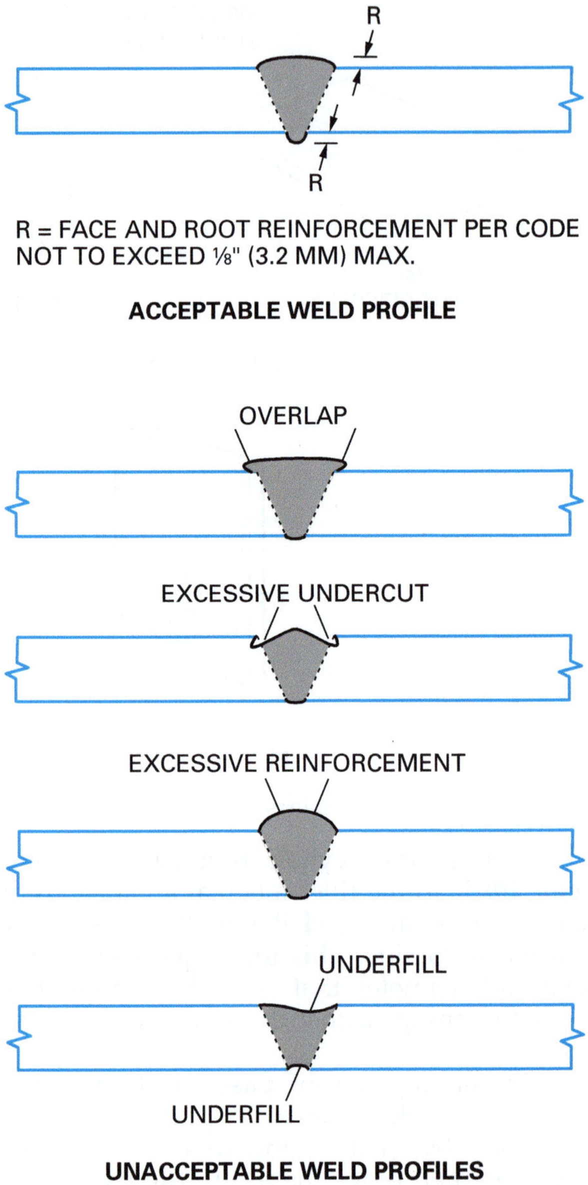

Figure 20 Acceptable and unacceptable groove weld profiles.

Step 3 Run the root pass using an appropriate diameter of carbon steel filler metal and either the on-the-wire or the keyhole technique. Use a 15- to 20-degree push angle, a 20-degree filler metal angle to the plate surface, and the work angle for Pass 1 (*Figure 21*).

Step 4 Clean the root pass.

Step 5 Run the remaining passes using either 3/32" or 1/8" (2.4 mm or 3.2 mm) carbon steel filler metal, a 15- to 20-degree push angle, and the work angles shown in *Figure 21*. Hold the filler metal about 20 degrees above the plate surface. Clean the weld between each pass.

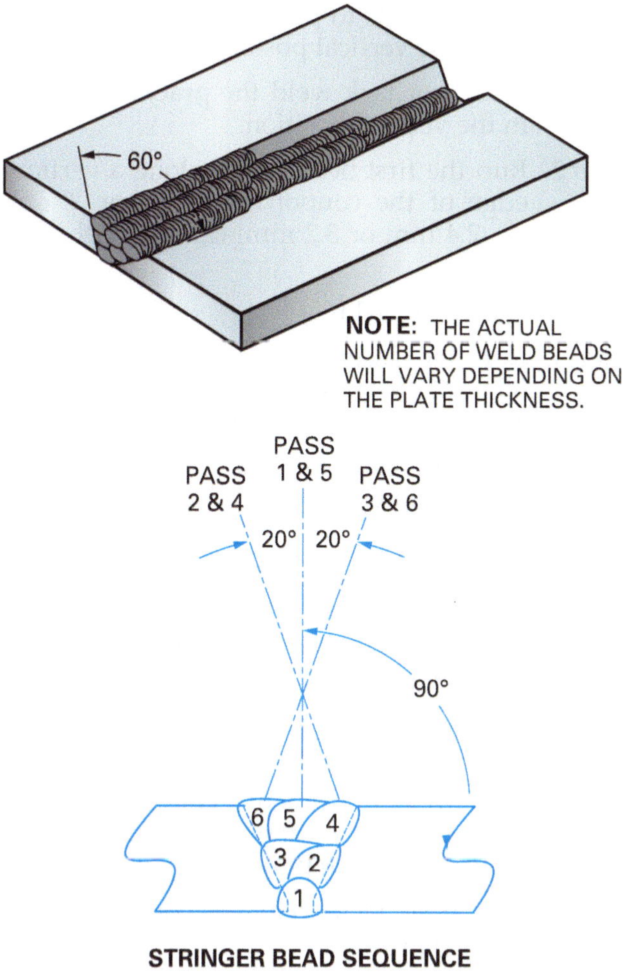

Figure 21 Multiple-pass 1G bead sequence and work angles.

Open V-groove welds can be run with a root and a single layer hot pass before beginning the split layer technique shown in *Figure 21.*

3.2.4 Practicing Horizontal Beads

Before welding a joint in the horizontal (2G) position, practice running horizontal stringer beads by building a pad in the horizontal position (*Figure 22*). Use 3⁄32" or 1⁄8" (2.4 mm or 3.2 mm) carbon steel filler metal to build the pad. Use a push angle of 15 to 20 degrees, and hold the filler metal in front of the weld at a 20-degree angle with the plate face.

Follow these steps to practice overlapping horizontal welding beads in the horizontal (2G) position:

Step 1 Clamp or tack weld the welding coupon in the horizontal position.

Step 2 Run the first pass along the bottom edge of the coupon using either 3⁄32" or 1⁄8" (2.4 mm or 3.2 mm) carbon steel filler metal.

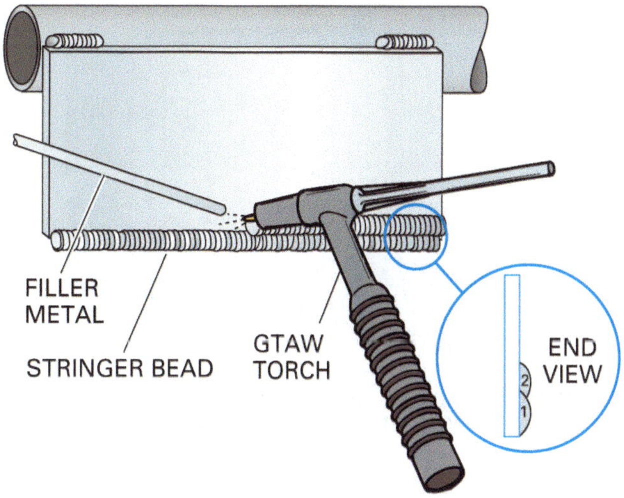

Figure 22 Building a pad in the horizontal position.

Step 3 Clean the weld.

Step 4 Run the second bead along the top edge, overlapping the first bead. Adjust the work angle to obtain a good tie-in.

Step 5 Continue running beads to complete the pad. Be sure to clean the weld between each pass.

3.2.5 Practicing Horizontal (2G) Position Open V-Groove Welds

Follow these steps to practice open V-groove welds in the horizontal (2G) position (*Figure 23*):

Step 1 Tack weld the practice coupon together as explained in the section related to weld coupon preparation. Use the standard or alternate weld coupon as directed by the instructor.

Step 2 Clamp or tack weld the coupon in the horizontal position.

Step 3 Run the root pass using either the on-the-wire or the keyhole welding technique, and using an appropriate diameter of carbon steel filler metal. Use a 15- to 20-degree push angle, a 20-degree filler metal angle with the plate surface, and the appropriate work angle for Pass 1 (*Figure 23*).

Step 4 Clean the weld.

Step 5 Run the remaining passes using either 3⁄32" or 1⁄8" (2.4 mm or 3.2 mm) carbon steel filler metal, a 15- to 20-degree push angle, and the work angles shown in *Figure 23.* Clean the weld between each pass.

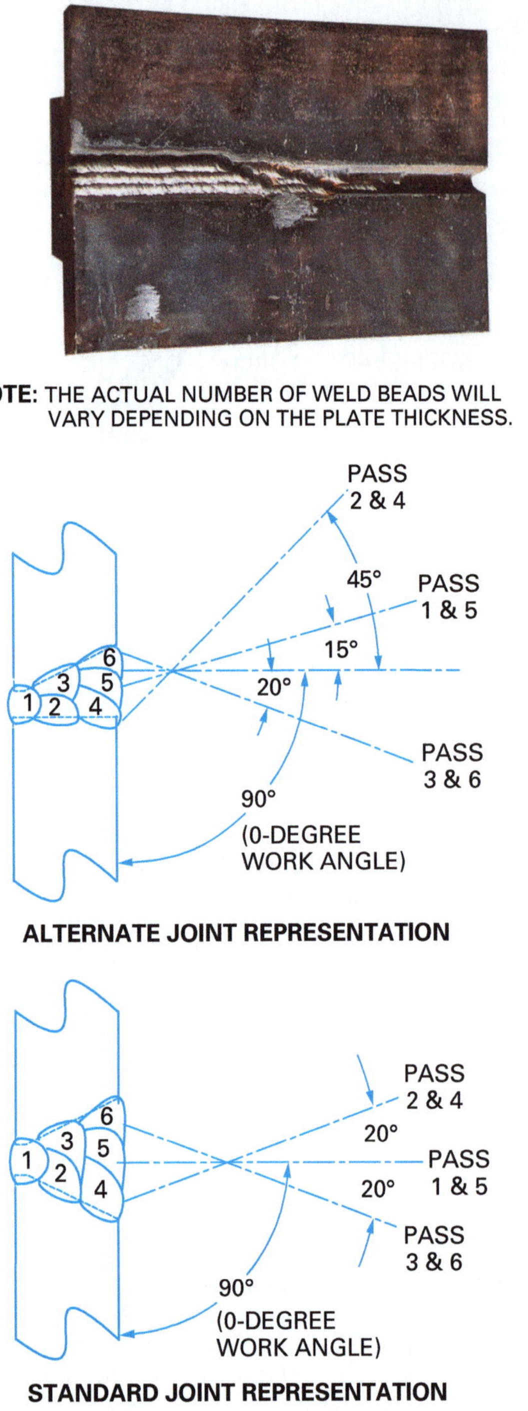

Figure 23 Multiple-pass 2G bead sequences and work.

3.2.6 Practicing Vertical Beads

Before welding a joint in the vertical position, practice running vertical stringer beads by building a pad in the vertical position. Use 3/32" or 1/8" (2.4 mm or 3.2 mm) carbon steel filler metal to build the pad. Weld vertically with a push angle of 15 to 20 degrees and a filler metal angle of about 20 degrees with the plate surface.

Follow these steps to practice overlapping vertical beads in the vertical position:

Step 1 Clamp or tack weld the practice coupon in the vertical position.

Step 2 Run the first bead uphill along a vertical edge of the coupon using either 3/32" or 1/8" (2.4 mm or 3.2 mm) carbon steel filler metal.

Step 3 Clean the weld.

Step 4 Run the second bead, overlapping the edge of the first bead. Adjust the work angle to obtain a good tie-in.

Step 5 Continue running beads to complete the pad. Clean the weld between each pass.

Figure 24 shows how to build a pad with stringer beads in the vertical position.

3.2.7 Practicing Vertical (3G) Position Open V-Groove Welds

Follow these steps to practice open V-groove welds in the vertical (3G) position (*Figure 25*):

Step 1 Tack weld the practice coupon together as explained in the section related to weld coupon preparation.

Step 2 Clamp or tack weld the coupon in the vertical position.

Step 3 Run the root pass uphill with either the on-the-wire or the keyhole welding technique. Use an appropriate diameter of carbon steel filler metal. Use a push angle of 15 to 20 degrees, a filler metal angle of 20 degrees with the plate surface, and the Pass 1 work angle (*Figure 25*).

Step 4 Clean the weld.

Step 5 Run the remaining passes uphill using either stringer or weave beads with 3/32"or 1/8" (2.4 mm or 3.2 mm) carbon steel filler metal, a 15- to 20-degree push angle, and the work angles shown in *Figure 25*. Clean the weld between each pass.

Backing and/or Backing Gas

Normally, backing and/or backing gas is not required for GTAW of carbon steel. However, it is required for stainless and most low-alloy steel and most non-ferrous metals.

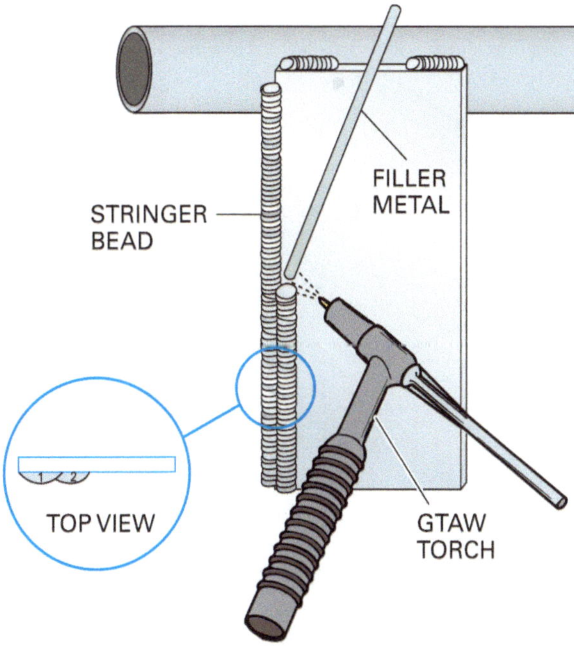

Figure 24 Building a pad in the vertical position.

3.2.8 Practicing Overhead Beads

Before welding a joint in the overhead position, practice running overhead stringer beads by building a pad in the overhead position (*Figure 26*). Use 3⁄32" or 1⁄8" (2.4 mm or 3.2 mm) carbon steel filler metal to build the pad. Weld overhead stringer beads with a push angle of 15 to 20 degrees. Hold the filler metal ahead of the weld at about 20 degrees with the plate surface. Follow these steps to practice overlapping vertical beads in the overhead position:

Step 1 Clamp or tack weld the coupon in the overhead position.

Step 2 Run the first bead along one edge of the coupon using either 3⁄32" or 1⁄8" (2.4 mm or 3.2 mm) carbon steel filler metal.

Step 3 Clean the weld.

Step 4 Run the second bead, overlapping the edge of the first bead. Adjust the work angle to obtain a good tie-in.

Step 5 Continue running beads to complete the pad. Be sure to clean the weld between each pass.

3.2.9 Practicing Overhead (4G) Position Open V-Groove Welds

Follow these steps to practice open V-groove welds in the overhead (4G) position (*Figure 27*):

Step 1 Tack weld the practice coupon together as explained in the section related to weld coupon preparation.

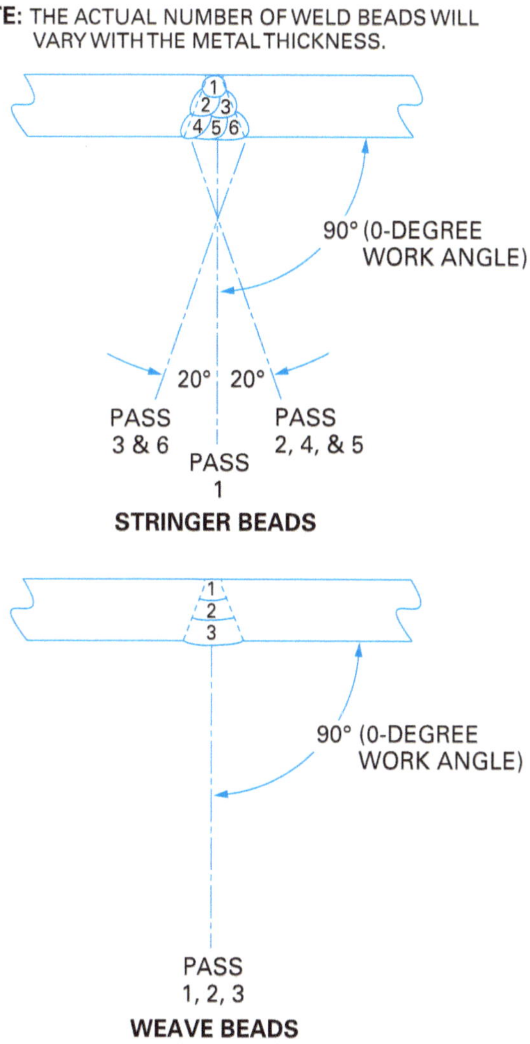

Figure 25 Multiple-pass 3G bead sequence and work angles.

Step 2 Clamp or tack weld the coupon in the overhead position.

Step 3 Run the root pass with either the on-the-wire or the keyhole welding technique, using an appropriate diameter carbon steel filler metal. Use a 15- to 20-degree push angle, a 20-degree filler metal angle to the plate surface, and the Pass 1 work angle (*Figure 27*).

Step 4 Clean the weld.

Step 5 Run the remaining passes using either 3⁄32" or 1⁄8" (2.4 mm or 3.2 mm) carbon steel filler metal, a 15- to 20-degree push angle, and the work angles shown in *Figure 27*. Clean the weld between each Pass.

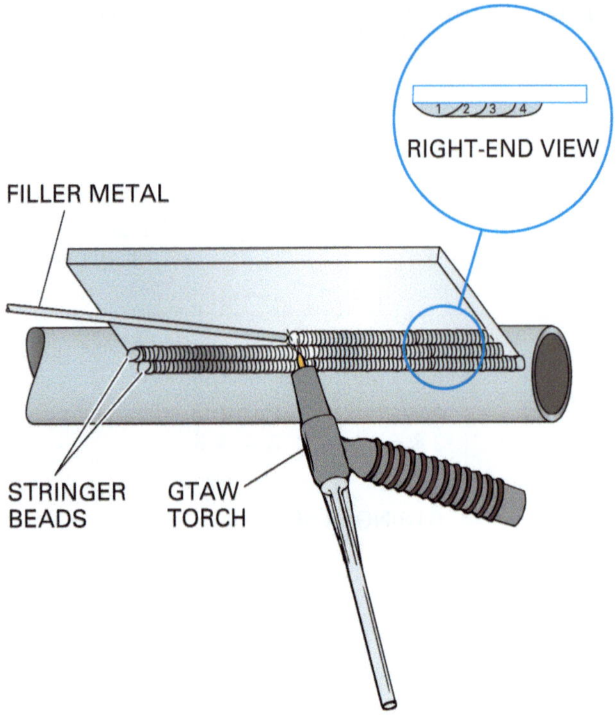

Figure 26 Building a pad in the overhead position.

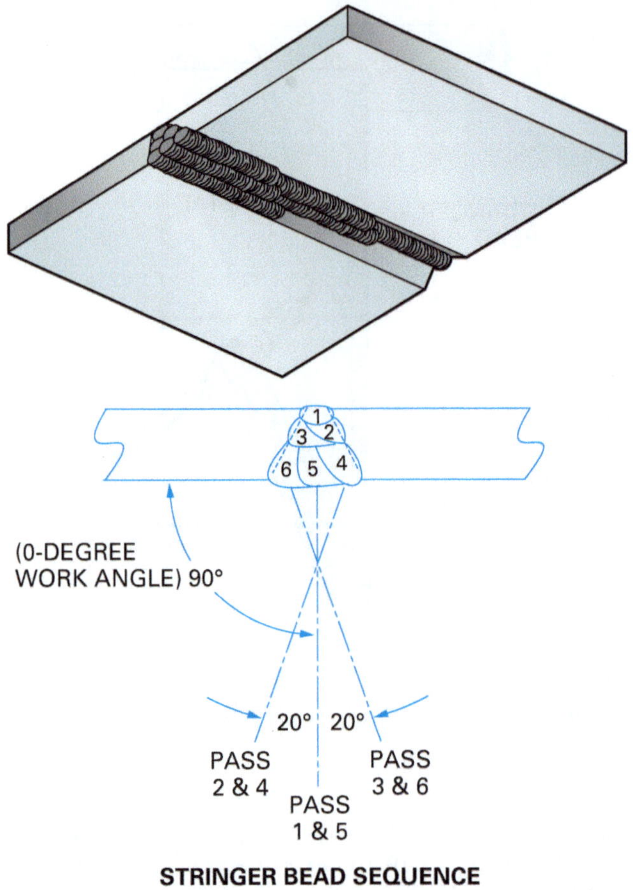

Figure 27 Multiple-pass 4G bead sequence and work angles.

Additional Resources

Lincoln Electric website: **www.lincolnelectric.com** offers sources for products and training.

OSHA Standard 1926.351, Arc Welding and Cutting. Current Edition. Washington, DC: Occupational Safety and Health Administration.

The Procedure Handbook of Arc Welding. Cleveland, OH: The Lincoln Electric Company, 2000.

Welding Handbook, Volume 1 (2001) and Volume 2 (2004). Miami, FL: The American Welding Society.

3.0.0 Section Review

1. The type of GTAW fillet weld that is normally made by welding uphill using a torch push angle is a(n) _____.
   a. horizontal position fillet weld
   b. flat position fillet weld
   c. vertical position fillet weld
   d. overhead position fillet weld

2. The most difficult part of welding an open V-groove butt joint is the _____.
   a. oscillation
   b. root pass
   c. up-cut
   d. termination

SUMMARY

Setting up GTAW equipment, preparing the welding work area, running stringer beads, and making acceptable fillet and groove welds in all positions are essential skills a welder must have to perform basic welding jobs or to progress to more difficult welding procedures. The ability to consistently produce acceptable welds requires an understanding of the necessary welding techniques as well as dedicated practice.

Review Questions

1. In GTAW, the filler metal is typically a _____.
   a. flux-cored wire
   b. handheld rod
   c. tungsten electrode
   d. base metal flux

2. GTAW produces high-quality welds without slag, corrosion, or _____.
   a. penetration
   b. ventilation
   c. termination
   d. oxidation

3. Any safety glasses, face shields, or helmet lenses used during GTAW activities must have the proper _____.
   a. light-reducing shade
   b. mirroring capabilities
   c. temperature tolerance
   d. electrical resistance

4. Before a tank or barrel is cut or welded, it should be cleaned and _____.
   a. sealed shut to prevent leaks or fumes
   b. filled with the proper shielding gas
   c. neutralized with compressed oxygen
   d. filled with water or purged with gas

5. When performing GTAW, open V-groove welds generally have a bevel angle of _____.
   a. 0 to 15 degrees
   b. 20 to 25½ degrees
   c. 30 to 37½ degrees
   d. 45 to 55 degrees

6. Torch travel speed affects the size of the GTAW weld puddle and the weld _____.
   a. convexity
   b. temperature
   c. appearance
   d. penetration

7. The two basic torch angles that must be controlled when performing GTAW are the _____.
   a. work angle and travel angle
   b. freehand angle and pull angle
   c. arc angle and work angle
   d. travel angle and keyhole angle

8. In the freehand technique, the torch electrode tip is manually held _____.
   a. against the base metal
   b. just above the weld puddle or base metal
   c. just below the weld puddle or base metal
   d. against the groove edges

9. Flat-position stringer beads should be run with a push angle of _____.
   a. 15 to 20 degrees
   b. 20 to 30 degrees
   c. 30 to 40 degrees
   d. 60 to 70 degrees

10. Compared to the puddle width used to produce it, a weave bead is much _____.
    a. deeper
    b. higher
    c. wider
    d. hotter

For Questions 11 through 13, refer to *Review Question Figure 1*.

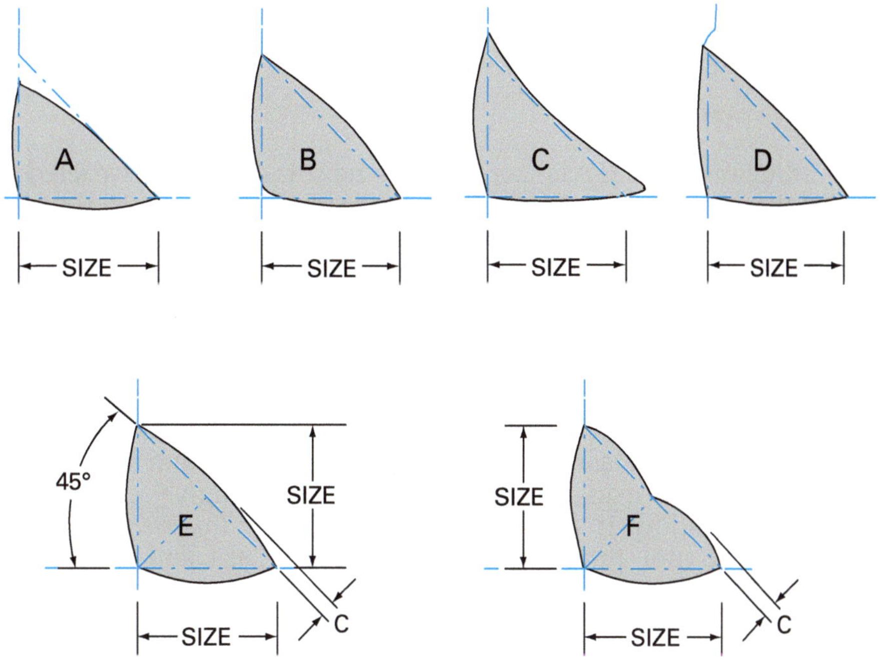

Figure 1

11. An example of an acceptable fillet weld is represented by _____
    a. Letter B
    b. Letter D
    c. Letter E
    d. Letter F

12. An example of an ideal fillet weld is represented by _____
    a. Letter A
    b. Letter C
    c. Letter E
    d. Letter F

13. An example of an unacceptable overlap fillet weld is represented by _____
    a. Letter B
    b. Letter C
    c. Letter D
    d. Letter E

14. When practicing flat (1G) position open V-groove welds, run the root pass using either the on-the-wire or the _____.
    a. side-to-side movement
    b. keyhole technique
    c. walking-the-cup angle
    d. freehand technique

15. When running stringer beads in the horizontal position, always start the first bead _____.
    a. along the top edge of the workpiece
    b. so that it overlaps the padding
    c. at the end of the tie-in weld
    d. along the bottom edge of the workpiece

Appendix

Performance Accreditation Tasks

The Performance Accreditation Tasks (PATs) correspond to and support the learning objectives in *AWS EG2.0, Curriculum Guide for the Training of Welding Personnel: Level 1 – Entry Welder.*

PATs provide specific acceptable criteria for performance and help to ensure a true competency-based welding program for students.

The following tasks are designed to evaluate your ability to run fillet and groove welds with GTAW equipment. Perform each task when you are instructed to do so by your instructor. As you complete each task, show it to your instructor for evaluation. Do not proceed to the next task until instructed to do so by your instructor.

NOTE to SENSE instructors: The AWS SENSE program also requires welding activities on aluminum and stainless steel base metals. This module does not support aluminum and stainless steel welding.

PATs 1 through 5 correspond to *AWS EG2.0, Module 7, Gas Tungsten Arc Welding (GTAW)*, Key Indicators 1, 3, 4, and 5.

PATs 6 through 9 correspond to *AWS EG2.0, Module 7, Gas Tungsten Arc Welding (GTAW)*, Key Indicators 1, 3, 4, and 6.

WELD A PAD ON CARBON STEEL PLATE IN THE FLAT POSITION USING GTAW STRINGER BEADS

As directed by the instructor, use GTAW process with carbon steel filler metal to build a pad on a carbon steel coupon in the flat position.

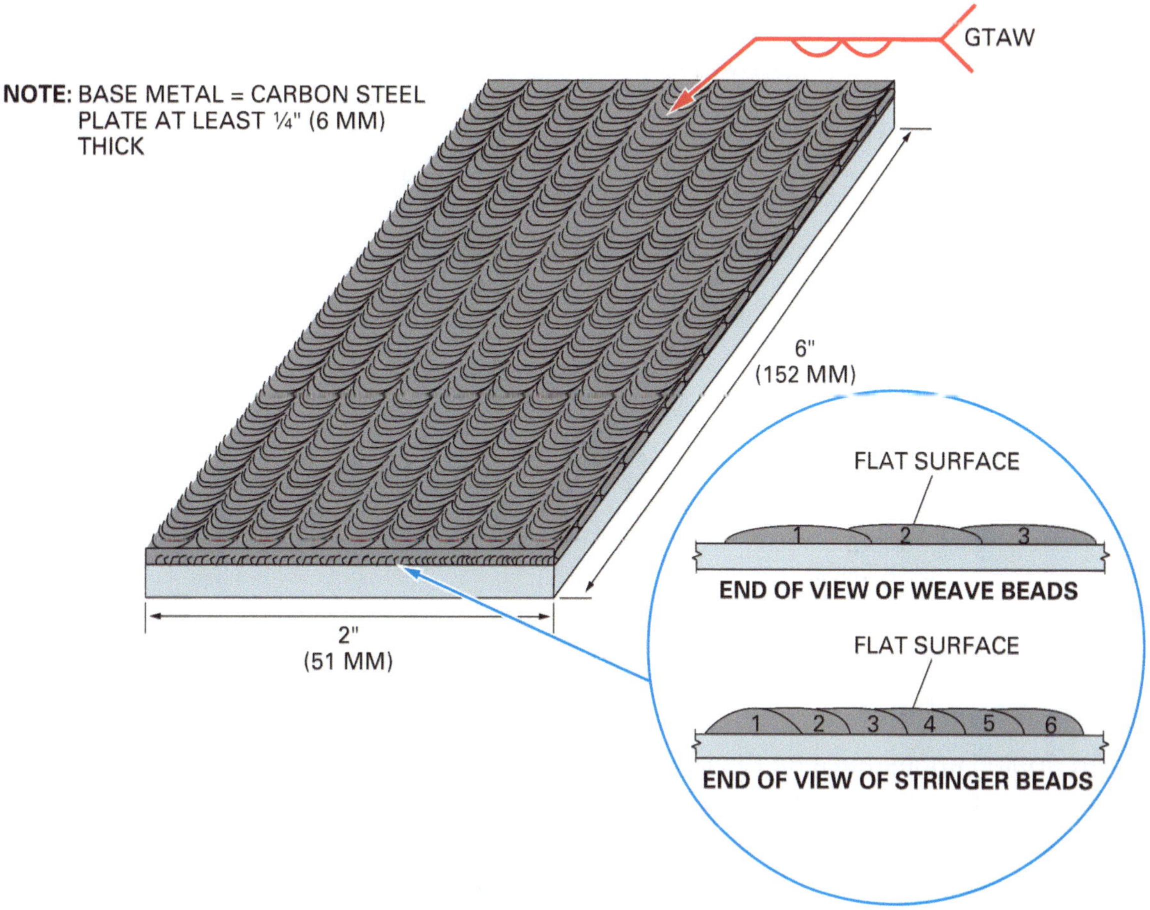

Criteria for Acceptance:

- Uniform rippled appearance on the bead face ____________
- Craters and restarts filled to the full cross section of the weld ____________
- Uniform weld width ±1⁄16" (1.6 mm) ____________
- Acceptable weld profile in accordance with the applicable code or standard ____________
- Smooth flat transition with complete fusion at the toes of the weld ____________
- No porosity ____________
- No undercut ____________
- No inclusions ____________
- No cracks ____________

MAKE MULTIPLE-PASS FILLET WELDS ON CARBON STEEL PLATE IN THE (1F) FLAT POSITION

As directed by the instructor, use GTAW process with carbon steel filler metal to make a six-pass fillet weld using stringer beads on carbon steel plate, as shown.

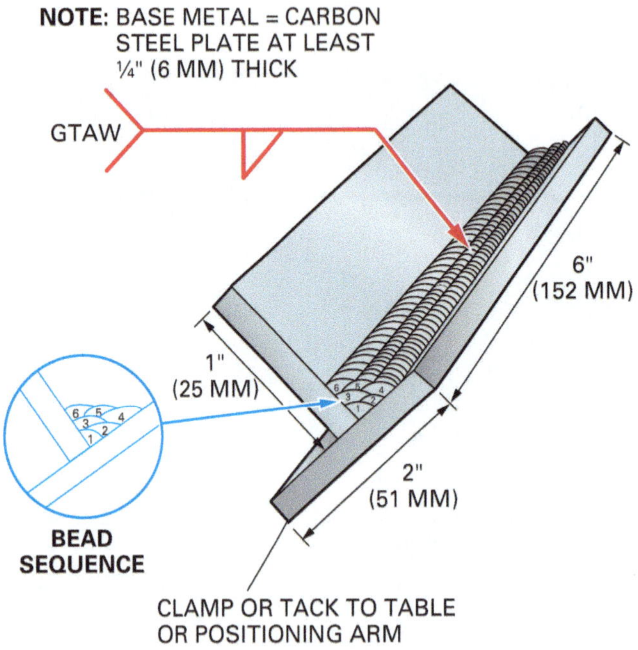

Criteria for Acceptance:

- Uniform rippled appearance on the bead face ____________
- Craters and restarts filled to the full cross section of the weld ____________
- Uniform weld width ±1⁄16" (1.6 mm) ____________
- Acceptable weld profile in accordance with the applicable code or standard ____________
- Smooth flat transition with complete fusion at the toes of the welds ____________
- No porosity ____________
- No undercut ____________
- No overlap ____________
- No inclusions ____________
- No cracks ____________

MAKE MULTIPLE-PASS FILLET WELDS ON CARBON STEEL PLATE IN THE (2F) HORIZONTAL POSITION

As directed by the instructor, use GTAW process with carbon steel solid filler metal to make a six-pass fillet weld using stringer beads on carbon steel plate, as shown.

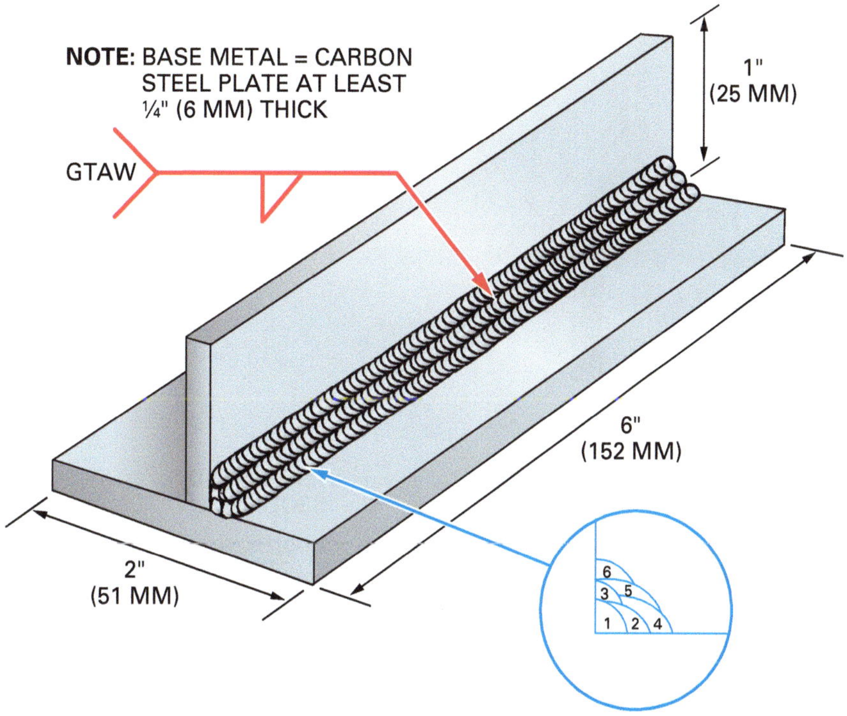

Criteria for Acceptance:

- Uniform rippled appearance on the bead face ________
- Craters and restarts filled to the full cross section of the weld ________
- Uniform weld width ±1/16" (1.6 mm) ________
- Acceptable weld profile in accordance with the applicable code or standard ________
- Smooth flat transition with complete fusion at the toes of the welds ________
- No porosity ________
- No undercut ________
- No overlap ________
- No inclusions ________
- No cracks ________

MAKE MULTIPLE-PASS FILLET WELDS ON CARBON STEEL PLATE IN THE (3F) VERTICAL POSITION

As directed by the instructor, use GTAW process with carbon steel filler metal to make a vertical fillet weld on carbon steel plate, as shown.

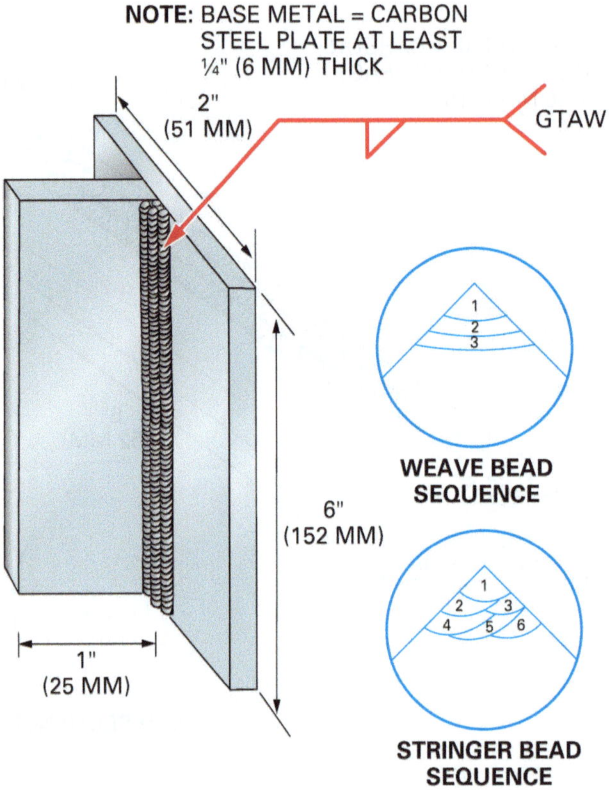

Criteria for Acceptance:

- Uniform rippled appearance on the bead face ______
- Craters and restarts filled to the full cross section of the weld ______
- Uniform weld width ±1⁄16" (1.6 mm) ______
- Acceptable weld profile in accordance with the applicable code or standard ______
- Smooth flat transition with complete fusion at the toes of the welds ______
- No porosity ______
- No undercut ______
- No inclusions ______
- No cracks ______

MAKE MULTIPLE-PASS FILLET WELDS ON CARBON STEEL PLATE IN THE (4F) OVERHEAD POSITION

As directed by the instructor, use GTAW process with carbon steel filler metal to make a six-pass fillet weld using stringer beads on carbon steel plate, as shown.

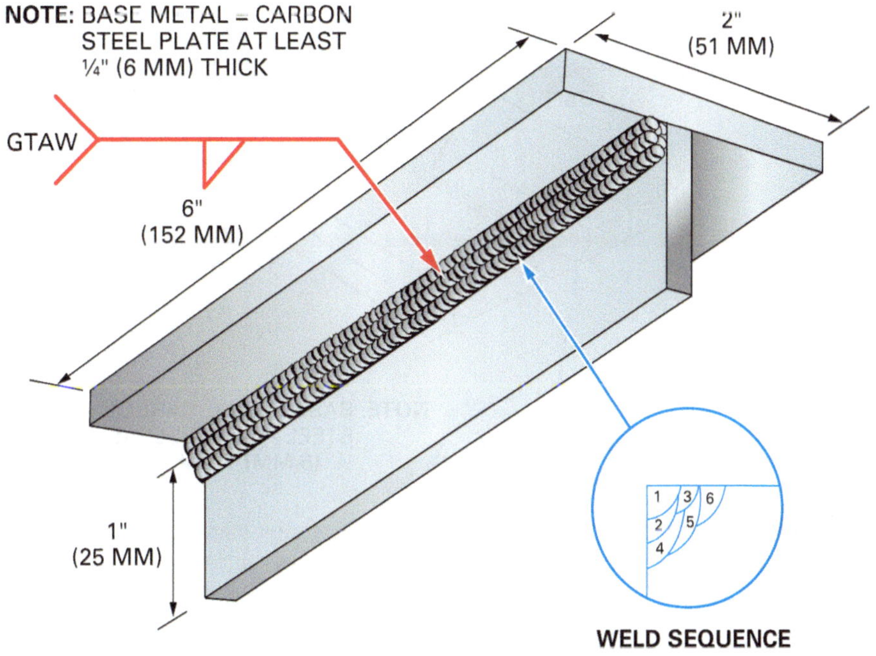

Criteria for Acceptance:

- Uniform rippled appearance on the bead face ____________
- Craters and restarts filled to the full cross section of the weld ____________
- Uniform weld width ±1⁄16" (1.6 mm) ____________
- Acceptable weld profile in accordance with the applicable code or standard ____________
- Smooth flat transition with complete fusion at the toes of the welds ____________
- No porosity ____________
- No undercut ____________
- No inclusions ____________
- No cracks ____________

MAKE MULTIPLE-PASS OPEN V-GROOVE WELDS ON CARBON STEEL PLATE IN THE FLAT (1G) POSITION

As directed by the instructor, use GTAW process with carbon steel filler metal to make multiple pass open V-groove welds on carbon steel plate in the flat (1G) position, as shown.

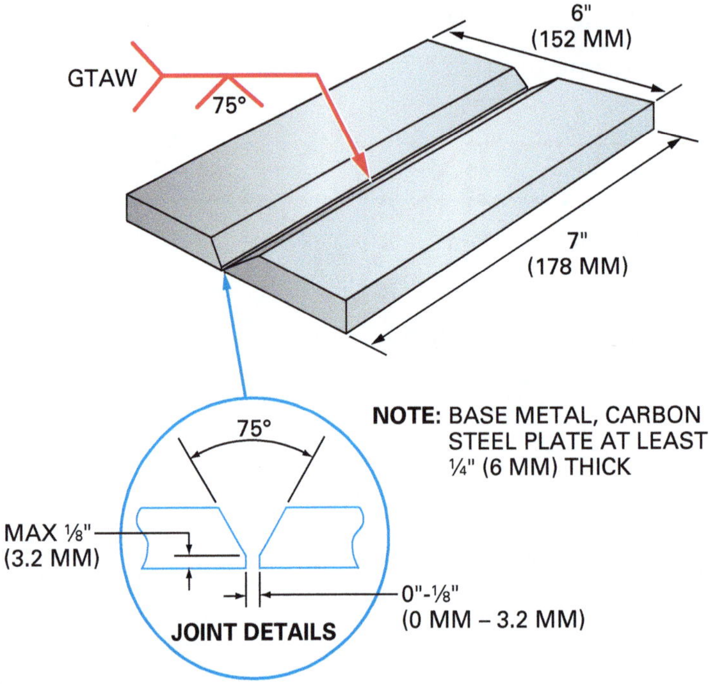

NOTE: BASE METAL, CARBON STEEL PLATE AT LEAST ¼" (6 MM) THICK

Criteria for Acceptance:

- Uniform rippled appearance on the bead face ____________
- Craters and restarts filled to the full cross section of the weld ____________
- Uniform weld size ±1⁄16" (1.6 mm) ____________
- Acceptable weld profile in accordance with the applicable code or standard ____________
- Smooth transition with complete fusion at the toes of the weld ____________
- Complete uniform root and face reinforcement at least flush with the base metal to a maximum buildup of ⅛" (3.2 mm) ____________
- No porosity ____________
- No overlap ____________
- No undercut ____________
- No inclusions ____________
- No cracks ____________
- No pinholes ____________

MAKE MULTIPLE-PASS OPEN V-GROOVE WELDS ON CARBON STEEL PLATE IN THE HORIZONTAL (2G) POSITION

As directed by the instructor, use GTAW process with carbon steel filler metal to make multiple-pass open V-groove welds on carbon steel plate in the horizontal (2G) position, as shown.

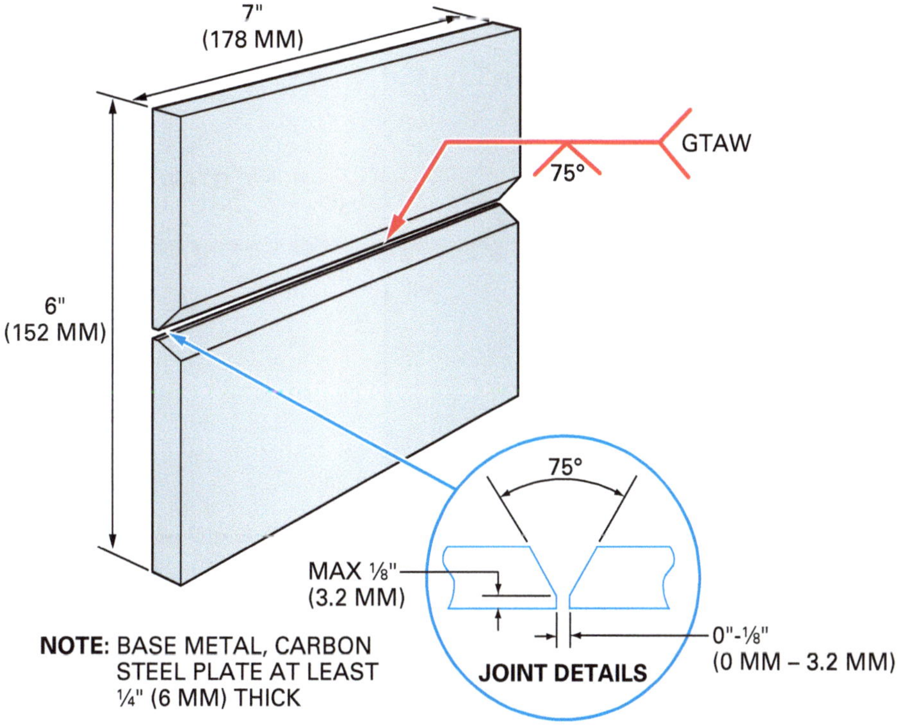

Criteria for Acceptance:

- Uniform rippled appearance on the bead face ____________
- Craters and restarts filled to the full cross section of the weld ____________
- Uniform weld size ±1⁄16" (1.6 mm) ____________
- Acceptable weld profile in accordance with the applicable code or standard ____________
- Smooth transition with complete fusion at the toes of the weld ____________
- Complete uniform root reinforcement at least flush with the base metal to a maximum buildup of ⅛" (3.2 mm) ____________
- No porosity ____________
- No overlap ____________
- No undercut ____________
- No inclusions ____________
- No cracks ____________
- No pinholes ____________

MAKE MULTIPLE-PASS OPEN V-GROOVE WELDS ON CARBON STEEL PLATE IN THE VERTICAL (3G) POSITION

As directed by the instructor, use GTAW process with carbon steel filler metal using stringer or weave beads to make multiple-pass open V-groove welds on carbon steel plate in the vertical (3G) position, as shown.

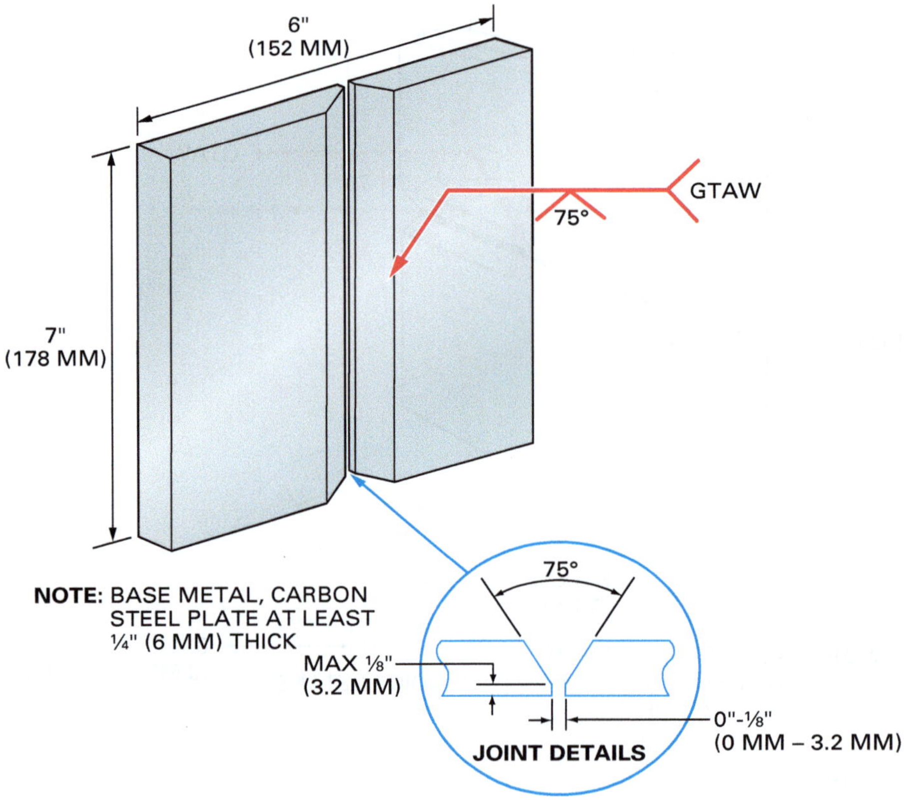

Criteria for Acceptance:

- Uniform rippled appearance on the bead face ____________
- Craters and restarts filled to the full cross section of the weld ____________
- Uniform weld size ±¹⁄₁₆" (1.6 mm) ____________
- Acceptable weld profile in accordance with the applicable code or standard ____________
- Smooth transition with complete fusion at the toes of the weld ____________
- Complete uniform root reinforcement at least flush with the base metal to a maximum buildup of ⅛" (3.2 mm) ____________
- No porosity ____________
- No overlap ____________
- No undercut ____________
- No inclusions ____________
- No cracks ____________
- No pinholes ____________

MAKE MULTIPLE-PASS OPEN V-GROOVE WELDS ON CARBON STEEL PLATE IN THE OVERHEAD (4G) POSITION

As directed by the instructor, use GTAW process with carbon steel filler metal to make multiple-pass open V-groove welds on carbon steel plate in the overhead (4G) position, as shown.

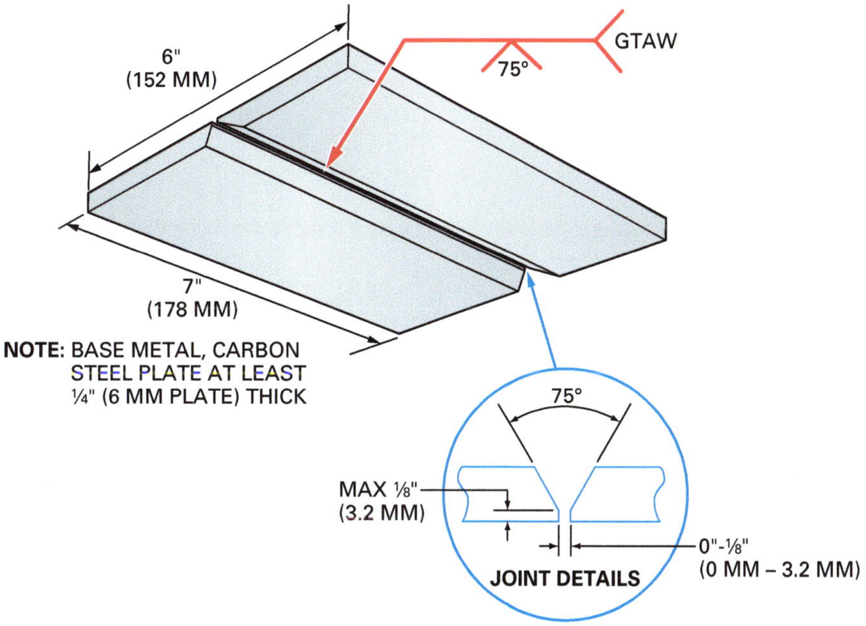

Criteria for Acceptance:

- Uniform rippled appearance on the bead face ____________
- Craters and restarts filled to the full cross section of the weld ____________
- Uniform weld size ±1⁄16" (1.6 mm) ____________
- Acceptable weld profile in accordance with the applicable code or standard ____________
- Smooth transition with complete fusion at the toes of the weld ____________
- Complete uniform root reinforcement at least flush with the base metal to a maximum buildup of ⅛" (3.2 mm) ____________
- No porosity ____________
- No overlap ____________
- No undercut ____________
- No inclusions ____________
- No cracks ____________
- No pinholes ____________

Trade Terms Introduced in This Module

On-the-wire: A root pass technique in which the filler metal is held in the bottom of the groove parallel to the plates and then melted and fused in place by moving the torch along the groove.

Walking-the-cup: A welding method in which the welder rests the edge of the torch nozzle (cup) against the base metal or groove edges to steady the torch and maintain a constant arc length.

Additional Resources

This module is intended as a thorough resource for task training. The following reference works are suggested for further study.

Lincoln Electric website: **www.lincolnelectric.com** offers sources for products and training.

OSHA Standard 1926.351, Arc Welding and Cutting. Current Edition. Washington, DC: Occupational Safety and Health Administration.

The Procedure Handbook of Arc Welding. Cleveland, OH: The Lincoln Electric Company, 2000.

Welding Handbook, Volume 1 (2001) and Volume 2 (2004). Miami, FL: The American Welding Society.

Figure Credits

The Lincoln Electric Company, Cleveland, OH, USA, Module opener

Zachry Industrial, Inc., Figure 7

Holley Thomas, SME, Figure 13

Terry Lowe, Figures 19 (photo), 20, 23 (photo)

Section Review Answer Key

Answer	Section Reference	Objective
Section One		
1. b	1.1.3	1a
2. d	1.2.3	1b
Section Two		
1. c	2.1.3	2a
2. a	2.2.1	2b
Section Three		
1. c	3.1.3	3a
2. b	3.2.1	3b

NCCER CURRICULA — USER UPDATE

NCCER makes every effort to keep its textbooks up-to-date and free of technical errors. We appreciate your help in this process. If you find an error, a typographical mistake, or an inaccuracy in NCCER's curricula, please fill out this form (or a photocopy), or complete the online form at **www.nccer.org/olf**. Be sure to include the exact module ID number, page number, a detailed description, and your recommended correction. Your input will be brought to the attention of the Authoring Team. Thank you for your assistance.

Instructors – If you have an idea for improving this textbook, or have found that additional materials were necessary to teach this module effectively, please let us know so that we may present your suggestions to the Authoring Team.

NCCER Product Development and Revision

13614 Progress Blvd., Alachua, FL 32615

Email: curriculum@nccer.org
Online: www.nccer.org/olf

❑ Trainee Guide ❑ Lesson Plans ❑ Exam ❑ PowerPoints Other ____________________

Craft / Level: Copyright Date:

Module ID Number / Title:

Section Number(s):

Description:

Recommended Correction:

Your Name:

Address:

Email: Phone:

Physical Characteristics and Mechanical Properties of Metals

Overview

In order for welders to accurately identify base metals and select the proper filler metals, they must have a thorough understanding of the composition, classification, physical characteristics, and mechanical properties of common metals and alloys. They must be familiar with metallurgical factors that need to be considered during welding activities and must also be able to recognize standard commercial steel shapes used for fabrication and construction.

Module 29203

Trainees with successful module completions may be eligible for credentialing through the NCCER Registry. To learn more, go to **www.nccer.org** or contact us at **1.888.622.3720.** Our website has information on the latest product releases and training, as well as online versions of our *Cornerstone* magazine and Pearson's product catalog.

Your feedback is welcome. You may email your comments to **curriculum@nccer.org,** send general comments and inquiries to **info@nccer.org,** or fill in the User Update form at the back of this module.

This information is general in nature and intended for training purposes only. Actual performance of activities described in this manual requires compliance with all applicable operating, service, maintenance, and safety procedures under the direction of qualified personnel. References in this manual to patented or proprietary devices do not constitute a recommendation of their use.

29203 V5

29203
Physical Characteristics and Mechanical Properties of Metal

Objectives

When you have completed this module, you will be able to do the following:

1. Describe the composition and classification systems for a variety of metals.
   a. Describe the composition and classification system for ferrous metals.
   b. Describe the composition and classification system for low-alloy steel.
   c. Describe the composition and classification system for common-grade stainless steel.
   d. Describe the composition and classification system for specialty-grade stainless steel.
   e. Describe the composition and classification system for nonferrous metals.
2. Describe the physical and mechanical characteristics of metals and explain how to identify base metals.
   a. Describe the physical characteristics of different metals.
   b. Describe the mechanical properties of different metals.
   c. Explain how to identify base metals in field conditions.
   d. Describe metallurgy-related considerations for welding.
3. Identify the common structural shapes of metal.
   a. Identify the most common structural steel shapes.
   b. Identify different structural beam shapes.
   c. Identify pipe and tubing types.
   d. Identify other common metal forms, including rebar.

Performance Tasks

This is a knowledge-based module; there are no performance tasks.

Trade Terms

Alloy
American Iron and Steel Institute (AISI)
American Society for Testing and Materials (ASTM International)
Austenitizing
Casting
Certified mill test report (CMTR)
Coefficient
Ductile
Ductility
Ferrous
Hardenability
Heat-affected zone (HAZ)
Hot shortness
Interpass temperature
Malleable
Mechanical properties
Metallic
Mill test report (MTR)
Nonferrous metal
Notch toughness
Postweld heat treatment (PWHT)
Quench
Residual stress
Sintered
Society of Automotive Engineers (SAE)
Stress-relief heat treatment
Tempering
Tensile strength
Underbead cracking
Unified numbering system (UNS)
Web
Wide flange (WF) beam
Wrought

Industry Recognized Credentials

If you are training through an NCCER-accredited sponsor, you may be eligible for credentials from NCCER's Registry. The ID number for this module is 29203. Note that this module may have been used in other NCCER curricula and may apply to other level completions. Contact NCCER's Registry at 888.622.3720 or go to **www.nccer.org** for more information.

Contents

Topics to be presented in this module include:

Figures and Tables

SECTION ONE

1.0.0 METAL COMPOSITION AND CLASSIFICATION

Objective

Describe the composition and classification systems for a variety of metals.

a. Describe the composition and classification system for ferrous metals.
b. Describe the composition and classification system for low-alloy steel.
c. Describe the composition and classification system for common-grade stainless steel.
d. Describe the composition and classification system for specialty-grade stainless steel.
e. Describe the composition and classification system for nonferrous metals.

Trade Terms

Alloy: A metal that has had other elements added to it that substantially change its mechanical properties.

American Iron and Steel Institute (AISI): An industry organization responsible for preparing standards for steel and steel alloys that are based upon the Society of Automotive Engineers (SAE) code system.

American Society for Testing and Materials (ASTM International): An organization that developed a code system for identifying and labeling steels.

Austenitizing: Heating a steel up to or above the transformation temperature range to achieve partial or complete transformation to austenite grain structure.

Casting: The metal object produced by pouring molten metal into a mold.

Certified mill test report (CMTR): A quality assurance report provided to a customer by a metal manufacturer that shows the chemical content and testing results of the metal being purchased; also called a mill test report (MTR).

Coefficient: A numerical measure of a physical or chemical property that is constant for a system under specified conditions, such as the coefficient of friction.

Ductile: The term used to describe a material that can be bent or shaped without breaking.

Ductility: The characteristic of metal that allows it to be stretched, drawn, or hammered without breaking.

Ferrous: Relating to iron or an alloy that is mostly iron; a metal containing iron.

Hardenability: A characteristic of a metal that enables it to become hard, usually through heat treatment.

Heat-affected zone (HAZ): The part of the base metal that has not been melted but has been altered structurally by the weld heat.

Hot shortness: The condition of metal when it proves to be very brittle and unbendable at red heat but can be bent without showing signs of brittleness when cold or at white heat. This condition is often a result of high sulfur and phosphorus content.

Interpass temperature: In a multiple-pass weld, the temperature of the base metal at the time the next weld pass is started; it is usually the same temperature that is used for preheating.

Malleable: Able to be hammered or pressed into another shape without breaking.

Mechanical properties: The characteristics or traits that indicate how flexible and strong a metal will be under stress.

Metallic: Having characteristics of metal, such as ductility, malleability, luster, and heat and electrical conductivity.

Mill test report (MTR): A quality assurance report provided to a customer by a metal manufacturer that shows the chemical content and testing results of the metal being purchased; also called a certified mill test report (CMTR).

Nonferrous metal: A metal, such as aluminum, copper, or brass that lacks sufficient quantities of iron to be affected by its properties.

Notch toughness: The impact load required to break a test piece of weld metal measured by a notch toughness test (Charpy V-notch test). This test may be performed on metal below room temperature, at which point it may be more brittle.

Postweld heat treatment (PWHT): The process of heat treating the base metal and weld after the weld has been made.

Quench: To rapidly cool a hot metal using air, water, or oil.

Society of Automotive Engineers (SAE): An automotive society responsible for creating an early system for classifying carbon steels.

Stress-relief heat treatment: A technique used to reduce welding stresses and to increase the ductility and toughness of the weld metal and the heat-affected zone (HAZ).

Tempering: The process of reheating quench hardened or normalized steel to a temperature below the transformation range and then cooling it at any rate desired.

Tensile strength: The maximum stress or force that a material can withstand without breaking.

Unified numbering system (UNS): An industry-accepted numbering system for identifying ferrous metals and alloys.

Web: The plate joining the flanges of a girder, joist, or rail.

Wide flange (WF) beam: A metal I-beam that has an extra wide flange at the top and bottom.

Wrought: Formed or shaped by hammering or rolling.

All living and nonliving things are made up of chemical elements that can be altered by changes in temperature. A steak left on a grill will eventually burn; a piece of wood thrown into a fire will also burn. Metals are no different. When they are heated to a high enough temperature they will melt, and some will burn.

Metals, like other substances, are a collection of chemical elements bound together by natural forces or by man-made production processes. The chemical elements in some metals make them stronger than other metals. Those same chemical elements allow some metals to withstand stress or force better than others.

The composition of common welding base metals varies widely, from metals made of essentially one **metallic** element, to **alloys** (mixtures) of metallic and nonmetallic components. Metals are classified into two basic groups: **ferrous** metals, which are composed mainly of iron, and **nonferrous metals**, which contain very little or no iron. Ferrous metals include all steel, cast iron, **wrought** iron, **malleable** iron, and **ductile** (nodular) iron. Nonferrous metals and their alloys include the light metals (aluminum, magnesium, titanium); the heavy metals (copper, nickel, lead, tin, zinc); and the precious metals (platinum, gold, silver).

1.1.0 Ferrous Metal Composition and Classification Systems

Ferrous metals contain mostly iron and some carbon. The amount of carbon ranges from almost none in commercially pure iron to about 1.4 percent. Steel contains less than 1.7 percent carbon. Above 1.4 percent carbon, steel develops brittle properties similar to those of cast iron. Carbon steel, the largest group of the ferrous-based metals, contains much smaller amounts of carbon than cast iron. As the carbon content increases in carbon steel, the steel becomes harder and more wear-resistant. The carbon content of a particular carbon steel is the largest factor in determining its use. Steel **castings** usually contain a higher percentage of carbon than rolled plate and other rolled shapes. The casting foundry can usually provide the alloy formula.

Steels also contain impurities and trace amounts of elements such as manganese, silicon, phosphorus, copper, and lead. Sometimes additional amounts of these elements are intentionally added to steel to change its characteristics, such as to increase its forgeability (the ability to be heated in a forge and then beaten or hammered into shape) or to decrease a tendency toward forming blowholes (the vents that form when air or gases escape).

Steels, including carbon steels, low-alloy carbon steels, alloy steels, and stainless steels, are classified by various systems, such as specification number and grade or manufacturer's trade name and number. Steel classification systems include those developed by the **American Iron and Steel Institute (AISI)**, the **American Society for Testing and Materials (ASTM International)**, and the **Unified Numbering System (UNS)**, as well as manufacturer's trade names and identification numbers. Stainless steels have their own classification systems. They are sometimes referred to by the percentages of their chromium and nickel content, such as 18/8, 25/20, and 18/10; but this system has largely been replaced by the AISI stainless steel classification system. Common classification systems for steels are explained in the following sections.

1.1.1 Carbon Steel Classification

The principal classification system for carbon steels is the AISI numerical designation of Standard Carbon and Alloy Steels. It was originally a four-digit system developed by the **Society of Automotive Engineers (SAE)** for carbon steels commonly used in structural shapes, plate, strip, sheet, and welded tubing. Five digits are now used for some alloys. The classification system may be referred to as the AISI, SAE, or AISI/SAE system.

The AISI and SAE designations are essentially the same, except that the AISI system sometimes uses a letter prefix to indicate the manufacturing process that produced the steel. The following are examples of letter designations:

- *A* – Open-hearth steel
- *B* – Acid Bessemer carbon steel
- *C* – Basic open-hearth carbon steel
- *D* – Acid open-hearth carbon steel
- *E* – Electric furnace steel

Carbon and Alloy Steels

Steel is an alloy of iron and carbon typically containing less than 1 percent carbon. All steels also contain varying amounts of other elements such as manganese, silicon, phosphorus, sulfur, and oxygen. In addition, some standard alloy steels can also contain elements such as nickel, chromium, and molybdenum. There are currently about 3,500 different grades of steel with many different properties. Most of the steels produced today are carbon and alloy steels. Ninety-five percent of the construction and fabrication metals used worldwide consist of these materials.

The absence of a letter prefix indicates basic open-hearth or acid-Bessemer carbon steel. The prefixing letter (if any) designates the manufacturing process. The first two numerical digits represent the series (type and class) of steel. The third and fourth (and sometimes fifth) numerical digits specify the approximate percentage of carbon content. *Table 1* shows the AISI/SAE system for designating carbon and alloy steel types and classes and includes the AISI/SAE number.

The following examples demonstrate the use of the AISI/SAE numbers for carbon steels:

- AISI number C1020:
  - C = Indicates basic open-hearth carbon steel
  - 10 = Carbon steel, nonresulfurized
  - 20 = Contains approximately 0.20 percent carbon
- AISI number E2512:
  - E = Indicates electric furnace steel
  - 25 = Designates steel alloyed with approximately 5 percent nickel
  - 12 = Designates steel containing approximately 0.12 percent carbon
- AISI number E52100:
  - E = Indicates electric furnace steel
  - 5 = Contains either approximately 0.50 percent, 1.00 percent, or 1.45 percent chromium, designated by the next digit: 0, 1, or 2, respectively
  - 2 = Designates approximately 1.45 percent chromium
  - 100 = Designates approximately 1 percent carbon

The common group classification of carbon steels is primarily based on carbon content. Plain carbon steels (AISI series 10XX through 11XX) are basically iron-carbon alloys. High-strength low-alloy (HSLA) carbon steels (AISI series 13XX through 98XX) have small amounts of alloying elements to improve strength, hardness, and toughness or to increase resistance to corrosion,

Table 1 AISI/SAE System for Designating Carbon and Alloy Steel Types and Classes

ASI/ASE Number	Carbon Steel Types and Classes
1XXX	Carbon steels
10XX	Plain (nonresulfurized) carbon steel grades
11XX	Free machining, resulfurized (screw stock)
12XX	Free machining, resulfurized, rephosphorized
13XX	Manganese 1.75 percent
15XX	High-manganese (1.00 to 1.65 percent) carburizing steels
2XXX	Nickel steels
23XX	3.50 percent nickel
25XX	5.00 percent nickel
3XXX	Nickel-chromium steels
31XX	1.25 percent nickel, 0.65 or 0.80 percent chromium
32XX	1.75 percent nickel, 1.07 percent chromium
33XX	3.50 percent nickel, 1.50 or 1.57 percent chromium
34XX	3.00 percent nickel, 0.77 percent chromium
4XXX	Molybdenum steels
40XX	0.20 or 0.25 percent molybdenum
41XX	0.50, 0.80, or 0.95 percent chromium, 0.12, 0.20, 0.25, or 0.30 percent molybdenum
43XX	1.82 percent nickel, 0.50 or 0.80 percent chromium, 0.25 percent molybdenum
46XX	0.85 or 1.82 percent nickel, 0.20 or 0.25 percent molybdenum
47XX	1.05 percent nickel, 0.45 percent chromium, 0.20 or 0.35 percent molybdenum
48XX	3.50 percent nickel, 0.25 percent molybdenum
5XXX	Chromium steels
50XX	0.27, 0.40, 0.50, or 0.65 percent chromium
51XX	0.80, 0.87, 0.92, 0.95, 1.00, or 1.05 percent chromium
5XXXX	1.00 carbon, 0.50, 1.02, or 1.45 percent chromium
6XXX	Chromium-vanadium steels
61XX	0.60, 0.80, or 0.95 percent chromium, 0.10 or 0.15 percent (minimum) vanadium
86XX	0.55 percent nickel, 0.50 percent chromium, 0.20 percent molybdenum
92XX	0.65, 0.82, or 0.85 percent manganese, 1.40 or 2.00 percent silicon, 0.00 or 0.65 percent chromium
93XX	3.25 percent nickel, 1.20 percent chromium, 0.12 percent molybdenum
94XX	0.45 percent nickel, 0.40 percent chromium, 0.12 percent molybdenum
97XX	0.55 percent nickel, 0.20 percent chromium, 0.20 percent molybdenum
98XX	1.00 percent nickel, 0.80 percent chromium, 0.25 percent molybdenum
XXBXX	Boron steels
XXLXX	Leaded steels

Steel Alloying Elements

Some alloying elements used with steel and the reasons they are added include the following:

- Manganese – Promotes forgeability
- Silicon – Prevents blowholes
- Phosphorus – Improves machinability
- Copper – Inhibits corrosion
- Lead – Improves machinability

heat, and environmental damage. The plain carbon steels are commonly grouped as follows:

- *Low-carbon* – 0.10 percent to 0.15 percent carbon, 0.25 percent to 1.50 percent manganese
- *Mild-carbon* – 0.15 percent to 0.30 percent carbon, 0.60 percent to 0.70 percent manganese
- *Medium-carbon* – 0.30 percent to 0.50 percent carbon, 0.60 percent to 1.65 percent manganese
- *High-carbon* – 0.50 percent to 1.00 percent carbon, 0.30 percent to 1.00 percent manganese

The HSLA carbon steels are alloyed with one or more of the elements manganese, nickel, chromium, or molybdenum to provide higher strength, better toughness, improved weldability, and in some cases, greater resistance to corrosion. As the percentage of these elements increases, changes in the weldability of the low-alloy steels make electrode selection and welding procedures more critical. HSLA steels include the following:

- Low-nickel steels
- Low-nickel chrome steels
- Low-manganese steels
- Low-alloy chromium steels
- Weathering steels

Another class of steel in the AISI series is sulfur steel. It is not an alloy steel but is still included in the series.

1.1.2 ASTM Steel Specification System

ASTM International develops and publishes specifications for use in material production and testing. Welding codes and procedures typically use the ASTM designation to identify metal types because it includes specifications for mechanical and physical characteristics of the material as well as chemical composition. Piping and tubing are the largest group of materials fabricated by welding covered in ASTM specifications.

Prefix letters are a general part of each specification and provide a general idea of the specification content. For example, A is used for ferrous metals, B for nonferrous metals, and E for miscellaneous subjects, such as examination and testing.

Sometimes an S is added before the prefix letter. This indicates that ASME International, publishers of the Boiler and Pressure Vessel Code, has adopted that specification.

With the exception of a few mild steel grades, such as SA-36 (mild steel plate) and SA-53 (mild steel pipe), ASTM specifications are three-digit numbers. Most specifications also have various grades, classes, or types. *Table 2* gives examples of some of the ASTM specifications for alloy steels.

NOTE

The complete collection of ASTM standards consists of 15 sections comprising 65 volumes, with a separate index. The table is provided only to illustrate sample ASTM numbers.

When companies buy steel, the manufacturer will provide the customer with a certificate indicating the quality of the steel. This document is often called a **mill test report (MTR)** or a **certified mill test report (CMTR)**. The report mainly shows the chemical content and testing results of the metal that was purchased (*Figure 1*). Such information is necessary in case the structure being built from the material fails. For liability reasons, this information must be tracked and kept on record.

Included in the mill test report are the ASTM specifications, size, grade, and length of the product, which in this case is **wide flange (WF) beams**. The report also shows the heat number, chemical analysis, and physical properties of the steel. These markings are also found on the materials covered by the report. *Figure 2* shows how the markings look on an actual beam.

UNS

The UNS used to identify metals was developed jointly by ASTM International and SAE as a way of relating the different numbering systems for metals and alloys. Its objective is to eliminate the confusion caused when more than one identification number is used to identify the same material or when the same identification number is used to identify two different materials. It is anticipated that this system will eventually replace the AISI and other systems.

Each UNS number has a letter prefix followed by five digits. Often the prefix letter relates to the family of metals. For example, UNS numbers A00001 through A99999 refer to aluminum and aluminum alloys, and UNS numbers C00001 through C99999 signify copper and copper alloys. When possible, the digits in the UNS sequence contain the same numbering sequences used in the other systems. For example, the UNS number G10200 corresponds to the AISI/SAE number 1020 for carbon steel. Most engineering and materials reference books contain tables that cross-reference AISI and SAE numbers for metals to the corresponding UNS numbers.

Table 2 ASTM Specification Standards

ASTM Number		Tensile	Type of Base Metal
A209	Grade T1	55,000	C-0.05 Mo tubes
	Grade T1a	60,000	C-0.05 Mo tubes
	Grade T1b	53,000	C-0.05 Mo tubes
A217	Grade T17	60,000	1.00 Cr-0.15 V tubes
A334	Grade 5	65,000	5.00 Ni-Steel pipe
A335	Grade P1	55,000	C-0.05 Mo pipe
	Grade P2	55,000	0.60 Cr-0.85 Mo pipe
	Grade P3	60,000	1.57 Cr-0.70 Mo pipe
	Grade P3b	60,000	2.00 Cr-0.55 Mo pipe
	Grade P11	60,000	1.25 Cr-0.55 Mo-Si pipe
	Grade P12	60,000	0.95 Cr-0.55 Mo pipe
	Grade P15	60,000	1.40 Si-0.55 Mo pipe
A517	Grade A	115,000	0.05 Cr-0.025 Mo-Si plates
	Grade B	115,000	0.05 Cr-0.02 Mo-V plates
	Grade P	115,000	0.125 Ni-0.10 Cr-0.05 Mo plates

C – Carbon Cr – Chromium Mo – Molybdenum Ni – Nickel Si – Silicon V – Vanadium

The markings show who made the steel, where it was made, its original dimensions, and its heat number or batch number. Heat or batch numbers are tracking numbers that the steel producer assigns to a particular ladle of steel. When a product, such as a beam or plate, is made from that ladle of steel, the tracking number is placed on that product in case of recall or problems with the chemistry. The following are explanations of the letters and numbers shown on the beam in *Figure 2*:

- *SDI* – Typically the name of the manufacturer.
- *USA* – Country of origin where the steel was melted.
- *A046196* – The heat number or batch number, which can be used to match the material to a material test report. Heat numbers are normally unique to each manufacturer, and they are usually meaningful only to the producer. The number provides a way to trace the product. The material test report lists the mechanical test results and chemistry. These can be checked against the material specification to ensure that the material test report results meet the requirements of the material specification.
- *W 16 × 36* – This beam has a depth of approximately 16 inches (actual depth 15⅞ inches or about 403 mm) and weighs 36 pounds per foot of length (53.6 kilograms per meter). If you looked up W 16 × 36 in the *American Institute of Steel Construction (AISC) Handbook*, you would find the beam's flange width and thickness, **web** thickness, beam depth, and many other design properties.

NOTE

One piece of information that is not included on the beam in *Figure 2* is the ASTM specification to which the beam was made (for example, A36, A441, A572, A588, etc.).

Plate steel also has mill markings, such as batch or heat numbers (*Figure 3*). The information at the top of the plate indicates that it was not heat treated after being rolled. The S7-09438 heat number is the batch number. The size information is self-explanatory. Grade A36 indicates that the plate was manufactured in accordance with the ASTM specifications for A36 steel plate. The other information at the bottom of the plate shows who made it and who bought it.

Steel Mills, Inc.
First Steel Mill
123 Rebar Way
Narvona, State 90426 USA

Mill Test Report

Chemical and Physical Test Report																		**Customer P.O. Number** ________
Shape & Size	**Grade**				**Specification**													**Sales Order** ____________
32mm Rebar (#10)	420(60)				ASTM A615/A615M-03													
Heat Identification Number	C	Mn	P	S	Si	Cu	Ni	Cr	Mo	V	Nb	B	N	Sn	Al	Ti	Zr	Ca	**C Equivalent**
123-123	0.32	0.91	0.015	0.067	0.32	0.38	0.13	0.16	0.016	0.000	0.014	0.000	0.000	0.003	0.000	0.000	0.000	0.0000	0.502

Mechanical Test					
Yield:79800 PSI, 550.2 MPA	Tensile:102000 PSI, 703.27 MA	% EL: 15.0/8 in, 15.0/200MM	Bend: Okay	Def: .073, 1.85MM	%I/h 3.3L

This material, including the billets, was produced and manufactured in the United States of America.

The above figures are certified extracts from the original chemical and physical test records, as contained in the permanent records of the company.

____
Signature of Quality Assurance Manager

____
Signature of Mill Metallurgist

Figure 1 Example of a mill test report.

Figure 2 Mill markings on a beam.

Figure 3 Mill markings on plate steel.

There is no set format for labeling steel, so steel producers have their own way of labeling their products. The mill markings on the beams and plates shown in *Figures 1, 2,* and *3* were stenciled or sprayed onto the steel. Other manufacturers attach paper labels to their products, as shown in *Figure 4*.

The mill markings on most steel will not be as clearly visible as those shown in these examples. Steel is often stored outside in the weather and may be exposed to other rough conditions or handling that can damage the markings.

When preparations are being made for a welding job, the job planner determines which steel to use. On some jobs, lay-down yard personnel may bring the materials to the work area. However, the fabricators (fitters and welders) must be able to ensure that the delivered steel is the same as that indicated on the welding procedure specification. *Figures 5A* and *5B* show a WPS with standard or English measurements. *Figure 6* shows a WPS with metric values.

When only a part of the original piece of steel is needed, welding personnel must transfer the heat number of the original piece to the piece they are cutting from the main piece. The transfer of numbers must be done before the part is cut away from the original piece. Transferring the heat number ensures that the removed piece of steel can be tracked when it is welded into place. Follow site-specific guidelines for transferring numbers.

On high-profile jobs, the process of tracking these numbers can be very involved because of liability concerns. Many jobs today require the welders to closely track the materials being used. The owner representatives (usually the quality control personnel) can use these heat or batch numbers to trace where the raw materials came from, the day and shift on which the material was produced, and who was on duty at the time. Most plants will not allow material to be used if the heat number has been ground off or rusted over. From a contractor's point of view, material that lacks a heat number should never be used in any critical application. If a failure occurred and the installer could not produce a valid heat number, then the contractor/installer would assume full responsibility, even though the steel was produced by someone else.

1.2.0 Low-Alloy Steel Groups

Low-alloy steels fall into the following general groups:

- High-strength low-alloy steels
- Quench-and-tempered steels
- Heat-treatable low-alloy steels
- Chromium-molybdenum steels

1.2.1 High-Strength Low-Alloy Steels

High-strength low-alloy (HSLA) steels are designed to meet specific mechanical requirements per ASTM specifications. The principal difference between plain-carbon steels and HSLA steels is the addition of alloys such as manganese, chromium, nickel, copper, molybdenum, and vanadium. Typically, these alloys total less than 2 percent. The carbon content of HSLA steels is about the same as that in low-carbon and mild steels (0.06 percent to 0.28 percent). Compared to mild steels, HSLA steels are designed to provide a combination of higher strength, better corrosion resistance, and improved **notch toughness**. The weldability of HSLA steel is similar to that of mild steel.

Figure 4 Paper labels attached to steel products.

HSLA steels can be divided into two general groups: Group A and Group B. Steels in Group A are designed for high strength, while steels in Group B are designed for high strength, improved atmospheric corrosion resistance, and notch toughness. Typical steels in Group A are A441, A572, and A633, which are used for structural applications such as pipelines, buildings, bridges, machinery, and railroad equipment. Other steels in Group A are A225 and A737, which are used for pressure vessels. Typical steels in Group B are A242 and A588, which are used for high-strength structural applications that may include unpainted applications. Group B steels specifically intended for low-temperature applications (the nickel content is as high as 9 percent) are A203 and A353.

CAUTION

The welding recommendations provided are general guidelines only. When welding HSLA steel, the site WPS or site quality standards must be followed. Failure to follow proper standards can result in rejected and/or defective welds.

ABC CONSTRUCTION

DATE OF ISSUE 7/20/13

WELDING PROCEDURE SPECIFICATION
IGMA-11-AF

REVISION	2 - 11/16/14	**SUPPORTING PQR NO. AND PAGE:** IGMA-11 PAGE B
PROCESS	GTAW / SMAW	
TYPE	MANUAL	

JOINTS (QW-402)

JOINT DESIGN	V OR U GROOVES AND FILLETS, PER THE PROJECT DRAWINGS AND SPECIFICATIONS
BACKING	NONE
BACKING MATERIAL	COMPATIBLE WITH BASE MATERIALS IF USED
JOINT FIT-UP REQUIREMENTS	1/8" +/- 1/16"
OTHER	NO RETAINERS (QW 402.11) ROOT SPACING PER DRAWINGS AND SPECIFICATIONS

BASE METALS (QW-403)

P-NO 1 GRP NO. ______ to P-NO 1 GRP NO. ______

BASE METAL SPEC. ______ to ______

THICKNESS RANGE QUALIFIED (QW-451)

BASE METAL GROOVE 3/16" TO 8" FILLET ALL

DEPOSITED WELD METAL GTAW - 3/4" MAX (No weld pass > 1/2" thick)

SMAW - 8" MAX (No weld pass > 1/2" thick)

OTHER ______

FILLER METALS/ELECTRODES (QW-404); PROGRESSION (QW-405) **POSITIONS: ALL**

LAYER	ROOT	PROCESS	GTAW	AWS CLASS	ER70S-2	PROGRESSION	UPHILL
SFA NO.	5.18	F NO.	6	A NO.	1	SIZES	1/16", 3/32", 1/8"
LAYER	WELDOUT	PROCESS	SMAW	AWS CLASS	E-7018	PROGRESSION	UPHILL
SFA NO.	5.1	F NO.	4	A NO.	1	SIZES	3/32", 1/8", 5/32"
LAYER		PROCESS		AWS CLASS		PROGRESSION	
SFA NO.		F NO.		A NO.		SIZES	

OTHER NO CONSUMABLE INSERTS (QW 404.22) NO PENETRATION FLUX (QW 404.50)

PREHEAT (QW - 406)

PREHEAT TEMPERATURE	75° F (175F when t > 1")
MAX INTERPASS TEMPERATURE	600° F
PREHEAT MAINTENANCE	NONE
OTHER	

POSTWELD HEAT TREATMENT (QW-407)

TEMPERATURE	1150° F
TIME RANGE	1HR/INCH ONE HOUR MINIMUM
MAX HEATING & COOLING RATES	PER THE APPLICABLE ASME DESIGN CODE

Figure 5A Sample WPS with standard measurements (1 of 2).

ABC CONSTRUCTION

REVISION 2 - 11/16/14

WELDING PROCEDURE SPECIFICATION
IGMA-11-AF

GAS (QW-408)

SHIELDING GAS	GTAW - 100% ARGON	FLOW RATE	15-35 CFH
PURGING GAS	NONE	FLOW RATE	N/A
TRAILING GAS	NONE	FLOW RATE	N/A
OTHER			

ELECTRICAL CHARACTERISTICS (QW-409)

PROCESS:	GTAW	PROCESS:	SMAW	PROCESS:	
ELECTRODE:	See Note 1	ELECTRODE:	E-7018	ELECTRODE:	
CURRENT:	DC	CURRENT:	DC	CURRENT:	
POLARITY:	SP	POLARITY:	RP	POLARITY:	

CHARACTERISTICS:				CHARACTERISTICS:				CHARACTERISTICS:			
ER70S-2			TRAVEL SPEED	E7018			TRAVEL SPEED				TRAVEL SPEED
SIZE	AMPS	VOLTS	IPM	SIZE	AMPS	VOLTS	IPM	SIZE	AMPS	VOLTS	IPM
1/16"	50-150	N/A	Manual	3/32"	50-120	N/A	Manual				
3/32"	75-180	N/A	Manual	1/8"	70-150	N/A	Manual				
1/8"	100-250	N/A	Manual	5/32"	90-200	N/A	Manual				

GTAW - PULSED CURRENT = NONE

TECHNIQUE (QW-410)

STRING OR WEAVE BEAD	USE STRINGER BEAD FOR POSITION 2G. USE EITHER STRINGER BEAD OR WEAVE BEAD FOR POSITIONS 1G, 3G, 4G, 5G, AND 6G.
ORIFICE OR GAS CUP SIZE	4 Thru 12
INITIAL AND INTERPASS CLEANING	BRUSHING AND/OR GRINDING
METHOD OF BACK GOUGING	THERMAL OR MECHANICAL IF REQUIRED (If Thermal grind back to bright metal)
OSCILLATION	N/A
CONTACT TUBE TO WORK DISTANCE	N/A
MULTIPLE OR SINGLE PASS PER SIDE	MULTIPLE
MULTIPLE OR SINGLE ELECTRODES	SINGLE
PEENING	NONE
OTHER	

NOTES NO AUTOGENOUS WELDING

1.) A potential health risk exists from inhaling grinding dust from EWTh-2 tungsten electrodes. As an alternative, EWCe-2 and/or EWLa-2 tungsten electrodes are acceptable.

Figure 5B Sample WPS with standard measurements (2 of 2).

Weld Procedure Number	30 P1 TIG 01 Issue A
Qualifying Welding Procedure (WPAR)	WPT17/A

Manufacturer:	National Fabs Ltd 25 Lane End Birkenshaw Leeds	Method Of Preparation and Cleaning:	Machine and Degrease
Location:	Workshop	Parent Metal Specification:	Grade 304L Stainless Steel
Welding Process:	Manual TIG	Parent Metal Thickness	3 to 8mm Wall
Joint Type:	Single Sided Butt Weld	Pipe Outside Diameter	25 to 100mm
		Welding Position:	All Positions
		Welding Progression:	Upwards

Joint Design

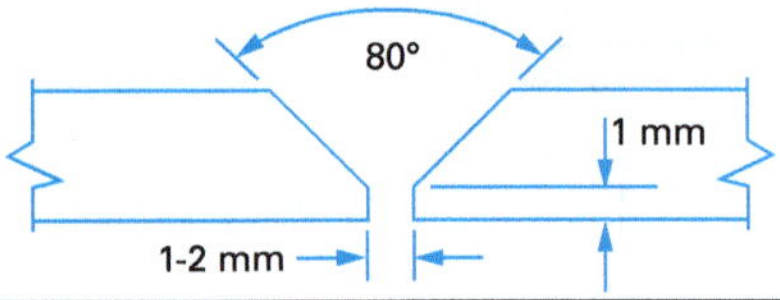

Welding Sequences

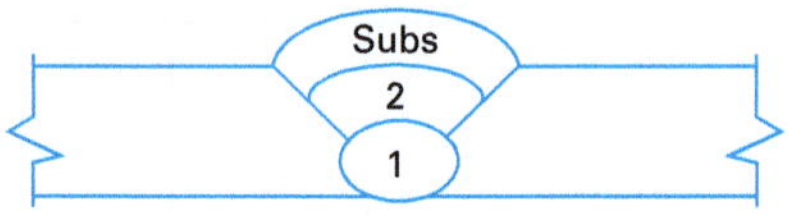

Run	Process	Size of Filler Metal	Current A	Voltage V	Type of Current/Polarity	Wire Feed Speed	Travel Speed	Heat Input
1 2 And Subs	TIG TIG	1.2 mm 1.6 mm	70 - 90 80 - 140	N/A	DC– DC–	N/A	N/A	N/A

Welding Consumables:-	
Type, Designation Trade Name:	BS 2901 Part 2 : 308S92
Any Special Baking or Drying:	No
Gas Flux:	Argon 99.99% Purity
Gas Flow Rate - Shield:	8 - 12 LPM
- Backing:	5 LPM
Tungsten Electrode Type/ Size:	2% Thoriated 2.4mm Dia
Details of Back Gouging/Backing:	Gas Backing
Preheat Temperature:	5°C Min
Interpass temperature:	200°C Max
Post Weld Heat Treatment Time, temperature, method: Heating and Cooling Rates*:	Not Required

Production Sequence

1. Clean weld and 25mm borders to bright metal using approved solvent.
2. Position items to be welded ensuring good fit up and apply purge.
3. Tack weld parts together using TIG, tacks to at least 5mm min length.
4. Deposit root run using 1.2mm dia. wire.
5. Inspect root run internally.
6. Complete weld using 1.6mm dia wire using stringer beads as required.
7. 100% Visual inspection of completed weld

Revision History

Date	Issue	Changes	Authorization
26/11/2005	A	First changes	Jack Straw *Jack Straw*

Figure 6 Sample WPS with metric measurements.

HSLA steels can be welded using the common welding processes, although successful welding requires consideration of preheating, **interpass temperature** , and control of hydrogen, including the use of low-hydrogen electrodes when applicable. This information can be found on the WPS. Typically, only the root pass, and possibly the hot pass, are made with the gas tungsten arc welding (GTAW) process on HSLA steels. Fill passes and cover passes are often made with faster processes such as shielded metal arc welding (SMAW) or gas metal arc welding (GMAW). One advantage of using the GTAW process is that hydrogen is easier to control because fluxes are not used. Minimum preheat and interpass temperatures are determined based on the alloy type and the thickness of the thinner part being welded.

CAUTION

Steels that are wet or below 32°F (0°C) should always be preheated to at least 70°F (21°C) and maintained at that temperature to remove moisture and prevent its formation. Failure to preheat metals may result in rejected and/or defective welds.

Table 3 provides minimum preheat and interpass temperatures for typical HSLA steels and also includes the ASTM steel numbers and grades.

Filler metal for welding HSLA steel depends on the alloy and the application. Mild steel filler metal can be used when joint strength is not critical or when alloy steel is being joined to mild steel. Alloy steel should be used when the strength of the joint must equal or approach that of the base metal. High-alloy stainless steel is used for special applications including welding dissimilar metals or welding high-nickel steel, such as A353.

Typically, HSLA steels are not postweld heat-treated; however, this may be required to improve ductility or to maintain the dimensions of the weldment during machining.

1.2.2 Quench-and-Tempered Steels

Quench-and-tempered (Q&T) steels are furnished in the heat-treated condition, which generally consists of austenitizing, quenching, and tempering. Some Q&T steels fall into the carbon steel classification, some into the low-alloy steel classification, and others into the alloy classification of the AISI/SAE system. Q&T steels combine high yield and tensile strength with good notch toughness, ductility, corrosion resistance, or weldability, depending on their intended use. Q&T steels have less than 0.25 percent carbon, with yield strengths of 50,000 to 180,000 psi (≈344,738 to 1,241,056 kPa).

Hot Tapping

Steel pipes are commonly hot-tapped to install a bypass around a repair area or to add a branch line. With hot tapping, threaded or welded taps (called threadolets and weldolets, respectively) are welded to the outside of the pipe while material or liquid is still flowing. Special tools can then be used to work through the installed fitting, penetrate the pipe, and install a shutoff valve without the loss of the pipe contents. In other words, the complete connection is made without draining the pipe contents. The pipe remains hot.

Low-alloy pipes cannot be hot-tapped because the material flowing through the pipe will cool the pipe rapidly and prevent proper preheat, interpass, and postheat temperature maintenance. This can result in defective welds, as well as pipe failure during welding. A site engineer should be consulted before any hot tap is made due to concerns related to the contents of the pipe.

Most Q&T steels are furnished as plate, although some are available as seamless pipe.

Q&T steels are used where high strength and minimum weight are essential. Many of the Q&T steels are covered by ASTM classifications. A common group of Q&T steels known as HY steels are covered by military specifications (MIL). The most common HY steel is HY-80, which has a yield stress of 80,000 psi (≈551,581 kPa). HY-100, HY-130, HY-150, and HY-180 are also available. Types and typical uses of Q&T steels include the following:

- *A514 and A517* – Used for earth-moving equipment, pressure vessels, bridges, TV towers, and ships.
- *A533 Grade B* – Used for nuclear pressure vessels.

Table 3 Minimum Preheat and Interpass Temperatures for Typical High-Strength Low-Alloy Steels Welded with Low-Hydrogen Electrodes

ASTM Steel	Thickness		Minimum Temperature
	(in)	(mm)	
A242	Up to 0.75	Up to 19	32°F (0°C)
A441	0.81 to 1.50	21 to 38	50°F (10°C)
A572, Gr 42, 50	1.56 to 2.50	40 to 64	150°F (66°C)
A588	Over 2.50	Over 63	225°F (107°C)
A572, Gr 60, 65	Up to 0.75	Up to 19	50°F (10°C)
A633, Gr E	0.81 to 1.50	21 to 38	150°F (66°C)
	1.56 to 2.50	40 to 64	225°F (107°C)
	Over 2.50	Over 64	300°F (140°C)

- *A543 Type B* – Used in nuclear reactor vessels, ships, and submarines. Classes 1 and 2 steels are similar to HY-80 and HY-100.
- *A553* – Used for cryogenic applications. It contains 8 to 9 percent nickel for high strength at extremely low temperatures.
- *A678* – Used for high-strength structural applications.

To prevent cracking, most Q&T steels can be welded by common welding processes using preheat and interpass temperatures as shown in *Table 4*. These temperatures vary depending on the type of steel being welded. Because Q&T steels obtain their strength from heat treatment, slow cooling can destroy the benefits of the original heat treatment. Slow cooling is generally caused by the following:

- Excess preheat
- Exceeding the recommended interpass temperature
- High heat input, such as high amperage with slow travel speed and the use of weave beads instead of stringers

CAUTION

When welding Q&T steels, pay close attention to the welding procedures concerning preheat and interpass temperature recommendations to prevent loss of strength or cracking.

On the other hand, cooling too quickly can cause a hard, brittle zone that may crack. Fast cooling is caused by the following:

- Insufficient preheat
- Insufficient interpass temperature
- Insufficient heat input during welding

When welding Q&T steels, always use low-hydrogen techniques and stringer beads to minimize heat input. Typically, the GTAW process is used only on thin sections or for the first one or two passes.

Most Q&T steels are designed to be used in the as-welded condition, with no postheat. Sometimes thick sections of A533 or A537 steels may require postheat treatment. If postheat treatment is required, use extreme care and follow the procedures exactly to prevent adversely affecting the base metal.

1.2.3 Heat-Treatable Low-Alloy Steels

Heat-treatable low-alloy (HTLA) steels are alloy steels that contain carbon in the range of about 0.25 to 0.45 percent, and small amounts of chromium, nickel, and/or molybdenum to enhance **hardenability**. They are generally welded in the annealed condition, and then the entire weldment is heat treated. The relatively high carbon content of HTLA steels allows them to be heat treated to a very high strength and hardness. Because HTLA steels must be heat treated after welding, they are typically used for smaller parts that can be placed in tempering ovens.

HTLA steels are produced to contain very low levels of impurities. During welding, low hydrogen techniques are used, and extreme care must be taken to keep the weld joint and filler metal absolutely clean and free of impurities. For this reason, GTAW is the preferred process for welding HTLA steels. Heat input is also very important, and welding procedures that take into account current, voltage, travel speed, and preheat temperatures must be followed.

HTLA steels generally require higher preheat and interpass temperatures than other types of alloy steels to prevent hardening of the weld zone and hot cracks. They also require **postweld heat treatment (PWHT)**. *Table 5* shows minimum preheat and interpass temperatures for HTLA steels and also includes the AISI steel numbers and thickness ranges.

HTLA steels should receive postweld heat treatment immediately after they are welded. The type of postweld heat treatment depends on the type of steel and the preheat and interpass temperature

Table 4 Minimum Preheat and Interpass Temperatures for Q&T Steels

Thickness Range		Temperature				
Inches	**Millimeters**	**A514/A517**	**A533**	**A537**	**A543**	**A678**
Up to 0.50	Up to 13	50°F (10°C)	50°F (10°C)	50°F (10°C)	100°F (38°C)	50°F (10°C)
0.56 to 0.75	14 to 19	50°F (10°C)	100°F (38°C)	50°F (10°C)	125°F (52°C)	100°F (38°C)
0.81 to 1.00	21 to 25	125°F (52°C)	100°F (38°C)	50°F (10°C)	150°F (66°C)	100°F (38°C)
1.10 to 1.50	27 to 38	125°F (52°C)	200°F (93°C)	100°F (38°C)	200°F (93°C)	150°F (66°C)
1.60 to 2.00	40 to 51	175°F (79°C)	200°F (93°C)	150°F (66°C)	200°F (93°C)	150°F (66°C)
2.10 to 2.50	53 to 64	175°F (79°C)	300°F (149°C)	150°F (66°C)	300°F (149°C)	150°F (66°C)
Over 2.50	Over 64	225°F (107°C)	300°F (149°C)	225°F (107°C)	300°F (149°C)	–

Table 5 Minimum Preheat and Interpass Temperatures for HTLA Steels

AISI Steel	Thickness Range		Minimum Preheat and Interpass Temperature
	Inches	Millimeters	
4027	Up to 0.50	Up to 13	50°F (10°C)
	0.60 to 1.0	15 to 25	150°F (66°C)
	1.1 to 2.0	27 to 51	250°F (121°C)
4037	Up to 0.50	Up to 13	100°F (38°C)
	0.60 to 1.0	15 to 25	200°F (93°C)
	1.1 to 2.0	27 to 51	300°F (149°C)
4130/ 5140	Up to 0.50	Up to 13	300°F (149°C)
	0.60 to 1.0	15 to 25	400°F (204°C)
	1.1 to 2.0	27 to 51	500°F (260°C)
4135/ 4140	Up to 0.50	Up to 13	350°F (177°C)
	0.60 to 1.0	15 to 25	450°F (232°C)
	1.1 to 2.0	27 to 51	500°F (260°C)
4340/ 8630	Up to 0.5	Up to 13	200°F (93°C)
	0.6 to 1.0	15 to 25	250°F (121°C)
	1.1 to 2.0	27 to 51	300°F (149°C)
8640	Up to 0.50	Up to 13	200°F (93°C)
	0.60 to 1.0	15 to 25	300°F (149°C)
	1.1 to 2.0	27 to 51	350°F (177°C)

maintained during welding. Allowing the weldment to cool after welding without some type of heat treatment generally results in cracks. Immediately after it is welded, HTLA steel should be stress-relieved following strict guidelines depending on the alloy and the preheat and interpass temperatures used.

1.2.4 Chromium-Molybdenum Steels

Chromium-molybdenum (Cr-Mo) steels contain 0.5 to 9 percent chromium and 0.5 to 1 percent molybdenum. The carbon content is generally less than 0.20 percent. Cr-Mo steels, which are sometimes called chrome-moly or heat-resisting alloy steels, are widely used in the petroleum industry and in steam-generating plants for elevated-temperature applications. The chromium provides improved resistance to oxidation and corrosion, and the molybdenum increases strength at elevated temperatures. Cr-Mo steels are also used in the aircraft and racing industries as tubing for highly stressed parts (*Figure 7*). Cr-Mo steels that are in the AISI/SAE 4100 series are available in forging, castings, plate, pipe, and tubing to various ASTM specifications.

Cr-Mo steels are hardenable, so a welding procedure that includes preheating, and sometimes postheating, must be used. The preheat temperature depends on the amount of alloy, the carbon content, the weld material, and base metal thickness. Generally, the higher the alloy content, carbon content, or thickness, the higher the preheat temperature. *Table 6* shows minimum preheat and interpass temperatures for Cr-Mo steels including their types and ASTM steel designations.

Once the Cr-Mo base metal has been preheated, welding should proceed without interrupting the heating cycle to avoid hard spots and cracking. If the base metal contains less than 9 percent chromium or is less than 1 inch (25 mm) thick, the heating cycle should be 800°F (427°C) per half hour per 1 inch (25 mm) of thickness. After being heated, wrap the metal and let it cool until it is at ambient temperature.

CAUTION

Check your welding procedures for any special precautions needed when interrupting a heating cycle. Failure to follow proper welding procedures will result in defective welds.

When the welding consumables are a potential source of hydrogen, such as with SMAW, it may be necessary to raise the preheat temperature and hold the weldment at this elevated temperature for a period of time. This allows the hydrogen to escape before the base metal cools, which reduces the risk of cracking. To avoid potential problems that can occur when the heating cycle of a Cr-Mo base metal is interrupted, always plan to complete the heating cycle once it has started.

Some Cr-Mo steels containing less than 1.25 percent Cr and 0.05 percent Mo can be placed in service as-welded if a high preheat temperature was used and the section is relatively thin. Other Cr-Mo steels should either be stress-relieved, annealed, or normalized and tempered.

Figure 7 Steel tubing used to construct a race car frame.

Table 6 Minimum Preheat and Interpass Temperatures for Cr-Mo Steels

ASTM Steel	Type	Up to 0.5 inch (13 millimeters)	0.5 to 1.0 inch (13 to 25 millimeters)	Over 1.0 inch (Over 25 millimeters)
A335-P2 A335-P12 A335-P11	½ Cr - ½ Mo 1 Cr - ½ Mo 1¼ Cr - ½ Mo	300°F (149°C)	300°F (149°C)	300°F (149°C)
A369-FP3b A335-P22	2 Cr - ½ Mo 2¼ Cr - 1Mo	300°F to 350°F (149°C to 177°C)	300°F to 350°F (149°C to 177°C)	350°F (177°C)
A335-P21	3 Cr - 1 Mo	300°F to 350°F (149°C to 177°C)	300°F to 350°F (149°C to 177°C)	400°F (204°C)
A335-P5 A335-P7 A335-P9	5 Cr - ½ Mo 7 Cr - ½ Mo 9 Cr - 1 Mo	400°F (204°C)	400°F (204°C)	500°F (260°C)

*Maximum carbon content of 0.15 percent. For higher content, preheat should be increased by 100°F to 200°F (38°C to 93°C)

Stress-relief heat treatment is used to reduce welding stresses and to increase the ductility and toughness of the weld metal and the heat-affected zone (HAZ). The stress-relief temperature ranges from about 1,150°F to 1,400°F (621°C to 760°C) and depends on the alloy being treated. The entire weldment or the weld zone is generally held at the stress-relief temperature for one hour per inch of thickness.

The procedures for annealing or normalizing and tempering are generally the same as those used for HTLA steels discussed earlier in this module.

1.3.0 Common Grade Stainless Steel Classifications

Stainless steels are typically classified as being common grade or specialty grade. This section covers common grade stainless steels. The specialty grades are discussed later in this module. Common grade stainless steels (corrosion-resistant steels) are iron-based alloys that normally contain at least 11 percent chromium. Other alloying elements, including nickel, carbon, manganese, and silicon, may be present in varying quantities, depending on the specific type of stainless steel, in order to enhance its physical and mechanical properties. The main physical characteristics of all stainless steels are their resistance to corrosion and heat. Some have good low- and high-temperature mechanical properties.

When compared with mild steels, stainless steels have the following characteristics:

- Lower coefficients of thermal conductivity that increase the chances of distortion
- Higher coefficients of thermal expansion that increase the chances of distortion
- Higher electrical resistances that increase the tendency to build up heat from welding current

Common grade stainless steels are classified by their grain structures. The type of grain structure is determined by the specific alloy content of the stainless steel and its heat treatment during manufacture. Based on their microcrystalline structures, the classification of common grade stainless steels is divided into these three groups:

- Austenitic
- Ferritic
- Martensitic

1.3.1 Austenitic Stainless Steels

Austenitic stainless steels are nonmagnetic in the annealed condition and non-hardenable by heat treatment. However, they can be hardened significantly by cold working. Austenitic stainless steels combine excellent corrosion and heat resistance with good mechanical properties over a broad temperature range. Austenitic stainless steels make up the largest of the three common grade stainless steel groups. For this reason, they are the type of stainless steel encountered most often by welders. They include the AISI 200 and 300 series stainless steels (*Table 7*) that all contain significant amounts of both chromium and nickel. Austenitic steels are sometimes further subdivided in two classifications based on their compositions: chromium-nickel (AISI 300 series) and chromium-nickel-magnesium (AISI 200 series). The range of applications for austenitic stainless steels includes housewares, containers, industrial piping and vessels, and architectural facades.

Austenitic stainless steels are also available in low-carbon (L-grade) and high-carbon (H-grade) types. The letter L after a stainless steel type, such as 304L, indicates a low-carbon content of 0.03 percent or under. Similarly, the letter H after a stainless steel type indicates a high-carbon content, ranging between 0.04 and 0.10 percent.

L-grades are typically used when annealing after welding is impractical, such as in the field where pipe and fittings are welded. H-grades are used when the steel will be subjected to extreme temperatures; the higher carbon content helps it retain strength.

1.3.2 Ferritic Stainless Steels

Ferritic stainless steels are always magnetic. They are hardened to some extent by cold working, not by heat treatment. Ferritic stainless steels combine corrosion resistance and heat resistance with fair mechanical properties over a narrower temperature range than austenitic steels. Ferritic stainless steels are straight chromium stainless steels containing 11.5 to 27 percent chromium, about 1 percent manganese, and little or no nickel. The carbon content is 0.20 percent or less. Examples of ferritic stainless steels include AISI types 405, 409, 430, 442, 444, and 446. They typically are used for decorative trim, sinks, and automotive applications, particularly exhaust systems.

1.3.3 Martensitic Stainless Steels

Martensitic stainless steels are magnetic and can be hardened by quenching and tempering. They are excellent for use in mild environments such as the atmosphere, freshwater, steam, and with weak acids. However, they are not resistant to severely corrosive solutions. Martensitic stainless steels comprise two groups: chromium-martensitic and martensitic. The chromium-martensitic stainless steels contain from 11.5 to 18 percent chromium, about 1 percent manganese, and in some cases 0 to 2.5 percent nickel. Examples of chromium-martensitic stainless steels are AISI types 403, 410, 414, 420, 422, 431, and 440. Martensitic stainless steels contain only 4 to 6 percent chromium and no nickel. For this reason, they are not considered true stainless steels, although their corrosion resistance is much greater than that of mild carbon steels, even at elevated temperatures. Examples of martensitic stainless steels are AISI types 501 and 502.

1.4.0 Specialty Grade Stainless Steel Classifications

In addition to the common grades of stainless steel, the following four specialty grades of stainless steel are also available:

- Precipitation-hardening
- Superaustenitic
- Superferritic
- Duplex

Table 7 AISI Classification Numbers and Composition for Selected Stainless Steels

AISI No.	C%	Mn%	Si%	Cr%	Ni%	Other Elements
Chromium-nickel-magnesium-austenitic (Nonhardenable)						
201	0.15 Max	5.5–7.5	1.0	16.0–18.0	3.0–5.5	N 0.25 Max
202	0.15 Max	7.5–10	1.0	17.0–19.0	4.0–6.0	N 0.25 Max
Chromium-nickel-austenitic (Nonhardenable)						
301	0.15 Max	2.0	1.0	16.0–18.0	6.0–8.0	—
302	0.15 Max	2.0	1.0	17.0–19.0	6.0–10.0	—
303	0.15 Max	2.0	1.0	17.0–19.0	8.0–10.0	Mo 0.0–0.6
304	0.08 Max	2.0	1.0	18.0–20.0	8.0–10.5	—
304L	0.03 Max	2.0	1.0	18.0–20.0	8.0–12.0	—
308	0.08 Max	2.0	1.0	19.0–21.0	10.0–12.0	—
309	0.20 Max	2.0	1.0	22.0–24.0	12.0–15.0	—
310	0.25 Max	2.0	1.5	24.0–26.0	19.0–22.0	—
316	0.08 Max	2.0	1.0	16.0–18.0	10.0–14.0	Mo 2.0–3.0
316L	0.03 Max	2.0	1.0	16.0–18.0	10.0–14.0	Mo 2.0–3.0
321	0.08 Max	2.0	1.0	17.0–19.0	9.0–12.0	Ti, 5xC Min
347	0.08 Max	2.0	1.0	17.0–19.0	9.0–13.0	Cb+Ta10xC Min
348	0.08 Max	2.0	1.0	17.0–19.0	9.0–12.0	Ti 0.10 Max
Chromium-martensitic (Hardenable)						
403	0.15 Max	1.0	0.5	11.5–13.0	—	—
410	0.15 Max	1.0	1.0	11.5–13.5	—	—
414	0.15 Max	1.0	1.0	11.5–13.5	1.25–2.5	—
416	0.15 Max	1.0	1.0	12.0–14.0	—	S 0.15 Min
420	0.15 Min	1.0	1.0	12.0–14.0	—	—
422	0.20–0.25	1.0	0.75	11.0–13.0	0.5–10	Mo 0.75–1.25
440A	0.60–0.75	1.0	1.0	16.0–18.0	—	W 0.75–1.25
440B	0.75–0.95	1.0	1.0	16.0–18.0	—	V 0.15–0.3
440C	0.95–1.2	1.0	1.0	16.0–18.0	—	Mo 0.75 Max
						Mo 0.75 Max
						Mo 0.75 Max
Chromium-ferritic (Nonhardenable)						
405	0.08 Max	1.0	1.0	11.5–14.5	—	Al 0.10–0.30
409	0.08 Max	1.0	1.0	10.5–11.75	—	Ti, 6xC Min
430	0.12 Max	1.0	1.0	16.0–18.0	—	—
442	0.20 Max	1.0	1.0	18.0–23.0	—	Ti, 0.20 + 4
444	0.025 Min	1.0	1.0	17.5–19.5	1.0	(%C+%N)
446	0.20 Max	1.50	1.0	23.0–27.0	1.0	Mo 1.75–2.5,
						N 0.035 Max
						(Nb+T2), 0.2 + 4
						(%C+%N) Min
						N 0.25 Max
Martensitic						
501	0.11 Max	1.0	1.0	4.0–6.0	—	Mo 0.4–0.65
502	0.10 Max	1.0	1.0	4.0–6.0	—	Mo 0.4–0.65

1.4.1 Precipitation-Hardening Stainless Steels

Precipitation-hardening (PH) stainless steels contain alloying additions, such as aluminum, that enable them to be hardened by a solution and aging heat treatment. Precipitation-hardening stainless steels were developed after World War II because the design and manufacture of jet aircraft required the use of stainless steels with a better weight-to-strength ratio. Precipitation-hardening stainless steels have designations such as 17-4PH, PH13-8MO, and AM350.

1.4.2 Superaustenitic Stainless Steels

Superaustenitic stainless steels have compositions that fall between 300-series stainless steels and nickel-alloy steels. They are highly resistant to corrosion and stress-corrosion cracking caused by chloride-containing solutions. Many of these alloys contain deliberate amounts of copper and/or nitrogen to increase their strength and corrosion resistance. These stainless steel types are widely used in process industry applications in place of common austenitic stainless steels that have a tendency to fail because of pitting, crevice corrosion, or chloride-induced stress corrosion cracking.

1.4.3 Superferritic Stainless Steels

Superferritic stainless steels differ from common-grade ferritic stainless steels in their lower carbon and nitrogen content and higher molybdenum and/or chromium content. This combination provides much higher corrosion resistance in oxidizing and chloride chemical environments and excellent resistance to corrosion in salt water. Examples of superferritic stainless steel designations include E-Brite®, Sea-Cure®, and AL 29-4-2®.

1.4.4 Duplex Stainless Steels

Duplex stainless steels are supplied with approximately equal amounts of ferrite and austenite, thus the name "duplex". They also contain about 24 percent chromium and 5 percent nickel. Developed around 1970, duplex steels are among the more recently developed stainless steels. They offer higher yield strength and superior resistance to pitting and stress cracking than the 300 series stainless steels. One example of duplex stainless steel is Type 2205, which accounts for more than 80 percent of duplex use. Duplex stainless steels are used for heat exchangers, tubes, and pipes for applications involving the production and handling of gas and oils, and for desalinization.

Highly alloyed, high-performance duplex stainless steels, called super-duplex stainless steels, are also made. These contain higher levels of chromium, molybdenum, and nitrogen. They were developed to provide outstanding resistance to acids, acid chlorides, and other caustic solutions typical of those encountered in the chemical, petrochemical, and pulp and paper industries.

1.5.0 Nonferrous Metals Classification Systems and Groups

Nonferrous metals include all metals except iron. Commonly used nonferrous metals include aluminum, magnesium, titanium, copper, nickel, zinc, tin, lead, and the precious metals. Nonferrous metals with densities lower than that of steel are considered light metals. They include, in order of increasing density, magnesium, beryllium, aluminum, and titanium. The heavier common nonferrous metals include copper, nickel, lead, tin,

Why Stainless Steel Doesn't Rust

When steel comes in contact with oxygen and water vapor in the atmosphere, a chemical reaction occurs and the steel begins to change to its original form of iron oxide. The scale that forms on the surfaces of metals when they are exposed to air is an oxide. The chromium added to the steel prevents it from rusting or staining, thus the name "stainless steel". The chromium reacts with the oxygen and water in the atmosphere to form a microscopic and transparent layer of chromium oxide called a passive film. The sizes of the chromium atoms and their oxides are about the same, so they join to form a tight, stable layer only a few atoms thick on the surface of the steel. If the stainless steel surface is cut or scratched, disrupting the passive film, it is self-healing because more chromium oxide immediately forms and covers the exposed surface, protecting it from iron oxide (rust) again.

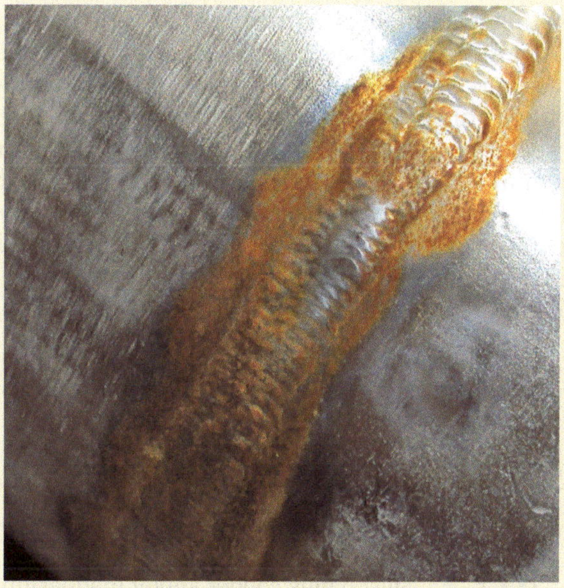

This is why after repeated use and sharpening, stainless steel cutlery remains bright and shiny. However, if the surface of stainless steel is contaminated by a ferrous metal, the ferrous material will rust. The rust in this photo was traced back to surface preparation with tools, such as grinding wheels and points, previously used to prepare carbon steel. This rust would not be present in a properly prepared joint, nor would it be evident immediately after the weld is completed.

and zinc, as well as the precious metals platinum, gold, silver, and their alloys. Each of the nonferrous metals and alloys is classified by a different and unrelated classification system.

1.5.1 Aluminum Alloys

Pure aluminum is a soft, lightweight metal that can be used in its nearly pure form. Because aluminum is very soft and ductile and has a low tensile strength, it is often alloyed with elements such as copper, manganese, silicon, magnesium, and zinc to make different alloys with specific properties for very different purposes. When aluminum is exposed to the atmosphere, a microscopic and transparent aluminum-oxide layer forms immediately that protects the metal from further oxidation. This characteristic gives aluminum its high resistance to corrosion.

Aluminum is lightweight, has one-third the weight of steel, and can be made very strong, ductile, and malleable. It is a good conductor of heat and electricity and is highly resistive to weathering and corrosion from many acids. However, aluminum can be corroded easily by alkalis because they attack the oxide layer. Aluminum is also noncombustible and nonmagnetic and is widely used in electrical shielding and near inflammable or explosive substances. It is decorative and easily formed, cast, and machined. Polished aluminum has the highest reflectivity of any material, including mirror glass.

Aluminum alloys are broadly classified as either casting alloys or wrought alloys for rolling, forging, or extruding. The Aluminum Association, Inc. (AA) uses a three-digit plus one decimal system to designate each type of casting alloy (*Table 8*). The first digit (1 through 9) shows the main alloying element or elements. For the 1XX.X series this is essentially pure aluminum (99.00 percent aluminum). With the exception of the 1XX.X series, the second and third digits identify the specific alloys. For the 1XX.X series only, the second and third digits give the degree of aluminum purity above 99.00 percent. For example, if the second and third digits are 60, the alloy contains a minimum of 99.60 percent aluminum. The decimal indicates if the alloy composition is for the final casting (0.0) or for ingot (0.1 or 0.2, depending on impurity limits). Modifications to the alloys are indicated by the use of prefix letters such as A, B, C.

Similarly, the AA uses a four-digit numbering system for wrought aluminum alloy types (*Table 9*) that is not related to the three-digit casting alloy designation system. In this wrought aluminum system, the first digit identifies the alloy group by the major alloying element. For the 1XXX group, the second digit designates modifications and impurity limits. A zero (0) indicates no special control on individual impurities. A second digit of 1 through 9 indicates special control over one or more impurities and is assigned consecutively. For groups 2XXX through 8XXX, the second digit indicates alloy modifications and is consecutively assigned. A second digit of 0 indicates the original alloy. The last two digits indicate the minimum percentage of aluminum.

Around the World

Recycled Steel

Steel is the world's most recycled material. According to the World Steel Association, more than 650 million tons of steel were recycled annually between 2009 and 2013.

Table 8 AA System Code Designations for Basic Cast Aluminum Alloys

Designation	Major Alloying Element
1XX.X	Near pure aluminum
2XX.X	Copper
3XX.X	Silicon plus copper and/or magnesium
4XX.X	Silicon
5XX.X	Magnesium
6XX.X	Unused series
7XX.X	Zinc
8XX.X	Tin
9XX.X	Other element

Table 9 AA System Code Designations for the Basic Wrought Alloy Groups

Designation	Major Alloying Element
1XXX	Non-aluminum elements are 1 percent or less
2XXX	Copper
3XXX	Manganese
4XXX	Silicon
5XXX	Magnesium
6XXX	Magnesium and silicon
7XXX	Zinc
8XXX	Other elements
9XXX	Unused series

Going Green

Recycling Aluminum

Aluminum can be recycled repeatedly with no loss of quality or its normal properties.

The following examples explain AA wrought aluminum numbers:

- AA Number 1075:
  - 1075 = Indicates aluminum content of 99.0 percent or greater
  - 1075 = Indicates no special control on the impurities
  - 1075 = Indicates an aluminum content of 99.75 percent
- AA Number 1180:
  - 1180 = Indicates aluminum content of 99.0 percent or greater
  - 1180 = Indicates impurities are limited
  - 1180 = Indicates an aluminum content of 99.80 percent

In general, aluminum and its alloys can be welded by the following methods: oxyfuel or air-fuel; electric arc; electrical resistance; and brazing.

1.5.2 Magnesium Alloys

Magnesium is mainly used in alloy form. It is the lightest of the structural metals, having less than one-quarter the density of steel. It is mainly used in the aerospace and automotive industries.

Why Aluminum Must Be Cleaned Before Welding

The aluminum-oxide coating protecting aluminum must be thoroughly cleaned from the metal before welding. The reason is that, while pure aluminum melts at about 1,200°F (about 650°C), the aluminum-oxide coating that protects the metal does not melt until it reaches the much higher temperature of 3,700°F (nearly 2,040°C).

All cleaning materials used must be contained and disposed of in accordance with local environmental laws. Cleaning something as large as a boat presents a greater challenge when it comes to containing the runoff of cleaning materials.

Magnesium is the eighth most common element in the world and the sixth most abundant metal, comprising about 2.5 percent of the earth's surface. Because seawater contains about 0.13 percent magnesium, some production facilities process seawater to obtain the magnesium. This is done after the precipitation of other salts from the seawater, which leaves a magnesium-rich brine. The uses of magnesium center on three properties of the metal: its ability to form intermetallic compounds with other metals, its high chemical reactivity, and its low density. Magnesium can be extruded, forged, rolled, and machined.

Magnesium alloys are broadly classified as casting alloys or wrought (rolled, forged, or extruded) alloys. Both of these classifications use the *ASTM Standard B951, Standard Practice for Codification of Unalloyed Magnesium and Magnesium-Alloys, Cast and Wrought.* Designations consist of one or two letters representing the alloying elements, followed by their respective percentages, and rounded off to the nearest whole number. The letters representing the alloying elements are as follows:

- *A* – Aluminum
- *B* – Bismuth
- *C* – Copper
- *D* – Cadmium
- *E* – Rare Earths
- *F* – Iron
- *G* – Magnesium
- *H* – Thorium
- *K* – Zirconium
- *L* – Beryllium
- *M* – Manganese
- *N* – Nickel
- *P* – Lead
- *Q* – Silver
- *R* – Chromium
- *S* – Silicon
- *T* – Tin
- *Y* – Antimony
- *Z* – Zinc

Magnesium alloy designations also contain information indicating the temper condition. This part of the designation, separated by a hyphen from the main designation of letters and numbers, consists of a letter or a letter and number. The letters and numbers that indicate the temper are as follows:

- *F* – As fabricated
- *O* – Annealed
- *H10, H11* – Strain hardened
- *H23, H24, H26* – Strain hardened and annealed
- *T4* – Solution heat treated
- *T5* – Artificially aged
- *T6* – Solution heat treated and artificially aged
- *T8* – Solution heat treated, cold worked, and artificially aged

In the alloy AZ92A-T6, the first A is aluminum and the Z is zinc. The 9 indicates that the aluminum percentage is between 8.6 and 9.4. The 2 shows that the zinc percentage is between 1.5 and 2.5. The final A denotes that this is the first alloy to receive the AZ92 designation. The letter and number T6 indicate solution heat treatment and artificially aged tempering.

Magnesium and its alloys are best welded by the GTAW process for thin sections and the GMAW process for thick sections. The composition of the filler metals should be compatible with the base metal. However, zinc above 1 percent increases **hot shortness** and can result in cracking. *Table 10* shows examples of magnesium alloys and their principal uses.

1.5.3 Titanium Alloys

Titanium is mainly used in alloy form. Its alloys are high-strength, lightweight (about one-half the density of steel), corrosion-resistant structural metals that retain their strength at elevated temperatures. It is used in aircraft engines and structures, chemical processing, surgical implants, sporting goods, and marine and other applications where corrosion resistance is required.

Titanium is a reactive metal. At high temperatures, it readily combines with oxygen, hydrogen, and nitrogen to form stable compounds. Like aluminum and magnesium, titanium forms a tight, protective, microscopic oxide film on clean surfaces at room temperature, when in contact with oxygen. This makes it highly resistant to corrosion at low temperatures. Titanium alloys are classified into three groups according to the crystal structures that are stable at room temperature.

- *Alpha alloys* – Are the usual crystalline structures of pure titanium. Aluminum is added to stabilize the crystalline structure. These alloys have excellent strength and are oxidation-resistant from 600°F to 1,100°F (about 316°C to 593°C). They have good ductility and weldability and can be hardened by working, but not by heat treatment.
- *Beta alloys* – Can contain vanadium, chromium, and aluminum. Iron and chromium are added to make them heat-treatable, high-strength alloys. They have fair weldability.
- *Alpha-beta alloys* – Contain one or more of the elements manganese, molybdenum, tin, vanadium, iron, and aluminum. They have a wide range of mechanical characteristics, but have poor weldability.

Table 10 Examples of Magnesium Alloys

ASTM Number	Nominal Alloy Composition (By %)						
	Al	Mn	Zn	Zr	Rare Earths	Th	Mg
Sand and Permanent Mold Castings							
AZ92A-T6	9.0	0.10	2.0	–	–	–	Balance
AZ63A-T6	6.0	0.15	3.0	–	–	–	Balance
AZ81A-T4	7.6	0.13 Min	0.7	–	–	–	Balance
AZ91C-T6	8.7	0.13	0.7	–	–	–	Balance
EZ33A-T5	–	–	2.6	0.8	3.3	–	Balance
HK31A-T6	–	–	0.3	0.7	–	3.3	Balance
HZ32A-T6	–	–	2.1	0.8	–	3.3	Balance
Die Castings							
AZ91A-F	9.0	0.13	0.7	–	–	–	Balance
AM60A-F	6.0	0.13	–	–	–	–	Balance
AZ91D-F	9.0	0.13	0.7	–	–	–	Balance
Extrusions							
AZ61A-F	6.5	0.15	1.0	–	–	–	Balance
MIA-F	–	1.20	–	–	–	–	Balance
AZ80A-T5	8.5	0.12	0.5	–	–	–	Balance
AZ10A-F	1.3	0.20	0.4	–	–	–	Balance
AZ31B-F	3.0	0.20	1.0	–	–	–	Balance
ZK60A-F	–	–	5.5	0.45	–	–	Balance
Sheet and Plate							
AZ31B-H24	3.0	0.20	1.0	–	–	–	Balance
HK31A-H24	–	–	–	0.7	–	3.3	Balance

Titanium is considered difficult to weld because the weld must be kept in an inert (nonreactive) gas atmosphere until it cools to below the reactive temperature. Small amounts of hydrogen or large amounts of oxygen or nitrogen can cause embrittlement in the heat-affected zone, which leads to breakability. Titanium is usually welded by the GTAW process.

About 34 different titanium alloys are currently being produced, including the ones shown in *Table 11*.

1.5.4 Beryllium Alloys

Beryllium (Be) is a high-strength, lightweight metal. It is only about two-thirds as heavy as aluminum, but is six times stiffer than steel. Beryllium's heat capacity is five times that of copper, meaning that one pound (0.45 kg) of it will absorb as much heat as five pounds (2.3 kg) of copper. For this reason, it is widely used in aerospace and nuclear reactor applications.

Titanium

Commercial production of titanium did not begin until the 1950s. At that time, titanium was recognized for its importance and was developed as a unique, lightweight, high-strength alloy for use in the engine and airframe components of high-performance jet aircraft. Today, titanium alloys are widely used in many applications that previously used metals such as stainless and specialty steels, copper alloys, and nickel-based alloys.

Table 11 Examples ofTitanium Alloys

Designation	Nominal Alloy Composition (By %)				
	Al	Sn	Zr	Mo	Other
Unalloyed Grades					
ASTM Grade 1	—	—	—	—	—
ASTM Grade 7	—	—	—	—	0.2 Pd
ASTM Grade 11	—	—	—	—	0.12–0.25 Pd
ASTM Grade 12	—	—	—	0.3	0.6–0.9 Ni
Alpha and Near Alpha Alloys					
Ti-5Al-2.5Sn	5.0	2.5	—	—	—
Ti-5Al-2.5Sn-ELI	5.0	2.5	—	—	—
Ti-6Al-2Sn-4Zr-2Mo	6.0	2.0	4.0	2.0	—
Ti-6Al-2Nb01Ta-0.8Mo	6.0	—	—	1.0	2 Nb, 1 Ta
Alpha-Beta Alloys					
Ti-6Al-4V	6.0	—	—	—	4.0 V
Ti-6Al-4V-ELI	6.0	—	—	—	6.0 V
Ti-6Al-6V-2Sn	6.0	2.0	—	—	0.75 Cu, 6.0 V
Ti-8Mn	—	—	—	—	8.0 Mn
Ti-7Al-4Mo	7.0	—	—	4.0	—
Ti-6Al-2Sn-4Z-6Mo	6.0	2.0	4.0	6.0	—
Beta Alloys					
Ti-13V-11Cr-3Al	3.0	—	—	—	11.0 Cr, 13.0 V
Ti-8Mo-8V-2Fe-3Al	3.0	—	—	8.0	8.0
Ti-11.5Mo-6Zr-4.5Sn	—	4.5	6.0	11.5	—

Beryllium is a reactive metal, and at high temperatures it readily combines with oxygen, hydrogen, and nitrogen to form stable compounds. At low temperatures, it is highly resistant to corrosion. Small amounts of nitrogen and hydrogen or large amounts of oxygen cause embrittlement. *Table 12* lists beryllium alloys and some of their mechanical properties, such as tensile and yield strength.

Beryllium is considered a difficult metal to weld because it must be kept in an inert gas atmosphere while it is being welded.

WARNING!

Dust, fumes, vapors, and fine machining particles of beryllium are cancer-causing agents. If inhaled, they can cause chronic beryllium disease. An independent air supply is required when working with beryllium.

Table 12 Examples of Beryllium Alloys and Mechanical Properties

Designation and Percent Be	Tensile Strength		Yield Strength		Commercial Form	Description or Properties
	psi	kPa	psi	kPa		
HP-20 (98.0 percent)	48,000	330,948	35,000	241,317	Block	Pressed powder
HP-8 (99.0 percent)	37,000	255,106	27,000	186,158	Block	Pressed powder
HP-12 (98.5 percent)	38,000	262,200	30,000	206,843	Block	Pressed powder
HP-40 (92.0 percent)	65,000	448,159	60,000	413,685	Block	Pressed powder
C-10 (99.0 percent)	20,000	137,895	—	—	Block	Pressed powder
PR-2 (98.0 percent)	77,000	530,896	55,000	379,212	Sheet/plate	Some form at 1,350°F (732°C)
IS-2 (99.0 percent)	45,000	310,264	31,000	213,737	Sheet/plate	High form at 700°F (371°C)
XT-20 (98.0 percent)	77,000	530,896	40,000	275,790	Extrusion	Good mechanical properties
XT-40 (92.0 percent)	90,000	620,528	50,000	344,738	Extrusion	High strength
Lockalloy	60,000	413,685	50,000	344,738	Extrusion	Service at 600°F to 800°F (316°C to 427°C)

Magnesium

Magnesium is widely used in specialty racing cars and some commercial cars as a lightweight material to minimize weight and fuel consumption because of the 20 to 25 percent weight saving over aluminum. It typically is used in the construction of transmission casings, intake manifolds, and cylinder head covers. Magnesium pistons and various other engine parts are also being used in some race cars. It is difficult to visually differentiate aluminum and magnesium. Use positive metal identification (PMI) test equipment to verify these materials.

1.5.5 Copper and Copper Alloys

Copper, like aluminum, is often used in nearly pure form for electrical and heat conductors. It is alloyed with zinc to make brass, and with tin to make bronze. Other bronzes are formed by alloying copper with aluminum, silicon, or beryllium. Many variations of alloys exist within each of the brass and bronze groups, made by varying the percentages of the alloying metals and by adding other metallic and nonmetallic elements. These include zinc, tin, nickel, silicon, aluminum, cadmium, and beryllium.

Beryllium

Beryllium allows the transmission of X-rays better than other metals or glass. Because of this property, along with its high melting point, it is widely used as the window portion of high-intensity X-ray tubes, such as those used in high-resolution X-ray machines and imaging equipment.

Copper and copper alloys are classified into the following major groups:

- *Coppers* – Coppers are metals that have a designated minimum copper content of 99.3 percent or higher.
- *High-copper alloys* – For the wrought-type products, these are alloys with a copper content of less than 99.3 percent but more than 96 percent that do not fall into any other copper alloy group. Cast-type high copper alloys have a copper content in excess of 94 percent, to which silver may be added to obtain special properties.
- *Brasses* – These alloys contain zinc as the principal alloying element, with or without other alloying elements. The wrought-type alloys are grouped into three categories of brasses:

copper-zinc alloys, copper-zinc-lead alloys (leaded brasses), and copper-zinc-tin alloys (tin brasses). Cast alloys are grouped into five categories of brasses: copper-tin-zinc alloys (red, semi-red, and yellow brasses); manganese bronze alloys (high-strength yellow brasses); leaded manganese bronze alloys (leaded high-strength yellow brasses); copper-zinc-silicon alloys (silicon brasses and bronzes); and cast copper-bismuth and copper-bismuth-selenium alloys.

- *Bronzes* – Bronzes are copper alloys in which the major alloying element is not zinc or nickel. For wrought-type alloys, there are four groups of bronzes: copper-tin-phosphorus alloys (phosphor bronzes); copper-tin-lead-phosphorus alloys (leaded phosphor bronzes); copper-aluminum alloys (aluminum bronzes); and copper-silicon alloys (silicon bronzes). Cast-type alloys are divided into four groups of bronzes: copper-tin alloys (tin bronzes), copper-tin-lead alloys (leaded and high leaded tin bronzes), copper-tin-nickel alloys (nickel-tin bronzes), and copper-aluminum alloys (aluminum bronzes). Alloys known as manganese bronzes, in which zinc is the major alloying element, are included in the brasses.
- *Copper-nickels* – Copper-nickel alloys have nickel as the principal alloying element, with or without other alloying elements.
- *Copper-nickel-zinc alloys* – These alloys contain zinc and nickel as the principal and secondary alloying elements, with or without other elements. They are commonly called nickel silvers.
- *Leaded coppers* – Leaded coppers comprise a series of cast alloys of copper with 20 percent or more lead, sometimes with a small amount of silver, but without tin or zinc.
- *Special alloys* – These are alloys whose chemical compositions do not fall into any of the above categories.

Two classification systems for copper alloys are the *ASTM Standard B224, Standard Classification of Coppers,* (*Table 13*) and the Copper Development Association, Inc. designation system (*Table 14*). The Copper Development Association system has been updated to fit the UNS, and includes all commercially available metals and alloys. The format for the UNS designation is CNNN00, in which NNN designates numbers. As shown in *Table 14,* the alloys are divided into wrought and cast alloy categories. For this reason, an

Statue of Liberty

The Statue of Liberty contains 179,000 pounds (81,193 kg) of copper. When copper oxidizes, it forms a blue-green self-protective coating, which is why the copper skin of the Statue of Liberty remained virtually intact after 100 years of being subjected to high sea winds, driving rains, and scorching sun. Close analysis showed that weathering and oxidation of the copper skin had caused only 0.005 inch (0.13 mm) of wear in a century. For this reason, the copper skin did not need to be significantly rebuilt when the statue was renovated for its centennial. However, high-alloy copper saddles and rivets were used during the restoration to fasten the copper skin to the framing underneath. The outline of the copper sheets can be seen in this image. This was done to ensure the structural integrity of the statue and guard against any galvanic reaction problems. Galvanic action is corrosion caused by an electrical current produced as the result of a chemical reaction between two dissimilar metals.

Table 13 ASTM Classification System for Copper Alloys

ASTM Designation	Alloy Type	Typical Uses
Tough-Pitch Coppers		
ETP	Electrolytic, tough-pitch	Bus bars, brazing rods, wire anodes, forgings
FRHC	Fire-refined, high-conductivity, tough-pitch	Mechanical applications
FRTP	Fire-refined, tough-pitch	Sheets, strips, plate
ATP	Arsenical, tough-pitch	Roofing, radiator cores
STP	Silver-bearing, tough-pitch	Pans, printing rolls, fasteners
SATP	Silver-bearing, arsenical, tough-pitch	Pans, printing rolls, fasteners
Oxygen-Free Coppers		
OFH	Oxygen-free (without residual deoxidants)	Tubing wave guides, starting anodes, wire
OFP	Oxygen-free, phosphorus-bearing	Welding rods, forging
OFTPE	Oxygen-free, tellurium- and phosphorus-bearing	Welding rods, forging
OFS	Oxygen-free, silver-bearing	Plate, sheets, rods, forging
OFTE	Oxygen-free, tellurium-bearing	Bars, free-machining
Deoxidized Coppers		
DHP	Phosphorized, high-residual phosphorus	Tubes, pipes, anodes, projectile rotating bands
DLP	Phosphorized, low-residual phosphorus	Tubes, wave guides, general use
DPS	Phosphorized, silver-bearing	Heat exchangers, steam lines, condenser tubes, tubes for general use
DPA	Phosphorized, arsenical	Heat exchangers, steam lines, condenser tubes, tubes for general use
DPTE	Phosphorized, tellurium-bearing	Free-machining

Table 14 UNS Designations for Classifying Copper Alloys

Copper Designation	Composition
Tough-Pitch Coppers	
C11X00	99.95% oxygen-free, high-conductivity copper
C12X00	99.88+% copper (tough-pitch copper)
C19X00	96+% copper (high-copper alloys)
C2XX00	Copper-zinc alloys (brasses)
C3XX00	Copper-zinc-lead alloys (leaded brasses)
C4XX00	Copper-zinc-tin alloys (tin brasses)
C51X00	Copper-tin alloys (phosphor bronzes)
C54X00	Copper-tin-lead alloys (leaded phosphor bronzes)
C62X00	Copper-aluminum alloys (aluminum bronzes)
C65X00	Copper-silicon alloys (silicon bronzes)
C70X00	Copper-nickel alloys
Cast Alloy Groups	
C80X00	99+% copper, copper alloys
C81X00	High-copper alloys (beryllium copper)
C83X00	Red brasses and leaded red brasses
C84X00	Yellow brasses and leaded yellow brasses
C86X00	Manganese and leaded manganese-bronze alloys
C87X00	Silicon bronzes and brasses
C90X00	Tin bronzes

alloy made in both a wrought and cast form can have different UNS numbers, depending on the method of manufacture.

Tough-pitch copper is oxygen bearing, and most of the impurities have been oxidized during refining to increase conductivity. Continuous casting methods have allowed the oxygen content to be reduced to 0.03 percent or less to achieve the desired results. During fire refining, the oxygen that comes from the air injected to cause the oxidation of impurities also combines with hydrogen to form steam. The steam creates gas pockets (porosity) in the metal during solidification. (In the past, high porosity was a characteristic of tough-pitch copper.) When tough-pitch copper is reheated by welding or annealing, the hydrogen trapped in the copper is diffused into the metal and reacts with cuprous oxide to form insoluble water vapor. The expansion associated with this reaction forces the grains apart, which causes embrittlement of the metal. For applications where copper must be resistant to gassing, the oxygen is sometimes eliminated by adding a deoxidizer such as phosphorus to the molten metal before casting. Such copper is specified as deoxidized copper.

For welding copper alloys, GMAW and GTAW are most commonly used.

1.5.6 Nickel and Nickel Alloys

Nickel, a corrosion-resistant metal, is used as an alloying element as well as for plating other metals. Nickel-clad steel and nickel-based alloys containing over 50 percent nickel are used extensively in industry. They have good corrosion resistance and elasticity, and unique magnetic thermal expansion and thermal conductivity. Many high-strength, high-temperature nickel super alloys have been developed for aerospace, automotive, and hot-die applications. Alloys of nickel and chromium (Nichrome®) are used to make heating elements of high electrical resistance.

UNS and international standard designations have been assigned to many nickel-alloys. However, trade names, such as Monel 400®, Inconel 600®, and Hastelloy 22® are more commonly used to identify them. *Table 15* lists some common nickel alloys and some of their uses and properties.

In general, nickel and nickel-based alloys can be welded by SMAW, GMAW, and GTAW processes. Welding nickel alloys is very similar to welding austenitic stainless steel, with the following exceptions:

- When welding nickel alloys, the surface oxide melts at a much higher temperature than the base metal, so surfaces should be thoroughly cleaned.
- Embrittlement can be caused by lead, sulfur, phosphorus, or other low-temperature metals. The surface should be free of all contaminants.
- Weld penetration in nickel-based alloys is shallower than with other metals and requires the use of very wide bevel and groove angles.

1.5.7 Lead, Tin, and Zinc

Lead, tin, zinc, and their alloys have very low melting points. For this reason, they are usually soldered and not welded. Zinc-plated, or galvanized, steels can be welded, but the fumes are toxic and must not be inhaled.

1.5.8 Platinum, Gold, Silver, and Other Precious Metals

The precious metals are osmium, iridium, rhodium, platinum, ruthenium, gold, palladium, silver, and their alloys. They are used industrially for electrical contacts, mirror coatings, catalysts, electrodes, high-temperature crucibles, and dental structures and fillings. They are usually brazed, soldered, or welded with special equipment.

Table 15 Common Nickel Alloys and Some of Their Uses

Alloy	Ni	C	Mn	Fe	S	Si	Cu	Cr	Al	Ti	Other	Typical Properties and Uses
Commercially Pure Nickel												
A-Nickel	99.4	0.1	0.2	0.15	0.005	0.05	0.1					Chemical industry, electroplating
D-Nickel	95.2	0.1	4.5	0.15	0.005	0.05	0.05					Resistance to sulfur attack
E-Nickel	97.7	0.1	2.0	0.10	—	0.05	0.05					Similar to D-Nickel
L-Nickel	99.4	0.02	0.2	0.15	0.005	0.05	0.1					Severe plastic forming operations
Z-Nickel	94.0	0.16	0.25	0.25	0.005	0.04	0.05					Age hardenable, springs, pump rods, shafts
Nickel-Copper Alloys												
Monel	66.0	0.12	0.90	1.35	0.005	0.15	31.50		2.80	0.50		Corrosion resistance, toughness, high strength
R-Monel	66.0	0.18	0.90	1.35	0.005	0.15	31.50					Free-machining
K-Monel	65.3	0.15	0.60	1.00	0.005	0.15	29.50					Age-hardenable, nonmagnetic
H-Monel	65.0	0.1	0.9	1.5	0.015	3.0	29.50					Age-hardenable, machinable
S-Monel	63.0	0.1	0.9	2.0	0.015	4.0	30.0					Age-hardenable, anti-galling, impellers pump liners
Nickel-Iron Alloys												
Invar	36.0	*	*	63.0	*							Very low expansion coefficient length standards
Platinite	46.0											Glass seal
Permalloy	78.5											High-magnetic permeability, submarine telegraph cables
Nickel-Chromium and Nickel-Iron-Chromium Alloys												
Inconel	76.4	0.04	0.20	7.20	0.007	0.2	0.10	15.85				Heat and corrosion resisting, nonmagnetic
Inconel 702	78.0	0.02	0.05	0.30	0.007	0.15	0.05	15.85	3.0	0.60		Age-hardenable, operates up to 2,400°F (1,316°C)
S-Inconel	68.0	0.20	1.0	8.5	0.008	5.0	0.5	15.5				Age-hardenable, anti-galling
Inconel-X-750	72.9	0.04	0.70	6.8	0.007	0.30	0.05	15.0	0.80	2.5	CB-0.9	Age-hardenable, used up to 1,500°F (816°C)
Nimonic 80A	74.5	0.05	0.55	0.55	0.007	0.20	0.05	20.0	1.30	2.50		Age-hardenable, similar to Inconel X
Nimonic 90	57.0	0.05	0.50	9.50	0.007	0.20	0.05	20.45	1.65	2.60	CO-17.00	Similar to Nimonic 80A
Nichrome	80.0	0.05	0.01	0.5	—	0.20	—	20.0				Heating elements
Chromel	90.0							10.0				Heating elements, thermocouples
Nicoloy 901	37.35	0.04	0.45	33.75	0.007	0.30	1.10	13.50	0.25		MO-5.90	Age-hardenable, gas turbine wheels
Ni-O-NEL	41.35	0.03	0.65	31.65	0.007	0.35	1.80	20.2	0.40	0.50	MO-310	Resistant to certain hot acids
60-12 Ni-Cr	62.45	0.55		25.0				12.0				Resistance heating element
65-15 Ni-Cr	65.0			24.00	0.007			15.00				Resistance heating element
67-17 Ni-Cr	72.45	0.55		10.0				17.00				Heat-resisting alloy

Additional Resources

AWS B1.10M/B1.10 Guide for the Nondestructive Examination of Welds. Miami, FL: American Welding Society.

AWS B1.11 Guide for the Visual Examination of Welds. Miami, FL: American Welding Society.

AWS D3.5-93R Guide for Steel Hull Welding. Miami, FL: American Welding Society.

AWS D3.6M Specification for Underwater Welding. Miami, FL: American Welding Society.

AWS D3.7 Guide for Aluminum Hull Welding. Miami, FL: American Welding Society.

Machinery's Handbook. Erik Oberg, Franklin D. Jones, and Christopher J. McCauley. New York, NY: Industrial Press, Inc. 2004.

Metals and How to Weld Them. T. B. Jefferson. Cleveland, OH: The James F. Lincoln Arc Welding Foundation.

OSHA 1926.351, Arc Welding and Cutting. Current Edition. Washington, DC: Occupational Safety and Health Administration (OSHA).

Stick Electrode Product Catalog. Cleveland, OH: The Lincoln Electric Company, 2008.

Stick Electrode Welding Guide. Cleveland, OH: The Lincoln Electric Company, 2004.

The Procedure Handbook of Arc Welding. 14th Edition. Cleveland, OH: The James F. Lincoln Welding Foundation, 2000.

Welding Handbook, 4 vols. Miami, FL: American Welding Society.

Lincoln Electric website: **http://www.lincolnelectric.com** offers sources for products and training.

1.0.0 Section Review

1. A document that a customer receives from a steel manufacturer that indicates the quality of the steel being purchased is a(n) ____.
   a. safety data sheet (SDS)
   b. ferrous metal classification bill (FMCB)
   c. certified mill test report (CMTR)
   d. evaluated properties rating index (EPRI)

2. The amount of alloy added to make high-strength low-alloy (HSLA) steel is typically ____.
   a. less than 2 percent
   b. about 5 percent
   c. between 6 and 8 percent
   d. about 10 percent

3. Ferritic stainless steels are always ____.
   a. martensitic
   b. high in nickel content
   c. hardenable by tempering
   d. magnetic

4. A form of highly alloyed stainless steel developed to provide outstanding resistance to acids and other caustic solutions is called ____.
   a. superferritic
   b. super duplex
   c. superaustenitic
   d. super martensitic

5. What nonferrous metal has the lowest density of the structural metals and is commonly used in the aerospace industry?
   a. Magnesium
   b. Aluminum
   c. Nickel
   d. Titanium

Section Two

2.0.0 Metal Characteristics

Objective

Describe the physical and mechanical characteristics of metals and explain how to identify base metals.

a. Describe the physical characteristics of different metals.
b. Describe the mechanical properties of different metals.
c. Explain how to identify base metals in field conditions.
d. Describe metallurgy-related considerations for welding.

Trade Terms

Residual stress: The strain that occurs in a welded joint after the welding has been completed, as a result of mechanical action, thermal action, or both.

Underbead cracking: Subsurface cracking in the base metal under or near the weld.

All metals have different characteristics and mechanical properties. These physical and mechanical properties can be used to identify a particular metal. In addition, a metal's properties must be taken into consideration prior to and during welding operations.

2.1.0 Physical Characteristics of Metals

Physical characteristics are distinctive for every metal. They include the following variable traits:

- Density (specific gravity)
- Electrical conductivity
- Thermal conductivity
- Thermal expansion
- Melting point
- Corrosion resistance

2.1.1 Density

Density is the mass (weight) of a specific volume of metal. Density may also be expressed as specific gravity or specific density. Specific gravity is the weight of a metal compared to the weight of an equal volume of water. The higher the density of a material, the heavier it will be.

Light metals, such as aluminum and magnesium, have low densities (low weights for a given volume) while ferrous metals, such as carbon steel and stainless steel, have much higher densities. An iron or steel item weighs nearly three times as much as an aluminum item of the same size. *Table 16* lists the densities of some common metals in pounds per cubic inch and grams per cubic centimeter.

2.1.2 Electrical Conductivity

Electrical conductivity is a metal's ability to conduct electricity. All metals can conduct electricity to some degree. Silver is one of the best conductors, but it is too expensive to use for commercial power transmission. Copper is not quite as good a conductor as silver, but for the same diameter conductor or wire, it is a better conductor than aluminum. However, for conductors weighing the same per unit of length, aluminum will carry more current than copper because it has a larger cross-section.

The electrical conductivity of a metal is calculated from its measured resistance to current flow (its resistivity). Resistivity is the inverse (opposite) of conductivity. The lower the resistivity value, the higher the conductivity value and the more current a metal can carry for its size. The resistivity of any metal increases and its conductivity decreases as its temperature rises. This means that metals with higher electrical resistances build up heat from welding current faster. They also do not transfer heat away as efficiently, resulting in higher distortion and stresses in the weld zone. *Table 17* lists the electrical resistivity of some common metals in microhms per cubic centimeter.

Table 16 Densities of Some Common Metals

Metal	Density	
	(lb/in^3)	(g/cm^3)
Aluminum	0.098	2.71
Copper	0.324	8.96
Gold	0.698	19.32
Iron or steel	0.284	7.87
Magnesium	0.063	1.74
Manganese	0.267	7.43
Molybdenum	0.369	10.20
Nickel	0.309	8.90
Tin	0.264	7.30
Titanium	0.163	4.54
Tungsten	0.697	19.30
Zinc	0.258	7.13

Table 17 Electrical Resistivity of Some Common Metals

Metal	Electrical Resistivity (microhms/cm^3)
Aluminum	2.92
Copper	1.67
Gold	2.19
Iron or steel	9.71
Magnesium	4.46
Manganese	185.00
Molybdenum	5.17
Nickel	6.84
Titanium	80.00
Zinc	5.91

2.1.3 Thermal Conductivity

Thermal conductivity is a metal's ability to conduct heat. Metals that are good conductors of electricity are also good conductors of heat. As temperature increases, a metal's resistance to the flow of heat increases. Because a metal with poor heat conductivity does not carry the weld heat away from the weld zone efficiently, the temperature rises and distortion is greater. Silver and copper are excellent heat conductors, iron and nickel are fair conductors, and titanium and manganese are poor conductors.

2.1.4 Thermal Expansion

Thermal expansion is the change in size that occurs in a material as its temperature changes. Solids expand in all dimensions when heated and contract from all dimensions when cooled. The increase in unit length when a solid is heated one degree is called its coefficient of linear expansion. The larger the coefficient of expansion, the greater the dimensional increase of the solid for each degree of temperature increase. Large dimensional changes cause distortion and stresses in weldments. *Table 18* lists the coefficients of linear expansion for a variety of metals. *Figure 8* shows the expansion of common metal pipes in inches per 100 feet for a range of temperatures in degrees Fahrenheit.

2.1.5 Melting Point

The melting point of a solid is the temperature at which the solid converts into a liquid. *Table 19* lists a number of metals and their melting points.

2.1.6 Corrosion Resistance

Metals and alloys are corroded by different agents at varying rates. Corrosion is important to consider because it can severely reduce the tensile strength of a metal. When selecting a metal for a particular use, its ability to resist corrosion by the atmosphere and by gaseous or liquid chemicals may be an important factor. Certain metals may not react to specific chemicals. For example, iron is often used to handle strong caustics. Other metals produce a protective oxide coating that retards further corrosion. Both aluminum and lead form weather-resistant oxide coatings in the atmosphere. The stainless steels and nickel-based alloys are designed to withstand corrosion from such things as the atmosphere, elevated temperatures, and strong chemicals.

2.2.0 Mechanical Properties of Metals

The mechanical properties of metals determine how they will react to or be affected by the external forces of tension, compression, torsion, shear, impacts, and cold shaping (working). The ability of metals to permanently change or deform is called plasticity. Plasticity is different from elasticity, which refers to the ability of a material to change temporarily and then revert to its original form.

2.2.1 Stress-Strain Relationship

Stress is the force or load applied to a specimen that causes it to deform. Strain is the magnitude of deformation caused by stress. Within certain limits, stress increases directly with strain. The four types of stress are as follows:

- *Tension* – Opposed, in-line forces pulling away from each other at opposite sides of a specimen
- *Compression* – Opposed, in-line forces pushing toward each other against opposite sides of a specimen
- *Torsion* – Opposed, separated, twisting forces acting on the same axis of a specimen
- *Shear* – Opposed, nonintersecting forces acting opposite each other and perpendicular to a specimen

Figure 9 shows the force directions that result from different types of stress. Often several different stress types act on a specimen at the same time.

Table 18 Coefficients of Linear Expansion of Some Common Metals

Metal	Coefficient (a) 10^{-6}/°F	Coefficient (a) 10^{-6}/°C
Aluminum	13.3	25.0
Copper	9.2	16.5
Iron	6.5	12.0
Magnesium	14.0	25.0
Manganese	12.0	22.0
Molybdenum	2.7	5.0
Nickel	7.4	12.8
Titanium	4.7	8.6
Zinc	22.1	39.7

2.2.2 Elasticity and Elastic Limit

Elasticity is the ability of a material to be deformed without the deformation becoming permanent. As long as a material remains within its elastic range, it will return to its original size and shape when the deforming force (stress) is removed. While within the elastic range, the strain will be directly proportional to the stress that produces it.

Every material has its elastic limit. When the elastic limit is exceeded, the strain is no longer proportional to the stress, and the material will not return to its original size or shape. The material has been distorted. Elastic limits are commonly measured by an extensometer.

2.2.3 Modulus of Elasticity

Every material strains (deforms) in direct proportion to the applied stress while it remains within its elastic range; but the amount each material strains for a given stress is not the same. The ratio of stress to strain is called the modulus of elasticity. It remains the same for any material under the same proportional conditions. For example, a material that will stretch 1 inch (25.4 mm) under a 10-pound (4.5-kg) tension will stretch 2 inches (50.8 mm) under a 20-pound (9-kg) tension, as long as the elastic limit is not exceeded. Similarly, a material that will stretch ¼ inch (6.4 mm) under a 10-pound (4.5-kg) tension will stretch ½ inch (12.7 mm) under a 20-pound (9-kg) tension as long as the elastic limit is not exceeded. The modulus of elasticity of a metal can be used to predict how much the metal will deform under a specified load, provided the elastic limit is not exceeded.

2.2.4 Yield Strength and Tensile Strength

Yield strength is the maximum tensile stress a material can withstand before irreversibly deforming. Once stressed beyond its yield strength, the material will not return to its original shape when the stress is removed. Tensile strength, on the other hand, is the maximum tensile stress a material can withstand without breaking. The maximum load on a tensile material when it fractures is called the ultimate load. Brittle materials tend to break at values near their yield strength, while more elastic materials may deform significantly before breaking. Both yield and tensile strength are expressed in force per cross-sectional unit area. Engineers are interested in both values during the design process.

2.2.5 Ductility

Ductility is the plasticity exhibited by a material under tension loading. Ductility is measured by dividing the change in length of a material by the original length at the point of failure. Ductility is the characteristic that allows metals to be drawn into thinner sections without breaking. Copper and aluminum are examples of common ductile metals.

2.2.6 Hardness

The hardness of a material can be defined as its resistance to indentation. It is related to its elastic and plastic properties. The greater the hardness, the greater the resistance to indentation, penetration, and wear. Hardness is measured by pressing a steel ball or diamond point against a surface with a calibrated force. The diameter or depth of the impression is then measured to determine the hardness. Standard hardness testers include the Brinell, Rockwell, and Vickers instruments. *Figure 10* shows a portable hardness tester suitable for testing on the job site.

The Brinell hardness test is performed by pressing a hardened 10-mm steel ball into the metal being tested. Force is applied by standard weights: 500 kg for aluminum, copper, and other soft materials; and 3,000 kg for iron, steel, and

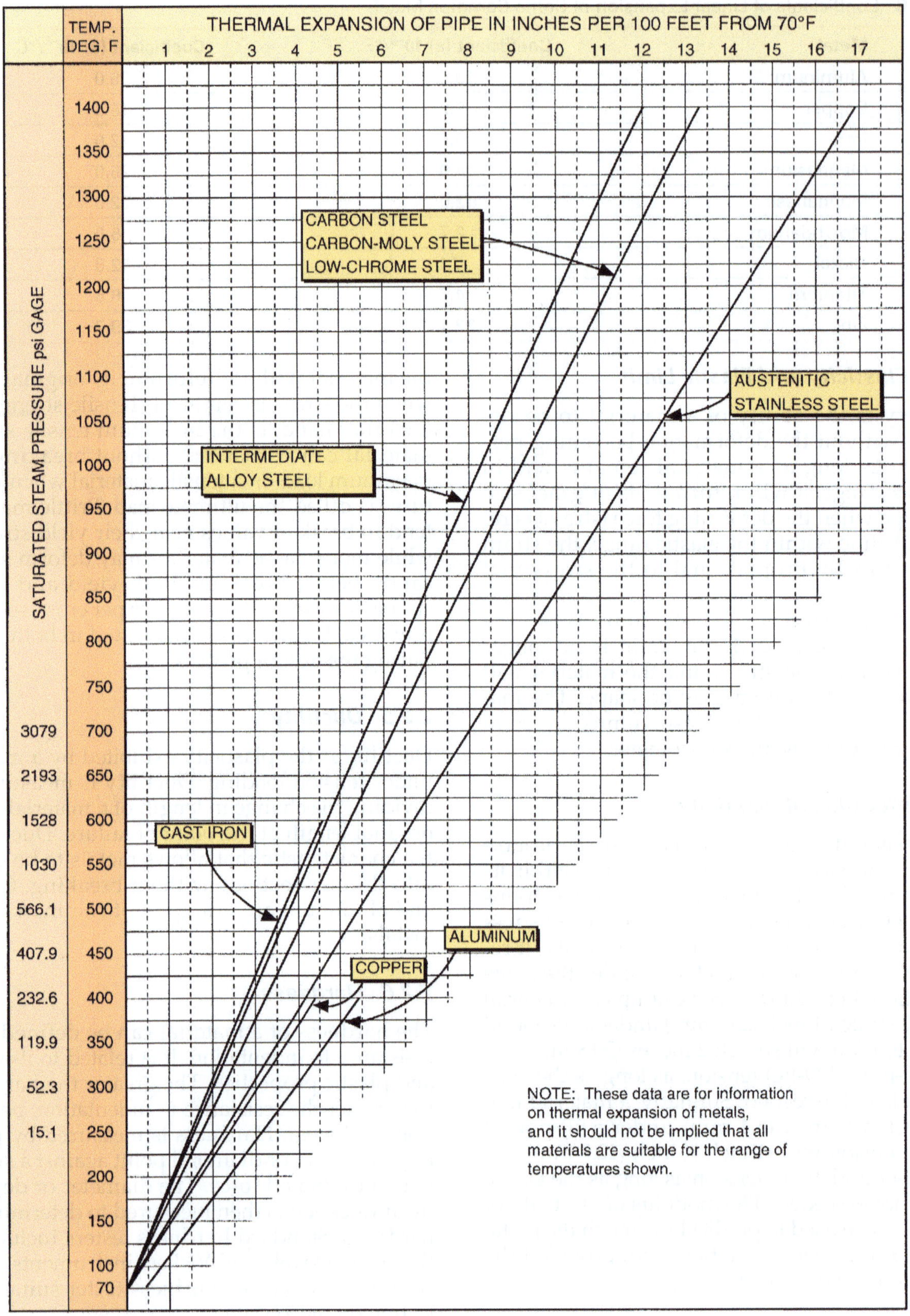

Figure 8 Expansion of different types of pipe.

Table 19 Melting Points of Some Common Metals

Metal	Melting Point
Aluminum	1,215°F (657°C)
Copper	1,981°F (1,083°C)
Iron	2,800°F (1,538°C)
Magnesium	1,202°F (650°C)
Manganese	2,273°F (1,245°C)
Molybdenum	4,760°F (2,627°C)
Nickel	2,651°F (1,455°C)
Titanium	3,300°F (1,816°C)
Zinc	787°F (419°C)

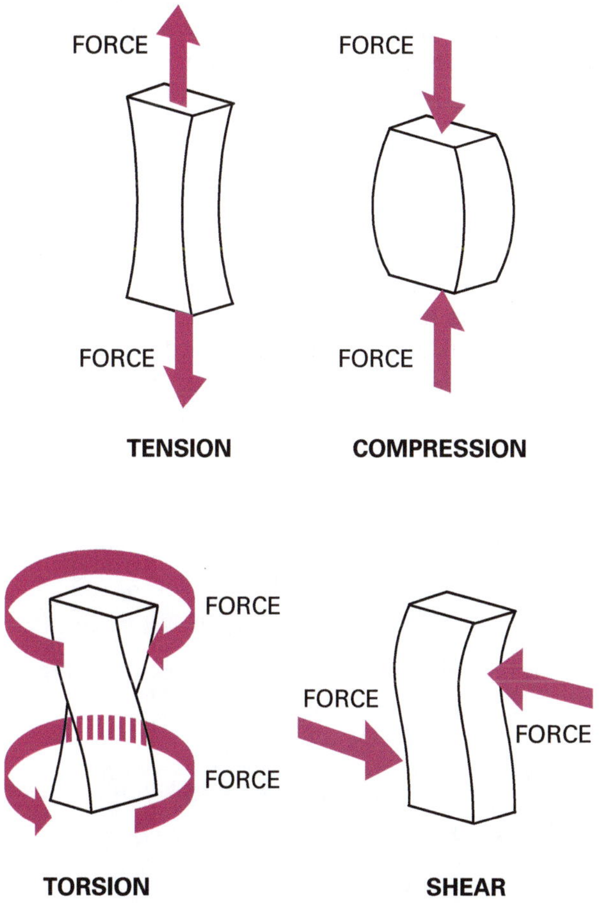

Figure 9 Force directions that result in each stress type.

other hard metals. The diameter of the indentation is measured under a microscope and then converted to the Brinell hardness number.

The Rockwell hardness tester uses a 1⁄16-inch (1.6-mm) hardened steel ball with a 100-kg weight on soft metals, and a diamond cone and 150-kg weight on hard metals. Depth of penetration is read directly from hardness scales on the tester.

The Vickers hardness tester uses a pyramid-shaped diamond penetrator with a variable 1- to 120-kg load. The hardness is read from standard tables by comparing the width of the dent with the pressing load.

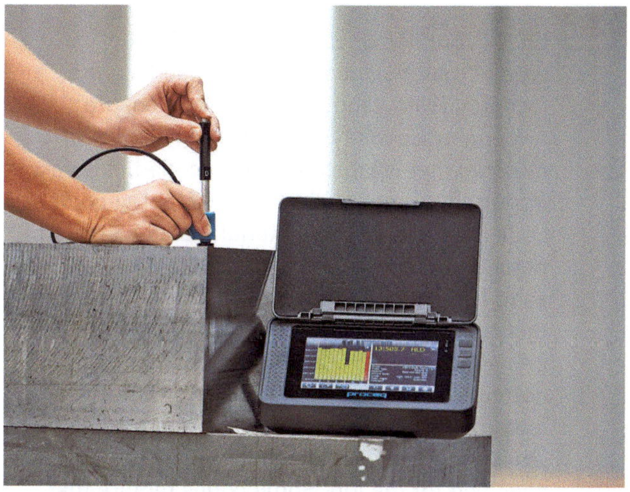

Figure 10 Portable hardness tester.

2.3.0 Field Identification of Base Metals

Many of the common base metals and alloys can be generally identified by simple tests and observation of their characteristic metallic features. This is especially true of the ferrous metals, which can be identified by magnetic attraction, surface appearance, structural form or shape, and grinding sparks.

2.3.1 Metal Labeling

Because it is difficult to identify base metal types, most sites have a system to label or color-code the various types of metals being used. Labeling is performed by printing the ASTM number on the material (shape or plate) with an indelible marker or paint stick. Color-coding involves assigning a color to a base metal type and then spraying that color on the material. When color-coding is used, a master chart is often posted near the storage area. The master chart identifies the color assigned to each base metal.

For stainless steel and nickel alloys, special care should be taken to mark only with low-chloride markers. Using markers other than the low-chloride markers will leave a residue on the metal that can contaminate the weld. Specialty markers, which usually contain white or yellow ink, are available for marking metals such as stainless steel (*Figure 11*).

Figure 11 A typical low-chloride marker for specialty metals.

NOTE

Check your site quality standards for the type of material labeling that is used at your location. Follow those standards to ensure material traceability. When cutting a piece of material that is labeled or color-coded, it is very important to make sure that the label or color code is visible on the material that is being left behind. If possible, make the cut in such a way as to leave the label or color code still visible. If this is not possible, be sure to mark the material being left behind to ensure that others can correctly identify the material. It is also a good idea to mark the material being removed so that everyone involved can identify it as the material specified for the job. Some codes require that a witness be present when marking the two pieces.

2.3.2 Identification by Magnet

All carbon steels, cast steels, and cast irons are attracted to a magnet. Of the stainless steels, only ferritic and martensitic stainless steels can be magnetized, and they have the bright, corrosion-free surface typical of stainless steel. The largest group of stainless steels—austenitic stainless steels—are nonmagnetic. Nonferrous metals and manganese steels are also nonmagnetic.

CAUTION

Magnetic testing is sometimes unreliable because of alloys or changes that can occur during welding. Do not rely solely on magnetic testing.

2.3.3 Identification by Appearance

Both hot-rolled and cold-finished carbon steel are milled in standard structural shapes, such as rods, bars, beams, channels, and angles. Unweathered, hot-rolled carbon steel has a blue-gray, smooth-to-scaly surface, but it oxidizes to a rough, red-brown rust when exposed to the weather. The corners of hot-rolled shapes are slightly rounded or chamfered.

Unweathered, cold-rolled carbon steel has a smooth, gray finish that oxidizes to a fine red-brown rust when weathered. Except for round shapes, the sides and edges are very flat and smooth, and the corners are well formed, not rounded or chamfered.

Cast steel and cast iron look similar on the surface. They both exhibit the shape characteristics typical of castings: curved surfaces and rounded edges; thin and thick regions; reinforcing ribs or raised areas; spokes; and irregular shapes. The cast surfaces are usually slightly textured and can contain mold seams. When broken, cast iron has a grainy appearance, while cast steel has a smooth texture. Cast iron is used to make everything from water-meter covers, to engine crankshafts, to fan blades (*Figure 12*).

There are three types of cast iron: gray, white, and malleable. The exterior surface of all the cast irons has a slightly irregular texture that comes from the use of sand molds. Malleable iron is sometimes forged. If it has been forged, it will have a smoother appearance and die marks.

Gray cast iron contains a high percentage of carbon and silicon. Gray cast iron is only formed by slow cooling, which forms large crystals and a soft, easily machined metal. During the slow cooling process, the high silicon content forces the carbon to separate out in the form of graphite flakes (free carbon). This gives a fresh fracture its characteristic grainy and gray appearance. If you rub a clean finger or white cloth against a freshly fractured gray cast-iron surface, the graphite will leave a dark smear. Gray cast iron can be welded by following special procedures.

White cast iron is chilled rapidly, which forms small crystals and a very hard and brittle metal. The carbon is united with the iron and is not free carbon. Freshly fractured cast iron has an irregular, fine, silvery-white, crystalline surface.

Figure 12 A compressor's cooling fan blade made of cast iron.

Identification of Nonferrous Metals

The electrical conductivity measuring instrument shown can be used to determine the composition of nonferrous metals, including titanium alloys and silver. By using eddy currents to determine the electrical conductivity of a nonferrous metal under test, conclusions can be made about the composition, structure, and condition of the material.

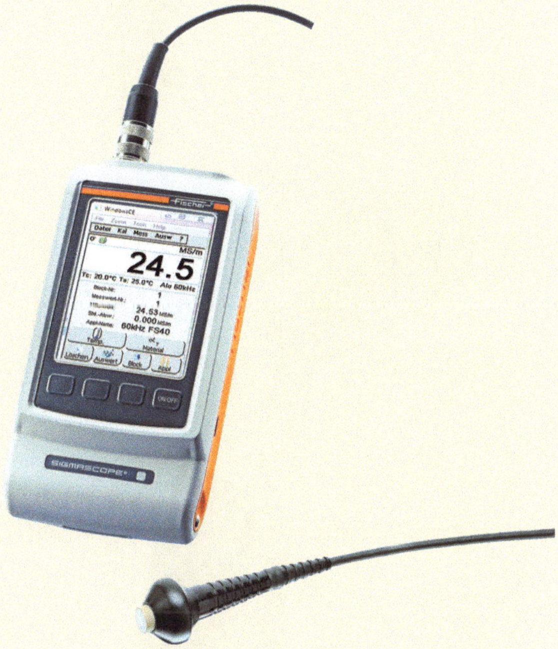

Because the weldability of white cast iron is poor, it is generally not welded.

Malleable cast iron is white cast iron that has been annealed. The annealing process results in a casting with an outer shell of white cast iron that grades into a core of gray cast iron. The resulting casting is tougher and stronger than gray cast iron and less brittle than white cast iron. Malleable cast iron can be welded, but special procedures must be used or it will revert back to white cast iron.

2.3.4 Identification by X-Ray Fluorescence Spectrometry

Portable metal analyzers that use X-ray fluorescence (XRF) spectrometry provide a fast and accurate analysis for metal and metal alloy identification. A fluorescent X-ray is created when an X-ray of sufficient energy, emitted from a miniaturized X-ray tube or from a small, sealed radioactive isotope, strikes an atom in the sample, dislodging an electron from one of the atom's inner orbital shells. The atom regains stability, filling the vacancy left in the inner orbital shell with an electron from one of the atom's higher energy orbital shells. The electron drops to the lower energy state by releasing a fluorescent X-ray, and the energy of this X-ray is equal to the specific difference in energy between two quantum states of the electron.

When a sample is measured using X-ray fluorescence, each element present in the sample emits its own unique fluorescent X-ray energy spectrum. By simultaneously measuring the fluorescent X-rays emitted by the different elements in the sample, handheld XRF analyzers can rapidly determine those elements that are present in the sample, along with their relative concentrations—in other words, the elemental chemistry of the sample. For samples with known ranges of chemical composition, such as common grades of metal alloys, the XRF analyzer also identifies most sample types by grade name, through comparing the analyzed concentrations with known values contained in an alloy library programmed within the analyzer. This entire process is typically completed in less than five seconds, and is completely nondestructive.

Thirty or more elements may be analyzed simultaneously by measuring the characteristic fluorescence X-rays emitted by a sample. *Figure 13* shows a handheld analyzer, its screen, and the analyzer in use.

2.4.0 Metallurgical Considerations for Welding

Many metallurgical variables must be considered when determining the effects of welding on both the base metal and the weld. Base metal composition and thickness, filler metal, and joint configuration all affect the weld quality. There are also several processes and design considerations used to change the metal characteristics in order

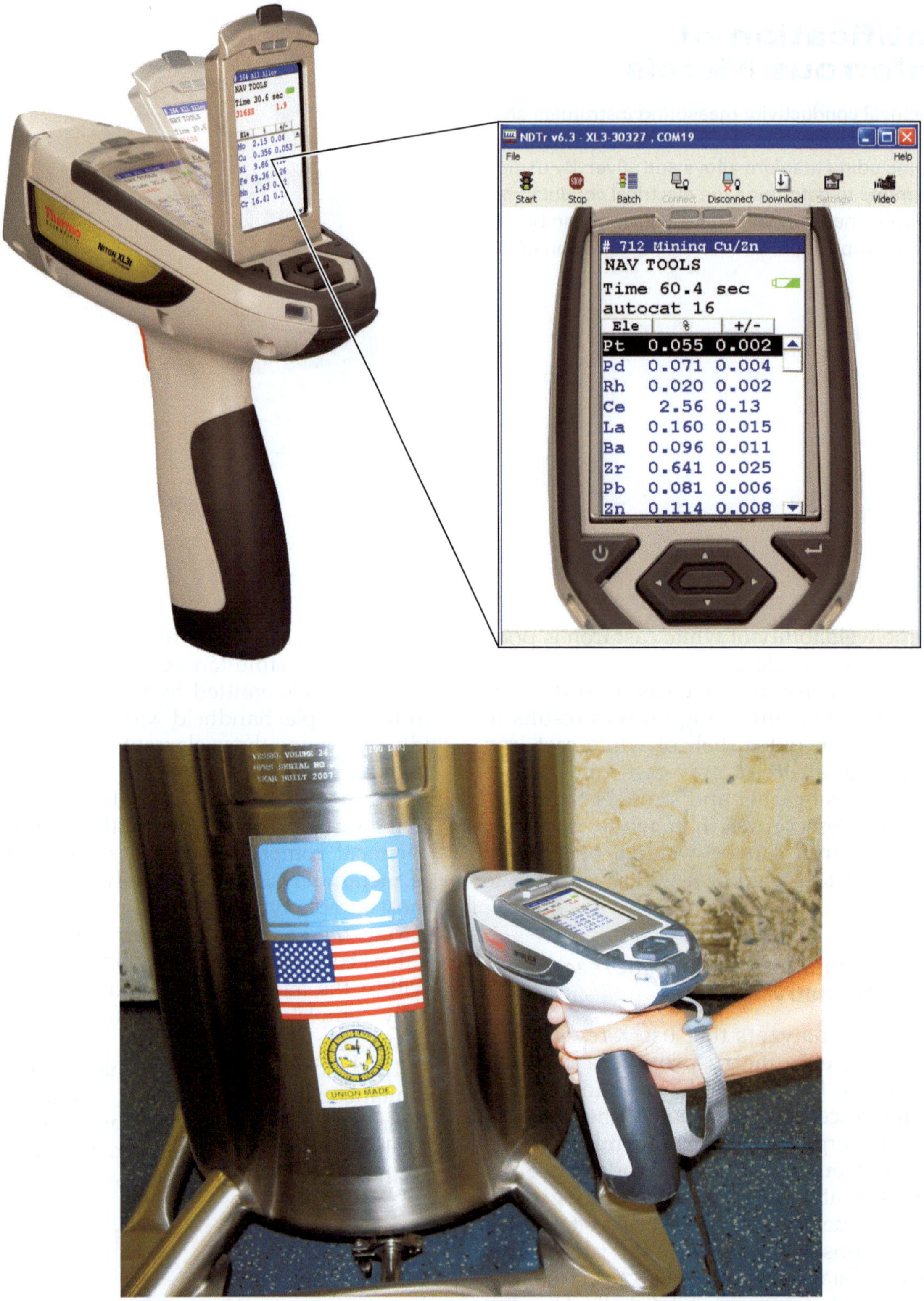

PHOTOS COURTESY OF THERMO FISHER SCIENTIFIC.

Figure 13 NITON XL3 handheld analyzer.

to produce a more desirable weld. The following sections explain the metallurgical factors that must be considered to make acceptable welds:

- Base metal preparation
- Joint design considerations
- Filler metal selection
- Filler metal and electrode selection considerations
- Preheating and interpass temperature control
- Postweld heat treatment

2.4.1 Base Metal Preparation

Because of the defects caused by surface contaminants and the possibility of toxic fumes, all welding codes require surface cleaning prior to any cutting or welding activity. In general, the codes state that the surface to be thermally cut or welded must be clean and free from paint, oil, rust, scale, and other material that would be detrimental to either the weld or the base metal when heat is applied.

The most common defect caused by surface contamination is porosity, which occurs when gas pockets or voids appear in the weld metal. When the porosity has a length greater than its width and is approximately perpendicular to the weld face, it is called piping porosity. Piping porosity is formed as the gas pocket floats toward the surface of the weld, leaving a void. The gas pocket is trapped as the weld metal solidifies, but as the next layer of weld is deposited, it continues to float up. Porosity and piping porosity may not be visible on the surface of the weld.

Removing surface contaminants from base metal is done using mechanical and/or chemical means. Mechanical cleaning is the most common method. It can involve the use of hand tools, power tools, and special sandblasting equipment.

WARNING!

When performing mechanical cleaning, be sure to wear safety glasses and a face shield for protection from the flying particles produced during the cleaning operation. In addition, special clothing is required for sandblasting. Respiratory protection may also be needed.

Several different types of chemicals can be used to clean metal before cutting or welding. Their primary purpose is to clean oil and grease from metal surfaces that require high-quality welds. The cleaning chemical to be used should be selected specifically for the type of base metal and the application. Using the wrong chemical can cause a reaction with the base metal that results in burning, pitting, or discoloration of the surface. All chemicals used for cleaning metal are very strong, which enables them to do their job quickly. Because they are so strong, they can be very hazardous if mishandled.

WARNING!

Never use a chemical without specific and recent training in its use. The mishandling of cleaning chemicals can cause severe burns to the skin, lungs, and eyes. Some chemicals can also cause death or blindness. Always use approved solvents. Procedures and methods for using, storing, and disposing of most chemicals are given in the SDS/MSDS for the specific chemical being used. These are available at the shop or job site.

2.4.2 Joint Design Considerations

Welded joints are selected primarily for the safety and strength required for the service conditions. When selecting the joint, groove, and weld to use, there are many considerations to take into account. The following sections explain those considerations.

Determining the Weight and Density of Materials

At 39°F (4°C), water is at its maximum density of 62.42 lb/ft^3 (≈1,000 kg/m^3). Since water is the standard to which other materials are compared, it is given a specific gravity value of 1.

If the specific gravity of a material is known, its weight can be calculated by multiplying its specific gravity by the weight of an equal volume of water. For example, the specific gravity of aluminum is 2.71. Its weight in pounds per cubic foot (lb/ft^3) can be found using the formula 2.71×62.42 lb/ft^3 = 169.16 lb/ft^3. In other words, one cubic foot of aluminum has a density (weight) of 169.16 pounds. In metric units, the calculation is $2.71 \times 1{,}000$ kg/m^3 = 2,710 kg/m^3.

Similarly, if the weight of a material is known, its specific gravity can be calculated by dividing its weight by the weight of an equal volume of water. For example, one cubic foot of aluminum weighs 169.16 pounds. The specific gravity of aluminum can then be determined using the formula 169.16 lb/ft^3 ÷ 62.42 lb/ft^3 = 2.71. In other words, aluminum has a specific gravity of 2.71. In metric units, the calculation is 2,710 kg/m^3 ÷ 1,000 kg/m^3 = 2.71.

A welding code is a detailed listing of the rules and principles that apply to specific welded products. Codes ensure that safe and reliable welded products will be produced and that the weldments will be reasonably safe. Clients specify which codes to follow when placing orders or letting contracts. These orders or contracts impose severe penalties for not conforming to the code(s) specified. In addition, when codes are specified, the use of these codes is mandated with the force of law by one or more government jurisdictions. Always check the contract, order, or project specification for the code(s) specified.

Codes require that a WPS be written for all critical welds. A WPS is a written set of instructions for producing sound welds. It includes the type of joint to use, as well as the groove preparation, if required. Each WPS is written and tested in accordance with a particular welding code or specification. All welding requires that acceptable industry standards are followed, but not all welds require a WPS. If a weld does require a WPS, it must be followed. The requirement to use a WPS is often listed on project drawings as a note or in the tail of the welding symbol. If you are unsure whether the welding being performed requires a WPS, do not proceed until you check with your supervisor.

One of the easiest joint preparations is the square groove joint because it only requires butting the edges of the plate together. Butt joints are welded with a square groove weld that can have a partial or complete joint penetration weld. A partial joint penetration weld has much less strength than a complete penetration weld. The *AWS D1.1 Structural Welding Code – Steel* imposes restrictions for complete penetration welds. With SMAW, the maximum base metal thickness is ¼ inch (6 mm plate) and welding from both sides is required. In addition, a root opening of half the thickness of the base metal is required, and the root of the first weld must be gouged before the second weld is made. For GMAW and FCAW, the maximum base metal thickness is ⅜ inch (10 mm plate) with the same requirements for welding from both sides and for back-gouging.

Fillet welds also require very little joint preparation. When using a fillet weld on outside corner welds, two types of fit-up can be used: corner to corner or half lap. The corner-to-corner joint is difficult to assemble because neither plate can support the other, and care must be taken not to burn through the corner when welding. The half-lap joint is easier to assemble, requires less weld metal, and has less danger of corner burn-through. The half-lap requires a second weld on the inside of the corner. If a half-lap fit-up is used, allowances in the plate dimensions for the lap must be made. Several methods of joint preparation are explained in the following items:

- *Single- and double-groove welds (Figure 14)* – When possible, use a double-groove joint in place of a single groove. The double groove requires half the weld metal of the single groove. Also, welding from both sides reduces distortion because the forces of distortion work against each other.
- *U- and J-groove preparation* – On thick joints, U- and J-grooves often require less weld metal than V or bevel joints. The major disadvantage of U- and J-grooves is that they require more preparation time. For this reason, V and bevel joints are used much more frequently than J- or U-grooves.
- *Groove angles and root openings (Figure 15)* – The purpose of the groove angle is to allow electrode access to the root of the weld. If the groove angle is larger than necessary, it requires additional weld metal to fill. This increases the time and cost to complete the weld and increases the chance of distortion. If the groove angle is too small, it can result in weld discontinuities. The root preparation is sized to control melt-through. Increasing or decreasing the root preparation (root opening and root face) results in excess melt-through or insufficient root penetration. As a general rule, as the groove angle decreases, the root opening increases to compensate. Root faces are used with open-root joints, but not when metal backing strips are used (except for aluminum).

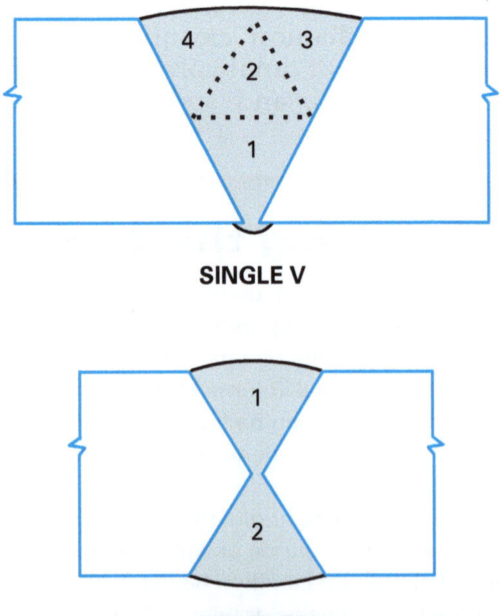

Figure 14 Single- and double-groove welds.

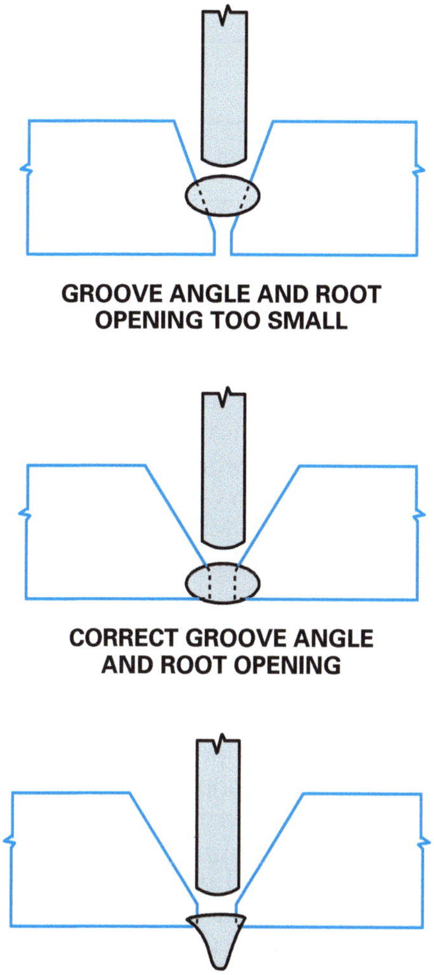

Figure 15 Proper and improper groove angles and root openings.

- *Groove welds with root openings* – For groove welds on plate, the groove angle is 60 degrees, and the maximum size of the root opening and root face is ⅛ inch (about 3.2 mm). For open root welds on pipe, the groove angle is 60 or 75 degrees, depending on the code or specifications used. The maximum size of the root opening and root faces is referred to in the WPS. *Figure 16* shows open-root joint preparation, including the groove angles.

2.4.3 Filler Metal Selection

Because so many different electrodes are available, the welder must be able to distinguish between them and select the correct electrode for the job. The following sections will explain how to select electrodes.

Electrodes are often classified into one of four groups based on their general characteristics:

- Fast-freeze

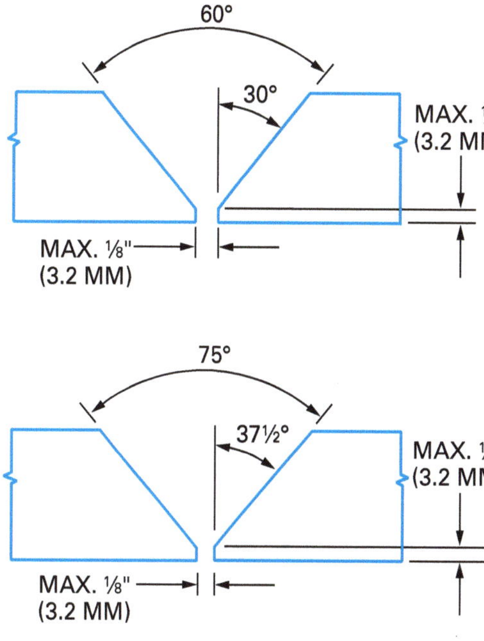

Figure 16 Open-root joint preparation.

- Fast-fill
- Fill-freeze
- Low-hydrogen

These four groups are identified by F-numbers. Refer to *Table 20* as you progress through the discussion of each group and their characteristics. The mild steel electrodes used with SMAW are classified as F1, F2, F3, and F4. Qualification with an electrode in a particular F-number classification qualifies the welder with all electrodes identified in that classification and in lower F-number classifications. For example, a typical AWS welder qualification test is a V-groove weld with metal backing in the 3G and 4G positions using an F4 electrode. Passing this test qualifies the welder to weld with F4 and all lower F-number electrodes, and to make groove and fillet welds in all positions.

Fast-fill electrodes (F1) have powdered iron in the flux, which gives them high deposition rates but requires that they only be used for flat welds and horizontal fillet welds. These electrodes have shallow penetration with excellent weld appearance and almost no spatter. The heavy slag covering is easy to remove. Electrodes in the fast-fill group will also have a 2 as the next-to-last number, as shown in *Table 20*. They include electrodes such as the following:

- E6027
- E7024
- E7028 (also in the low-hydrogen F4 group)

Typical applications for fast-fill electrodes are production welds on plate more than ¼" thick and flat and horizontal fillets and lap welds.

Table 20 AWS F-Number Electrode Classification

Group	AWS Electrode Classification				
F4	EXX15	EXX16	EXX18		Low-Hydrogen
F3	EXX10	EXX11			Fast-Freeze
F2	EXX12	EXX13	EXX14		Fill-Freeze
F1	EXX20	EXX24	EXX27	EXX28	Fast-Fill

Fill-freeze electrodes (F2), also called fast-follow electrodes, are all-position electrodes that are commonly used for vertical and flat welding. They have medium deposition rates and penetration and are excellent for sheet metal. The weld bead ranges from smooth and ripple-free to even with distinct ripples. The medium slag covering is easy to remove.

Electrodes in the fill-freeze group include the following:

- E6012
- E6013
- E7014

Typical applications for fill-freeze electrodes are the following:

- Vertical fillet or lap welds
- Sheet metal lap and fillet welds
- Joints having poor fit-up in the flat position
- General purpose welding

Fast-freeze electrodes (F3) are all-position electrodes that provide deep penetration. However, the arc creates a lot of fine weld spatter. The weld bead is flat with distinct ripples and a light slag coating that can be difficult to remove. The arc is easy to control, making fast-freeze electrodes good for overhead and vertical welding. These more challenging positions often make it harder for the welder to maintain a precise arc gap. Electrodes in the fast-freeze group include the following:

- E6010
- E7010
- E6011

Typical applications for fast-freeze electrodes are as follows:

- Vertical and overhead plate welding
- Pipe welding (cross-country and in-plant)
- Welds made on galvanized, plated, or painted surfaces
- Joints requiring deep penetration, such as square butts
- Sheet metal welds

WARNING!

Many surface coatings, such as galvanized coatings, give off very toxic fumes when heated, cut, or welded. Fumes from certain types of electrodes may also be hazardous. Breathing these fumes can cause damage to the lungs and possibly other organs. Identify any hazard for the coatings with which you are working by referring to the SDS/MSDS for that particular coating. Also, read the SDS/MSDS contained in the electrode package. Use proper ventilation and take the necessary precautions to protect yourself and other workers in the immediate area.

Low-hydrogen electrodes (F4) are designed for welding high-sulphur, phosphorus, and medium- to high-carbon steels, which have a tendency to develop porosity and underbead cracking. These problems are caused by hydrogen that is absorbed during welding. Low-hydrogen electrodes are also available with fast-fill and fill-freeze characteristics. They produce dense, X-ray-quality welds with excellent notch toughness and ductility. Electrodes in the low-hydrogen group include the following:

- E7015
- E7016
- E7018
- E7028
- E7048

Typical applications for low-hydrogen electrodes are the following:

- X-ray quality welds
- Welds requiring high mechanical properties
- Crack-resistant welds in medium- to high-carbon steels
- Welds that resist hot-short cracking in phosphorus steels
- Welds that minimize porosity in sulphur steels
- Welds in thick or highly restrained mild or alloy steels to minimize the danger of weld cracking
- Welds in alloy steels requiring a strength of 70,000 psi (≈4,826 bars) or more
- Multiple-pass welds in all positions

If hydrogen is present in the weld zone, it will create defects. The greatest source of hydrogen is airborne water vapor, which is composed of oxygen and hydrogen. Low-hydrogen electrodes are carefully manufactured to eliminate moisture in the flux coating. Low hydrogen is defined as being less than 16 milliliters (ml) per 100 grams of weld metal. The amount of hydrogen in these electrodes varies from 4 to 16 ml per 100 grams of metal weld.

Low-hydrogen electrodes have two unique suffixes. The M suffix is for compliance. Most military codes require an M suffix. When the suffix H and a number are added, the numbers relate to the amount of hydrogen in the electrodes. As a result, the electrodes will have suffixes such as H4, H8, and H16. These three suffixes indicate the dryness of the electrodes, identifying the milliliters of hydrogen per 100 grams of metal weld they contain. An additional suffix may also be added to these dryness suffixes. A suffix such as H4R would be one such additional suffix. The R means indicates they are moisture resistant. Without an R suffix, the electrodes may only be exposed to the atmosphere for a maximum of 4 hours. Those with an R suffix have an allowable exposure of up to 8 hours.

If low-hydrogen electrodes are exposed to the atmosphere, they will begin to absorb moisture and will no longer be low-hydrogen as a result. This is the reason for their short atmospheric exposure limits. For this reason, all low-hydrogen electrodes are shipped in hermetically sealed metal containers. Once opened, the electrodes must be stored in a heated oven. The temperature range of this oven is 100°F to 550°F (38°C to 228°C). However, it is important to note that placing the oven in areas below freezing may not allow it to maintain its maximum temperature, if that setting is necessary. Additional guidelines for the storage and handling of low-hydrogen electrodes are covered later in this module.

CAUTION

The welding codes establish strict guidelines for the storage, handling, and re-drying of low-hydrogen electrodes. These procedures will be referenced in your site quality standards and the WPS. Severe penalties will result from violation of these standards. Improperly storing these electrodes will lead to poor-quality welds and the possibility of having numerous welds rejected. Be sure you read and understand your site standards before working with low-hydrogen electrodes.

2.4.4 Filler Metal and Electrode Selection Considerations

If there is a WPS for the weld to be made, the filler metal will be specified. It will be given as an AWS classification and/or manufacturer's standard and will sometimes include a manufacturer's name.

When selecting filler metals and electrodes not specified in a WPS, consider the factors explained in the following sections.

The filler metal should be compatible with the base metal's chemical composition and mechanical properties. The chemical composition and mechanical properties of the base metal can be obtained from the base metal mill specifications. The mill specifications can then be compared to various tables that list base metal and filler metal acceptability.

A weld made on mild steel with a high-tensile electrode, such as E12018, is weaker than one made with E7018. The high-tensile weld is so strong that when stress is placed on the weld, it does not give. It transfers all stress to the heat-affected zone along the weld. This can cause cracking. A weld made with E7018 will give along with the base metal and distribute the stress across the weld, which makes a stronger joint. The opposite occurs when mild-steel electrodes are used to weld high-strength steel. All the stress occurs in the weld because the surrounding base metal is so much stronger. This causes the weld to fail prematurely.

Always use an electrode smaller in diameter than the thickness of the metal to be welded. For sheet metal, 3⁄32-inch (2.4 mm) E6013 is recommended because it has shallow penetration. To minimize the joint preparation, a deep penetration E6010 or E7010 electrode can be used for the root pass. Before welding, always clean the base metal surface to remove surface corrosion, plating, coatings, and paint.

The welding position is important because some electrodes, such as E7024 and E7028, can only be used in the flat and horizontal fillet weld positions. In addition, never attempt to weld in vertical or overhead positions with electrodes larger than 3⁄16 inch (4.8 mm), except for low-hydrogen electrodes. The recommended limitation for low-hydrogen is 5⁄32 inch (4 mm).

The type of joint to be welded is also important to consider when selecting an electrode. When lap and T joints can be positioned flat, they are usually welded with E7024 or E7028. Weld butt joints that need deep penetration using E6010. If the joint has poor fit-up, use a shallow-penetrating electrode such as E6012 or E6013. If low-hydrogen electrodes are used for open-root joints, back-gouging may be required. To eliminate the need for back-gouging and to make the root pass easier to run, use metal backing, or put the root in with E6010 and then fill out with low-hydrogen.

2.4.5 Preheating and Interpass Temperature Control

Preheating and interpass temperature control procedures are sometimes performed to preserve weldment strength, ductility, and weld quality. Proper preheating and interpass temperature control can reduce shrinkage, prevent excessive hardening, and prevent underbead cracking. Interpass temperature control is also required during certain welding procedures to prevent or reduce localized hardening and shrinkage. If certain metals cool too quickly between weld passes, undesirable changes occur in the base and weld metals, and the heat from successive passes will not reverse these changes. Preheating and interpass temperature heat treatment of metals is described in detail in module titled, *Preheating and Postheating of Metals*.

2.4.6 Postweld Heat Treatment

Postweld heat treatment (PWHT) is the heating of a weldment or assembly after welding has been completed. It can be used on most types of base metals. Postweld heat treatment may be necessary to reduce **residual stresses** and to improve corrosion resistance, dimensional stability, fatigue resistance, and impact and low-temperature strength. Postweld heat treatment equipment and techniques include furnaces, banding, and localized heat treatment. Postweld heat treatments include stress relieving, annealing, normalizing, and tempering. Postweld heat treatment of metals is described in detail in *Preheating and Postheating of Metals*.

Additional Resources

AWS B1.10M/B1.10 Guide for the Nondestructive Examination of Welds. Miami, FL: American Welding Society.

AWS B1.11 Guide for the Visual Examination of Welds. Miami, FL: American Welding Society.

AWS D3.5-93R Guide for Steel Hull Welding. Miami, FL: American Welding Society.

AWS D3.6M Specification for Underwater Welding. Miami, FL: American Welding Society.

AWS D3.7 Guide for Aluminum Hull Welding. Miami, FL: American Welding Society.

Lincoln Electric website: **http://www.lincolnelectric.com** offers sources for products and training.

Machinery's Handbook. Erik Oberg, Franklin D. Jones, and Christopher J. McCauley. New York, NY: Industrial Press, Inc. 2004.

Metals and How to Weld Them. T. B. Jefferson. Cleveland, OH: The James F. Lincoln Arc Welding Foundation.

OSHA 1926.351, Arc Welding and Cutting. Current Edition. Washington, DC: Occupational Safety and Health Administration (OSHA).

Stick Electrode Product Catalog. Cleveland, OH: The Lincoln Electric Company, 2008.

Stick Electrode Welding Guide. Cleveland, OH: The Lincoln Electric Company, 2004.

The Procedure Handbook of Arc Welding. 14th Edition. Cleveland, OH: The James F. Lincoln Welding Foundation, 2000.

Welding Handbook, 4 vols. Miami, FL: American Welding Society.

2.0.0 Section Review

1. The mass, or weight, of a specific volume of metal is commonly expressed as ____.
   a. purity
   b. hardness
   c. density
   d. resistivity

2. A type of stress in which opposed, in-line forces pull away from each other at opposite sides of a specimen is ____.
   a. torsion
   b. compression
   c. shear
   d. tension

3. A handheld analyzer that can be used to accurately identify most types of metal relies on the use of ____.
   a. X-ray fluorescence spectrometry
   b. magnetic resonance imagery
   c. the modulus of elasticity
   d. tension loading ductility

4. The type of electrode that is best suited for steel that has a tendency to develop porosity and underbead cracking is ____.
   a. fast-freeze
   b. low-hydrogen
   c. fill-freeze
   d. fast-fill

Section Three

3.0.0 Common Structural Steel Shapes

Objective

Identify the common structural steel shapes of metal.

a. Identify the most common structural steel shapes.
b. Identify different structural beam shapes.
c. Identify pipe and tubing types.
d. Identify other common metal forms, including rebar.

Trade Term

Sintered: A process of making parts from a powdered metal in which the metal is molded and then heated in a furnace without melting to fuse it into a solid metal.

Metals are commercially available in many shapes and sizes. They can be shaped by casting, rolling, drawing, forging, cutting, machining, extruding, and spinning. Structural steel and other common structural shapes are formed by rolling hot or cold metal between a succession of rollers. The rollers are specially configured for the shapes they will produce.

3.1.0 Common Structural Shapes

Structural steel includes a wide range of steel types, classes, and shapes. To ensure uniform standards, the ASTM specifies the properties of strength, weight, corrosion resistance, and weldability for various steel classifications. *Table 21* lists some typical structural steel types and classifications. Note that the minimum yield stress is expressed in the table as ksi. This is one of the standard notations used for structural steel. The term *ksi* is defined as kips per square inch, where kips means kilo-pounds (1,000 pounds) of dead load. Hence, a yield stress of 36 ksi means 36,000 pounds per square inch of dead load.

The shape or type of structural steel is identified on drawings by a symbol or abbreviation. Sizes and dimensions are always given in a specified order. The specification format for designating shape and size is shown with the relevant shape in the following sections.

NOTE

It is important to remember that dimensions for various structural steel shapes that are given in US customary units (such as inches) do not directly correspond to metric dimensions for those same shapes. In other words, a steel plate that measures ¼" × 4' × 10' will not have an exact metric equivalent. An actual metric steel plate size that is close to the same size might measure 6 mm × 1,250 mm × 3,000 mm.

3.1.1 Plate

Plate is rolled metal with a uniform thickness equal to or greater than 3⁄16 inch (4.8 mm), and its identifying symbol is PL. Its thickness is given in inches or millimeters, and its length and width are typically listed in feet or meters.

Example specification:

PL ¼" × 4' × 10'

3.1.2 Sheet Metal

Sheet metal is rolled metal with a uniform thickness less than 3⁄16 inch (4.8 mm). Thickness is given in inches or millimeters, or indicated by an American Wire Gauge (AWG) number. Length and width are given in feet or a metric equivalent. Sheet metal is often rolled into coils.

Example specification:

Sheet No. 12 AWG × 4' × 8'

3.1.3 Bars

Bars are rolled to a variety of cross-sectional shapes and sizes (*Figure 17*). Standard bar shapes include the following:

- *Round* – Round bar is specified by BAR, followed by the bar diameter, often with a diameter symbol (a circle with a slash through it), and then the length.
- *Square* – Square bar is specified by BAR, followed by the face width, often with a face width symbol (a square with a slash through it), and then the length.
- *Bar (flat)* – Flat bar is specified by BAR, followed by the width dimension, the thickness dimension, and then the length.
- *Z-bar* – Z-bar is specified by Z, followed by the flange width, the web depth, the flange and web thickness, and then the length. The flange and the web are always the same thickness.
- *Hexagonal* – Hexagonal bar is specified by HEX, followed by the bar thickness (measured across the flats), and then the length.

Table 21 Common Types and Classifications of Structural Steel

Steel Type	ASTM Class	Minimum Yield Stress (ksi)	Form		Remarks
Carbon	A 36	36	Plate Shapes Bars	Sheets, strips Rivets Bolts, nuts	For buildings and general structures; available in high-toughness grades
	A529	42	Plate Shapes	Bars	For buildings and similar construction
High-strength	A440	42 to 50	Plate Shapes	Bars	Lightweight and superior corrosion resistance
High-strength, low-alloy	A 441	40 to 50	Plates Shapes	Bars	Primarily for lightweight welded buildings and bridges
	A 572	42 to 65	Several types Plate	Shapes Bars	Lightweight high-toughness for buildings, bridges, and similar structures
Corrosion-resistant, high-strength, low-alloy	A 242	42 to 50	Plate Shapes	Bars	Lightweight and added durability weathering grades available
	A 588	42 to 50	Plate Shapes	Bars	Lightweight, durable in high thickness; weathering grades available
Quenched and tempered alloy	A 514	90 to 100	Several types Plate	Shapes Bars	Strength varies with thickness and type

- *Octagonal* – Octagonal bar is specified by OCT, followed by the bar thickness (measured across the flats), and then the length.

3.1.4 Angles

Angles are L-shaped bars that have either equal-sized or unequal-sized legs that are always at 90 degrees to each other. *Figure 18* shows both types of angle bar stock and the specification formats for each type. The dimensions for the angle bar stock examples are given in US customary units and metric units. Remember, though, these two units of measurement do not directly correspond. The specification format represents the size and thickness of the angle bar stock legs. For example, a specification of L2 × 2 × ⅜ would indicate a 2" × 2" angle of ⅜" thickness in US customary units. For unequal leg stock, the longest leg is listed first, as shown in *Figure 18*.

3.1.5 Channels

Channels are U-shaped forms made of two flanges connected by a common web. The flanges extend from the same side of the web. The flanges can be of uniform thickness or tapered toward the outer edges.

There are several channel variations, each having a unique specification format designation. *Figure 19* shows two different channels and their related American Standard Channels specification format designations. It also shows similar channels with metric dimensions. The channel C8 × 11.5 has a nominal depth of 8 inches and a weight of 11.5 pounds per foot of length. Similarly, the channel C8 × 18.75 has a nominal depth of 8 inches and a weight of 18.75 pounds per foot of length. The metric channels both have a nominal depth of 200 mm. The thinner channel has a weight of 22 kg per meter of length, while the thicker channel has a weight of 25.30 kg per meter of length. As shown in the figure, the channel C8 × 11.5 has a flange width of 2¼ inches and a web thickness of ¼ inch; while the C8 × 18.75 channel has a flange width of 2¼ inches and a web thickness of ½ inch. The metric channels both have a flange width of 75 mm. The thinner channel has a web thickness of 10.5 mm, while the thicker channel has a web thickness of 11.5 mm. The specific dimensions for the beam depth, flange width, and web thickness for all common channel designations can be found in related tables of dimensions and properties in most structural engineering and architectural standards reference books.

SHAPE		SPECIFICATION FORMAT
ROUND	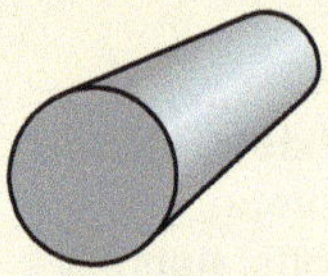	BAR (DIAMETER) ×(LENGTH) BAR ¾" ∅ × 8'3" OR BAR 19 ∅ × 3 M
SQUARE	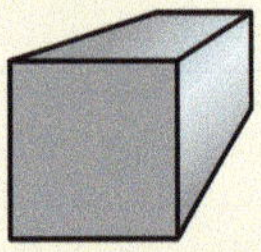	BAR (FACE WIDTH ⧄) ×(LENGTH) BAR ¾" ×8'3" OR BAR 19⧄ ×3 M
BAR	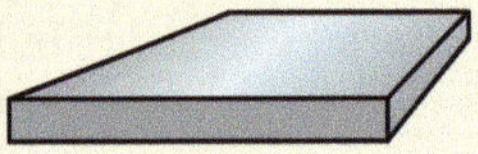	BAR (WIDTH) ×(THICKNESS) ×(LENGTH) BAR 2" ×¼"×8'3" OR BAR 50 MM ×6 MM ×3 M
Z-BAR	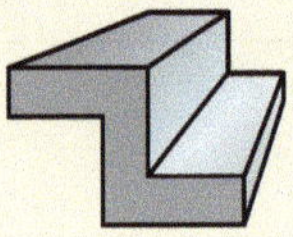	Z (FLANGE WIDTH) ×(WEB DEPTH) ×(FLANGE AND WEB THICKNESS) ×(LENGTH) Z 2" ×2"×½"×8'3" OR Z 50 MM ×50 MM×13 MM ×3 M
HEXAGONAL	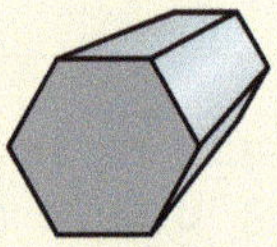	HEX (BAR THICKNESS) ×(LENGTH) HEX ⅞"×10' OR HEX 22 MM ×3 M
OCTAGONAL	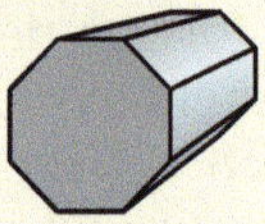	OCT (BAR THICKNESS) ×(LENGTH) OCT 1"×6' OR OCT 25 MM ×2 M

Figure 17 Bar shapes and specification formats.

3.2.0 Beams and Shapes from Beams

Beams (*Figure 20*) are made in I-shaped, H-shaped, and T-shaped cross sections, and they are made with flat or tapered flanges. In addition to different flange widths and web depths, the thickness of the flanges and webs varies with beam size. Also, beam weight (per linear foot or meter) for a given dimension can be increased by adding thickness to the web and flanges with very little change in beam width or depth. The specification designator describes the beam type, its nominal depth, and its weight per foot of length or kilogram per meter. An example specification designator for a beam is S24 × 120, where S24 represents the beam type (S) and the nominal depth in inches (24). The 120 value is the weight in pounds per foot of length. A comparable metric beam might have a 300 mm nominal depth and have a weight of 42.2 kilograms per meter. The specific dimensions for the depth, web thickness, and flange thickness for the various designations of beams are given in related tables of dimensions and properties in most structural engineering and architectural standards books.

I-Beams

The older American Standard I-beams have been replaced by the American Institute of Steel Construction (AISC) wide flange (WF) beams. These wide flange shapes are now used for most beams and columns. The wider flange provides a stronger beam than the older I-beam with the same web width.

Beams are identified with a standard set of symbols. The I-beam and H-beam letter symbols include the following:

- *S (American Standard)* – I-shaped beams with tapered flanges
- *WF (Wide flange)* – Wider flanges than S beams with thinner webs and nontapered flanges
- *M* – Similar to WF beams but with short flanges
- *HP* – H-shaped beams with nontapered flanges

T-beams include structural tees and tee shapes. Structural tees are made by cutting S, W, and M beams down the center, usually by shearing. The conventional symbol for a structural tee is the original beam letter followed by a T. For example, a tee cut from an S8 × 18.4 beam would be identified as ST4 × 9.2 in US customary units.

T-shapes are rolled into their final tee shape. The symbol for tee shapes is a T, without any other letter. The specification gives the nominal depth, flange width, thickness, and length. An example specification is T3 × 7.8. T-shapes are short pieces, commonly used as connections or supports.

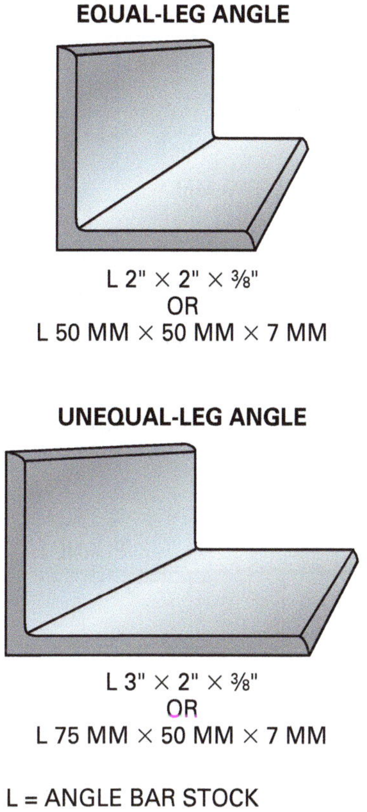

Figure 18 Angle bar stock and specification formats.

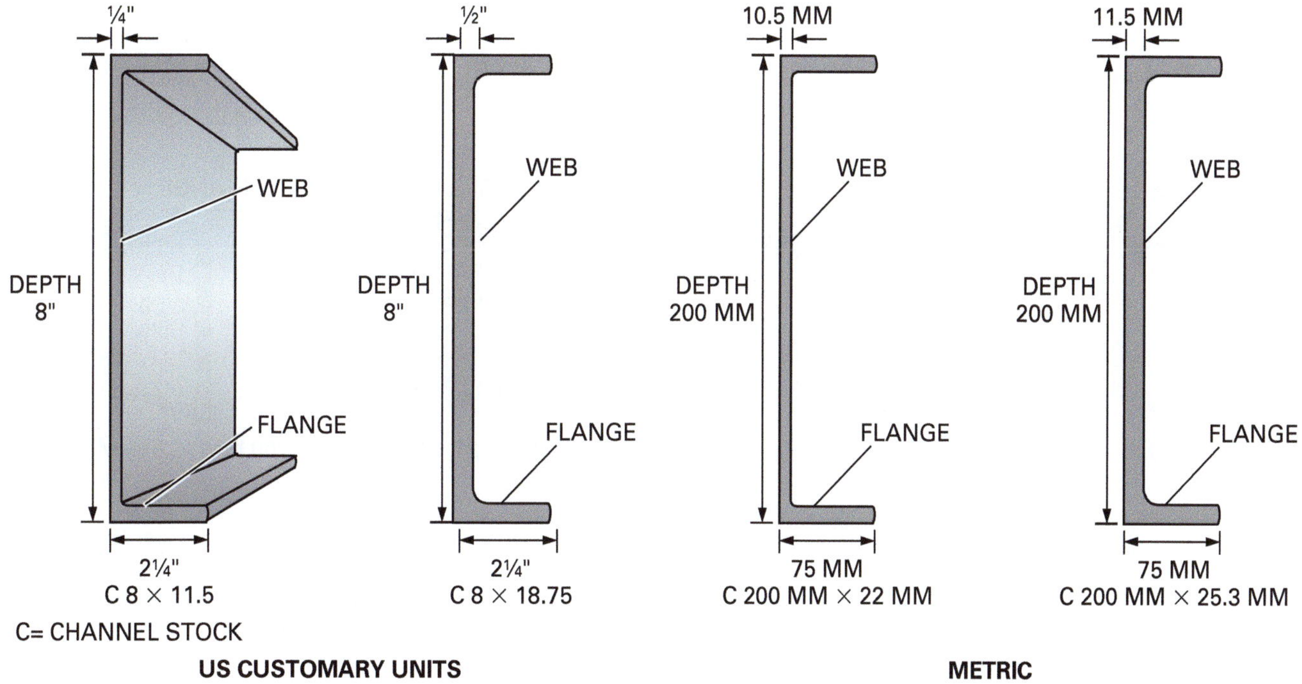

Figure 19 Channels and specification formats.

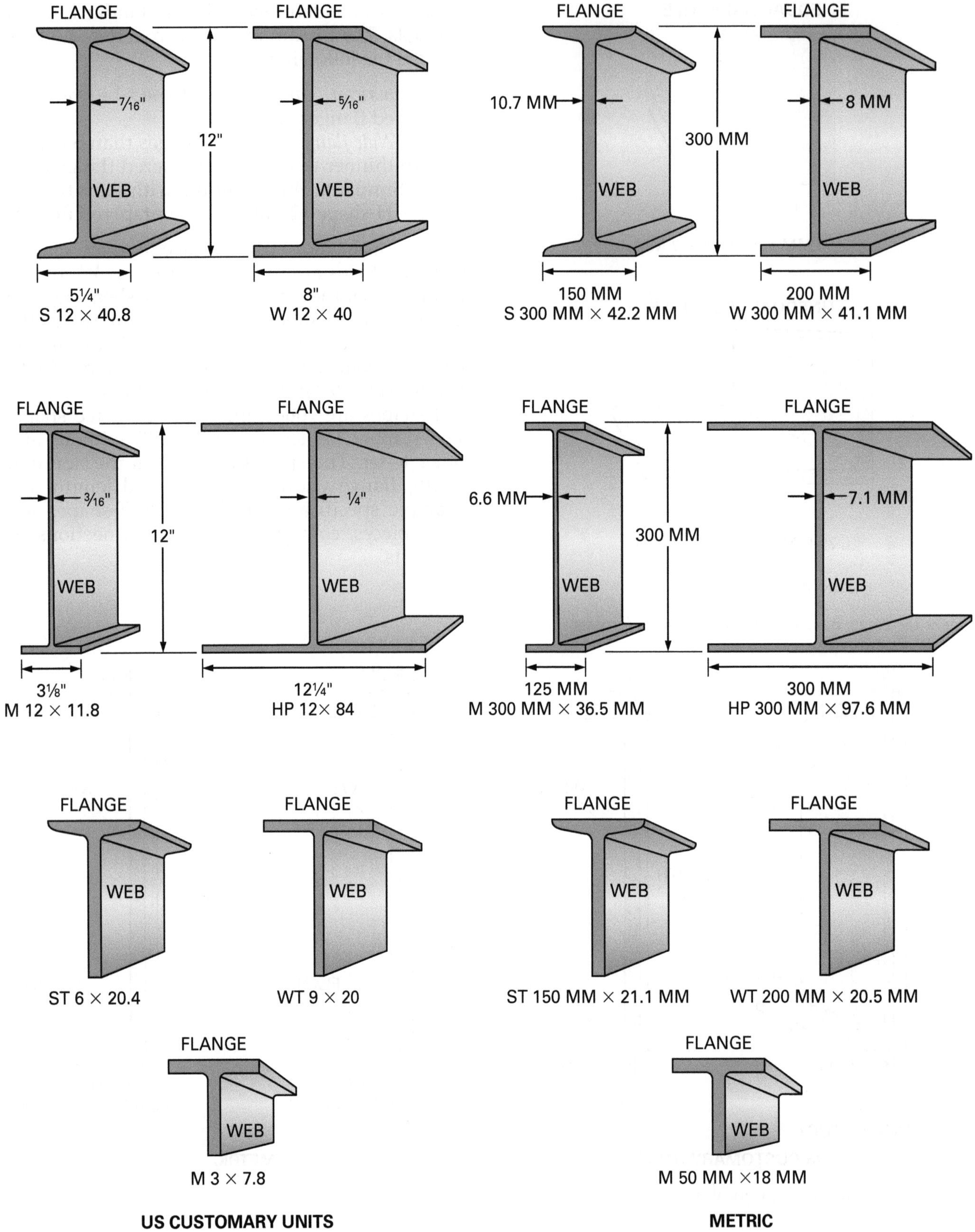

Figure 20 Structural beam and T-beam shapes.

3.3.0 Pipe and Tubing

Pipe and tubing share some basic similarities, but they are identified using different measuring methods and standards. This section focuses on various sizes of pipe, some common tubing shapes, and different methods for producing pipe and tubing.

3.3.1 Pipe

Steel pipe size is specified as a nominal inside diameter (ID) up through 12 inches (or 300 mm). The word *nominal* is used to identify the design size, but the actual inside diameter will not be exactly that given size. Pipe 14 inches (350 mm) and larger is specified by the actual outside diameter (OD). Pipe weight, which is determined by the wall thickness, is specified by a schedule number or strength. Common schedule numbers are 5, 10, 20, 30, 40, 60, 80, 100, 120, 140, and 160. Strengths are specified as standard (STD), extra strong (XS), and extra-extra strong (XXS). Standard falls between Schedules 30 and 40. Extra-strong falls between Schedules 60 and 80. Extra-extra strong is Schedule 160 or higher.

Because the OD of a given pipe size is always the same, the ID becomes smaller as the pipe schedule increases. The additional wall thickness is taken from the ID while the OD remains fixed. *Table 22* shows common pipe sizes and schedules with nominal wall thicknesses in US customary units and similarly sized metric pipe measurements. As with other steel shapes, nominal size dimensions for pipe that are given in US customary units do not directly correspond to metric pipe. For instance, a 12-inch pipe is not identical to a 300-millimeter pipe.

3.3.2 Tubing

Tubing is manufactured in square, rectangular, and round cross sections. Round tubing can be distinguished from pipe by its dimensions. Unlike pipe, standard round tubing is measured by its actual OD.

The cross sections of square and rectangular tubing have even outside dimensions and slightly rounded corners. *Figure 21* shows three standard tubing shapes and their specification formats in inches, as well as three similarly sized metric tubing shapes.

3.3.3 Seamed and Seamless Tubing and Pipe

Both standard and nonstandard shapes can also be produced by other processes. For example, tubing can be produced by extrusion or by rolling and welding. Extrusions can be produced in an almost limitless variety of shapes. When tubing is produced by rolling and welding, the most common welding process used is electronic resistance welding (ERW).

3.4.0 Other Common Forms

In addition to common structural steel shapes, beams, pipe, and tubing, there are other forms of steel that are commonly used in industry. These other forms include reinforcing bars, forged objects, cast objects, and powdered metals.

Why Doesn't the Pipe Size Match Pipe ID or OD?

When pipe was first made, it was made from lead, which required a thick wall to hold the specific pressure for a given ID. When the switch was made from lead pipe to steel pipe, a decision had to be made whether to maintain the OD so the steel pipe would fit existing fittings, or to change the OD and keep the ID size. The decision was made to maintain the OD to provide direct replacement with the existing fittings. At that point, neither nominal ID nor OD corresponded to size.

Table 22 Standard Pipe Sizes and Schedules with Nominal Wall Thicknesses

		Nominal Wall Thickness (in)												
Nom. Pipe Size	Outside Diam.	Sched. 10	Sched. 20	Sched. 30	STD	Sched. 40	Sched. 60	XS	Sched. 80	Sched. 100	Sched. 120	Sched. 140	Sched. 160	XXS
⅛	0.405	–	–	–	0.068	0.068	–	0.095	0.095	–	–	–	–	–
¼	0.540	–	–	–	0.088	0.088	–	0.119	0.119	–	–	–	–	–
⅜	0.675	–	–	–	0.091	0.091	–	0.126	0.126	–	–	–	–	–
½	0.840	–	–	–	0.109	0.109	–	0.147	0.147	–	–	–	0.188	0.294
¾	1.050	–	–	–	0.113	0.113	–	0.154	0.154	–	–	–	0.219	0.308
1	1.315	–	–	–	0.133	0.133	–	0.179	0.179	–	–	–	0.250	0.358
1¼	1.660	–	–	–	0.140	0.140	–	0.191	0.191	–	–	–	0.250	0.382
1½	1.900	–	–	–	0.145	0.145	–	0.200	0.200	–	–	–	0.281	0.400
2	2.375	–	–	–	0.154	0.154	–	0.218	0.218	–	–	–	0.344	0.436
2½	2.875	–	–	–	0.203	0.203	–	0.276	0.276	–	–	–	0.375	0.552
3	3.500	–	–	–	0.216	0.216	–	0.300	0.300	–	–	–	0.438	0.600
3½	4.000	–	–	–	0.226	0.226	–	0.318	0.318	–	–	–	–	–
4	4.500	–	–	–	0.237	0.237	–	0.337	0.337	–	0.438	–	0.531	0.674
5	5.563	–	–	–	0.258	0.258	–	0.375	0.375	–	0.500	–	0.625	0.750
6	6.625	–	–	–	0.280	0.280	–	0.432	0.432	–	0.562	–	0.719	0.864
8	8.625	–	0.250	0.277	0.322	0.322	0.406	0.500	0.500	0.594	0.719	0.812	0.906	0.875
10	10.750	–	0.250	0.307	0.365	0.365	0.500	0.500	0.594	0.719	0.844	1.000	1.125	1.000
12	12.750	–	0.250	0.330	0.375	0.406	0.562	0.500	0.688	0.844	1.000	1.125	1.312	1.000
14 OD	14.000	0.250	0.312	0.375	0.375	0.438	0.594	0.500	0.750	0.938	1.094	1.250	1.406	–
16 OD	16.000	0.250	0.312	0.375	0.375	0.500	0.656	0.500	0.844	1.031	1.219	1.438	1.594	–
18 OD	18.000	0.250	0.312	0.438	0.375	0.562	0.750	0.500	0.938	1.156	1.375	1.562	1.781	–
20 OD	20.000	0.250	0.375	0.500	0.375	0.594	0.812	0.500	1.031	1.281	1.500	1.750	1.969	–
22 OD	22.000	0.250	0.375	0.500	0.375	–	0.875	0.500	1.125	1.375	1.625	1.875	2.125	–
24 OD	24.000	0.250	0.375	0.562	0.375	0.688	0.969	0.500	1.218	1.531	1.812	2.062	2.344	–
26 OD	26.000	0.312	0.500	–	0.375	–	–	0.500	–	–	–	–	–	–
28 OD	28.000	0.312	0.500	0.625	0.375	–	–	0.500	–	–	–	–	–	–
30 OD	30.000	0.312	0.500	0.625	0.375	–	–	0.500	–	–	–	–	–	–
32 OD	32.000	0.312	0.500	0.625	0.375	0.688	–	0.500	–	–	–	–	–	–
34 OD	34.000	0.312	0.500	0.625	0.375	0.688	–	0.500	–	–	–	–	–	–
36 OD	36.000	0.312	0.500	0.625	0.375	0.750	–	0.500	–	–	–	–	–	–
42 OD	42.000	–	–	–	0.375	–	–	0.500	–	–	–	–	–	–

3.4.1 Reinforcing Bars

Reinforcing bars (rebars), sometimes called re-rods, are used for concrete reinforcement. They are available in several grades. These grades vary in yield strength, ultimate strength, percentage of elongation, bend-test requirements, and chemical composition. Reinforcing bars can be coated with different compounds, such as epoxy, for use in concrete where corrosion can be a problem.

The ASTM International has established standard specifications for reinforcing bars. These grades appear on bar-bundle tags, in color coding, in rolled-on markings on the bars, and/or on bills of materials. The ASTM specifications are as follows:

- *A615, Standard Specification for Deformed and Plain Billet-Steel Bars for Concrete Reinforcement*
- *A706, Standard Specification for Low-Alloy Steel Deformed and Plain Bars for Concrete Reinforcement*
- *A996, Standard Specification for Rail-Steel and Axle-Steel Deformed Bars for Concrete Reinforcement*

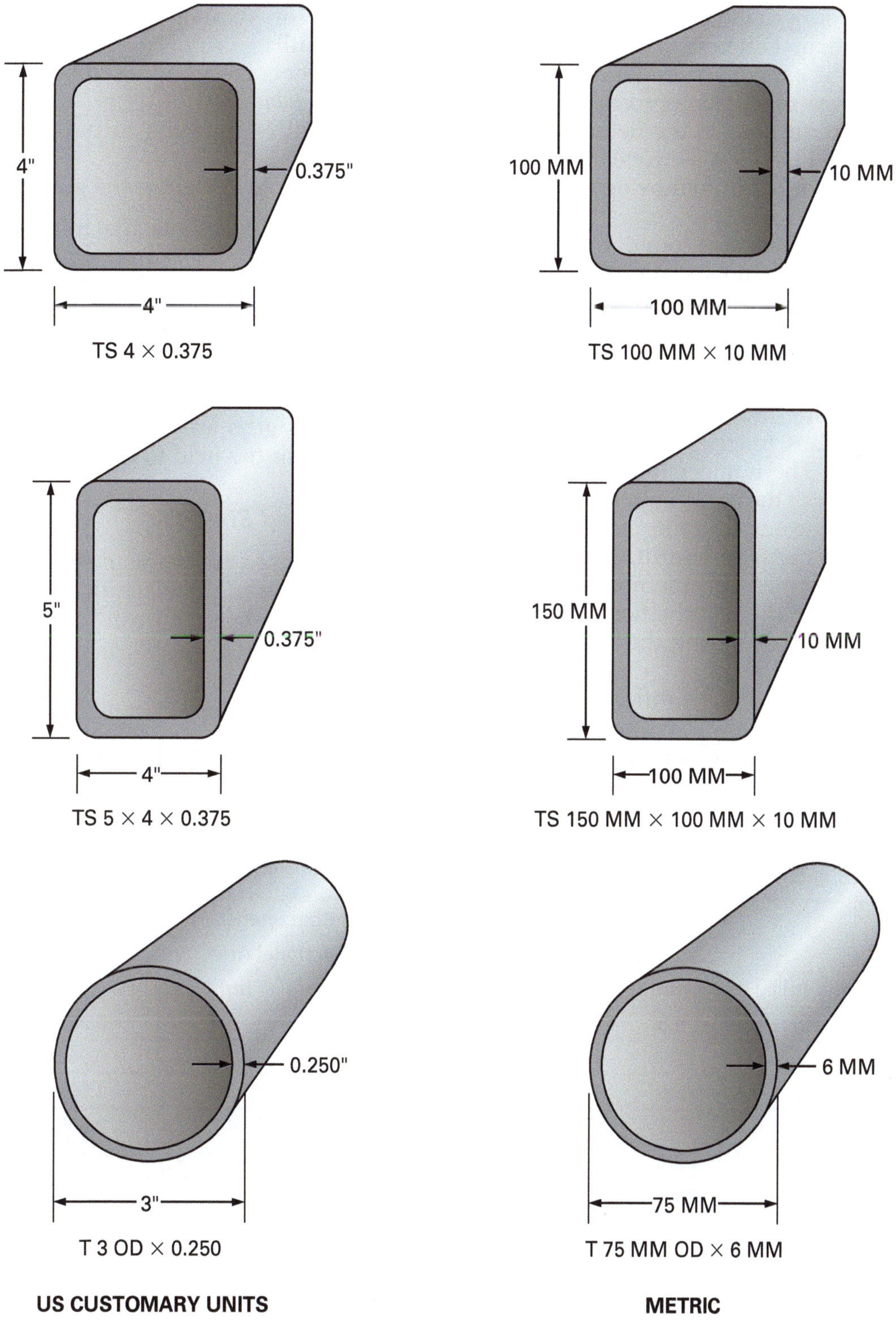

Figure 21 Standard tubing shapes and specification formats.

The standard configuration for reinforcing bars is the deformed bar. Different patterns may be impressed upon the bars depending on the manufacturer, but all are rolled to conform to ASTM specifications. The purpose of the deformation is to improve the bond between the concrete and the bar and to prevent the bar from moving within the concrete.

Plain bars are smooth and round without deformations on them. Plain bars are used for special purposes: dowels at expansion joints, where the bars must slide in a sleeve; for expansion and contraction joints in highway pavement; and for column spirals.

Deformed bars are designated by a number in eleven standard sizes (metric or inch-pound) as shown in *Table 23*. The number indicates the approximate diameter of the bar in eighths of an inch or millimeters. For example, a #16 metric bar has an approximate diameter of 16 mm. The nominal dimension of a deformed bar (the nominal diameter does not include the deformation) is equivalent to that of a plain bar having the same weight per foot.

As shown in *Figure 22*, bar identification is made by ASTM specifications; these require that each bar manufacturer roll the following information onto the bar:

- A letter or symbol to indicate the manufacturer's mill
- A number corresponding to the size number of the bar (*Table 23*)
- A symbol or marking to indicate the type of steel (*Table 24*)
- A marking to designate the grade (*Table 25*)

The grade represents the minimum yield or tension strength, measured in kips per square inch (ksi) or megapascals (MPa), that the type of steel will withstand before it permanently stretches (elongates). Today, Grade 420 is the most commonly used rebar. Rebars are normally supplied from the mill bundled in 60-foot lengths (12-meter lengths are common for metric rebar) that are cut in the field to the required length.

3.4.2 Forged Shapes

Forging is used to produce specialized shapes where high strength is required. An example is forged high-pressure pipe fittings such as flanges and elbows.

3.4.3 Cast Shapes

Casting is another method of producing shapes. Cast iron and steel are used for machine parts, automobile components, and large castings.

Table 23 ASTM Standard Metric and Inch-Pound Reinforcing Bars

Bar Size		Nominal Characteristics					
		Diameter		Cross-Sectional Area		Weight	
Metric	[in-lb]	mm	[in]	mm	[in]	kg/m	[lbs/ft]
#10	[#3]	9.5	[0.375]	71	[0.11]	0.560	[0.376]
#13	[#4]	12.7	[0.500]	129	[0.20]	0.944	[0.668]
#16	[#5]	15.9	[0.625]	199	[0.31]	1.552	[1.043]
#19	[#6]	19.1	[0.750]	284	[0.44]	2.235	[1.502]
#22	[#7]	22.2	[0.875]	387	[0.60]	3.042	[2.044]
#25	[#8]	25.4	[1.000]	510	[0.79]	3.973	[2.670]
#29	[#9]	28.7	[1.128]	645	[1.00]	5.060	[3.400]
#32	[#10]	32.3	[1.270]	819	[1.27]	6.404	[4.303]
#36	[#11]	35.8	[1.410]	1006	[1.56]	7.907	[5.313]
#43	[#14]	43.0	[1.693]	1452	[2.25]	11.38	[7.65]
#57	[#18]	57.3	[2.257]	2581	[4.00]	20.24	[13.60]

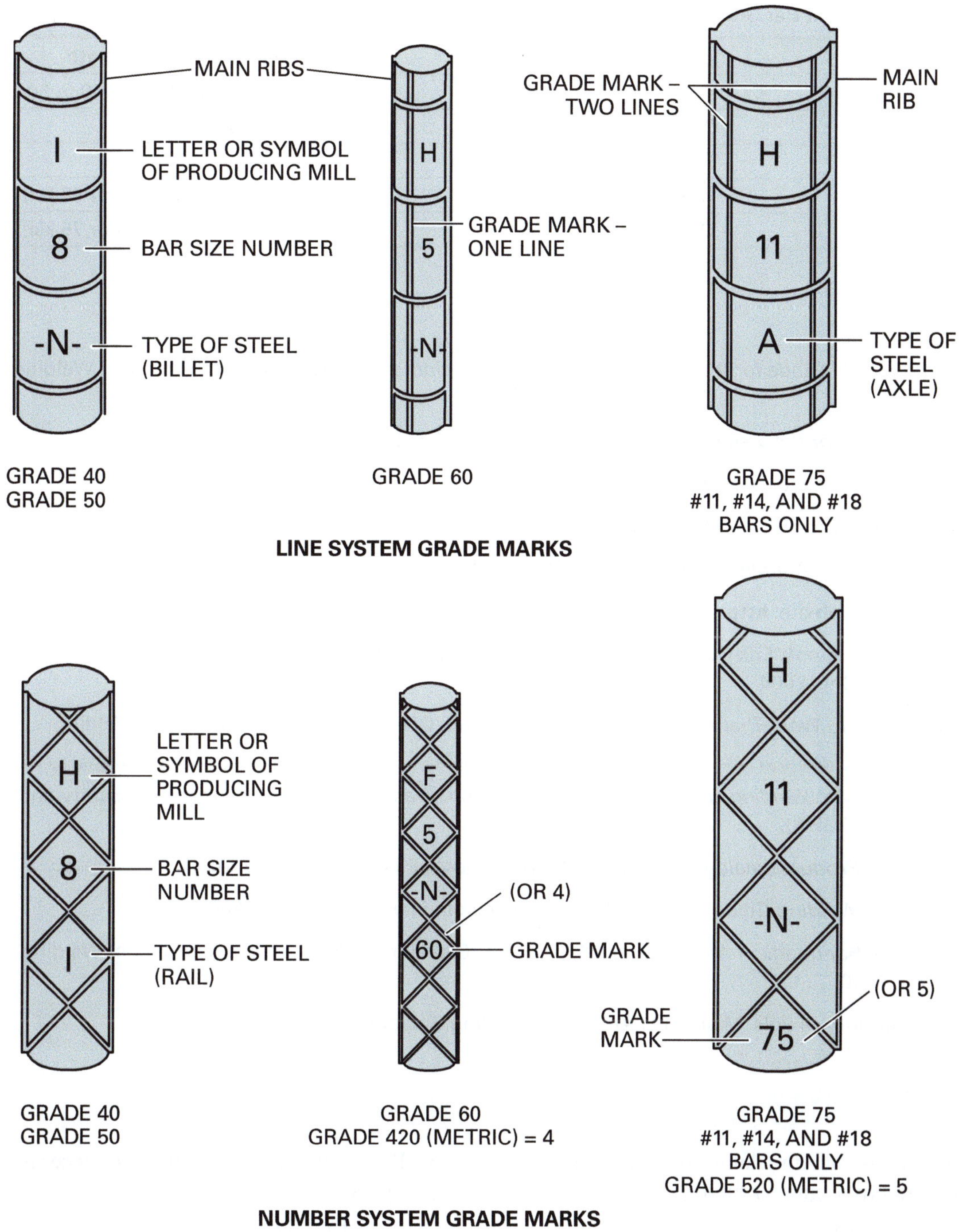

Figure 22 Reinforcement bar identification.

Table 24 Reinforcement Bar Steel Types

Symbol/Marking	Type of Steel
A	Axle *(ASTM A996)*
S or N	Billet *(ASTM A615)*
I or IR	Rail *(ASTM A996)*
W	Low-alloy *(ASTM A706)*

3.4.4 Powdered Metals

Commercially prepared powdered metals are used for welding fillers, surface coatings, and mold castings. When cast to make parts, the molded powder is usually **sintered** (heated in a furnace without melting) to fuse it into a solid metal. This is a common technique for manufacturing complex-shaped or porous parts. It is also used with high-temperature metals that cannot be liquid-cast.

Table 25 Reinforcement Bar Grades

Grade	Identification	Mimimum Yield Strength
40 and 50	None	40,000 to 50,000 psi (40 to 50 ksi)
60	One line or the number 60	60,000 psi (60 ksi)
75	Two lines or the number 75	75,000 psi (75 ksi)
420	The number 4	420 MPa (60,000 psi or 60 ksi)
520	The number 5	520 MPa (75,000 psi or 75 ksi)

Additional Resources

AWS B1.10M/B1.10 Guide for the Nondestructive Examination of Welds. Miami, FL: American Welding Society.

AWS B1.11 Guide for the Visual Examination of Welds. Miami, FL: American Welding Society.

AWS D3.5-93R Guide for Steel Hull Welding. Miami, FL: American Welding Society.

AWS D3.6M Specification for Underwater Welding. Miami, FL: American Welding Society.

AWS D3.7 Guide for Aluminum Hull Welding. Miami, FL: American Welding Society.

Lincoln Electric website: **http://www.lincolnelectric.com** offers sources for products and training.

Machinery's Handbook. Erik Oberg, Franklin D. Jones, and Christopher J. McCauley. New York, NY: Industrial Press, Inc. 2004.

Metals and How to Weld Them. T. B. Jefferson. Cleveland, OH: The James F. Lincoln Arc Welding Foundation.

OSHA 1926.351, Arc Welding and Cutting. Current Edition. Washington, DC: Occupational Safety and Health Administration (OSHA).

Stick Electrode Product Catalog. Cleveland, OH: The Lincoln Electric Company, 2008.

Stick Electrode Welding Guide. Cleveland, OH: The Lincoln Electric Company, 2004.

The Procedure Handbook of Arc Welding. 14th Edition. Cleveland, OH: The James F. Lincoln Welding Foundation, 2000.

Welding Handbook, 4 vols. Miami, FL: American Welding Society.

3.0.0 Section Review

1. Steel plate is rolled metal with a uniform thickness equal to or greater than ____.
   a. ⅛" (0.04 mm)
   b. 3⁄16" (0.06 mm)
   c. ¼" (0.08 mm)
   d. 5⁄16" (0.10 mm)

2. The three steel beam cross-section shapes are ____.
   a. A, H, and I
   b. I, L, and T
   c. H, T, and V
   d. I, H, and T

3. Pipe that is 14" (350 mm) and larger is specified by its ____.
   a. actual ID
   b. nominal ID
   c. actual OD
   d. nominal OD

4. The steel form that is most likely to be used for expansion and contraction joints in highway pavement is ____.
   a. forged high-pressure pipe
   b. plain reinforcing bars
   c. sintered powdered metal
   d. deformed rebar rods

SUMMARY

Being successful as a welder depends not only on the ability to make satisfactory welds, but also on having a thorough understanding of base metals, alloys, and filler metals. Welders must know about the composition, classification, physical characteristics, and mechanical properties of metals. They must also understand the numerous metallurgical variables that must be considered during welding activities, and they need to be able to identify different forms and shapes of structural metal. The information presented in this module should provide an adequate foundation for career advancement in welding.

Review Questions

1. The largest group of the ferrous-based metals includes the ____.
   a. cast irons
   b. alloy steels
   c. heavy metals
   d. carbon steels

2. Quench-and-tempered steels and chromium-molybdenum steels are examples of ____.
   a. common-grade stainless steels
   b. low-alloy steels
   c. high-alloy steels
   d. specialty-grade stainless steels

3. A lower coefficient of thermal conductivity along with a higher coefficient of thermal expansion and a higher electrical resistance are all characteristics of ____.
   a. nickel alloy
   b. carbon steel
   c. stainless steel
   d. aluminum alloy

4. A type of stainless steel that provides excellent resistance to salt water corrosion is ____.
   a. superferritic
   b. duplex
   c. superaustenitic
   d. precipitation-hardening

5. The principal alloying element in brasses is ____.
   a. aluminum
   b. zinc
   c. beryllium
   d. bronze

6. Corrosion resistance is an important characteristic of a metal because corrosion can severely reduce a metal's ____.
   a. tensile strength
   b. interpass ductility
   c. thermal expansion
   d. residual stress

7. The ability of a material to be strained (deformed) without permanent deformation is known as ____.
   a. ductility
   b. modulus
   c. elasticity
   d. tensile strength

8. The resistance to indentation of a material is called ____.
   a. ductility
   b. hardness
   c. tensile strength
   d. modulus of elasticity

9. The three types of cast iron are malleable, white, and ____.
   a. ductile
   b. green
   c. blue
   d. gray

10. Malleable cast iron is white cast iron that has been ____.
    a. annealed
    b. cooled slowly
    c. forged
    d. chilled rapidly

11. The most common defect caused by surface contamination of a metal is ____.
    a. rust
    b. porosity
    c. corrosion
    d. toxic fumes

12. A channel that has a designation of C12 × 30 is a (n) ____.
    a. American Standard Channel with a depth of 12 inches
    b. Miscellaneous Channel with a depth of 30 inches
    c. Miscellaneous Channel with a weight of 12 pounds per foot
    d. American Standard Channel with a weight of 12 pounds per foot

13. Shearing an S, W, or M beam down the center creates a(n) ____.
    a. wide-flange beam
    b. H beam
    c. structural tee
    d. seamed I

14. Tubing can be produced by rolling and welding or by ____.
    a. casting
    b. compression
    c. annealing
    d. extrusion

15. The standard configuration for reinforcing bars is the ____.
    a. dowel bar
    b. deformed bar
    c. column bar
    d. plain bar

Trade Terms Introduced in This Module

Alloy: A metal that has had other elements added to it that substantially change its mechanical properties.

American Iron and Steel Institute (AISI): An industry organization responsible for preparing standards for steel and steel alloys that are based upon the Society of Automotive Engineers (SAE) code system.

American Society for Testing and Materials (ASTM International): An organization that developed a code system for identifying and labeling steels.

Austenitizing: Heating a steel up to or above the transformation temperature range to achieve partial or complete transformation to austenite grain structure.

Casting: The metal object produced by pouring molten metal into a mold.

Certified mill test report (CMTR): A quality assurance report provided to a customer by a metal manufacturer that shows the chemical content and testing results of the metal being purchased; also called a mill test report (MTR).

Coefficient: A numerical measure of a physical or chemical property that is constant for a system under specified conditions, such as the coefficient of friction.

Ductile: The term used to describe a material that can be bent or shaped without breaking.

Ductility: The characteristic of metal that allows it to be stretched, drawn, or hammered without breaking.

Ferrous: Relating to iron or an alloy that is mostly iron; a metal containing iron.

Hardenability: A characteristic of a metal that enables it to become hard, usually through heat treatment.

Heat-affected zone (HAZ): The part of the base metal that has not been melted but has been altered structurally by the weld heat.

Hot shortness: The condition of metal when it proves to be very brittle and unbendable at red heat but can be bent without showing signs of brittleness when cold or at white heat. This condition is often a result of high sulfur and phosphorus content.

Interpass temperature: In a multiple-pass weld, the temperature of the base metal at the time the next weld pass is started; it is usually the same temperature that is used for preheating.

Malleable: Able to be hammered or pressed into another shape without breaking.

Mechanical properties: The characteristics or traits that indicate how flexible and strong a metal will be under stress.

Metallic: Having characteristics of metal, such as ductility, malleability, luster, and heat and electrical conductivity.

Mill test report (MTR): A quality assurance report provided to a customer by a metal manufacturer that shows the chemical content and testing results of the metal being purchased; also called a certified mill test report (CMTR).

Nonferrous metal: A metal, such as aluminum, copper, or brass, that lacks sufficient quantities of iron to be affected by its properties.

Notch toughness: The impact load required to break a test piece of weld metal measured by a notch toughness test (Charpy V-notch test). This test may be performed on metal below room temperature, at which point it may be more brittle.

Postweld heat treatment (PWHT): The process of heat treating the base metal and weld after the weld has been made.

Quench: To rapidly cool a hot metal using air, water, or oil.

Residual stress: The strain that occurs in a welded joint after the welding has been completed, as a result of mechanical action, thermal action, or both.

Sintered: A process of making parts from a powdered metal in which the metal is molded and then heated in a furnace without melting to fuse it into a solid metal.

Society of Automotive Engineers (SAE): An automotive society responsible for creating an early system for classifying carbon steels.

Stress-relief heat treatment: A technique used to reduce welding stresses and to increase the ductility and toughness of the weld metal and the heat-affected zone (HAZ).

Tempering: The process of reheating quench hardened or normalized steel to a temperature below the transformation range and then cooling it at any rate desired.

Tensile strength: The maximum stress or force that a material can withstand without breaking.

Underbead cracking: Subsurface cracking in the base metal under or near the weld.

Unified numbering system (UNS): An industry-accepted numbering system for identifying ferrous metals and alloys.

Web: The plate joining the flanges of a girder, joist, or rail.

Wide flange (WF) beam: A metal I-beam that has an extra wide flange at the top and bottom.

Wrought: Formed or shaped by hammering or rolling.

Additional Resources

This module is intended as a thorough resource for task training. The following reference works are suggested for further study.

AWS B1.10M/B1.10 Guide for the Nondestructive Examination of Welds. Miami, FL: American Welding Society.

AWS B1.11 Guide for the Visual Examination of Welds. Miami, FL: American Welding Society.

AWS D3.5-93R Guide for Steel Hull Welding. Miami, FL: American Welding Society.

AWS D3.6M Specification for Underwater Welding. Miami, FL: American Welding Society.

AWS D3.7 Guide for Aluminum Hull Welding. Miami, FL: American Welding Society.

Machinery's Handbook. Erik Oberg, Franklin D. Jones, and Christopher J. McCauley. New York, NY: Industrial Press, Inc. 2004.

Metals and How to Weld Them. T. B. Jefferson. Cleveland, OH: The James F. Lincoln Arc Welding Foundation.

OSHA 1926.351, Arc Welding and Cutting. Current Edition. Washington, DC: Occupational Safety and Health Administration (OSHA).

Stick Electrode Product Catalog. Cleveland, OH: The Lincoln Electric Company, 2008.

Stick Electrode Welding Guide. Cleveland, OH: The Lincoln Electric Company, 2004.

The Procedure Handbook of Arc Welding. 14th Edition. Cleveland, OH: The James F. Lincoln Welding Foundation, 2000.

Welding Handbook, 4 vols. Miami, FL: American Welding Society.

Lincoln Electric website: **http://www.lincolnelectric.com** offers sources for products and training.

Figure Credits

Robert I. Carr, Ph.D., P.E., Module opener

Concrete Reinforcing Steel Institute, Figure 1

Topaz Publications, Inc., Figures 2–4, 11

Zachry Industrial, Inc., Figures 5A, 5B

The Lincoln Electric Company, Cleveland, OH, USA, Figure 6, SA03, SA04

Photo Courtesy of Proceq SA Schwerzenbach, Switzerland, Figure 9

W.W. Grainger, Inc., Figure 10

Thermo Fisher Scientific, Figure 12

Photo Courtesy of U.S. Army, SA01

American Welding Society®, SA02

U.S. National Park Service, SA05

Fischer-Technology, Inc., SA06

Section Review Answer Key

Answer	Section Reference	Objective
Section One		
1. c	1.1.2	1a
2. a	1.2.1	1b
3. d	1.3.2	1c
4. b	1.4.4	1d
5. a	1.5.2	1e
Section Two		
1. c	2.1.1	2a
2. d	2.2.1	2b
3. a	2.3.4	2c
4. b	2.4.3	2d
Section Three		
1. b	3.1.1	3a
2. d	3.2.0	3b
3. c	3.3.1	3c
4. b	3.4.1	3d

NCCER CURRICULA — USER UPDATE

NCCER makes every effort to keep its textbooks up-to-date and free of technical errors. We appreciate your help in this process. If you find an error, a typographical mistake, or an inaccuracy in NCCER's curricula, please fill out this form (or a photocopy), or complete the online form at **www.nccer.org/olf**. Be sure to include the exact module ID number, page number, a detailed description, and your recommended correction. Your input will be brought to the attention of the Authoring Team. Thank you for your assistance.

Instructors – If you have an idea for improving this textbook, or have found that additional materials were necessary to teach this module effectively, please let us know so that we may present your suggestions to the Authoring Team.

NCCER Product Development and Revision

13614 Progress Blvd., Alachua, FL 32615

Email: curriculum@nccer.org
Online: www.nccer.org/olf

❑ Trainee Guide ❑ Lesson Plans ❑ Exam ❑ PowerPoints Other ______________________

Craft / Level: Copyright Date:

Module ID Number / Title:

Section Number(s):

Description:

Recommended Correction:

Your Name:

Address:

Email: Phone:

Preheating and Postheating of Metals

Overview

Welders need to understand how heat affects metals and how to control heating during each stage of a welding operation in order to make strong, enduring welds. This module explains the relationship between heat and metal and describes various devices and methods that welders use to heat metals and measure temperature.

Module 29204

Trainees with successful module completions may be eligible for credentialing through the NCCER Registry. To learn more, go to **www.nccer.org** or contact us at **1.888.622.3720**. Our website has information on the latest product releases and training, as well as online versions of our *Cornerstone* magazine and Pearson's product catalog.

Your feedback is welcome. You may email your comments to **curriculum@nccer.org,** send general comments and inquiries to **info@nccer.org**, or fill in the User Update form at the back of this module.

This information is general in nature and intended for training purposes only. Actual performance of activities described in this manual requires compliance with all applicable operating, service, maintenance, and safety procedures under the direction of qualified personnel. References in this manual to patented or proprietary devices do not constitute a recommendation of their use.

29204 V5

29204
Preheating and Postheating of Metals

Objectives

When you have completed this module, you will be able to do the following:

1. Describe the relationship between heat and metal and identify preheating methods.
   a. Describe the relationship between heat and metal.
   b. Identify and describe methods used to preheat metal prior to welding.
   c. Identify and describe devices and products used to measure temperature.
2. Describe interpass temperature control and postheating processes.
   a. Describe interpass temperature control.
   b. Describe various postheating processes.

Performance Task

Under the supervision of your instructor, you should be able to do the following:

1. Preheat base metal to 350°F (177°C) and verify preheat using a temperature-indicating device.

Trade Terms

Annealing
Constituents
Critical temperature
Crystalline
Discontinuity
Heat-affected zone (HAZ)
Interpass temperature control
Martensite
Preheat weld treatment (PHWT)
Strain hardening
Work hardening

Industry Recognized Credentials

If you are training through an NCCER-accredited sponsor, you may be eligible for credentials from NCCER's Registry. The ID number for this module is 29204. Note that this module may have been used in other NCCER curricula and may apply to other level completions. Contact NCCER's Registry at 888.622.3720 or go to **www.nccer.org** for more information.

Contents

Topics to be presented in this module include:

Figures and Tables

Section One

1.0.0 Metal and Preheating

Objective

Describe the relationship between heat and metal and identify preheating methods.

a. Describe the relationship between heat and metal.
b. Identify and describe methods used to preheat metal prior to welding.
c. Identify and describe devices and products used to measure temperature.

Performance Task

1. Preheat base metal to 350°F (177°C) and verify preheat using a temperature-indicating device.

Trade Terms

Constituents: The elements and compounds, such as metal oxides, that make up a mixture or alloy.

Critical temperature: The temperature at which iron crystals in a ferrous-based metal transform from being face-centered to body-centered. This dramatically changes the strength, hardness, and ductility of the metal.

Crystalline: Like a crystal; having a uniform atomic structure throughout the entire material.

Heat-affected zone (HAZ): The portion of the base metal workpiece that is not melted but has had its properties affected by welding or other heat-intensive operation such as oxyfuel cutting.

Interpass temperature control: The process of maintaining the temperature of the weld zone within a given range during welding.

Martensite: A solid solution of carbon in alpha-iron that is formed when steel is cooled so rapidly that the change from austenite to pearlite is suppressed; responsible for the hardness of quenched steel.

Preheat weld treatment (PHWT): The process of heating the base metal before making a weld.

Preheat weld treatment (PHWT) is the controlled heating of the base metal immediately before welding begins. The main purpose of PHWT is to keep the weld and the base metal along the weld line in the heat-affected zone (HAZ) from cooling too fast. In some cases, the whole structure is preheated. It may not always be possible or practical to heat the whole structure. When this is the case, only the section near the weld is preheated.

1.1.0 Temperature and Metal Relationship

At room temperature, most metals are solid and have a crystalline structure (*Figure 1*). They become liquids if heated to a high enough temperature. Temperature is a measure of molecular activity. When the temperature is high enough, the forces that hold the molecules in their crystalline structures break down, and the metal becomes a liquid. At even higher temperatures, metals become gases. The temperature at which a metal becomes liquid is its melting point. The melting point is different for every metal. In the case of alloys (metal mixtures), the various constituents melt at different temperatures. Some, like carbon, may not melt at all.

Below the melting point, there is a temperature zone at which crystalline changes take place. Depending on the metal or alloy, these changes can seriously affect the metal's physical structure and mechanical characteristics. Quick cooling (quenching) of a metal while it is in this temperature zone can freeze some of the modified crystalline structure. As a result, the mechanical properties and physical traits of the cooled metal will be very different from those of the same metal if it had been cooled more slowly. Quenching does not have much of an effect on low-carbon and mild steels because they do not contain much carbon. However, quenching medium- and high-carbon steels produces a hard and brittle crystalline structure, which reduces the metal's toughness. *Table 1* shows the minimum preheat temperatures for various base materials, and it includes their welding processes and material thickness.

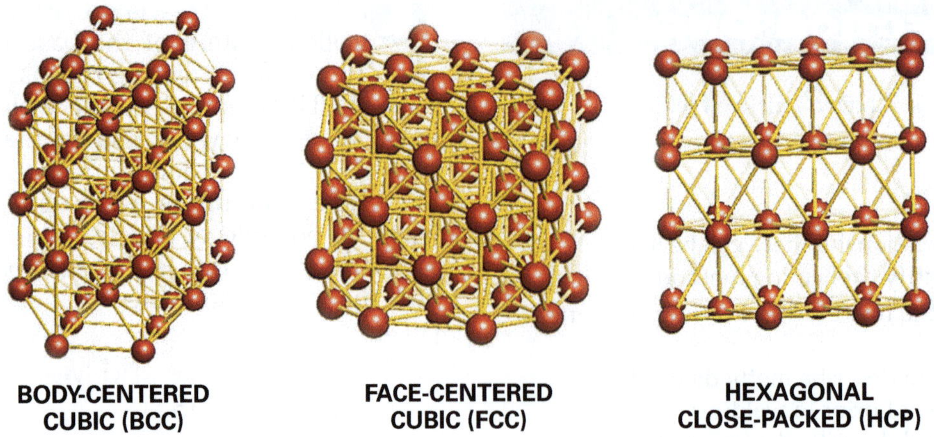

Figure 1 Crystal structures of iron.

Table 1 Minimum Preheat Temperature for Various Base Metals

Base Materials	Welding Processes	Material Thickness	Mimimum Preheat Temperature
ASTM A36 ASTM A53 GR. B ASTM A106 GR. B API 5L GR. B API 5L GR. X42	Shielded Metal Arc Welding With Other Than Low Hydrogen Electrodes	≤¾″ (19 mm) Between ¾″ and 1½″ (19 mm and 38 mm) Between 1½ and 2½″ (38 mm and 64 mm) >2½″ (64 mm)	None (See Note) 150°F (66°C) 225°F (107°C) 300°F (149°C)
ASTM A36 ASTM A53 GR. B ASTM A106 GR. B ASTM A572 GR. 42 ASTM A572 GR. 50 ASTM A586 API 5L GR. B API 5L GR. X42	Shielded Metal Arc Welding With Low Hydrogen Electrodes, Submerged Arc Welding, Gas Metal Arc Welding, and Flux-Cored Arc Welding	≤¾″ (19 mm) Between ¾″ and 1½″ (19 mm and 38 mm) Between 1½″ and 2½″ (38 mm and 64 mm) >2½″ (64 mm)	None (See Note) 50°F (10°C) 150°F (66°C) 225°F (107°C)
ASTM A572 GR. 60 ASTM A572 GR. 65 ASTM 5L GR. X52	Shielded Metal Arc Welding With Low Hydrogen Electrodes, Submerged Arc Welding, Gas Metal Arc Welding, and Flux-Cored Arc Welding	≤¾″ (19 mm) Between ¾″ and 1½″ (19 mm and 38 mm) Between 1½″ and 2½″ (38 mm and 64 mm) >2½″ (64 mm)	50°F (10°C) 150°F (66°C) 225°F (107°C) 300°F (149°C)
ASTM A514 ASTM A517	Shielded Metal Arc Welding With Low Hydrogen Electrodes, Submerged Arc Welding, Gas Metal Arc Welding, and Flux-Cored Arc Welding	≤¾" (19 mm) Between ¾" and 1½" (19 mm and 38 mm) Between 1½" and 2½" (38 mm and 64 mm) >2½" (64 mm)	50°F (10°C) 125°F (52°C) 175°F (79°C) 275°F (135°C)

NOTE: If below 32°F (0°C), preheat to 70°F (21°C)

NOTE

Tables in the Trainee Guide are for educational purposes only. These tables should not be relied upon as a resource for work outside of the learning environment. Always consult job specifications and other named project resources for current and accurate guidance.

When medium-carbon and high-carbon steel are welded, the temperature in the weld zone always reaches the melting point (above the critical temperature or transformation temperature), and the crystalline structure in the weld zone changes. If the metal has not been preheated enough, the large mass of the cool base metal will quench the weld zone and cause localized

Heat Treatments

Different forms of heat treatments are used to alter the properties of metals. Soft metals can be made hard, brittle, or strong using the correct heat treatment. They can be returned to their original soft form with heat treatment. Welding specifications often require that welded joints be heat-treated before welding or after fabrication. To avoid mistakes, heat treatments made to metals of any kind must be done in a planned and systematic way. This is because the use of temperatures that are too low or too high can result in doing more harm than good.

hardness and brittleness. If the metal has been adequately preheated, the weld zone will cool much more slowly, which will allow the crystalline structure to return to the crystalline form of the surrounding base metal.

Preheating may be required to do the following:

- Reduce shrinkage in the weld and the adjacent base metal
- Prevent excessive hardening and reduced ductility of the weld filler and base metals in the heat-affected zone
- Reduce hydrogen gas in the weld zone
- Ensure compliance with required welding procedures

The critical temperature of a metal is somewhere below its melting point. The temperature range that exists between a metal's critical temperature and its melting point is called the transformation temperature range. This range, generally considered to lie between 932°F and 1,040°F (500°C and 560°C), is the lowest temperature point at which the molecular structure of the metal will change during preheating, welding, and postheating. The melting point, strength, and hardness of steel are determined by the amount of carbon in the steel. *Table 2* shows the classifications of steel based on carbon content.

When preheating metal, apply heat to the region of the intended weld long enough to sufficiently heat the area. Insufficient heating will allow a relatively large weld to cool too quickly. To ensure proper preheating, measure the temperature of the preheated region again after waiting 15 to 20 minutes.

Table 2 Classification of Carbon Steel

Common Name	Carbon Content
Low-carbon steel	0.15 percent maximum
Mild steel	0.15 to 0.30 percent
Medium-carbon steel	0.30 to 0.50 percent
High-carbon steel	0.50 to 1.00 percent

1.1.1 Reducing Shrinkage

Because restrained assemblies are held firmly in position to maintain their shape, they can build up residual stresses following welding. The section that is welded expands due to the heat. When it cools, it shrinks, and stresses may occur if the surrounding area does not shrink along with it. With steel weldments, this shrinkage can cause distortion, warping, and even cracking. With cast-iron weldments, shrinkage usually causes the base metal to crack or the weld to fail. For example, if a broken cast-iron pulley spoke is being repaired, the rim and other spokes must be adequately preheated or they will be placed under heavy strain by the thermal expansion of the welded spoke. When the weld cools, the residual strains could cause the rim to crack or the welded spoke to crack again. If the rim and spokes near the weld are adequately preheated, they will shrink along with the welded spoke, reducing or eliminating most of the stress on the rim. *Figure 2* shows the areas of heavy strain resulting from inadequate preheating of a cast-iron pulley.

1.1.2 Preventing Excessive Hardening and Reduced Ductility

All welding processes use or generate temperatures higher than the melting point of the base

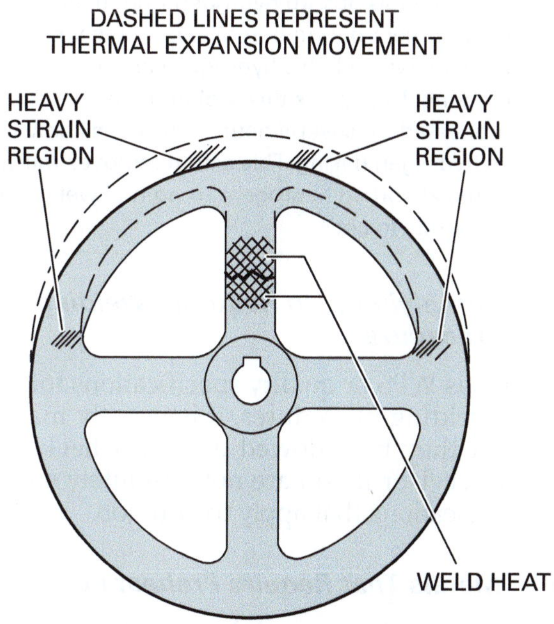

Figure 2 Strain caused by inadequate preheating of a cast-iron pulley.

metal. If the welded metal is a high-carbon alloy steel or cast iron that has been quenched, a hard, brittle metal called **martensite** will form in the weld zone. This causes the entire heat-affected zone parallel to the weld fusion line to become hard and brittle (*Figure 3*). The large temperature difference also causes differential thermal expansion and creates stresses in the weld region. Heavy plate and castings have a high heat absorption capacity that will cause the weld metal and adjacent base metal to quench unless they are adequately preheated.

1.1.3 Reducing Hydrogen Gas in the Weld Zone

The welding arc can cause moisture in the weld zone to separate into hydrogen and oxygen. The hydrogen dissolves in the molten base metal, and the oxygen combines with other elements to form slag. As the metal begins to cool, the hydrogen forms gas bubbles in the base metal along the weld boundary. These bubbles create pressure that strains the metal and causes underbead cracking along the weld boundary and at the toe. Preheating helps to remove moisture from the weld zone. This results in slower cooling, which allows the hydrogen more time to diffuse up through the filler metal and down into the adjacent base metal. The more time the hydrogen has to diffuse, the less pressure it causes, and the less chance there is for underbead cracking to occur. *Figure 4* is a simplified diagram showing underbead cracking caused by hydrogen bubbles.

> **NOTE**
> Some metals absorb hydrogen from their environment during normal use. Later, when they are welded, this hydrogen can cause cracking. To address this problem, the metal is preheated for several hours, a process known as hydrogen baking. This action removes much of the absorbed hydrogen, so normal welds are less likely to crack.

1.1.4 Complying with Required Welding Procedures

If a site has WPS or quality specifications for particular welding procedures, it is usually mandatory that they be followed exactly. Check with your supervisor if you are not absolutely sure of the specifications that apply to your job.

1.1.5 Metals That Require Preheating

Determining whether metal assemblies and conditions require preheating can be very difficult.

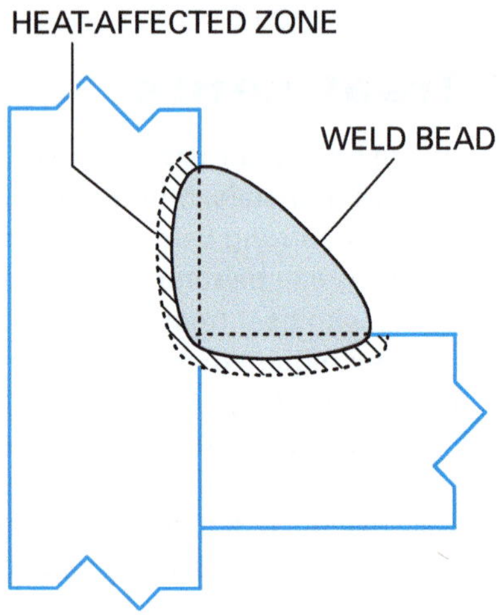

Figure 3 Heat-affected zone (HAZ).

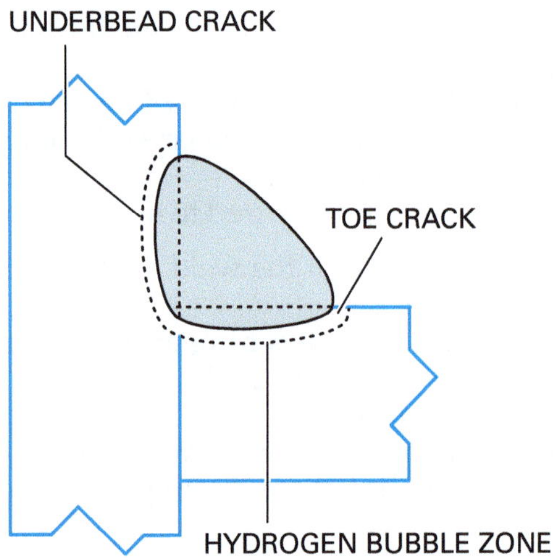

Figure 4 Underbead cracking caused by hydrogen bubbles.

Preheating and **interpass temperature control** depend on the base metal composition, thickness, and degree of restraint. Mild carbon steel, for example, requires preheating if its carbon content exceeds 0.30 percent. If the carbon content is lower than 0.30 percent, it must be heated to a temperature above 70°F (21°C). Some base metals require preheating only when they exceed a certain thickness or are too cold. For example, metals exposed to freezing winter temperatures usually require some preheating. Welding assembles and castings with complex shapes usually require preheating to avoid severe stresses or warping. Some alloys, such as high-carbon or alloy steels, require preheating to avoid brittleness and cracking. Not all metals require preheating. Low-carbon steels rarely need preheating. Always consult the WPS for specifics. *Table 3* shows the metals that usually require preheating and their conditions and forms.

Preheating requirements are affected by the alloy composition and thickness of the base metal,

Table 3 Metals That Usually Require Preheating

Metal or Alloy	Conditions or Forms
Aluminum	Large or thick section castings
Copper	All (prevents too-rapid heat loss)
Bronze, copper-based	All (prevents too-rapid heat loss)
Mild carbon steels	Restricted joints, complex shapes, freezing temperatures, and carbon content over 0.30 percent
Cast irons	All types and all shapes
Cast steels	Higher carbon content or complex shapes
Low-alloy steels	Thicker sections or restricted joints
Low-alloy steels	Heating temperature dependent on carbon or alloy content
Manganese steels	Heating temperature dependent on carbon or alloy content
Martensitic stainless steels	Carbon content over 0.10 percent
Ferritic stainless steels	Thick plates

the welding process to be used, and the ambient temperature. The hardenability of a steel is directly related to its carbon content and alloying elements. Because different types of steel have different amounts of carbon and alloying elements, they have different preheat requirements. As a general rule, the higher the carbon content, the higher the preheating temperature required. *Table 4* lists preheat requirements for steels, including the steel type, shapes or conditions, and preheat temperature.

1.2.0 Preheating Methods

Preheating is most effective when the entire welding assembly is preheated (general heating) in an oven or preheating furnace because the heating is even. However, because of site conditions or the

Underbead Cracking

Always consider preheating and postheating a weld when there is the danger of underbead cracking, also called hydrogen cracking. The risk of underbead cracking increases when welding carbon and low-alloy steels. Underbead cracking can occur when there is a combined presence of the following conditions:

- Diffusible (atomic) hydrogen in the steel
- A steel with a partly or wholly martensitic microstructure
- Tensile stress at the weldment and heat-affected zone
- A temperature below 300°F (149°C)

Proper preheat, interpass temperature maintenance, and postweld heat treatment, together with the proper electrode/shielding gas combination, can eliminate this problem.

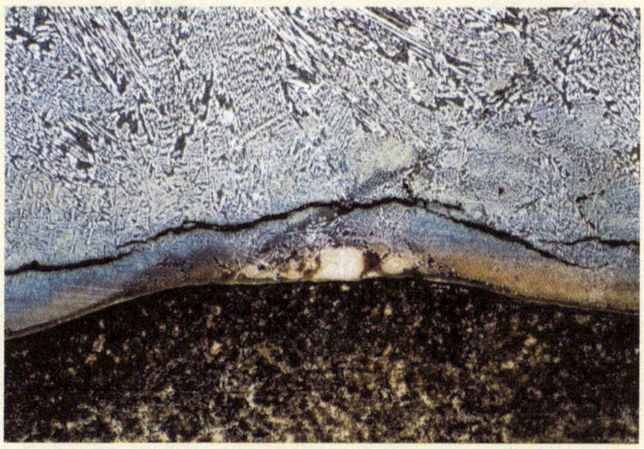

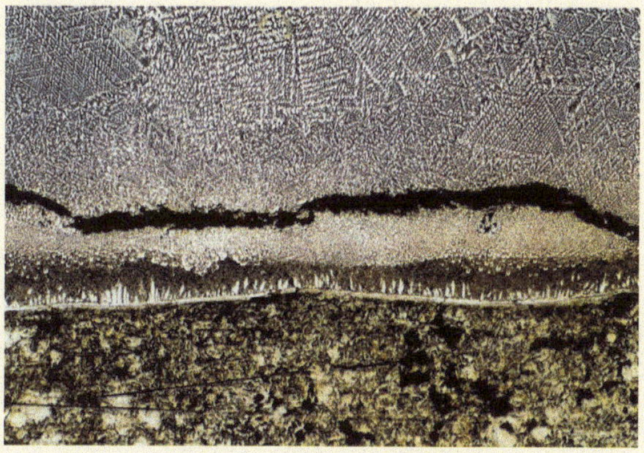

Table 4 Preheat Requirements for Steels

Steel Type	Shapes or Conditions	Preheat Temperature
Low-carbon steel (0.10–0.30 percent C)	Freezing temperatures	Above 70°F (21°C)
Low-carbon steel (0.10–0.30 percent C)	Complex large shapes	100°F–300°F (38°C–149°C)
Medium-carbon steel (0.30–0.45 percent C)	Complex large shapes	300°F–500°F (149°C–250°C)
High-carbon steel (0.45–0.80 percent C)	All sizes and shapes	500°F–800°F (250°C–427°C)
Low-alloy steel	Lower C or alloy percent	100°F–500°F (38°C–250°C)
Low-alloy steel	Higher C or alloy percent	500°F–800°F (250°C–427°C)
Manganese steel	Lower C or alloy percent	100°F–200°F (38°C–93°C)
Manganese steel	Higher C or alloy percent	200°F–500°F (93°C–250°C)
Cast iron	All shapes	200°F–400°F (93°C–204°C)
Martensitic stainless steel	0.10–0.20 percent C	500°F (250°C)
Martensitic stainless steel	0.20–0.50 percent C	500°F (250°C) and postheat
Ferritic stainless steel	¼" (6.4 mm) and thicker	200°F–400°F (93°C–204°C) and postheat

size or shape of the welded assembly, preheating by general heating is not always possible. Therefore, a variety of heating devices can be used for preheating. Localized heating of specific regions can be done with gas torches, blowtorches, electric resistance heaters, radio frequency induction heaters, or partial ovens that are built only around the region to be heated. Parts of the oven or enclosure can be removed so that the welder can have access and still maintain the preheating temperature.

Most welding shops have one or more types of preheating devices, such as the following:

- Oxyfuel torches
- Portable preheating torches
- Open-top preheaters
- Electric resistance heaters
- Induction heaters
- Enclosed furnaces

1.2.1 Oxyfuel Torches

If properly monitored, oxyfuel torches equipped with specialized heating tips can be used for localized preheating. *Figure 5* shows two types of torch handles to which heating tips can be mounted.

The specialized heating tips are designed to produce blowtorch flame patterns. Heating tip attachments with multiflame heating heads are often referred to as rosebuds. They are designed to mount directly to torch handles like welding tips. Heating tips designed to replace the cutting tips on straight cutting torches or combination cutting torch cutting attachments are also available. *Figure 6* shows several types of heating tips.

WARNING!

Always use the correct type of tip for the fuel gas being used. Using a tip with the wrong gas can cause the tip to explode. Observe the pressures and flow rates recommended by the manufacturer.

1.2.2 Portable Preheating Torches

Portable preheating torches are designed to burn either oil or gas combined with compressed air. Some use only gas and have an air blower to increase the intensity of the flame. *Figure 7* shows an air torch suitable for preheating large weldments. The combustion chamber head can easily be removed and an extension added to increase the length. This torch operates on various fuel gases combined with compressed air.

Preheating Steel in Low-Temperature or High-Humidity Conditions

Steel should not be welded if the steel temperature is below 32°F (0°C). If below 32°F (0°C), steel should be heated to at least 70°F (21°C) prior to welding. Under humid conditions, steel should be heated to a higher preheat temperature in order to dry any moisture from its surface.

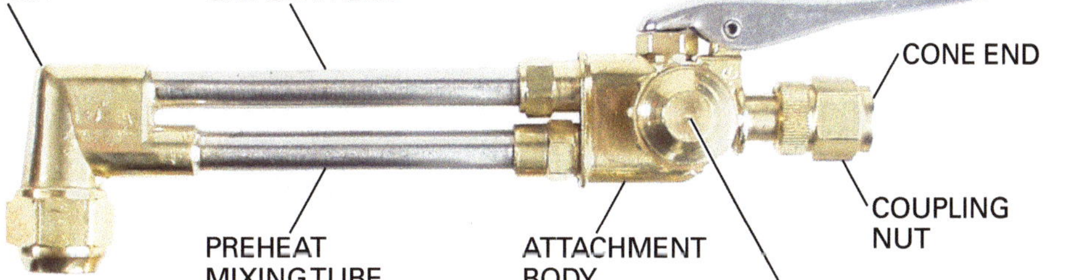

CUTTING TORCH ATTACHMENT

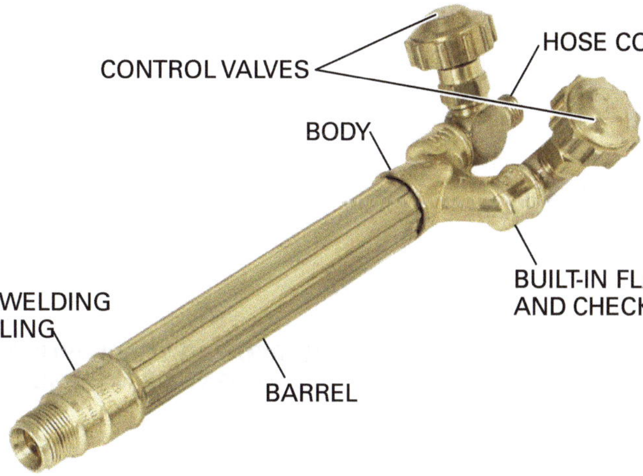

COMBINATION TORCH HANDLE

Figure 5 Combination torch handle and cutting torch attachment.

1.2.3 Open-Top Preheaters

Open-top preheaters come in several basic designs. The simplest is a large, horizontal gas burner with an adjacent or overhead support structure. The object to be heated is mounted on the support structure directly over the gas burners. A variation is the open flat-top preheater, which consists of a horizontal grate or series of bars with gas burners underneath (*Figure 8*). The item to be preheated is placed on the grating. This type of preheater works similar to a gas-powered kitchen stove.

Air Tube Burner

The gas fuel/compressed air tube burner can be used for the heat treatment of large weldments. The tube section has evenly spaced multiple heating tips along its length. A burner of this type provides a more even heat distribution to a large weldment.

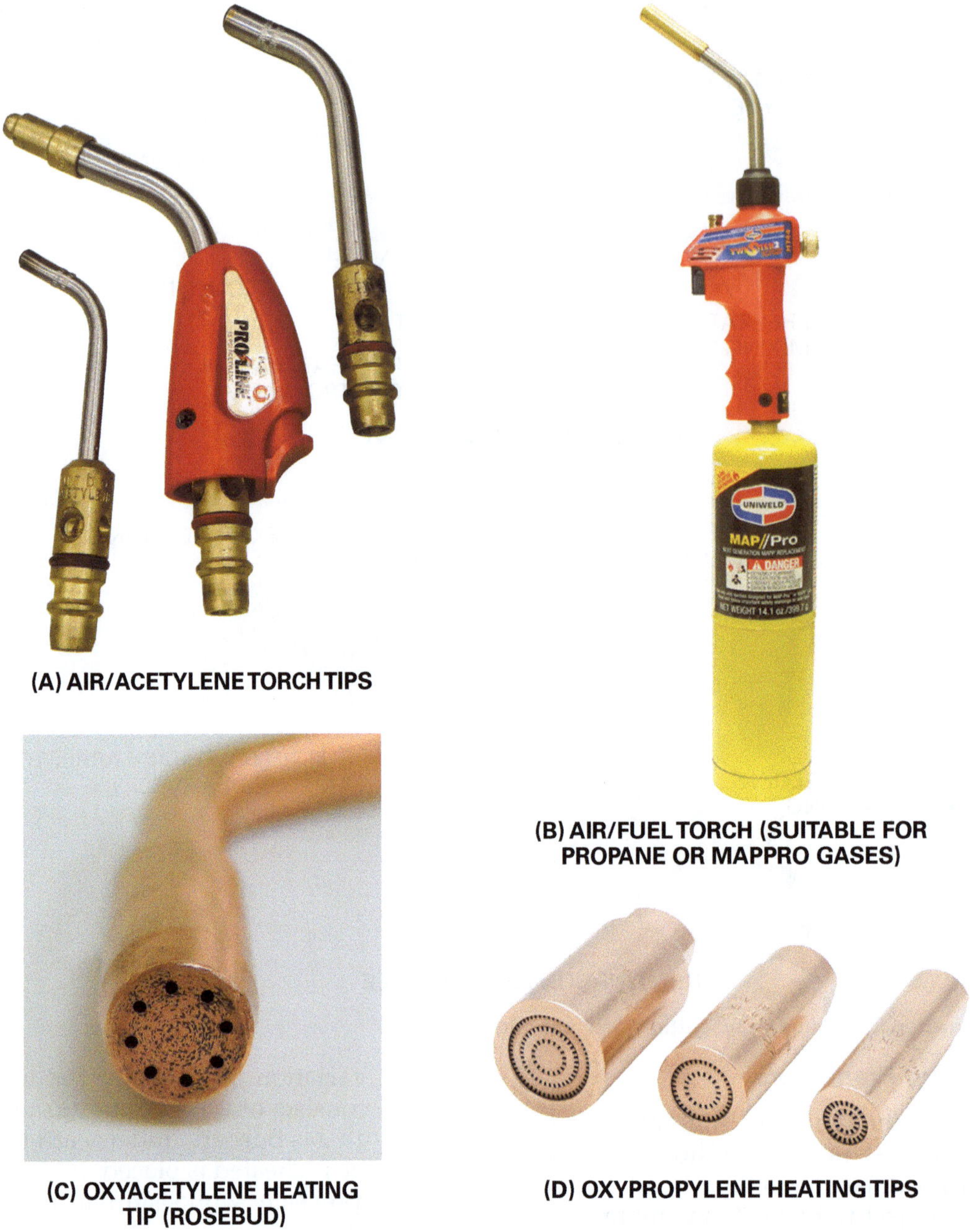

Figure 6 Types of heating tips.

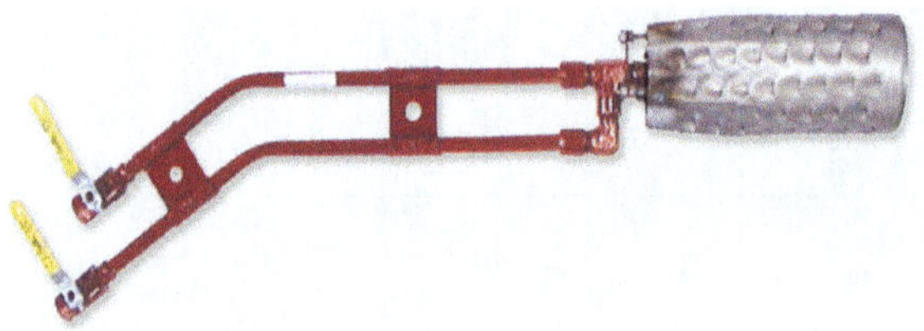

Figure 7 Gas preheating torch.

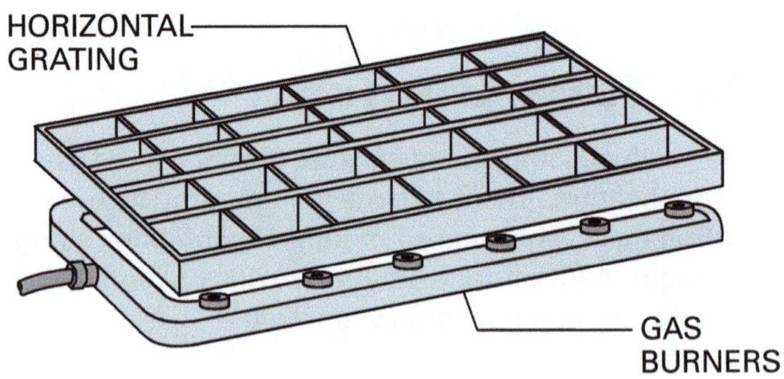

Figure 8 Open flat-top preheater.

1.2.4 Electric Resistance Heaters

Electric resistance heaters have tubular resistance elements that are formed in several shapes or styles to fit various applications (*Figure 9*). These elements are temporarily fastened against the surface of the metal to be heated. Electricity passing through the elements creates heat, which is transferred to the weldment.

1.2.5 Induction Heaters

In induction heating, an alternating magnetic field is generated within the metal body being heated. A magnetic coil is placed around the metal object, and a strong alternating current is passed through the coil. The resulting alternating magnetic field generated within the base metal creates circulating electrical currents known as eddy currents. The interaction of these currents with the electrical resistance of the base metal produces an electrical heating effect, raising the temperature of the base metal. *Figure 10* shows an air-cooled induction heating system consisting of a power supply, induction blanket, and the associated power and control cables. Air-cooled systems are typically used for preheat applications up to 400°F (204°C). This type of system typically uses a controller to monitor and automatically control the temperature.

1.2.6 Enclosed Furnaces

Enclosed furnaces or ovens include both permanent and temporary structures or partial enclosures that must be made of firebrick and clay, sheet metal, or some other noncombustible material. The item to be heated may be totally or partially enclosed. With very large items, it may not be practical or possible to enclose the entire item. Furnaces and ovens are heated by gas or oil burners, torches, or electric resistance heaters. *Figure 11* shows an example of a gas-fired box furnace.

1.3.0 Measuring Temperature

Control of preheating, interpass, and postheating temperatures depends on accurately determining the temperature of the weld zone or assembly. Both underheating and overheating can negatively affect a metal's physical and mechanical characteristics. There are a number of ways to determine the surface temperature of an item being heated. Some involve complex manufactured products, while others use less sophisticated methods. Devices and materials that are commonly used to determine surface temperatures include thermometers and temperature-sensitive indicators.

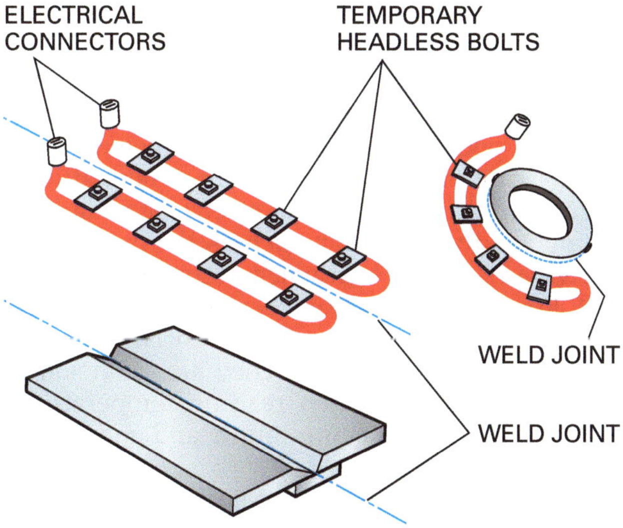

Figure 9 Resistance heating elements.

1.3.1 Thermometers

Electronic thermometers (*Figure 12*) now dominate industrial temperature measurement. These thermometers display temperature on either an analog meter or a digital liquid crystal display (LCD) readout. They generally use either a thermocouple or thermistor type of temperature probe, or both, to sense the heat and generate a temperature indication.

Thermocouple probes use a sensing device made of two dissimilar wires welded together at one end called a junction. When the junction is heated, it generates a low-level DC voltage that produces a proportional temperature reading on the electronic thermometer indicator. Thermocouple probes tend to be rugged and less expensive than thermistor probes. The probe shown in *Figure 12* is designed to clamp securely to small pipe; many probe styles are available. Thermistor probes use a semiconductor electronic element in which resistance changes in proportion to the change in temperature.

Often, several different probes are used with the same electronic thermometer to allow temperature measurement in different applications. Many electronic thermometers can also accommodate two or more probes at the same time, so that simultaneous measurements can be made at several locations. Most thermometers of this type can calculate and display the difference in temperature between the locations being measured.

Electronic thermometers are precise measuring instruments. Be sure to read and follow the manufacturer's instructions for operating electronic thermometers. Also, be sure to follow the manufacturer's instructions for calibration of the instrument.

Dial thermometers (*Figure 13*) are available in various forms. They are popular because they are rugged, small, and typically inexpensive. They come in many stem lengths and dial sizes. Because pocket-style dial thermometers have smaller calibrated scales, it is sometimes difficult to get accurate readings. An electronic version of the pocket thermometer, such as the one shown in *Figure 13*, works best when accuracy is important. Remember that most thermometers, regardless of the type, are generally more accurate in the mid-section of their range. They are often somewhat less accurate when reading temperatures that are close to the minimum and maximum values in their range.

Handheld infrared thermometers (*Figure 14*) are also very popular. Infrared thermometers allow the user to measure temperature without making contact with the surface of an object. For this reason, they are also called non-contact thermometers. They work by detecting infrared energy emitted by the object at which they are pointing. Laser sighting devices allow the user to aim the thermometer like a pistol at the surface to be measured. The temperature is displayed instantly. This type of thermometer is useful for taking temperature readings in hard-to-reach areas. However, it may not be effective on highly reflective surfaces. The unit may have an adjustable setting which can be set to allow the measurement of reflective surfaces. A thermometer with no such adjustment can still be used to measure the temperature of a shiny surface by applying a matte paint or non-reflective tape to the surface.

Induction Heaters

Because water cools more efficiently than air, water-cooled induction heating systems can be used for high-temperature preheating and stress relieving applications. This type of system is normally equipped with a temperature controller and temperature recorder. The controller and recorder are important components in stress-relieving applications.

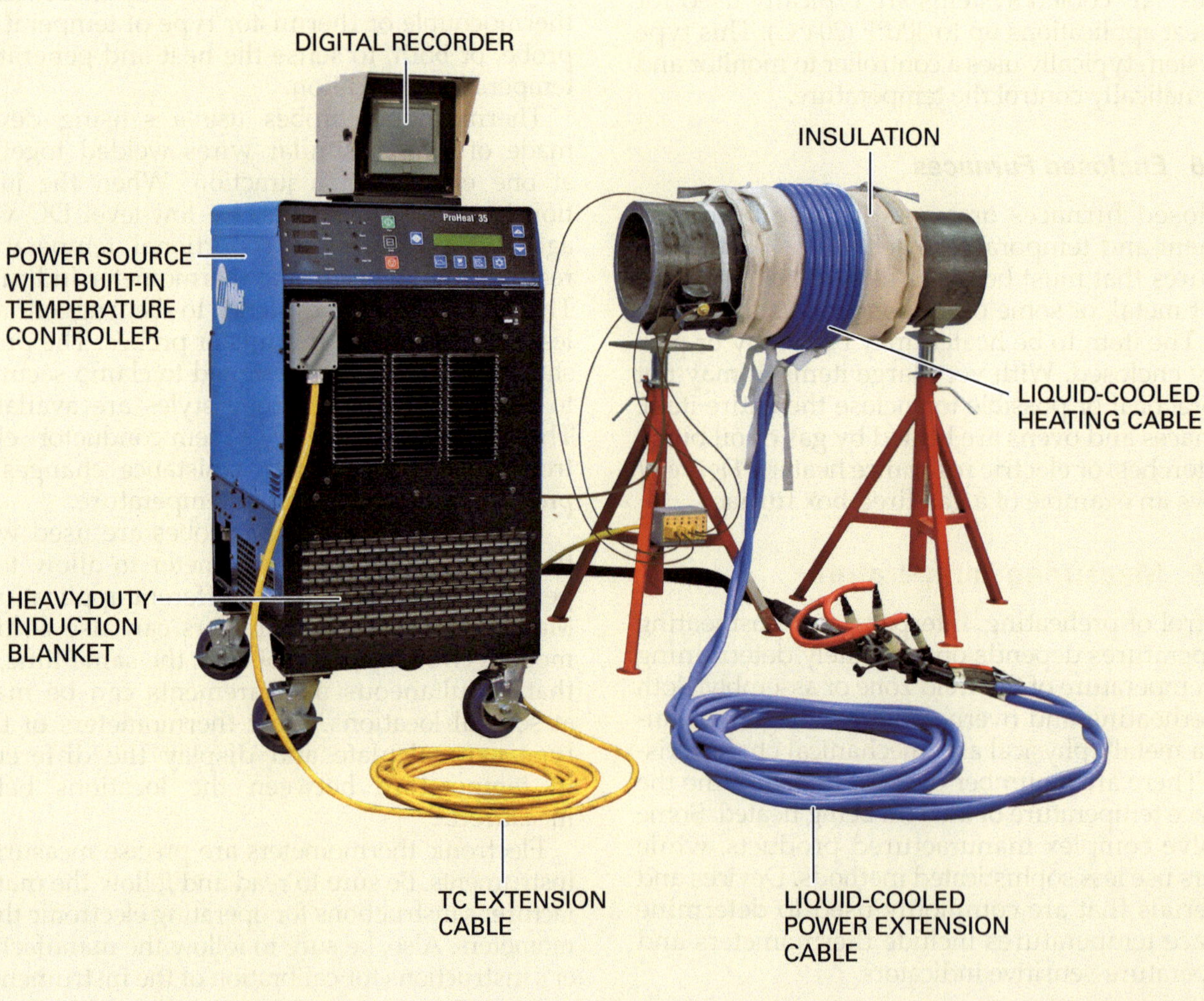

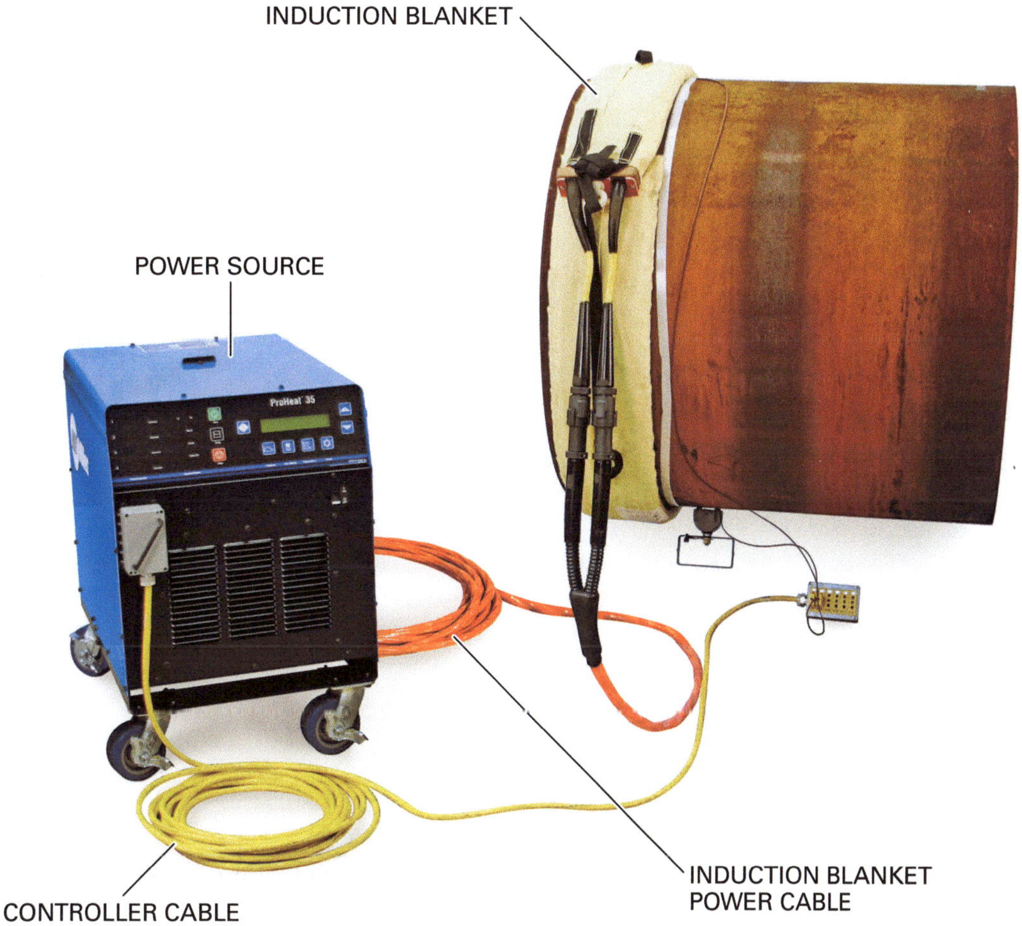

Figure 10 Induction heating system.

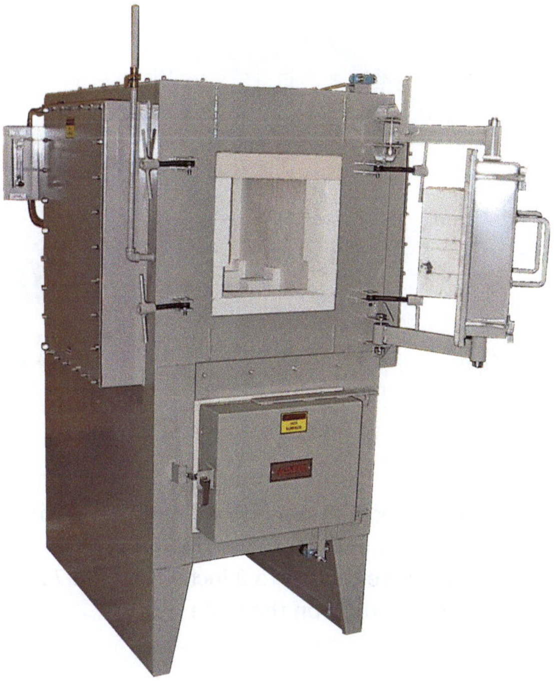

Figure 11 Box furnace.

It is important to understand the distance-to-spot ratio (D:S) of infrared thermometers. The D:S is the ratio of the distance to the object being measured compared to the diameter of the temperature measurement area. For example, if the D:S ratio is 12:1, taking a measurement from 12 inches (30.5 cm) away will give the average temperature of a spot 1" (25 mm) in diameter (*Figure 15*). The greater the distance, the larger the spot becomes. When using an infrared unit with a 12:1 D:S from 12 feet away, the spot becomes 1 foot in diameter. As a result, the temperature measurement may be inaccurate, unless the surface being measured is at least as large as the spot. Otherwise, the device will average the temperature of two or more unrelated surfaces.

1.3.2 Temperature-Sensitive Indicators

Temperature-sensitive indicators are devices that change form or color at specific temperatures or temperature ranges. These include commercially made materials and devices such as crayon sticks

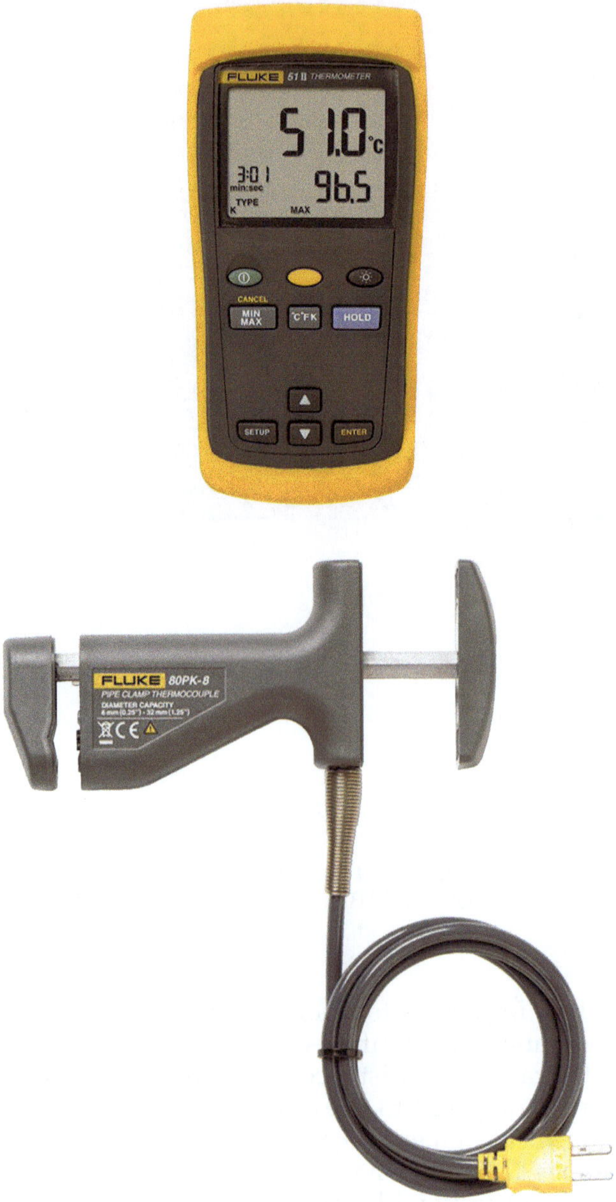

Figure 12 Electronic thermometer.

Figure 13 Dial and electronic pocket thermometers.

Figure 14 An infrared non-contact thermometer.

(*Figure 16 [A]*); liquids (*Figure 16 [B]*); and chalks, powders, pellets, and labels (*Figure 16 [C]*). These indicators are designed to change color or form at precise temperatures, often within a 1 percent tolerance.

Temperature sticks are among the most widely used temperature-sensitive indicators. The sticks consist of a series of graduated temperature-indicating crayons that cover the temperature range from 100°F to 1,150°F (38°C to 621°C). These sticks are used to mark on the base metal. At a specific temperature, the mark melts or changes color. *Table 5* is an example of a preheating chart that could be used when selecting temperature-indicating crayons for various metals.

CAUTION

Do not mark with the crayon in the weld zone. The recommended practice is to mark from the toe of the weld out 1 to 3 inches (25.4 to 76.2 mm), depending on the code.

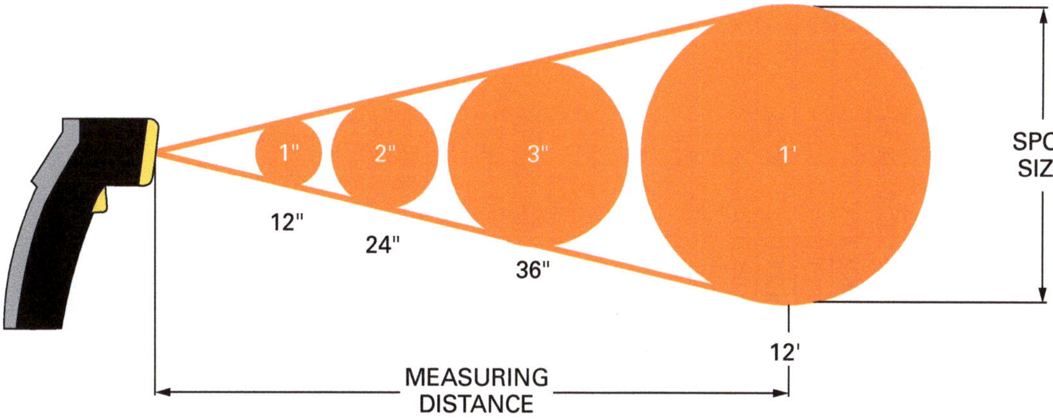

Figure 15 The D:S ratio illustrated.

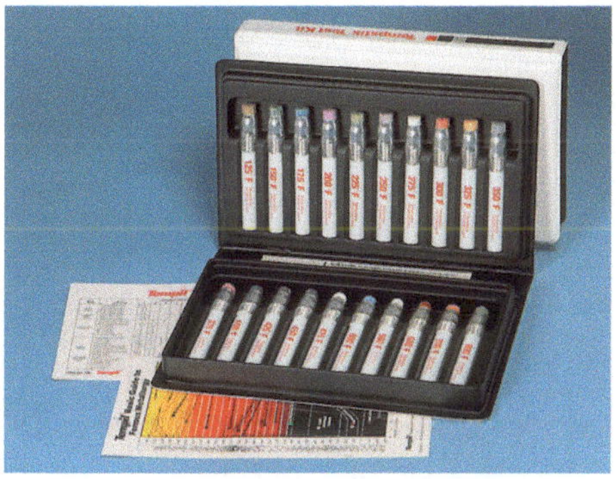
(A) TEMPERATURE STICKS

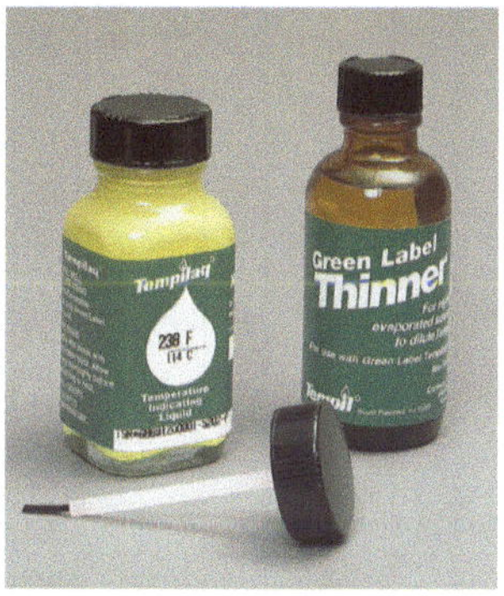

(B) TEMPERATURE LIQUID

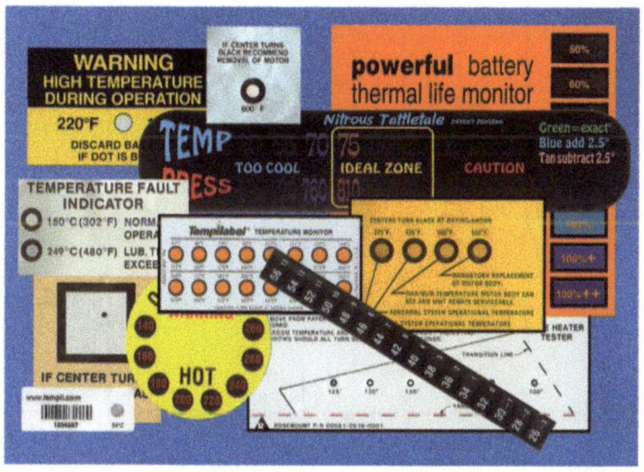

(C) TEMPERATURE LABELS

Figure 16 Temperature-sensitive indicators.

Furnace Thermocouples

The rate of temperature rise, holding time at temperature, and rate of cooling are critical to proper heat treatment. For this reason, thermocouples used in furnaces must measure the temperature of the metal being heat-treated, not the ambient temperature inside the furnace. To do this, the thermocouples must, of course, be in contact with the metal.

Table 5A Preheating Chart for Selecting Temperature-Indicating Crayons (1 of 2)

Metal Group	Metal Description	Approximate Composition Percentage								Recommended Preheat Temperature
		C	Mn	Bi	Cr	Ni	Mo	Cu	Other	
Plain Carbon Steels	Plain Carbon Steel	Below 0.20								Up to 150°F (Up to 65°)
	Plain Carbon Steel	0.20–0.30								150°F–300°F (65°C–150°C)
	Plain Carbon Steel	0.30–0.45								250°F–450°F (120°C–230°C)
	Plain Carbon Steel	0.45–0.80								450°F–750°F (230°C–400°C)
Plain Moly Steels	Carbon Moly Steel	0.10–0.20					0.50			150°F–250°F (65°C–120°C)
	Carbon Moly Steel	0.20–0.30					0.50			200°F–400°F (95°C–205°C)
Manganese Steels	Silicon Structural steel	0.35	0. 80	0.25						300°F–500°F (150°C–260°C)
	Medium Manganese Steel	0.20–0.25	1.0–1.75							300°F–500°F (150°C–260°C)
	SAE 1330 Steel	0.30	1.75							400°F–800°F (205°C–425°C)
	SAE 1340 Steel	0.40	1.75							500°F–800°F (260°C–425°C)
	SAE 1045 Steel	0.50	1.75							800°F–900°F (425°C–480°C)
	12% Manganese Steel	1.25	12.0							Usually not Required
	Manganese Moly Steel	0.20	1.65	0.20			0.35	0.40		300°F–500°F (150°C–260°C)
	Jalten Steel	0.35 max	1.50	0.30				0.20		400°F–800°F (205°C–425°C)
	Manten Steel	0.30 max	1.35	0.30				0.35 min		400°F–800°F (205°C–425°C)
High Tensile Steels	Armco High Tensile Steel	0.12 max				0.50 min	0.50 min	0.50–1.50		Up to 200°F (Up to 95°C)
	Double Strength #1 Steel	0.12 max				0.50–1.25	0.10 min	0.50–1.50		300°F–800°F (150°C–425°C)
	Double Strength #1A Steel	0.30 max				0.50–1.25	0.10 min	0.80		400°F–700°F (205°C–370°C)
	Mayarir Steel	0.12 max	1.25	0.35	0.2–1.0	0.25–0.75		0.50 max		Up to 300°F (Up to 150°C)
	Otiscoloy Steel	0.12 max		0.10 max	0.10 max			0.25 max		200°F–400°F (95°C–205°C)
	Nax High Tensile Steel	0.15–0.25		0.75	0.60	0.17			Zr. 12	Up to 300°F (Up to 150°C)
	Cromansil Steel	0.14 max		0.75	0.50		0.15 max	0.40		300°F–400°F (150°C–205C°)
	A.W. Dyn-el Steel	0.11–0.14						0.40		Up to 300°F (Up to 150°C)
	Corten Steel	0.12 max		0.25–1.0	0.5–1.5	0.55 max				200°F–400°F (95°C–200°C)

Table 5B Preheating Chart for Selecting Temperature-Indicating Crayons (2 of 2)

Metal Group	Metal Description	Approximate Composition Percentage								Recommended Preheat Temperature
		C	Mn	Bi	Cr	Ni	Mo	Cu	Other	
High Tensile Steels (Cont.)	Chrome Copper Nickel Steel	0.12 max	0.75		0.75	0.75		0.55		400°F–600°F (205°C–315°C)
	Chrome Manganese Steel	0.40	0.90		0.40			1.0		200°F–400°F (95°C–205°C)
	Yoloy Steel	0.05–0.35	0.3–1.0	0.3 max		1.75		0.9–1.25		200°F–600°F (95°C–315°C)
	Hi-Steel	0.12 max	0.6			0.55				200°F–500°F (95°C–260°C)
Nickel Steels	2¼% Nickel Steel	0.25				2.25				200°F–400°F (95°C–205°C)
	3½% Nickel Steel	0.23				3.50				200°F–400°F (95°C–205°C)
Moly Bearing Chromium and Nickel Steels	SAE 4140 Steel	0.40			0.90		0.20			600°F–800°F (315°C–425°C)
	SAE 4340 Steel	0.40			0.80	1.85	0.25			700°F–900°F (370°C–480°C)
	SAE 4615 Steel	0.15				1.80	0.25			400°F–800°F (205°C–425°C)
	SAE 4820 Steel	0.20				3.50	0.25			600°F–800°F (315°C–425°C)
Low Chrome Moly Steels	1¼% Cr – ½% Mo Steel	0.17 max			1.25		0.50			250°F–400°F (120°C–205°C)
	2¼% Cr – 1% Mo Steel	0.15 max			2.25		1.0			300°F–500°F (150°C–260°C)
Medium Chrome Moly Steels	5% Cr – ½% Mo Steel	0.15 max			5.0		0.5			400°F–800°F (205°C–425°C)
	7% Cr – ½% Mo Steel	0.15 max			7.0		0.5			400°F–800°F (205°C–425°C)
	9% Cr – 1% Mo Steel	0.15 max			9.0		1.0			400°F–800°F (205°C–425°C)
Plain High Chromium Steels	11½-13% Cr Type 410	0.15 max			12.0					400°F–800°F (205°C–425°C)
	16-18% Cr Type 430	0.12 max			17.0					300°F–500°F (150°C–260°C)
	23-27% Cr Type 448	0.20 max			25.0					300°F–500°F (150°C–260°C)
	18% Cr 8% Ni Type 304	0.07			18.0	8.0				Usually do not require preheat but it may be desirable to remove chill
Chrome Nickel Stainless Steels	25-12 Type 309	0.07			25.0	12.0				
	25-20 Type 310	0.10			25.0	20.0				
	18-8 Cb Type 347	0.07			18.0	8.0			Cb 10xc	
	18-8 Mo Type 316	0.07			18.0	8.0	2.5			
	18-8 Mo Type 317	0.07			18.0	8.0	3.5			
Irons	Cast Iron									700°F–900°F (370°C–480°C)
Nonferrous	Ni Resist									500°F–1,000°F (260°C–540°C)
	Nickel, Monel, Inconel					Preheat not usually required for thin sections. 70°F–400°F				
	Aluminum-Copper					(21°C–205°C) preheat may be desirable for thick sections.				

Determining Metal Temperatures by Color

The approximate temperature of a metal heated in a furnace can be determined by the different colors exhibited by the metal. Reference charts are available that welders can use to interpret metal temperatures based on observing the color of the metal.

The chart shown below is an example only; it should not be relied upon for actual temperature measurement. Also note that the colors shown here will appear differently in various environments. Sunlight, for example, will cause the colors to appear differently than they would in a darkened room.

COLOR		APPROXIMATE TEMPERATURE	
		°F	°C
FAINT RED		930	500
BLOOD RED		1075	580
DARK CHERRY		1175	635
MEDIUM CHERRY		1275	690
CHERRY		1375	745
BRIGHT CHERRY		1450	790
SALMON		1550	845
DARK ORANGE		1630	890
ORANGE		1725	940
LEMON		1830	1000
LIGHT YELLOW		1975	1080
WHITE		2200	1205

Additional Resources

The Procedure Handbook of Arc Welding, 14th Edition. Cleveland, OH: The James F. Lincoln Arc Welding Foundation, 2000.

Galvery, William L. Jr. and Frank M. Marlow. *Welding Essentials: Questions and Answers.* 2nd Edition. New York, NY: Industrial Press, 2007.

Welding Handbook. Vol. 2, Part 1: Welding Processes. Miami, FL: American Welding Society, 2004.

Welding Handbook. Vol. 3, Part 2: Welding Processes. Miami, FL: American Welding Society, 2004.

1.0.0 Section Review

1. The temperature range that exists between a metal's critical temperature and its melting point is called its ____.
   a. transformation temperature range
   b. preheat weld treatment span
   c. heat-affected zone (HAZ)
   d. discontinuity arc

2. Open-top preheaters work similarly to a(n) ____.
   a. convection oven
   b. oxyfuel torch
   c. box furnace
   d. kitchen gas stove

3. Handheld infrared thermometers are also called ____.
   a. temperature sticks
   b. finger elements
   c. non-contact thermometers
   d. optical thermocouples

SECTION TWO

2.0.0 TEMPERATURE MAINTENANCE AND POSTHEATING PROCESSES

Objective

Describe interpass temperature control and postheating processes.

a. Describe interpass temperature control.
b. Describe various postheating processes.

Trade Terms

Annealing: A postweld heat treatment for stress relieving weldments that is done at temperatures approximately 100°F (38°C) above the critical temperature with a prolonged holding period.

Discontinuity: The deformation, or interruption, of the crystal structure inside a metal.

Strain hardening: A method of increasing the strength of a metal by cold-working it.

Work hardening: The increase in strength that sometimes accompanies plastic deformation of a metal.

In addition to proper preheating, it is equally important to maintain proper temperature of a weldment or assembly both during welding operations and after the welding has been completed. This section introduces several postheating processes that help prevent weld failures by relieving stresses, improving corrosion resistance, restoring ductility, and reducing brittleness.

2.1.0 Interpass Temperature Control

Interpass temperature control is used to maintain the temperature of the weld zone during welding. It is used when the cooling rate is too high or too low to maintain the correct temperature of the weldment between weld passes. The minimum interpass temperature is usually the same as the preheat temperature. If welding is ever interrupted, the temperature of the weldment must be brought back to the interpass temperature before welding is continued. Usually, there is also a maximum interpass temperature, which is especially important in preventing overheating from heat buildup in smaller weldments.

When preheating is important, interpass temperature for multiple-pass welds is usually also important. Interpass temperature control is often simply a continuation of preheating. Heating equipment and techniques are the same.

With large, high-mass weldments and infrequent weld passes, the base metal can cool below the required preheat temperature if additional heat is not supplied. When the base metal cools below the specified minimum preheat temperature, it must be reheated to the preheat temperature before welding continues.

2.2.0 Postheating

Postweld heat treatment is the heating of a weldment or assembly after welding has been completed. It can be performed on most types of base metals. Postheat equipment and techniques are the same as those used for preheating.

Postweld heat treatments include the following the following:

- Stress relieving
- Annealing
- Normalizing
- Tempering

2.2.1 Stress Relieving

Stress relieving is the most commonly used postheat treatment. It is done below the lower transformation temperature, usually in the range of 1,050°F to 1,200°F (566°C to 649°C).

Stress relief treatment is done for the following reasons:

- To reduce residual (shrinkage) stresses in weldments and castings. This is especially important with highly restrained joints.
- To improve resistance to corrosion and stress corrosion cracking.
- To improve dimensional stability during machining operations.
- To improve resistance to impact loading and low-temperature failure.

Importance of Maximum Interpass Temperatures

Austenitic stainless steel typically has a 350°F (177°C) maximum interpass temperature because it is subject to intergranular migration of alloying elements (sensitization) between 800°F and 1,600°F (427°C and 871°C). Migration decreases the corrosion resistance of stainless steel.

Some codes cover stress relieving. These codes specify the heating rate, holding time, and cooling rate. The heating rate is usually 300°F to 400°F (149°C to 204°C) per hour. The holding time is typically one hour for each inch (25.4 millimeters) of thickness. The cooling rate is usually 300°F to 500°F (149°C to 260°C) per hour. Stress relieving requires the use of temperature indicators and temperature-control equipment.

2.2.2 Annealing

Annealing relieves much more stress than a stress relieving postheat treatment can, but it results in a steel of lower strength and higher ductility. Annealing is done at temperatures approximately 100°F (38°C) above the critical temperature, with a prolonged holding period. The heating is followed by slow cooling in the furnace, or by covering, wrapping, or burying the item. Austenitic stainless steel is an exception because it requires rapid cooling. Annealing is used to relieve the residual stresses associated with welds in carbon-molybdenum pipe and to relieve stresses in welded castings that contain casting strains. Generally, annealing is considered to leave the metal in its softest condition, with good ductility.

2.2.3 Normalizing

Normalizing is a process in which heat is applied to remove strains and reduce grain size. Normalizing is done at temperatures and holding periods comparable to those used in annealing. However, the cooling is usually done in still air outside the furnace, and the cooling rate is slightly faster than the slow cooling used in annealing. Normalizing usually results in higher strength and less ductility than annealing. It can be used on mild steel weldments to form a uniform austenite solid solution, to soften the steel, and to make it more ductile. Normalized metal is not as soft and free of stresses as fully annealed metal.

2.2.4 Tempering

Tempering, also called drawing, increases the toughness of quenched steel and helps avoid breakage and failure of heat-treated steel. It reduces both the hardness and the brittleness of hardened steel. Tempering is done at temperatures below the critical temperature, much lower than those used for annealing, normalizing, or stress relieving. Tempering is commonly used on tool steel.

2.2.5 Hardening

The hardness of a material is its resistance to indentation. In metals, hardness is related to their elastic and plastic properties. These properties can be altered when mechanical forces or temperature changes exert stresses or strains on the metal. The deformation, or interruption, of the crystal structure inside the metal is called **discontinuity**. Discontinuity is also called dislocation.

When a piece of metal is hit with a hammer, deformation occurs with each stroke. Hammering metal can form it into a new shape, but the amount of metal will still be the same. The reshaping of metal at ambient temperature is called cold forming or cold working. During cold forming, the metal is strained beyond its yield point, which is the point at which plastic deformation occurs and it will not return to its original form. When a metal is strained or stressed beyond its yield point at ambient temperature, it is said to have been work hardened. **Work hardening**, also called **strain hardening**, tends to make the component stronger (harder) and less prone to failure. Parts that have been hardened using the cold forming process may look brighter in appearance.

Plastic deformation is the permanent distortion of a metal after its elastic limits have been exceeded because of stress. Metals that have been cold worked have undergone plastic deformation. The ability of metals to flow in a plastic manner without breaking apart is the basis for all metal

Stress Relief

The most commonly used method for stress relief of weldments is postweld heat treatment. Stress relief can also be accomplished mechanically. One method involves attaching a mechanical vibrating device to the weldment during or immediately after welding. When activated, the device vibrates the weldment at a specific resonant frequency. The sustained force of the vibrations deforms but does not fracture the shape of the metal's grains. This is called plastic deformation. The plastic deformation of the metal's grains evens out the stress distribution within the weldment. The weight of the weldment generally determines the length of time that the weldment will be subjected to the vibrations.

Tempering Temperatures

If carbon steel is heated in an oxidizing atmosphere, a film of oxide forms on the clean surface, and the color of the surface changes as the temperature increases. The colors can be seen with the unaided human eye under low light conditions. Each color or shade is related to a specific temperature. The colors for plain carbon steel as described in the example chart provided here have been widely used in the trade in the past to determine the correct amount of temper. These colors are affected to some extent by the composition of the metal being heated; therefore, they should be used as a guide only.

COLOR		APPROXIMATE TEMPERATURE	
		°F	°C
VERY PALE YELLOW		430	221
LIGHT YELLOW		440	227
PALE STRAW-YELLOW		450	232
STRAW-YELLOW		460	238
DEEP STRAW-YELLOW		470	243
DARK YELLOW		480	249
YELLOW-BROWN		490	254
BROWN-YELLOW		500	260
SPOTTED RED-BROWN		510	266
BROWN PURPLE		520	271
LIGHT PURPLE		530	277
FULL PURPLE		540	282
DARK PURPLE		550	288
FULL BLUE		560	293
LIGHT BLUE		640	338

forming processes. Plastic flow is the process in which metals are stretched or compressed permanently without breaking or cracking. The resistance to continuing plastic flow as a metal is cold worked is also called work hardening.

When metal is heated to a point just below its melting point, the crystals in it are forced to deform to adapt to the strain caused by the heat. If the crystals in a metal are deformed during the welding process, the weldment will gain strength and become harder. In some situations, metals are intentionally heated again after they have been welded to give them more strength.

2.2.6 Heat Treatment Devices

Metals can be heated by a variety of methods and devices before or after welding. Properly controlling heat is a major consideration. Water-cooled induction heaters are used in some applications, but they are not always available or may not be the process required for the application. *Figure 17* shows a portable induction heater on top of its power supply/control box. A rear view of the power supply is also shown.

Depending on the length of the leads between the power supply and the handheld unit (work head), the operator can move the induction coils to the object rather than move the object to the coils. These portable induction units can be used to heat or cool the object. They can also be programmed to operate at different power levels and times. This variety of options allows better control of the preheat or postheat treatment.

Ceramic heating pads are electrical devices that are often used to heat and relieve stress in metals. In a ceramic heating pad (*Figure 18*), the individual ceramic components are threaded together as needed to cover the metal being heated. The heating pads can be placed on plate steel or laid over pipe. Electrical wires are threaded through the individual ceramic pieces, generating heat when power is applied. Each individual piece is rated at a specific amperage level. The number of pieces used is dependent upon the desired heating output and physical size of pad needed.

The wires from the heating pads must be connected to an electrical device that will control the amount of electrical current flowing through

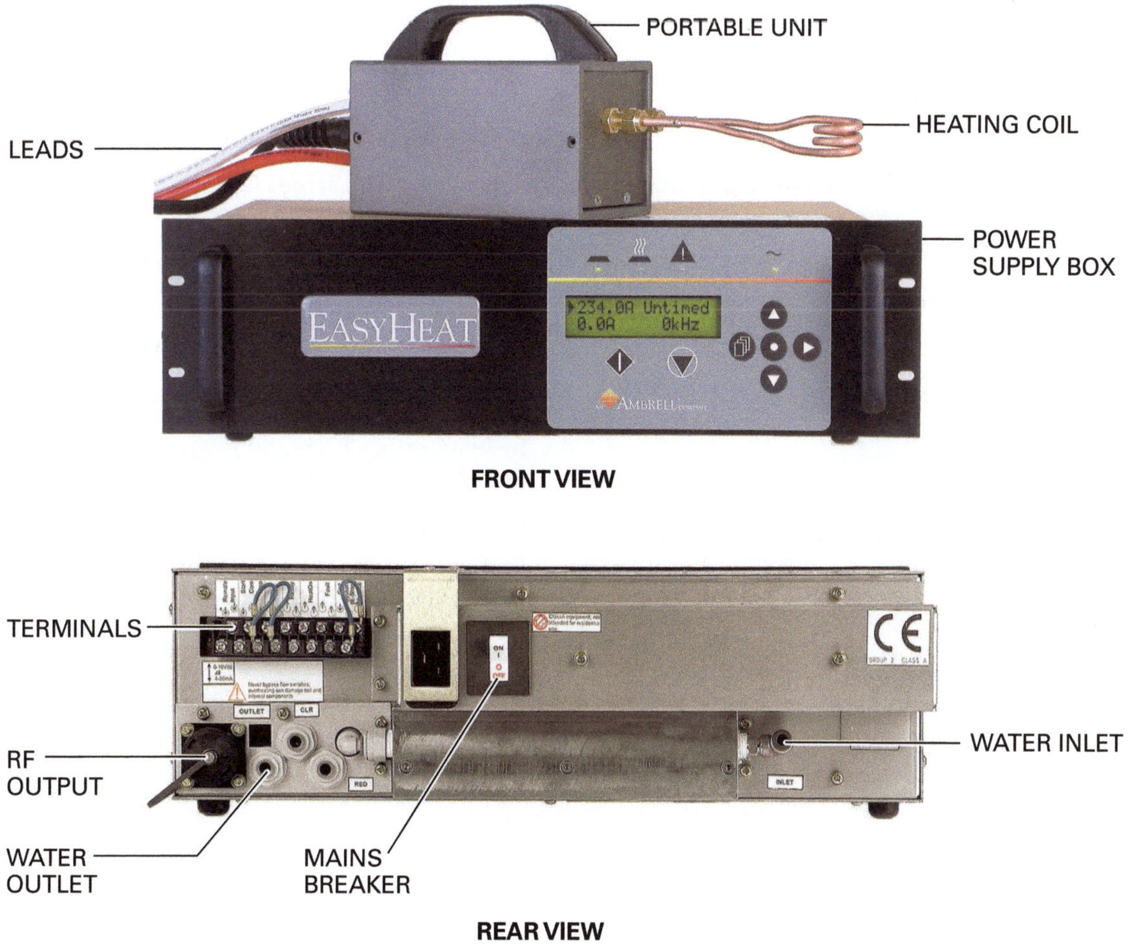

Figure 17 Portable induction heater and its power supply/control box.

the wires. Temperature-sensing thermocouples attached to the metal let the controlling device detect the temperature of the object being heated. The controller compares the thermocouple's signal against a preprogrammed temperature setting and determines if the metal needs to be warmer or cooler. If the controller senses that the metal is getting too hot or getting too hot too quickly, it can reduce the electrical power flowing through the ceramic pads of the heating device. On the other hand, if the controller determines that the metal is not heating fast enough, the controller can increase the current flow through the pads. This causes the pads to become warmer and the metal to heat more quickly. Ceramic heating pads are electrical, so they may be plugged into the AC outlet of a welding power supply.

2.2.7 Time-at-Temperature Considerations

In most heat treatment procedures, temperature and time-at-temperature are both specified by code. Time is very important because metallurgical changes are sluggish at temperatures below the critical temperatures. The maximum temperature is determined by the metal alloy. The holding time at the maximum temperature is based on the metal's thickness, usually one hour for each inch (25.4 mm) of thickness. The cooling rate is determined by the specific treatment or by any applicable code. Temperature control must also be used for postheating. Postheating temperatures are determined in the same manner as preheating temperatures.

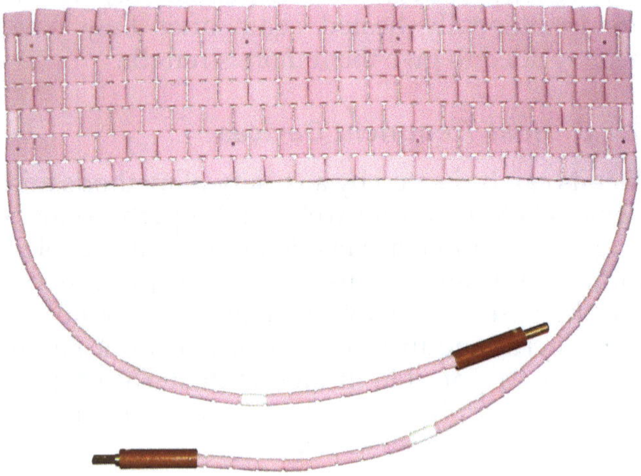

Figure 18 Ceramic heating pads.

Additional Resources

The Procedure Handbook of Arc Welding, 14th Edition. Cleveland, OH: The James F. Lincoln Arc Welding Foundation, 2000.

Galvery, William L. Jr. and Frank M. Marlow. *Welding Essentials: Questions and Answers*. 2nd Edition. New York, NY: Industrial Press, 2007.

Welding Handbook. Vol. 2, Part 1: Welding Processes. Miami, FL: American Welding Society, 2004.

Welding Handbook. Vol. 3, Part 2: Welding Processes. Miami, FL: American Welding Society, 2004.

2.0.0 Section Review

1. The interpass temperature for small weldments and frequent weld passes is likely to be specified as a ____.
   a. modal value
   b. maximum temperature
   c. calculated distribution
   d. minimum temperature

2. For stress relief on metals being welded, the holding time is typically how many hour(s) per inch (2.5 cm) of thickness of the metal?
   a. 14
   b. 12
   c. 1
   d. 2

SUMMARY

With certain alloy steel weldments and castings, preheating, interpass temperature control, and postheating are very important for relieving residual stresses and avoiding weld failures. Heating requirements vary with alloy composition, metal thickness, ambient temperature, and weldment configuration or shape. Proper preheating will preserve the weldment's strength, ductility, and weld quality. Postheating relieves residual stresses, improves corrosion resistance, restores ductility, and reduces brittleness. These processes result in stronger, tougher metal. Temperature control during preheating and postheating is very important because a temperature that is too high can weaken the base metal, and a temperature that is too low will not meet the heating objectives.

Review Questions

1. The controlled heating of the base metal immediately before welding begins is called ____.
   a. progressive base quenching
   b. thermal stress conditioning
   c. interpass temperature control
   d. preheat weld treatment

2. The temperature at which a metal turns liquid is its ____.
   a. melting point
   b. preheating point
   c. welding temperature
   d. postheating temperature

3. Hydrogen gas in the weld zone will cause ____.
   a. slow melting of the base metal
   b. underbead cracking
   c. hardening
   d. slag

4. The hardenability of steel is directly related to its alloying elements and ____.
   a. thickness
   b. carbon content
   c. preheat temperature
   d. welding temperature

5. Preheating is most effective when the welding assembly is heated ____.
   a. rapidly
   b. with a torch
   c. in an oven
   d. with induction heaters

6. Air-cooled induction heating systems are typically used for preheat applications up to ____.
   a. 300°F (149°C)
   b. 400°F (204°C)
   c. 600°F (316°C)
   d. 1,000°F (538°C)

7. Temperature-indicating crayons can be used to indicate temperatures ranging from 100°F (38°C) to ____.
   a. 400°F (204°C)
   b. 980°F (527°C)
   c. 1,150°F (621°C)
   d. 1,350°F (732°C)

8. What temperature is usually the same as the interpass temperature?
   a. Cutting
   b. Annealing
   c. Welding
   d. Preheat

9. For stress relief on metals being welded, the heating rate is usually how many degrees per hour?
   a. 100°F to 200°F (38°C to 93°C)
   b. 200°F to 300°F (93°C to 149°C)
   c. 300°F to 400°F (149°C to 204°C)
   d. 400°F to 500°F (204°C to 260°C)

10. Annealing is done at temperatures that are approximately how many degrees above the critical temperature?
    a. 100°F (38°C)
    b. 200°F (93°C)
    c. 300°F (149°C)
    d. 400°F (204°C)

Trade Terms Introduced in This Module

Annealing: A postweld heat treatment for stress relieving weldments that is done at temperatures approximately 100°F (38°C) above the critical temperature with a prolonged holding period.

Constituents: The elements and compounds, such as metal oxides, that make up a mixture or alloy.

Critical temperature: The temperature at which iron crystals in a ferrous-based metal transform from being face-centered to body-centered. This dramatically changes the strength, hardness, and ductility of the metal.

Crystalline: Like a crystal; having a uniform atomic structure throughout the entire material.

Discontinuity: The deformation, or interruption, of the crystal structure inside a metal.

Heat-affected zone (HAZ): The portion of the base metal workpiece that is not melted but has had its properties affected by welding or other heat-intensive operation such as oxy-fuel cutting.

Interpass temperature control: The process of maintaining the temperature of the weld zone within a given range during welding.

Martensite: A solid solution of carbon in alpha-iron that is formed when steel is cooled so rapidly that the change from austenite to pearlite is suppressed; responsible for the hardness of quenched steel.

Preheat weld treatment (PHWT): The process of heating the base metal before making a weld.

Strain hardening: A method of increasing the strength of a metal by cold-working it.

Work hardening: The increase in strength that sometimes accompanies plastic deformation of a metal.

Additional Resources

This module is intended as a thorough resource for task training. The following reference works are suggested for further study.

The Procedure Handbook of Arc Welding, 14th Edition. Cleveland, OH: The James F. Lincoln Arc Welding Foundation, 2000.

Welding Essentials: Questions and Answers. 2nd Edition. William L. Galvery, Jr. and Frank M. Marlow. New York, NY: Industrial Press, 2007.

Welding Handbook. Vol. 2, Part 1: Welding Processes. Miami, FL: American Welding Society, 2004.

Welding Handbook. Vol. 3, Part 2: Welding Processes. Miami, FL: American Welding Society, 2004.

Figure Credits

The Lincoln Electric Company, Cleveland, OH, USA, Module opener, Figure 6D

Carnegie Institution of Washington, Figure 1

Courtesy of Uniweld Products, Figures 5, 6B

Topaz Publications, Inc., Figures 6A and C, 14

Belchfire Corp., Figure 7, SA02

Miller Electric Mfg. Co., Figure 10, SA03

Lucifer Furnaces, Inc., Figure 11

Fluke Corporation, reproduced with permission, Figure 12

Courtesy of SPX Service Solutions, Figure 13

Tempil, Inc., Figure 16

Amerithern Inc., Figure 17

Bolttech Mannings Inc., Figure 18

Alloy Steel North America, SA01

Section Review Answer Key

Answer	Section Reference	Objective
Section One		
1. a	1.1.0	1a
2. d	1.2.3	1b
3. c	1.3.1	1c
Section Two		
1. b	2.1.0	2a
2. c	2.2.7	2b

NCCER CURRICULA — USER UPDATE

NCCER makes every effort to keep its textbooks up-to-date and free of technical errors. We appreciate your help in this process. If you find an error, a typographical mistake, or an inaccuracy in NCCER's curricula, please fill out this form (or a photocopy), or complete the online form at **www.nccer.org/olf**. Be sure to include the exact module ID number, page number, a detailed description, and your recommended correction. Your input will be brought to the attention of the Authoring Team. Thank you for your assistance.

Instructors – If you have an idea for improving this textbook, or have found that additional materials were necessary to teach this module effectively, please let us know so that we may present your suggestions to the Authoring Team.

NCCER Product Development and Revision

13614 Progress Blvd., Alachua, FL 32615

Email: curriculum@nccer.org

Online: www.nccer.org/olf

❑ Trainee Guide ❑ Lesson Plans ❑ Exam ❑ PowerPoints Other ____________

Craft / Level: Copyright Date:

Module ID Number / Title:

Section Number(s):

Description:

Recommended Correction:

Your Name:

Address:

Email: Phone:

Glossary

Alloy: A metal that has had other elements added to it that substantially change its mechanical properties.

American Iron and Steel Institute (AISI): An industry organization responsible for preparing standards for steel and steel alloys that are based upon the Society of Automotive Engineers (SAE) code system.

American Society for Testing and Materials (ASTM International): An organization that developed a code system for identifying and labeling steels.

Annealing: A postweld heat treatment for stress relieving weldments that is done at temperatures approximately 100°F (38°C) above the critical temperature with a prolonged holding period.

Arrow line: The line drawn at an angle from the reference line (either end or both ends) to an arrowhead at the location of the weld.

Austenitizing: Heating a steel up to or above the transformation temperature range to achieve partial or complete transformation to austenite grain structure.

Axially: In a straight line along an axis.

Axis: The center line of the weld or prepared base metal.

Blind hole: A drilled hole that does not go all the way through the object.

Casting: The metal object produced by pouring molten metal into a mold.

Certified mill test report (CMTR): A quality assurance report provided to a customer by a metal manufacturer that shows the chemical content and testing results of the metal being purchased; also called a mill test report (MTR).

Coefficient: A numerical measure of a physical or chemical property that is constant for a system under specified conditions, such as the coefficient of friction.

Composite metal-core wire electrode: A filler wire electrode with a hollow core containing powdered materials, primarily metals.

Constituents: The elements and compounds, such as metal oxides, that make up a mixture or alloy.

Counterbore: To enlarge a portion of a drilled hole so that a socket head fastener will seat flush or below the surface.

Countersink: A hole with tapered sides and a wider opening that allows a flat-head fastener to seat flush to the surface of the material in which the hole is made.

Critical temperature: The temperature at which iron crystals in a ferrous-based metal transform from being face-centered to body-centered. This dramatically changes the strength, hardness, and ductility of the metal.

Crystalline: Like a crystal; having a uniform atomic structure throughout the entire material.

Degassing: The escape of gases formed during the welding process that can cause imperfections in the weld.

Discontinuity: The deformation, or interruption, of the crystal structure inside a metal.

Ductile: The term used to describe a material that can be bent or shaped without breaking.

Ductility: The characteristic of metal that allows it to be stretched, drawn, or hammered without breaking.

Elevation: A height measurement above sea level or other identified point; also, the vertical sides of an object in an orthographic view.

Feather: The process of grinding a tack weld to a tapered edge.

Ferrous: Relating to iron or an alloy that is mostly iron; a metal containing iron.

Globular transfer: A GMAW process in which the electrode wire does not touch the base metal, and the filler metal is transferred randomly in the arc as large, irregular globules.

Hardenability: A characteristic of a metal that enables it to become hard, usually through heat treatment.

Heat-affected zone (HAZ): The part of the base metal that has not been melted but has been altered structurally by the weld heat.

Heat-affected zone (HAZ): The portion of the base metal workpiece that is not melted but has had its properties affected by welding or other heat-intensive operation such as oxy-fuel cutting.

Hot shortness: The condition of metal when it proves to be very brittle and unbendable at red heat but can be bent without showing signs of brittleness when cold or at white heat. This condition is often a result of high sulfur and phosphorus content.

Inductance: An electrical circuit property, adjustable on some FCAW power supplies, that influences the pinch effect.

Interpass temperature control: The process of maintaining the temperature of the weld zone within a given range during welding.

Interpass temperature: In a multiple-pass weld, the temperature of the base metal at the time the next weld pass is started; it is usually the same temperature that is used for preheating.

Isometric view: A drawing depicting a three-dimensional view of an object.

Lift-arc technology: A type of GTAW technology in which the tungsten electrode may be touched to the work, but the arc will not strike until the tip is lifted away to an appropriate distance.

Malleable: Able to be hammered or pressed into another shape without breaking.

Mark: A system of letters and numbers used with structural steel to identify the type of member and where it is placed.

Martensite: A solid solution of carbon in alpha-iron that is formed when steel is cooled so rapidly that the change from austenite to pearlite is suppressed; responsible for the hardness of quenched steel.

Mechanical properties: The characteristics or traits that indicate how flexible and strong a metal will be under stress.

Metallic: Having characteristics of metal, such as ductility, malleability, luster, and heat and electrical conductivity.

Mill test report (MTR): A quality assurance report provided to a customer by a metal manufacturer that shows the chemical content and testing results of the metal being purchased; also called a certified mill test report (CMTR).

Nonferrous metal: A metal, such as aluminum, copper, or brass, that lacks sufficient quantities of iron to be affected by its properties.

Notch toughness: The impact load required to break a test piece of weld metal measured by a notch toughness test (Charpy V-notch test). This test may be performed on metal below room temperature, at which point it may be more brittle.

On-the-wire: A root pass technique in which the filler metal is held in the bottom of the groove parallel to the plates and then melted and fused in place by moving the torch along the groove.

Orthographic projection: A method of developing multiple flat views of an object by projecting lines out from the object at right angles from the face.

Pinch effect: The tendency of the end of the wire electrode to melt into a droplet during short-circuit GMAW welding.

Pitch: The center-to-center distance between welds.

Postweld heat treatment (PWHT): The process of heat treating the base metal and weld after the weld has been made.

Potentiometer: A three-terminal resistor with an adjustable center connection; used for variable voltage control.

Preheat weld treatment (PHWT): The process of heating the base metal before making a weld.

Pulsed transfer: A GMAW welding process in which the welding power is cycled from a low level (background level) to a higher level (welding level), resulting in lower average voltage and current; this is a variation of the spray transfer process.

Quench: To rapidly cool a hot metal using air, water, or oil.

Reference line: The horizontal line in the center of the welding symbol from which all elements of the welding symbol are referenced. The reference line is one of the most important elements of the welding symbol.

Residual stress: The strain that occurs in a welded joint after the welding has been completed, as a result of mechanical action, thermal action, or both.

Root face: The portion of the edge of a part to be joined by a groove weld that has not been beveled or grooved; previously referred to as the land.

Scale: A method of drawing an object larger or smaller than its actual size while keeping the proportions the same.

Short-circuiting transfer: A GMAW welding process in which the electrode wire shorts against the base metal. The welding current repeatedly and rapidly melts the wire and establishes an arc, which is extinguished when the wire again touches the base metal.

Sintered: A process of making parts from a powdered metal in which the metal is molded and then heated in a furnace without melting to fuse it into a solid metal.

Society of Automotive Engineers (SAE): An automotive society responsible for creating an early system for classifying carbon steels.

Spray transfer: A GMAW welding process in which the electrode wire does not touch the base metal, and the filler metal is transferred axially (in a straight line) in the arc as fine droplets.

Stickout: The length of electrode that projects from the contact tube, measured from the arc tip to the contact tube.

Strain hardening: A method of increasing the strength of a metal by cold-working it.

Stress-relief heat treatment: A technique used to reduce welding stresses and to increase the ductility and toughness of the weld metal and the heat-affected zone (HAZ).

Synergic: Welding technology that controls voltage, current, and wire feed using a program optimized for each welding situation.

Tempering: The process of reheating quench hardened or normalized steel to a temperature below the transformation range and then cooling it at any rate desired.

Tensile strength: The maximum stress or force that a material can withstand without breaking.

Through hole: A drilled hole that goes completely through an object.

Tolerance: The allowance over or under the specified dimension, value, or other parameter.

Trim: A control found on synergic welding machines that is similar to the voltage control on regular machines and acts as a kind of override to the program.

Truncated: Having the pointed top (vertex) of a conical shape cut off, leaving only a plane section parallel to the base of the cone.

Tungsten: A nonstandard common name for the tungsten electrode used in GTAW; also, the principal metal from which the electrodes are constructed.

Underbead cracking: Subsurface cracking in the base metal under or near the weld.

Unified numbering system (UNS): An industry-accepted numbering system for identifying ferrous metals and alloys.

Walking-the-cup: A welding method in which the welder rests the edge of the torch nozzle (cup) against the base metal or groove edges to steady the torch and maintain a constant arc length.

Web: The plate joining the flanges of a girder, joist, or rail.

Weld symbol: A graphic character connected to the reference line of a welding symbol specifying the weld type.

Welding symbol: A graphical representation of the specifications for producing a welded joint; includes a reference line and arrow line and many also include a weld symbol.

Wide flange (WF) beam: A metal I-beam that has an extra wide flange at the top and bottom.

Work hardening: The increase in strength that sometimes accompanies plastic deformation of a metal.

Wrought: Formed or shaped by hammering or rolling.

